Applied Electrical Systems for Construction

Applied Electrical Systems for Construction

ROBERT CLIDERO • KENNETH SHARPE

Library of Congress Catalog Card Number 81-21939
ISBN 0-442-21660-2

Printed in the United States of America

Designed by Graphics International
50 Wolf Road
Albany, New York 12205

Published by Van Nostrand Reinhold Company
135 West 50th Street
New York, NY 10020

16 15 14 13 12 11 10 9 8 7 6 5 4 3 2 1

Library of Congress Cataloging in Publication Data

Clidero, Robert K.
Applied electrical systems for construction.

Bibliography: p. 365
Includes index.
1. Electric wiring, Interior. I. Sharpe, Kenneth H. II. Title.
TK3271.C56 621.319'24 81-21939
ISBN 0-442-21660-2 AACR2

Preface

Applied Electrical Systems for Construction is an adaptation for the American market of a best-selling Canadian text, *Applications of Electrical Construction,* by Robert K. Clidero and Kenneth H. Sharpe.

Although there are few dissimilarities between the Canadian Electrical Code and the United States' National Electrical Code, this text has been thoroughly revised and edited to reflect any differences which affect electrical applications in this country. While many devices and materials illustrated in the text are of Canadian origin, they are international products typically used interchangeably in either Canada or the United States, under various trade names of manufacturers.

Applied Electrical Systems for Construction introduces the nonmathematical theories behind electrical devices and equipment. Modern electrical products and their applications in electrical circuits are explained in simple terms, with emphasis on the reasons for electrical installation practices.

Neither the electrical industry nor the National Electrical Code had fully converted to the SI (metric) system of measurement at the time this book was printed. In preparation for the changes to be made in the near future, metric equivalents and useful conversions for English measurements have been used throughout the book. You will find it beneficial to refer to the most recent edition of the National Electrical Code as it applies to the electrical installations discussed in this book.

The authors of *Applied Electrical Systems for Construction,* Robert K. Clidero and Kenneth H. Sharpe, both entered the electrical trade after separately completing their high school education in the technical program at Western Technical and Commercial School, Toronto, where they majored in electricity. Mr. Clidero served a four-year apprenticeship, and worked as an electrician for eight years. He worked one year with the Ontario Hydro as an electrical inspector of residential wiring. Mr. Sharpe served a four-year apprenticeship, and worked as an electrician for 15 years. In 1961, both men entered the University of Toronto, College of Education, where they first met.

Both authors are now electricity teachers at Western Technical and Commercial School where each has served on the technical staff for almost twenty years. The authors are associates in the training of senior students entering the electrical field as apprentices or other related skill areas. Both men are members of the Ontario Electrical League.

Contents

Chapter 1 The Three-wire Distribution System 1
Information, sketches, and diagrams describing the method used to bring power into the home.

Chapter 2 Lighting-control Switches 11
Construction, types, ratings, applications, and circuit diagrams.

Chapter 3 Lampholders 29
Construction, sizes, materials, and installation requirements.

Chapter 4 Receptacles 34
Construction, types, location, and installation requirements.

Chapter 5 Conductors 45
Construction, types, sizes, insulation, and applications.

Chapter 6 Cord Fittings 58
Types, shapes, and installation techniques.

Chapter 7 Electrical Outlet Boxes 63
Sizes, shapes, wire capacity, installation methods, and applications.

Chapter 8 Nonmetallic-sheathed Cable Wiring 77
Types, sizes, fastening devices, preparation, connection to boxes, and applications.

Chapter 9 Solderless Connectors 88
Types, ratings, and methods of application; insulating tapes, materials, and techniques.

Chapter 10 Heat-control Switches 106
Circuit diagrams, multiposition units, and infinite-heat control switches; oven controls, self-cleaning ovens and circuits.

Chapter 11 Armored Cable 118
Construction, types, preparation, connection to boxes, requirements, and applications.

Chapter 12 Aluminum-sheathed Cable 125
Construction, sizes, preparation, installation techniques, and applications.

Chapter 13 Mineral-insulated (MI) Cable 135
Sizes, preparation, handling, and applications.

Chapter 14 Conduit Wiring 142
Rigid, flexible, thinwall, and plastic conduit; tools, related equipment, sizes, allied fittings, and requirements.

Chapter 15 Residential Service Wiring 173
Types, sizes, grounding, related equipment, installation methods, and requirements.

Chapter 16 Industrial Services 200
Wiring diagrams, three-phase systems, 3- and 4-wire systems, meter connections, grounding, circuit breakers, regulations, ground-fault protection devices and their operation.

Chapter 17 Fuses 219
Types, sizes, ratings, and applications.

Chapter 18 Residential Electric Heating and Cooling 234
Heating systems, types, sizes, insulation, heat-load calculations, and cost analysis, cooling systems, types, parts, and basic principles.

Chapter 19 Discharge Lighting Sources 268
Types, circuit diagrams, lamp construction, fluorescent and other types of mercury vapor lighting, tungsten-halogen lighting, and applications; energy conservation and lamp maintenance.

Chapter 20 Motor Control 302
Motor protection, across-the-line starting, reduced-voltage starting, starter construction and operation, magnetic starters, and circuit diagrams for control circuits, overload protection devices, types, and ratings.

Chapter 21 Fastening Devices 336
Screw fasteners, wood and metal types, bolts, masonry fasteners, types and methods of installation, powder-actuated fasteners, associated tools, accessories, and requirements.

Acknowledgments 364
For Reference 365
Index 367

Chapter 1

The Three-wire Distribution System

RESIDENTIAL VOLTAGES

Electrical power to the home is supplied by the local electric utility. Because so many appliances are available today, two voltages are needed. Lighting and receptacles for such small appliances as radios, toasters, coffee makers, frypans, and electric drills require a 120-volt (V) alternating-current (ac) supply. Large appliances, such as electric stoves, clothes dryers, some air conditioners, and electric heaters operate on a 240-V ac supply.

The *frequency* of an alternating-current system is the number of cycles of current that pass along the conductors in one second, measured in hertz (Hz). A frequency of 60 Hz is the standard frequency in North America. On a 60-Hz system, there are 120 pulses per second. One positive and one negative pulse make up one cycle, or hertz, Figure 1-1.

Residential voltages operate on a single-phase system. Simply stated, *single-phase* voltage results from one transformer having one single winding on the primary side, and one winding on the secondary side serving the premises. This secondary winding may be center-taped for the grounded neutral conductor. This results in the two-voltage supply to the home.

OBTAINING TWO VOLTAGES FROM THREE WIRES

When two dry-cell batteries of 1.5 V each are connected in series, 3 V are obtained, Figure 1-2. Connecting a wire midway between the two cells creates a 3-wire system providing two voltages. The single-phase distribution transformer used to supply a house or group of houses is wired in a similar manner.

DISTRIBUTION TRANSFORMER

The local electric utility uses a series of transformers to lower its voltages, in

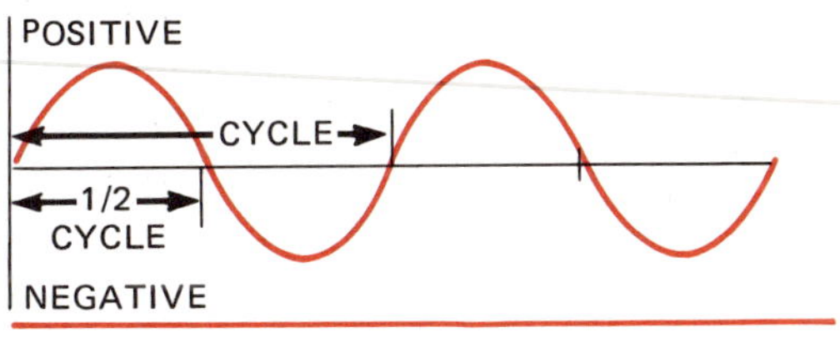

Fig. 1-1 Alternating-current pulse.

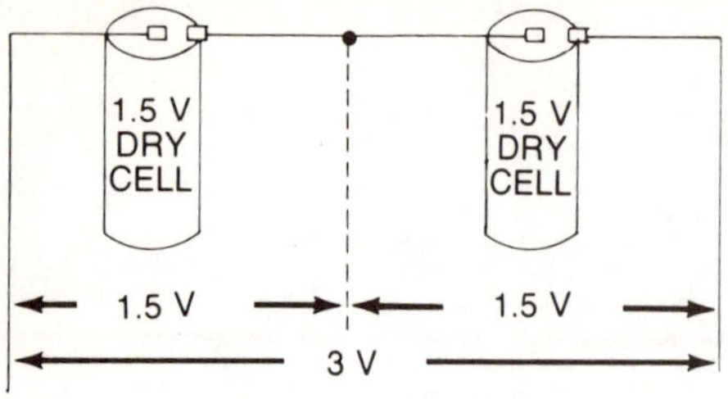

Fig. 1-2 Two voltages from three wires.

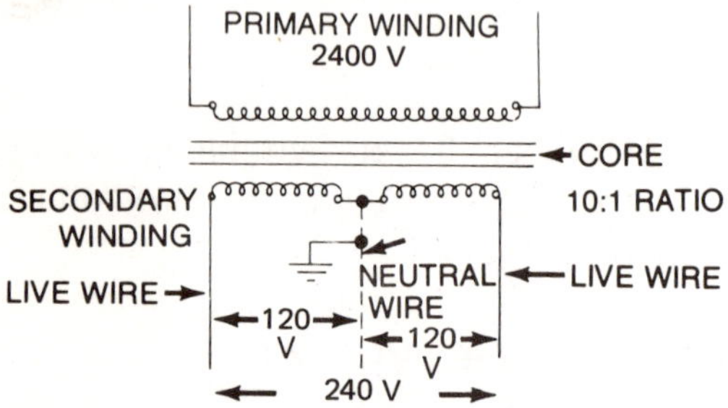

Fig. 1-3 Primary coil supplied with 2400 V.

stages, from the power substations to the premises. Each locality may differ slightly in the actual voltage that arrives at the distribution transformer on the street. One very common voltage is 2400 V, Figure 1-3.

High-tension (high-voltage) lines supply the primary coil of the transformer with 2400 V. The transformer reduces the voltage on a 10:1 ratio, giving a secondary, or output, of 240 V. A wire is connected to the midway point of the secondary winding, dividing its 240 V in half. This middle, or *neutral,* wire provides two voltages on a 3-wire system. Although two voltages are obtained, this is still considered to be a single-phase system. One-half of the transformer is used for 120 V to the neutral, and the complete secondary winding is used for 240 V.

The two outer wires of the secondary winding are known as the *live* wires. It should be noted that in a residential wiring system, the neutral wire is white or gray in color, while the live wires are usually black. For safety purposes, live wires are switch controlled and have fuses connected in series with them. The neutral wire is grounded (connected to the earth) at the transformer and in the residential switch box, and is never fused or switched.

RESIDENTIAL OVERHEAD SUPPLY SYSTEM (SERVICE DROP)

Figure 1-4 shows the method used to supply power to many of the houses in a community. In some areas, a fourth wire is brought from the electric pole to the house for the purpose of supplying a flat-rate, hot-water heater system. (This subject is covered in more detail in Chapter 15.)

RESIDENTIAL UNDERGROUND SUPPLY SYSTEM (SERVICE LATERAL)

Modern community planning has led to the development of an underground supply system. The main advantage of this system is to give the community an uncluttered look by placing the wires and cables below ground. Figure 1-5 shows a system of this type.

SWITCHING THE LIVE WIRE

On reaching the house, the service-entrance lines pass through the kilowatt-hour meter. A *kilowatt-hour meter* records the amount of power used in kilowatt-hours. The supply lines are then carried into the house by conduit, or service-entrance cable,

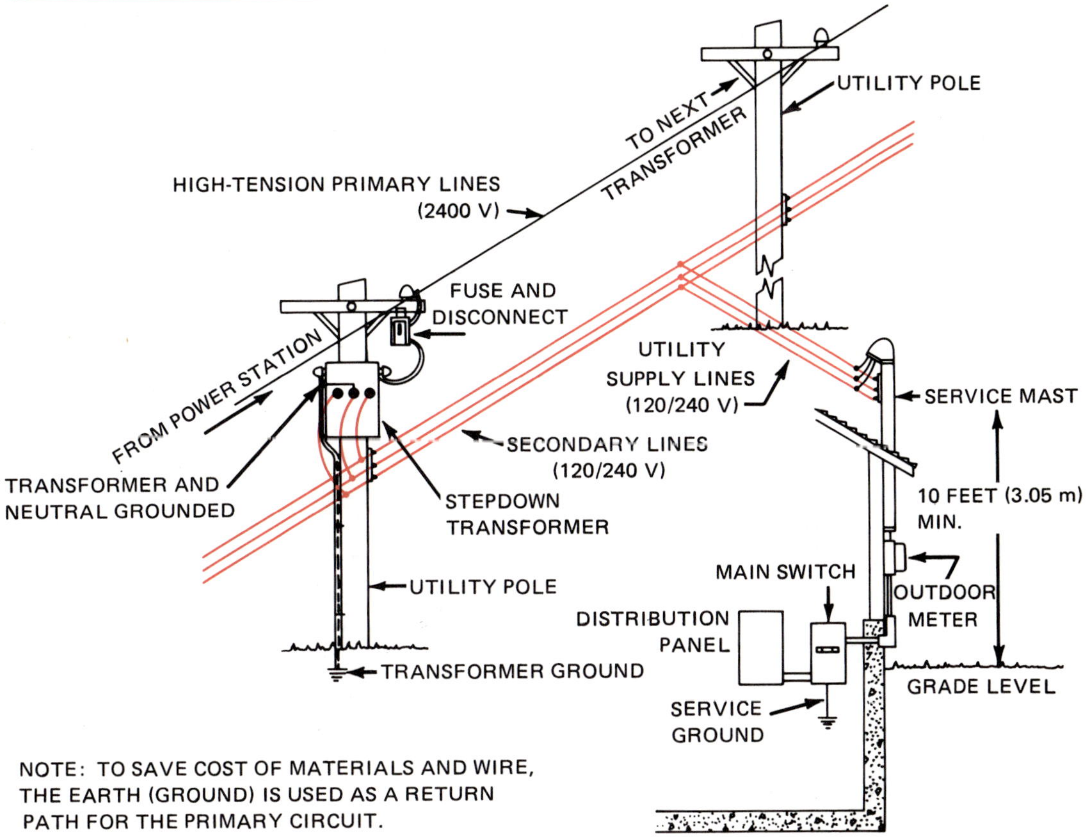

Fig. 1-4 Typical 3-wire distribution system.

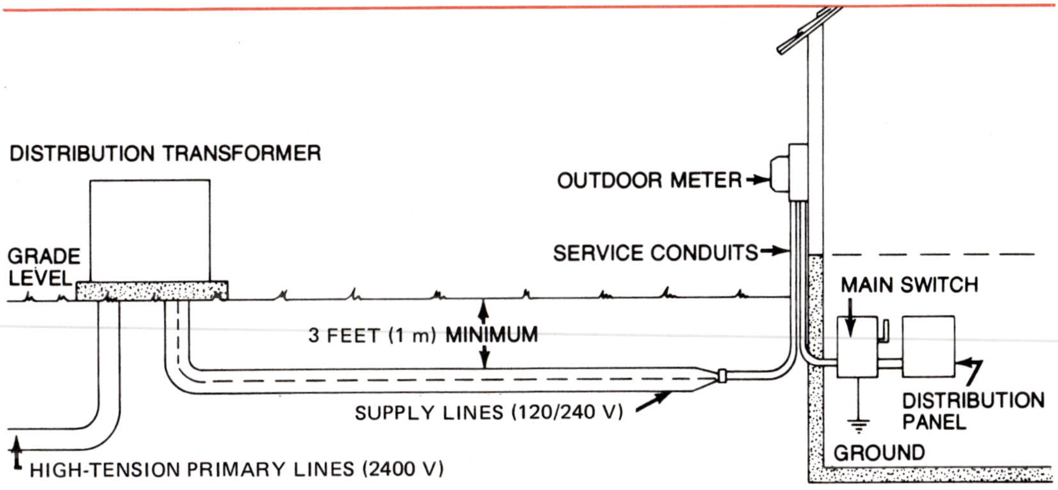

Fig. 1-5 Typical underground distribution system.

which takes them directly to the main disconnect switch. As shown in Figure 1-6, the supply lines are connected at the top of the switch, which is standard practice. The upper portion of the switch is called the *line side,* and most manufacturers have the word *line* printed somewhere near the line terminals.

Should an electrical emergency arise, turning this switch off will disconnect all of the power supply to all parts of the house.

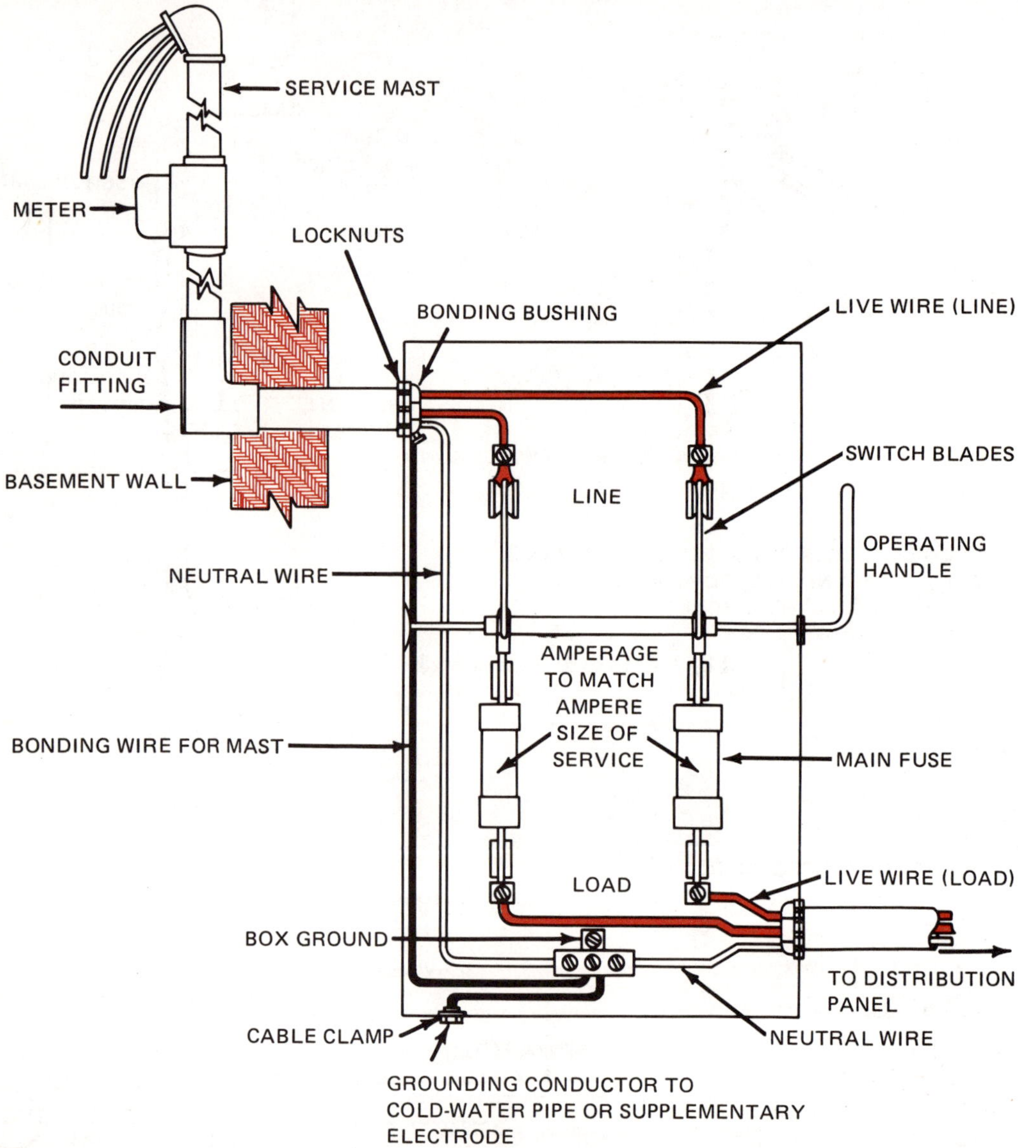

Fig. 1-6 Electrical connections in the main disconnect switch.

Power may be restored by turning the main switch on again. Once the main switch has been *pulled,* or turned off, all contents of the main switch box are safe to handle, with the exception of the two terminal connections that receive the incoming electrical supply lines. Fuses may be replaced or checked, repairs to the switch made, and any of the wires at the bottom, or *load,* side of the switch may be handled in safety.

FUSING THE LIVE WIRE

The purpose of any fuse is to limit the amount of electrical current *(amperes)* that can pass through a wire. If more current is passed through a wire than it is designed to carry, the wire will heat up and eventually start to burn its insulation covering. A fuse is designed to heat up, melt, and open the line before the wire it protects is damaged.

The fuses in the main disconnect switch are designed to protect those conductors going from the load side of the switch to the distribution panel. The *voltage* rating of the fuse is matched to the *voltage* of the supply system, while the *current rating* of the fuse is matched to the *current-carrying* capacity (ampacity) of the wire. The interrupting rating of the fuse must also keep the supply line open once the fuse has melted, thus preventing arcing.

The amount of current entering the house on one of the live wires must leave the house on the other live wire and/or the neutral wire. Therefore, a fuse is needed on *each* of the *live* wires. Figure 1-6 shows the location of the fuses in the main disconnect switch.

WHY THE NEUTRAL WIRE IS NOT FUSED

When the 3-wire distribution system first came into use, a fuse was placed in series with the neutral wire, as well as with the live wires. It was soon discovered that this practice was *dangerous.*

A 120-V lighting circuit was protected by two fuses: one on the live wire and one on the neutral wire. Since both wires were the same size, both fuses had the same current rating. If a fault causing a short-circuit condition occurred, and excessive current started to flow, the fuse would open the circuit and prevent damage to the wire. Since both fuses in the circuit were the same size, several things could happen.

Situation A. *Both fuses* could blow at the same time and prevent any further current flow. Both the live and neutral wires (conductors) would be safe to handle, and repairs could be made to the circuit without the danger of electrical shock, Figure 1-7.

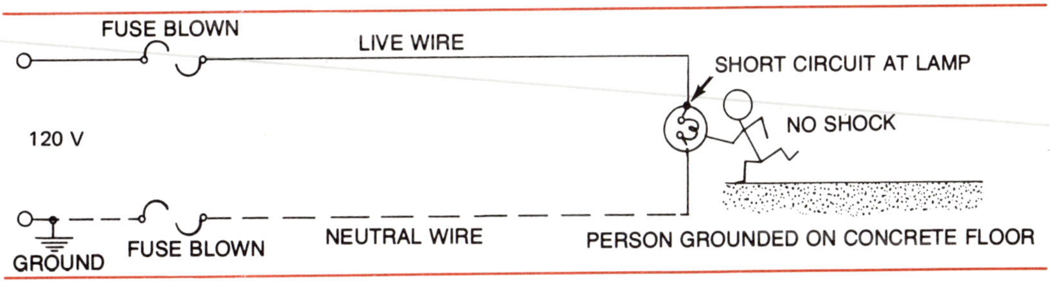

Fig. 1-7 Both fuses blow (Situation A).

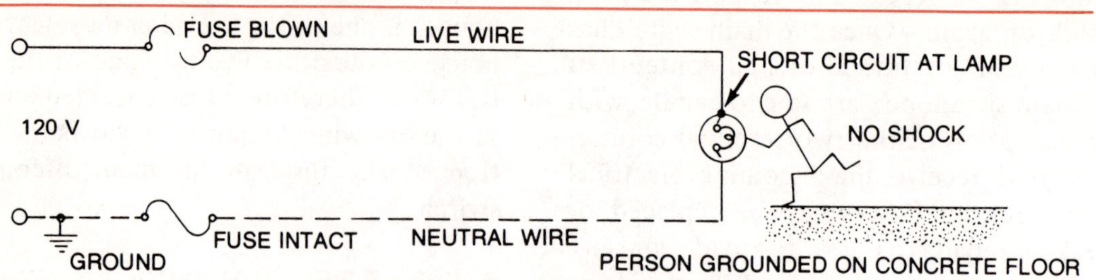

Fig. 1-8 Fuse on live wire blows (Situation B).

Situation B. The *fuse* on the *live wire* (conductor) could blow. Both fuses have the same rating, but some small difference in their construction could make one fuse weaker than the other, causing it to open first. When this happened, the conductors were protected by cutting the current flow, making the circuit safe to work on, Figure 1-8.

Situation C. The *fuse* on the *neutral* conductor could blow. Since there is no way to tell in advance which fuse might be weaker, chance alone determined which of the two fuses would be the one to open first. If the neutral fuse burned out first, it prevented further current flow from damaging the conductors. (The current entering a circuit is the same as the current leaving; so, the neutral fuse may open the circuit as well.) The live conductor circuit, however, would still be intact. **If grounded, a person attempting to repair the circuit could suffer a 120-V shock from the live wire.** The chances of such a shock occurring are high, because all electrical boxes in modern wiring systems are grounded, and because plumbing fixtures, damp concrete floors, and similar conditions are common in the home, Figure 1-9.

To eliminate the danger of shock, a fuse is no longer installed in the neutral wire. Modern service installations are designed to fuse *only* the *live* wires.

GROUNDING THE NEUTRAL WIRE

The majority of electrical service parts, wiring boxes, conduits, and similar fittings are made of metal. The insulation on con-

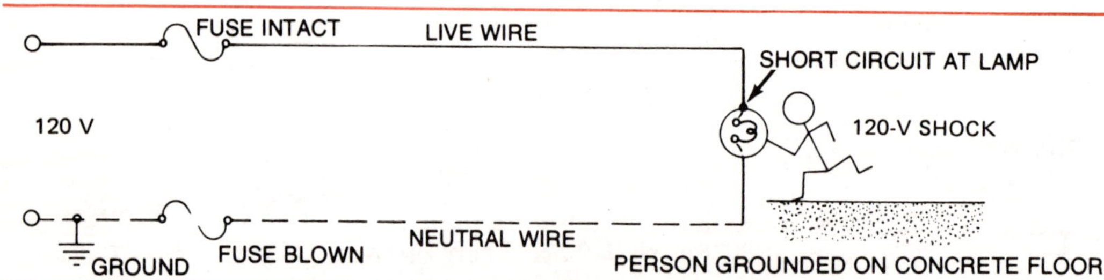

Fig. 1-9 Fuse on neutral wire blows (Situation C).

ductors is sometimes damaged during the initial installation of the circuit. It is also possible for the insulation to become worn over a period of years, allowing one of the wires to come in contact with the metal boxes. For these reasons, a completely grounded system is needed.

Service masts, which rise above the roof of many single-story houses, can be targets for lightning during thunderstorms. To prevent this metal conduit from attracting lightning, the service conduit and boxes are connected to the ground by driving grounding electrode rods into the earth and fastening a grounding conductor to the cold-water pipe where it enters the house. This installation provides a continuous electrical ground on the system.

If one of the two live wires in the 3-wire system comes into contact with the grounded service box, **a dangerous situation is established.** The risk is high for a person in a house to become grounded. The potential dangers are many. For example, if the second live wire of the 3-wire system comes into contact with the frame of a faulty power tool, and the first live wire touches the metallic service box, **a grounded person could suffer a 240-V shock – a shock strong enough to kill.** A person fixing a light over the kitchen sink could receive a 240-V shock by touching the sink and the second live wire in the fixture. Cleaning a laundry room fixture with a damp cloth while standing on the basement floor could also result in a 240-V shock, Figure 1-10.

In modern service installations, the main switch box is grounded, and a terminal block is placed in the lower portion of the switch where the neutral wire is intentionally grounded. With the neutral wire grounded, the maximum shock a person can suffer from any one live wire is 120 V. Thus, a grounded person working with power tools, repairing equipment, or cleaning fixtures can sustain a shock not higher than the voltage between the neutral and a live wire, or 120 V. **Under certain conditions, this voltage can kill,** but the danger is greatly reduced, Figure 1-11.

A short circuit results when one of the live wires accidentally touches a metal box or fitting with the neutral wire grounded. The fuse on the live wire blows (opens) and protects the circuit. The neutral wire is safe to handle even when a person is grounded, since there is no difference in the voltage between it and ground.

UNBALANCED SYSTEM

A 100-ampere, single-phase, 3-wire service supplies up to 100 amperes on each of the conductors.

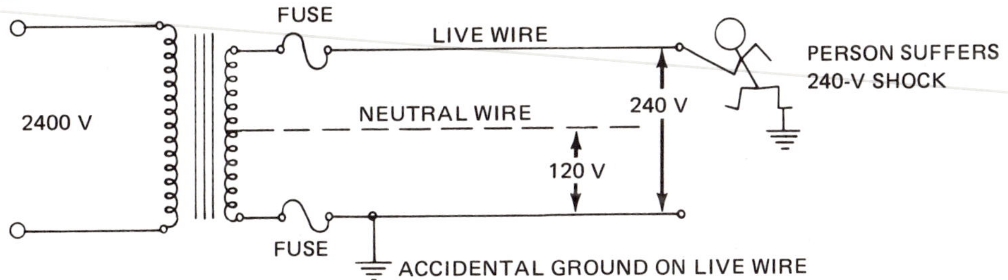

Fig. 1-10 Accidental grounding of the live wire creates a 240-V shock hazard.

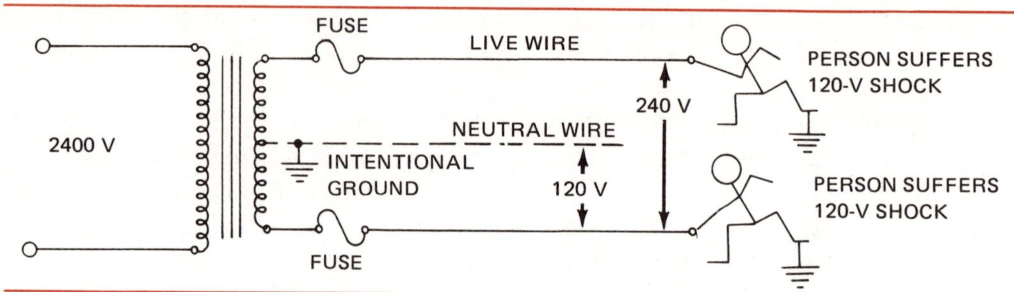

Fig. 1-11 Grounding the neutral wire limits the shock hazard to 120 V.

When a distribution panel is installed in a house, some attempt is usually made to distribute the load *evenly* on each of the live wires supplying the panel. However, it is unlikely that an equal number of lights or devices in a dwelling would be turned on in order for the load to be balanced on each side of the panel. The *neutral wire* in the system is very important in this situation, because it returns the unbalanced amount of current (the difference between the two live-wire currents) to the transformer.

A broken or disconnected neutral wire is a danger to the electrical equipment in the circuit. This is because the side of the system with the greater number of devices turned on has the greater number of parallel circuit paths. The more paths in a parallel circuit, the lower the total resistance of the circuit. The difference in electrical resistance between the two sides of the panel is determined by the number of devices operating on each side.

If the neutral wire is disconnected, the circuit is no longer a 120-V circuit. The two sides are then *in series* with each other, and have an applied voltage of 240 V. Since voltage is divided in a series circuit, the side of the panel with the fewer devices turned on (and, therefore, having a higher resistance) receives a voltage that is much higher than normal. The other side of the panel, with its lower resistance, receives the balance of the 240 V. Light bulbs glow very brightly with the increased voltage, and such sensitive equipment as stereos and television sets may be seriously damaged. When connecting service conductors, therefore, the *neutral* wire must be *disconnected last* and *reconnected first* when power is to be restored.

The danger of an unbalanced condition is another reason for *never fusing the neutral wire*. A blown fuse on the neutral wire results in an unbalanced situation.

As current passes through a circuit, each lamp forms part of the circuit and offers some resistance to the current flow, Figure 1-12. Three factors are involved in current flow: *line voltage, amperage,* and *resistance.* The value of each factor is determined according to *Ohm's Law,* which states that:

$$\text{current} = \frac{\text{voltage}}{\text{resistance}}$$

In terms of units, this is:

$$\text{amperes} = \frac{\text{volts}}{\text{ohms}}$$

As a mathematical formula, this is stated in symbols as:

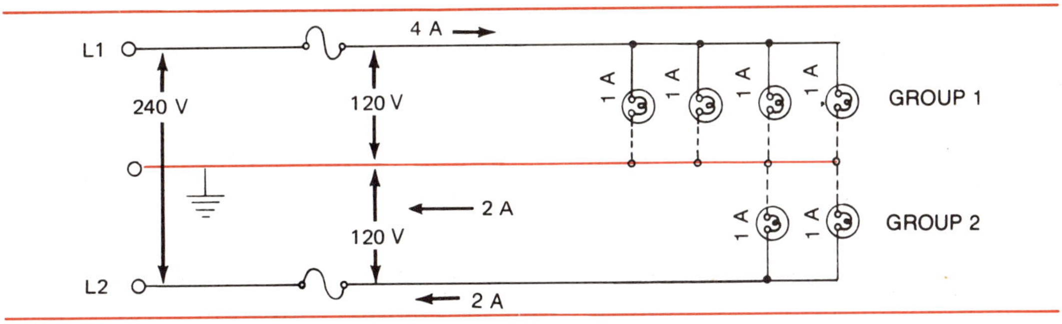

Fig. 1-12 Each lamp is the same size (amperage). The current flows are shown by the arrows.

$I = \frac{E}{R}$ where: I = amperes
E = voltage
R = resistance

To make calculations easier, each lamp shown in Figure 1-12 is the same size, receives 120 V, and draws 1 A of current. The resistance of *one* lamp is found by using Ohm's Law:

$$I = \frac{E}{R}$$

Transposed, this equals:

$$R = \frac{E}{I}$$

Since R represents resistance, measured in ohms which are represented by the Greek letter omega (Ω):

$$R\ (\text{ohms}) = \frac{E\ (\text{volts})}{I\ (\text{amperes})}$$

Therefore, the resistance of *one* lamp equals

$$\frac{120\ V}{1\ A}$$

That is, R equals 120 Ω.
The resistance of *one* lamp is 120 Ω.

The total resistance of *all* lamps of the same size in a parallel circuit is equal to the resistance of one lamp divided by the number of lamps in the circuit. Resistance of the lamps in Group 1:

$$R = \frac{120\ \Omega}{4\ \text{lamps}} = 30\ \Omega$$

Resistance of the lamps in Group 2:

$$R = \frac{120\ \Omega}{2\ \text{lamps}} = 60\ \Omega$$

When the neutral wire is broken, the lamps in Groups 1 and 2 are *in series* with one another, Figure 1-13. The *total resistance* for the whole circuit is now the *sum* of the resistances in *Groups 1* and *2:*

$$30\ \Omega + 60\ \Omega = 90\ \Omega$$

Since the neutral wire is broken, the line voltage is now 240 V, and the total current in the circuit is:

$$I = \frac{E}{R} = \frac{240\ V}{90\ \Omega} = 2.67\ A$$

The voltage in a series circuit is divided between the groups of lamps, according to Ohm's Law:

$$I = \frac{E}{R}$$

Transposed, this equals:

$$E\ (V) = I\ (A) \times R\ (\Omega) \text{ or } E = I \times R$$

The voltage of Group 1 is:

$$E = 2.67\ A \times 30\ \Omega$$
$$= 80.1\ V$$

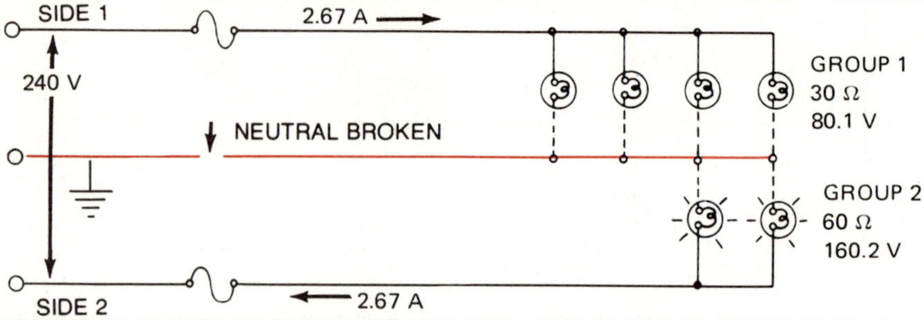

Fig. 1-13 Each lamp is the same size (amperage). The current flow is altered when the neutral wire is broken.

This low voltage causes the lamps in Group 1 to glow dimly.

The voltage of Group 2 is:

$$E = 2.67 \text{ A} \times 60\ \Omega$$
$$= 160.2 \text{ V}$$

This higher voltage causes the lamps in Group 2 to glow very brightly. The lamps' filaments now have an excessive amount of amperes (current) flowing through them. This may also cause the lamps to burn out, since they are rated to operate on 120 V ± 10 percent.

REVIEW

1. What voltage is used in houses for lighting, receptacles, and portable equipment? What voltage is used for heavy-duty equipment?
2. What is the electrical frequency most often used in North America?
3. Explain how two separate voltages are obtained from a 3-wire distribution system. Include a diagram.
4. What does the term *neutral* mean when applied to a conductor?
5. What is the purpose of the neutral wire?
6. What are three differences between the neutral and live wires of a circuit, other than color?
7. Explain (a) why the neutral wire is grounded, and (b) why the neutral wire is never fused.
8. What method is used to ground the neutral wire in a house?
9. Why are the incoming electrical supply lines connected to the upper terminals of the main switch?
10. What voltage is available (a) between two live wires, (b) between a live and neutral wire, (c) between a live wire and ground, and (d) between the neutral wire and ground?

Chapter 2

Lighting-control Switches

CONTROL OF CURRENT FLOW

The flow of electrical current in the various circuits of a building must be controlled. Current flow is controlled by using a variety of switches capable of opening and closing the circuits. Switches are installed in the live (ungrounded) conductors; *never* in the neutral (grounded) conductor.

TYPES OF SWITCHES

Switches are divided into two main categories, according to the method of installation.

Category 1. The *surface-mounted* type of switch is mounted on the face of the wall, with the entire body of the switch visible, Figure 2-1.

Category 2. The *flush-mounted* type of switch is mounted within an electrical box. The box is recessed into the wall, so that only the operating handle of the switch and the faceplate are visible, Figure 2-2. This type of switch looks neater than the surface-mounted type, and is used more extensively.

OPERATING MECHANISMS

Many different types of switches are available to control a wide variety of electrical devices. Many different types of operating mechanisms also are available to suit individual circuit needs. Figure 2-3 shows some of the more common switch operating mechanisms.

INTERNAL CONSTRUCTION

The basic function of a switch is to open and close a circuit. To do this, switches have an internal combination of fixed and

Fig. 2-1 Surface-mounted switch.

Fig. 2-2 Flush-mounted switch.

movable contacts, which are usually made of brass. Figure 2-4 shows two types of designs and methods for moving the contacts.

Switches designed to carry a high current have larger and stronger contacts than do switches designed to operate only one or two lights. Nearly all switches, however, have some form of internal spring to open and close the contacts quickly. If the contacts are slow to open and close, there is danger of the current arcing. *Arcing* occurs when a spark jumps across the contacts as they part, burning the contacts.

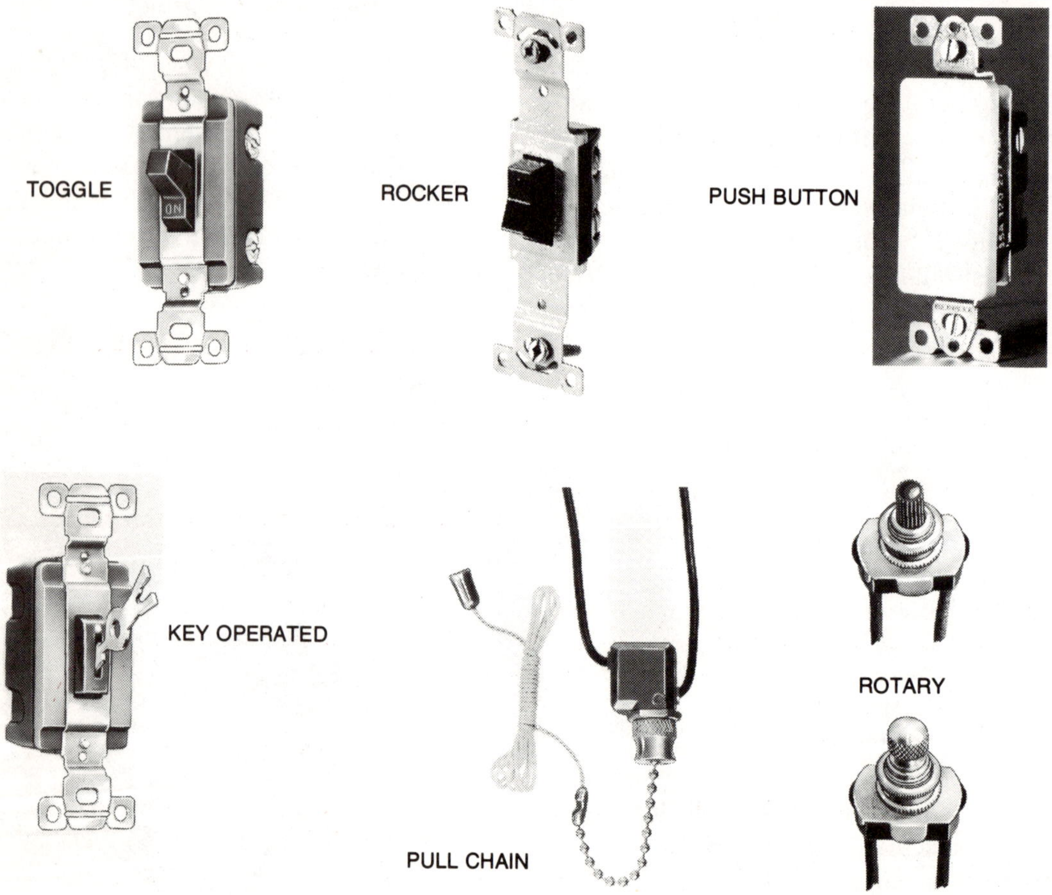

Fig. 2-3 Common switch operating mechanisms.

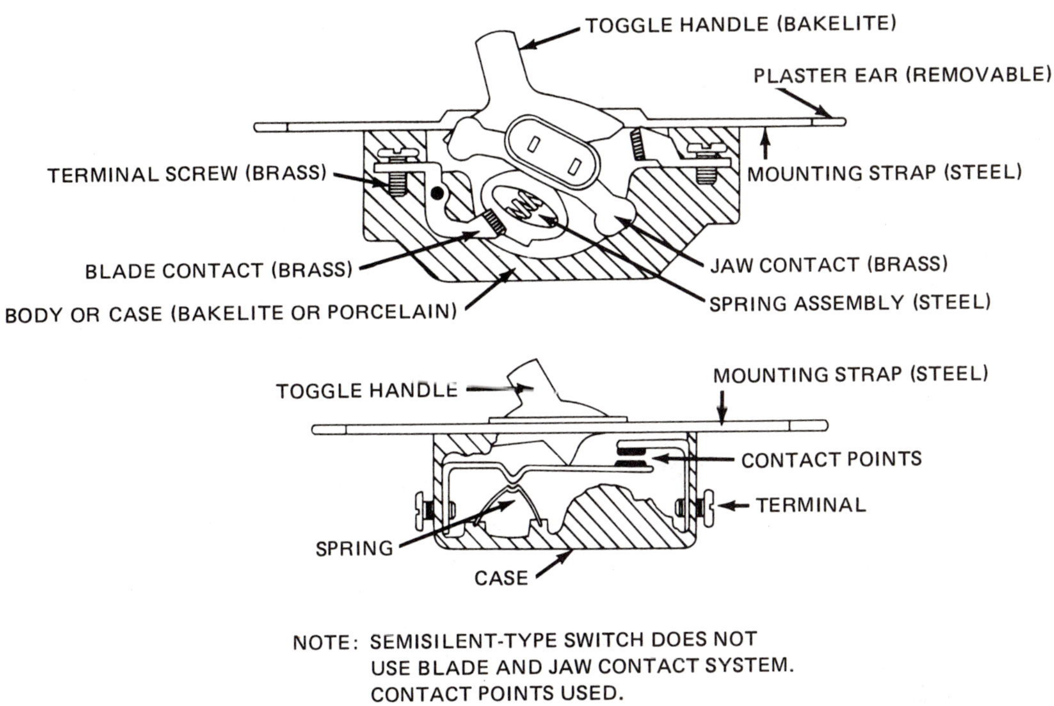

Fig. 2-4 Switch internal construction.

The main difference between a high-quality (expensive) switch and one of low quality is the size and strength of the contacts and spring. Switches are available in several grades. Switches are graded for specification, intermediate, hospital, or industrial uses. Take care to choose a switch that is capable of handling the intended load in amperes on the circuit, and rated for the voltage supplied.

SWITCH RATINGS

Manufacturers usually specify electrical ratings on the mounting bracket of the switch. Six ratings are available to assist a person in matching the switch to the job: voltage, amperes, test-laboratory listing, "T" rating, ac or dc, and horsepower (kW), Figure 2-5. The voltage, amperes, and listing ratings *must* be placed on *all* switches.

Voltage. The voltage rating refers to the electrical pressure the switch can control safely. Lighting switches should have a rating of 120 V to 125 V for control of *one* live wire, and a rating of 240 V to 250 V for control of *two* live wires.

Amperes. The rating in amperes refers to the ability of the contacts to carry current. The higher the current rating, the greater the load that can be controlled by the switch. The current ratings increase in units of five; for example, 10 A, 15 A, 20 A, and so forth.

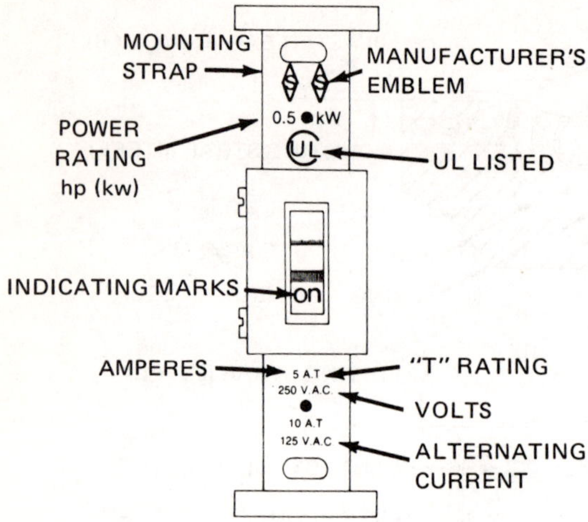

Fig. 2-5 Common switch ratings.

Test-laboratory Listing. All electrical products for sale or use in the United States must be submitted to a certified testing laboratory for testing and listing. Most electrical products made in the United States are tested and listed by the Underwriters' Laboratories, Inc., and bear the mark *UL* or *Und. Lab.* When tested and listed by the Underwriters' Laboratories, the products may carry the UL mark. The UL mark is a guarantee to the user that the manufacturer's ratings are correct.

Switches designed for special applications are marked with a "T" rating, an ac or dc rating, or a horsepower rating, as described next.

RATINGS AND USES OF SNAP SWITCHES

Snap switches must be used within their ratings, and according to the following information.

AC General-use Snap Switch. An ac general-use snap switch is a form of general-use snap switch suitable for use *only* on alternating-current circuits for controlling:

1. Resistive and inductive loads, including electric-discharge lighting, not exceeding the ampere rating of the switch at the voltage involved.
2. Tungsten-filament lamp loads not exceeding the ampere rating of the switch at 120 volts.
3. Motor loads not exceeding 80 percent of the ampere rating of the switch at its rated voltage.

A motor requires three to five times as much current when starting as it does when running. This high starting current, for example, is what causes room lights to dim briefly when a large motor, such as a refrigerator or air conditioner, starts up. The additional 20 percent rating of the switch is used for handling this high inrush starting current safely.

AC-DC General-use Snap Switch. An ac-dc general-use snap switch is a form of general-use snap switch suitable for use on *either* ac or dc circuits for controlling:

1. Resistive loads not exceeding the ampere rating of the switch at the voltage applied.
2. Inductive loads not exceeding 50 percent of the ampere rating of the switch at the applied voltage. Switches rated in horsepower are suitable for controlling motor loads within their rating at the voltage applied.
3. Tungsten-filament lamp loads not exceeding the ampere rating of the switch at the applied voltage if "T" rated.

Incandescent lamps (standard household lamps) have a tungsten filament. When the lamp is off and at normal room temperature, the filament's electrical resistance is very low. The low resistance of a cold filament allows a high inrush of current (8 to 10 times the normal operating current) to flow into the lamp for the length of time it takes to reach full brilliancy. Once the lamp is at operating temperature, filament resistance is high, and normal current flows through the lamp. Switches designed to control a group of incandescent lights must have a heavy-duty spring and contacts to withstand these operating conditions safely.

Horsepower-rated Switches. Switches are also used to control large motors. Recall that a motor requires three to five times as much current when starting as it does when running. To be capable of handling these high-current situations, a switch must be equipped with rugged contacts and a strong spring.

Since motors are rated in horsepower, the power rating indicates the maximum motor size that the switch can control safely. NEC *Article 430* requires that motor switches be marked in horsepower rating.

DC-rated Switches. Some industrial equipment is designed to operate on direct current. Switches for control of this equipment must have a dc rating. Once an arc is started between direct-current switch contacts, it follows the contacts, continues to burn, and destroys the switch within seconds. (Arcing between alternating-current switch contacts is usually less severe and damaging.)

Switches designed for direct-current use must have a very strong spring and contacts, together with adequate insulation.

CO/ALR Snap Switches. Snap switches rated at 20 amperes, or less, directly connected to aluminum conductors, shall be listed and marked CO/ALR.

DIMMER SWITCH RATING

A dimmer switch consists of an electronic circuit sensitive to the amount of load placed upon it. When installing a dimmer switch, (1) the electric power to the circuit *must* be turned *off,* and (2) care must be taken *not* to exceed the wattage rating on the unit. The total wattage of the lamps being controlled by the dimmer must be not greater than the wattage rating marked on the switch. A dimmer switch for incandescent lamps cannot be used for fluorescent lighting. However, dimmers for fluorescent lighting are readily available.

LIGHTING-LOAD CALCULATIONS

To determine the ampere rating of a lighting circuit, first *add* the wattage of all the lamps to be controlled by the switch. Then, *divide* the total by the circuit voltage to calculate the amperage (current flow) for the circuit.

For example, if a room has six lighting fixtures equipped with 100-W lamps on a 120-V circuit, the total wattage is 600 W (6 lights × 100 W). The current flow is 600 W divided by 120 V, which is 5 A. A 10-A, T-rated, 120-V switch would, therefore, be adequate for this installation. The current ratings for fluorescent lights, however, are usually marked on the ballast of each fixture. Adding these ampere ratings will give the current rating without the need of further calculations.

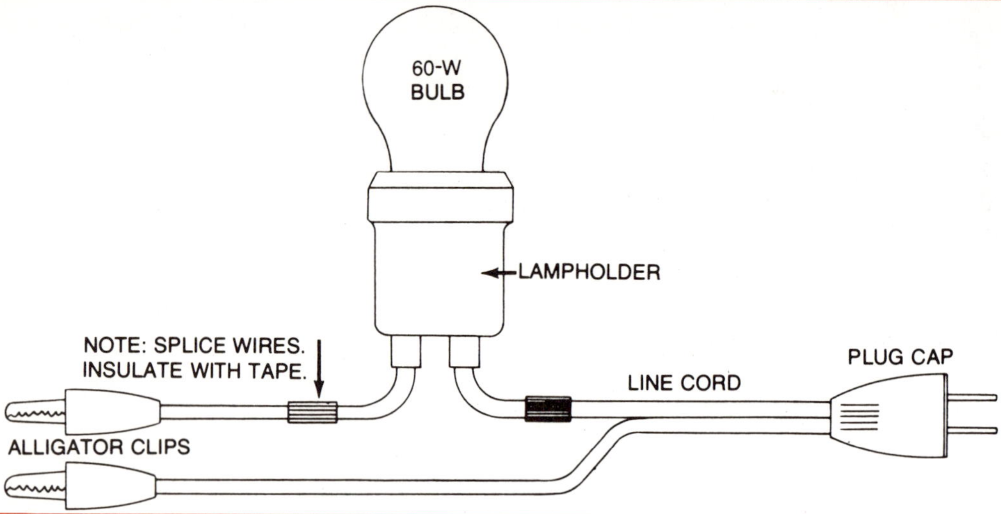

Fig. 2-6 Series-lamp tester (120-V device).

EQUIPMENT FOR TESTING SWITCHES

Switch mechanisms are usually completely enclosed, and the internal connections are hidden from view. Test meters are available for checking voltages, amperages, and resistances. Some meters, called *multimeters,* accomplish all of these tests. However, simple test equipment for locating the internal connections is available.

Both of the testers described in the next two paragraphs may be checked by touching the test clips together.

Series-lamp Tester. Figure 2-6 shows a tester made from a lamp socket, light bulb, and line cord. After the clips are fastened to the switch terminals, and if the lamp is on, the circuit is a complete, or closed, circuit.

A series-lamp tester should *not* be used on a switch connected to a circuit having its own source of voltage supply. The operator also must be careful *not* to touch the metal portion of the test clips, since this unit operates on 120 V.

Buzzer Tester. Figure 2-7 shows a tester made with a pair of dry-cell batteries, and a small buzzer or bell which signals when the circuit is closed. This simple unit is portable, and may be carried in a tool box.

The buzzer tester is to be used *only* after the unit to be tested has been removed from the live or energized circuit.

SWITCH WIRING SYMBOLS

A system of electrical symbols is often used to represent various types of switches on wiring diagrams. Figure 2-8 shows symbols commonly used on electrical drawings.

SWITCH APPLICATIONS

Switches are available in a variety of types for a number of uses. *Single-pole* and *double-pole* switches are used to control

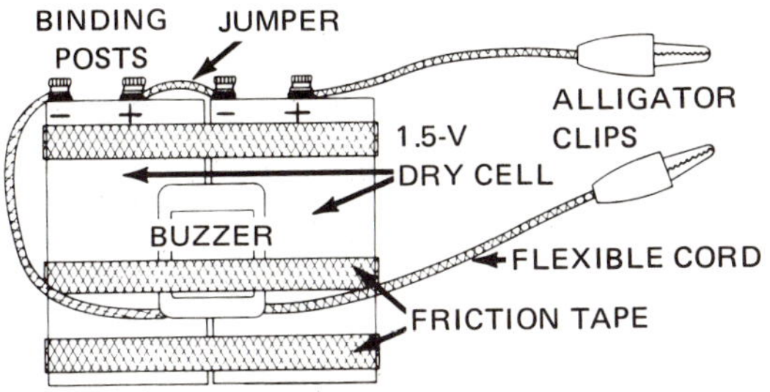

Fig. 2-7 Buzzer tester (battery powered). (Not to be used on live circuit.)

lights or equipment from one location. *Three-way* switches are used to control lights in such areas as rooms with two entrances, in hallways, at stairways, or wherever it is convenient to have control from either of two locations. *Four-way* switches are used where multiple-switch control is needed, such as in large houses, apartment buildings, or office buildings that have large rooms, three or more entrances, or stairways rising three or more floors.

Single-pole Switch. A single-pole switch is most commonly used for residential and commercial lighting circuits for such installations as fractional horsepower (kilowatt) motors, portable appliances, and portable tools.

A single-pole switch is designed to control a 120-V circuit (one live wire) from one location. The two terminal screws and the indicating marks (*on* and *off*) on the operating handle make it easy to recognize. Some manufacturers use a small dot on the operating handle to indicate the *on* position. Single-pole switches are available in a variety of styles and operating mechanisms, with current ratings to suit any installation for which they are designed.

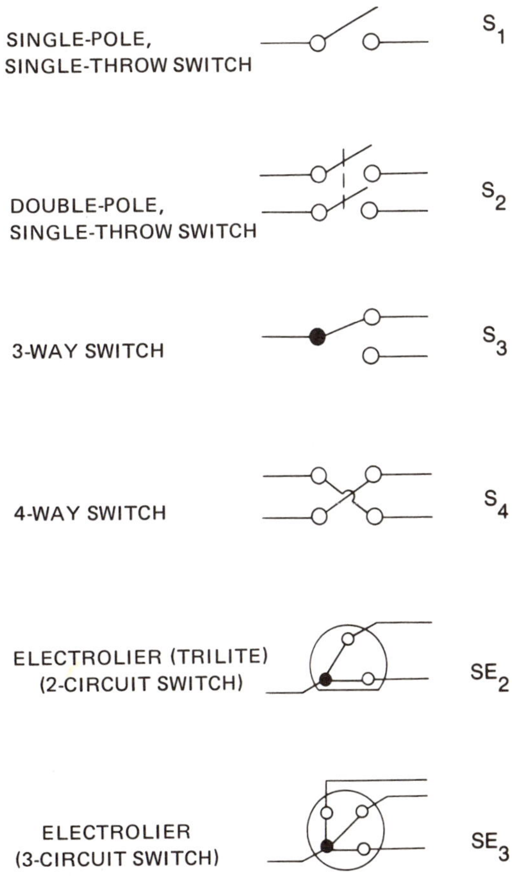

Fig. 2-8 Switch wiring symbols.

When installing a single-pole switch with a toggle or rocker mechanism, it is standard procedure to mount the switch with the operating handle in the *up* position (when the switch is turned on). Figure 2-9 is a schematic diagram showing a single-pole switch controlling two lights. Figure 2-10 shows how to connect this type of switch in a residential wiring circuit.

Remember that *only* the *live wire* is connected to the switch. (See "Switching the Live Wire," Chapter 1.) Figure 2-11

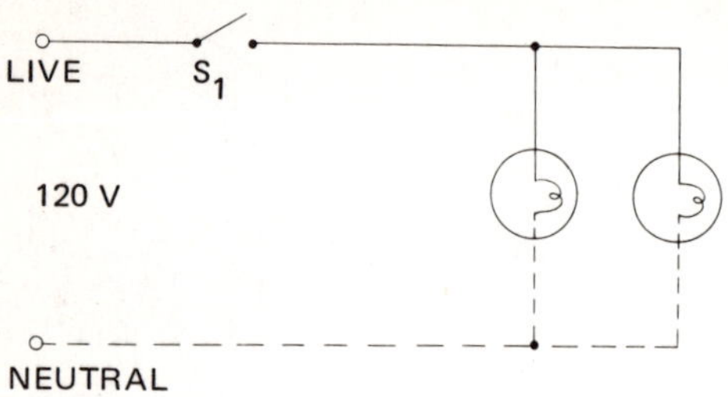

Fig. 2-9 Schematic wiring diagram showing two lamps controlled from one location.

shows the *danger* of using the neutral wire to control the circuit. In either case, the light may be turned off, but *the live wire is still dangerous* at the light fixture.

Double-pole Switch. Double-pole switches are available in a variety of current ratings, but most often in the heavy-duty range for such installations as electric heaters, air conditioners, and some motors operating on 240 V. This voltage requires two live wires and a switch capable of opening or closing simultaneously both live conductors of the circuit. The double-pole switch is, in simple terms, similar to a pair of single-pole switches in the same case or body. The double-pole switch is easily recognized by its four terminal screws and the indicating marks on the operating handle.

When installing a double-pole switch, the circuit conductors must be connected to the proper terminals. This prevents short-circuit damage to the switch contacts. The diagram in Figure 2-12 shows a double-pole switch controlling an electric heater. Figure 2-13 shows an improper connection which will seriously damage the switch the instant it is turned on. Figure 2-14 shows the proper

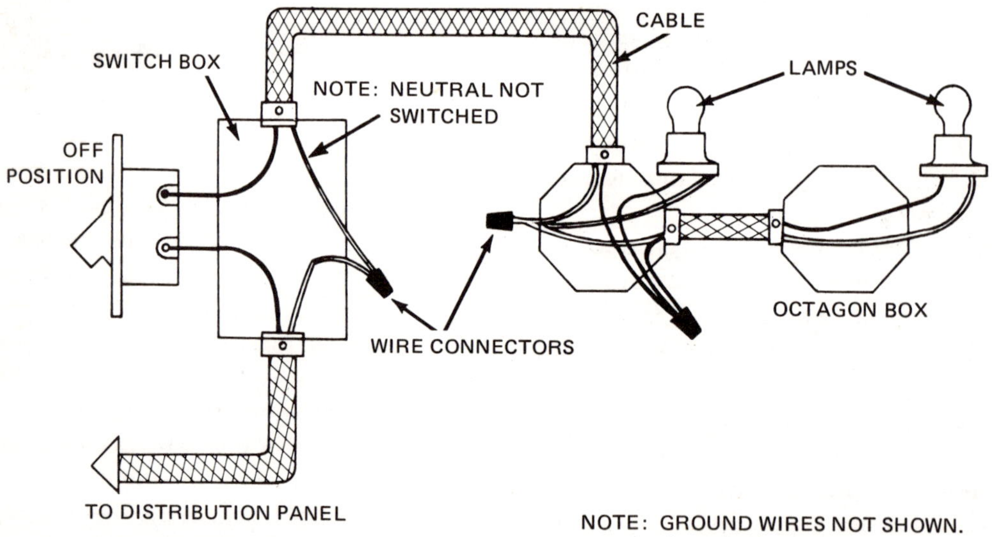

Fig. 2-10 Cable wiring diagram showing two lamps controlled from one location.

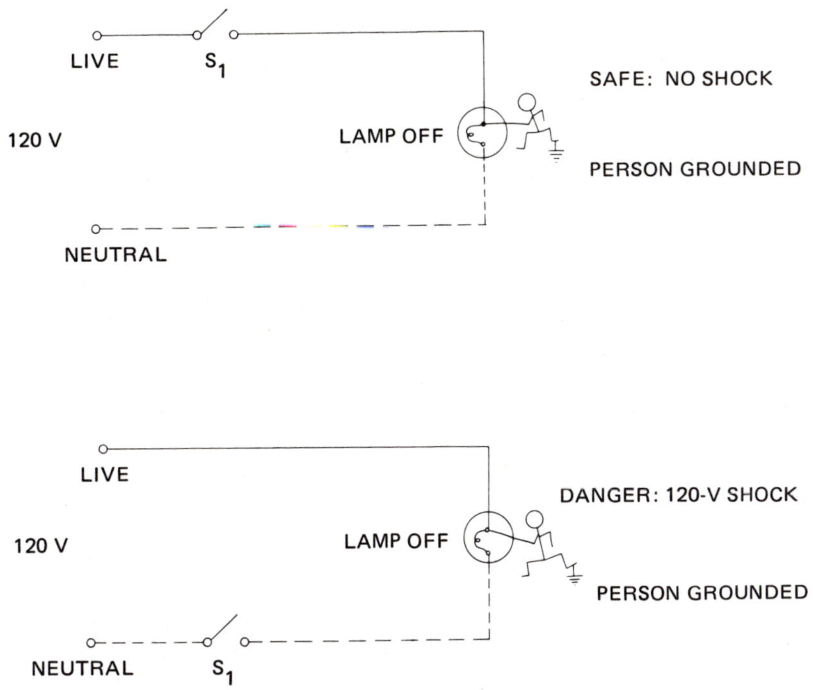

Fig. 2-11 Danger in connecting a switch in the neutral wire.

Fig. 2-12 Schematic wiring diagram of a double-pole switch.

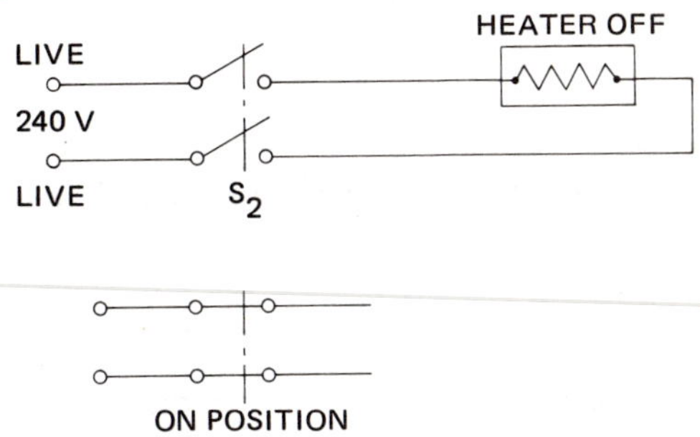

Fig. 2-13 Contact damage results if a double-pole switch is connected improperly.

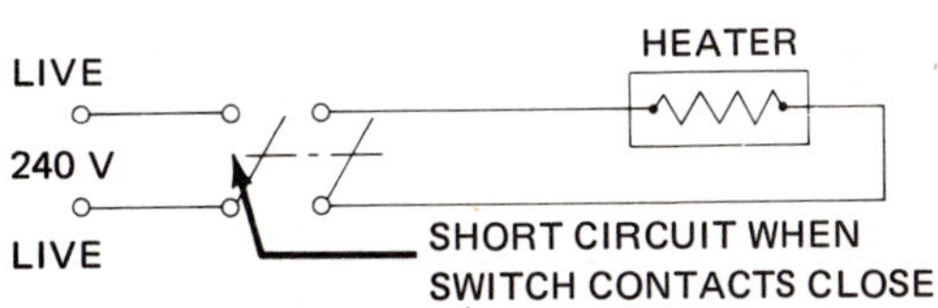

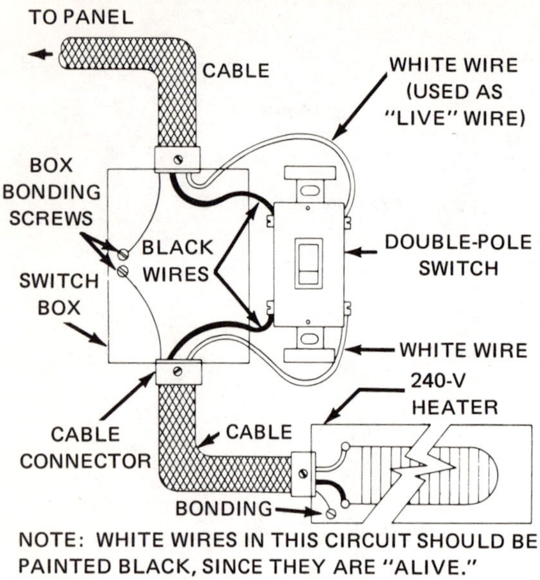

Fig. 2-14 Cable wiring diagram of a double-pole switch.

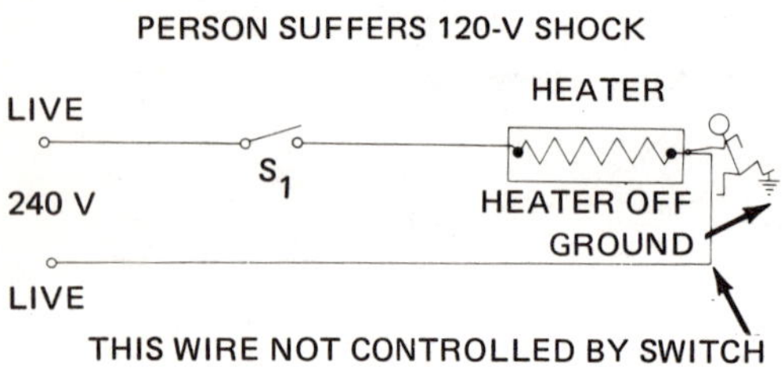

Fig. 2-15 Never use a single-pole switch on a 240-V circuit.

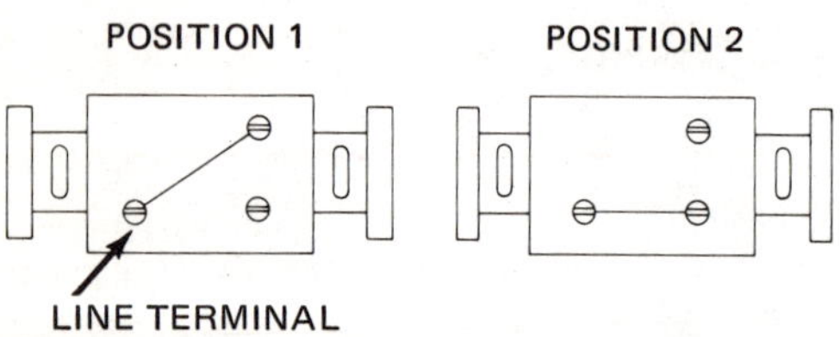

Fig. 2-16 Internal connections of the three-way switch.

way to connect this type of switch in a residential wiring circuit. Remember that a single-pole switch must not be used on a 240-V circuit. Figure 2-15 shows the *danger* of such a connection.

Three-way Switch. Three-way switches are used in pairs, usually on 120-V circuits, to control one live wire in order to provide independent control of a light or a group of lights from either of two locations. A three-way switch is easily recognized by its three terminals, one of which is marked as a *line* terminal, and by its lack of indicating marks. (The word *line* may be printed beside the terminal, or the terminal may be colored for easy recognition.)

The internal design of a three-way switch allows current to flow through the switch in either of its two positions. For this reason, the switch has no *on/off* marks on the operating handle. The circuit is controlled by using two switches as a team. Figure 2-16 shows the internal switch positions and also how one terminal – the common, or *line,* terminal – is used for both switch positions. The diagram in Figure 2-17 shows the three-way switch control of one light.

Three-way switches are available in a variety of styles and operating mechanisms,

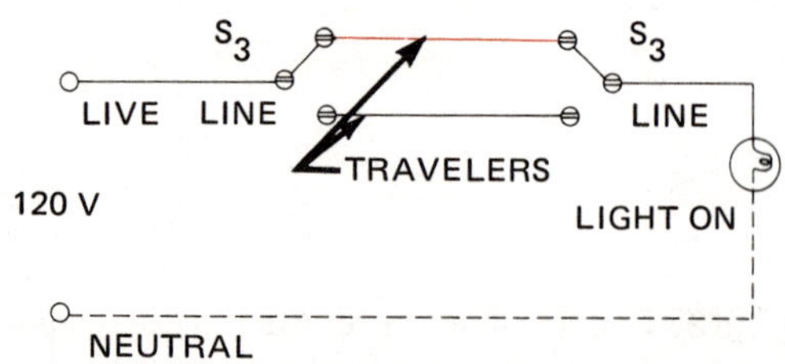

Fig. 2-17 Schematic wiring diagram showing one light controlled from two locations.

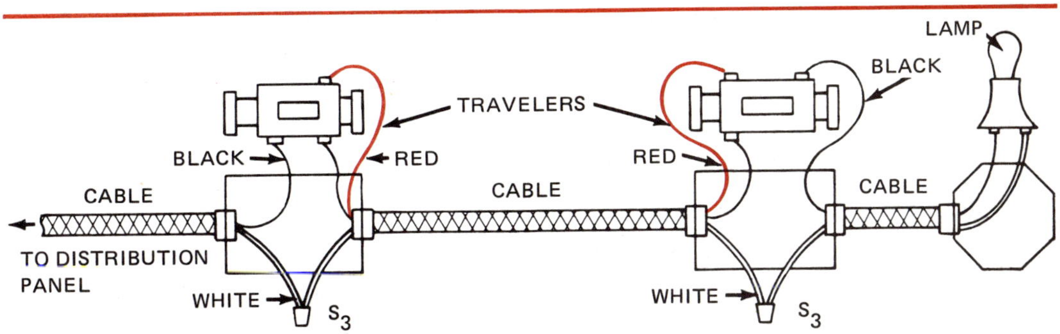

Fig. 2-18 Cable wiring diagram of 3-way switch control. (Bonding wire not shown.)

with current ratings to match the load being controlled. Figure 2-18 shows how to connect this switch in a residential wiring circuit. The two conductors running between the switches are often called *travelers,* or *messenger wires.* If the need arises, a three-way switch may also be used safely as a single-pole switch. Figure 2-19 illustrates such a circuit.

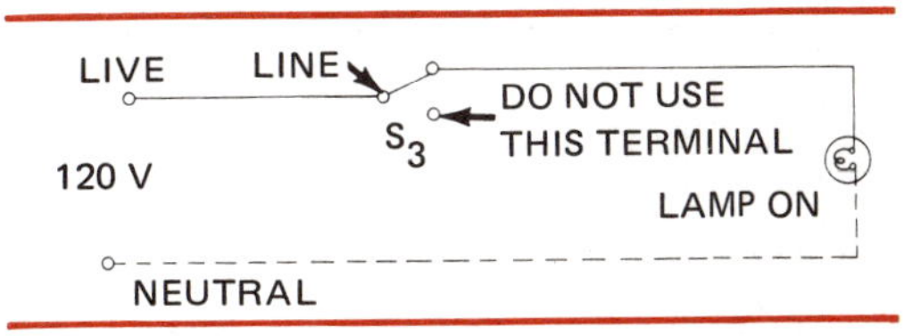

Fig. 2-19 Schematic wiring diagram showing how a 3-way switch can be used safely in place of a single-pole switch.

Four-way Switch. Multiple-switch control from more than two locations in large rooms or buildings is arranged by using four-way switches in combination with a pair of three-way switches in 120-V circuits. This combination is particularly useful in such areas as large rooms with three or more entrances, and in stairwells of buildings having three or more floors. Control of the light from any of the switch positions may be obtained by operating any one of the switches in the group.

The four-way switch is a *non-indicating* switch, because the lights can be controlled by any one switch in the circuit. It can be recognized by its four terminals and the lack of indicating marks on the operating handle. Without proper test equipment, the only way to tell the difference between the four-way switch and the double-pole switch is to remember that the four-way switch has *no* indicating marks, while the double-pole switch does.

Four-way switches are available in a variety of styles and operating mechanisms to suit any application. They may be more expensive than other types, however, because their internal mechanisms are more complicated and there is less consumer demand for them. The external terminal hookup of four-way switches differs among manufacturers. The wiring diagram usually furnished with the switch should be consulted.

Figure 2-20 shows the internal switch positions of one manufacturer, and Figure 2-21 shows the wiring for one light controlled from three locations. Figure 2-22 shows the wiring for one light controlled from five

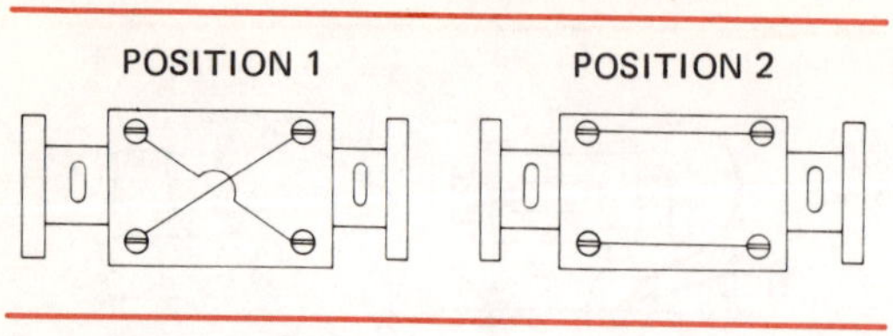

Fig. 2-20 Internal connections of the 4-way switch.

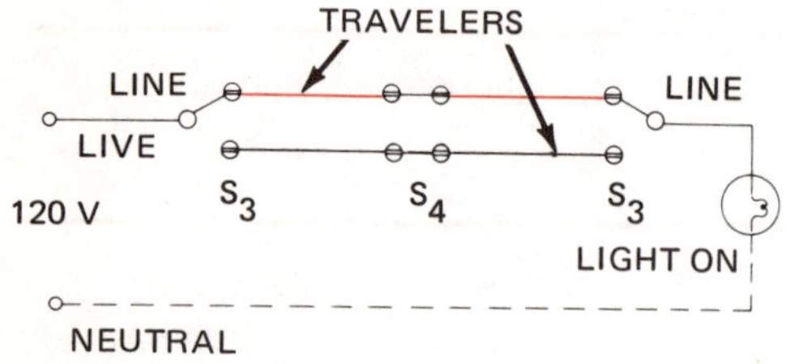

Fig. 2-21 Schematic wiring diagram showing one light controlled from three locations.

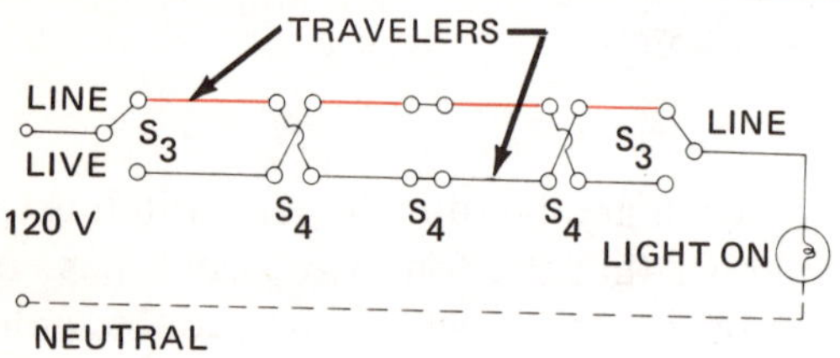

Fig. 2-22 Schematic wiring diagram showing one light controlled from five locations.

locations, and Figure 2-23 shows the cable wiring for three- and four-way switch control. *Note:* A three-way switch is *always* placed at each end of the circuit with one (or more) four-way switch in between.

ELECTROLIER SWITCHES

Two-circuit Electrolier. The two-circuit electrolier is more commonly known as the *trilite* (or three-way) switch. Used with the dual filament lamp, it provides three levels of light (low, medium, and high) for many types of table lamps, floor lamps, pole lamps, and hanging, or swag, lamps, Figure 2-24. The switch unit and bulb, in combination, give this circuit its versatility.

The trilite switch has three terminals, or leads, and is almost always a *rotary* switch. It usually has no indicating marks, since the installer may use either of the two line terminals, depending upon the choice of sequence. Trilite switches are used mainly for 120-V lighting units.

Figures 2-25 and 2-26 show the four positions of the trilite switch. One terminal, called the *common,* or *line,* terminal, is used in all three of the *on* positions. To turn the bulb off, the blades of the switch are simply moved away from the line terminal. A

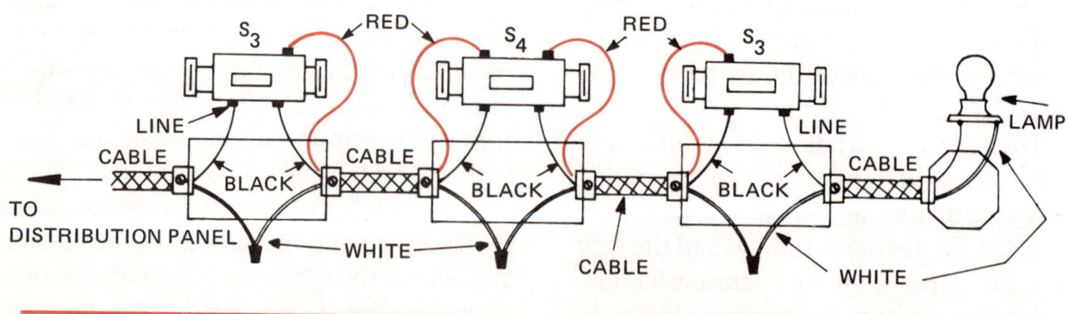

Fig. 2-23 Cable wiring diagram showing 3-way and 4-way switch control. (Bonding wire not shown.)

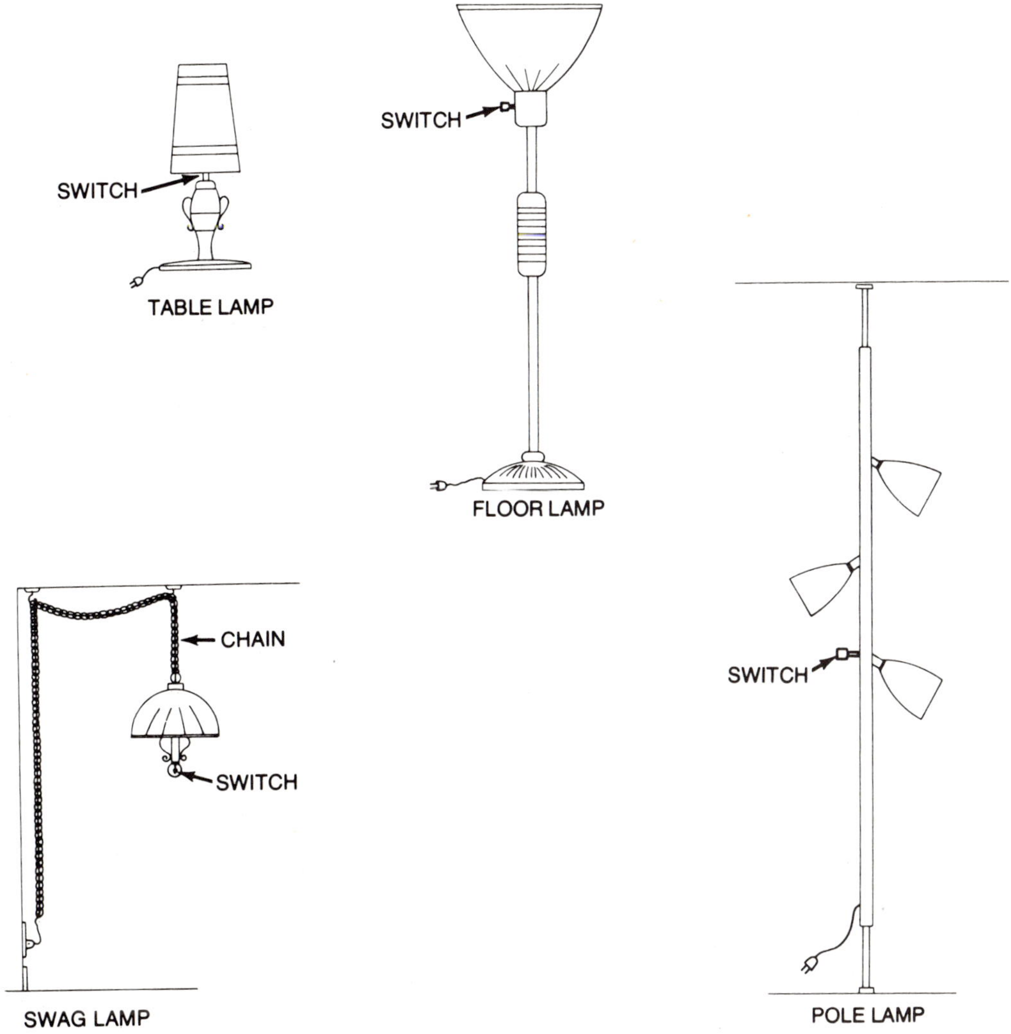

Fig. 2-24 Typical 2-circuit electrolier switch applications.

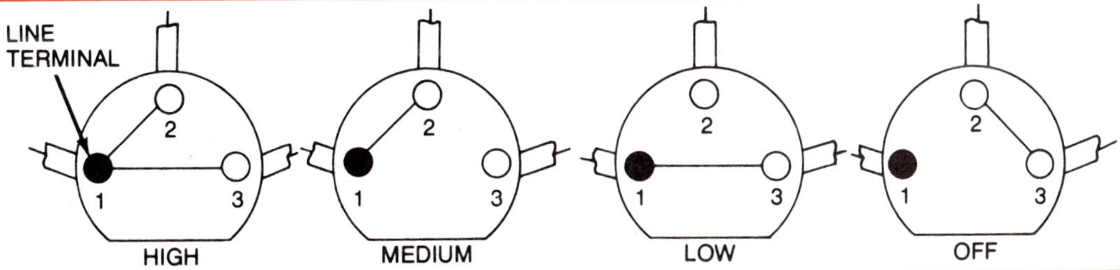

Fig. 2-25 Trilite switch positions for high to low sequence.

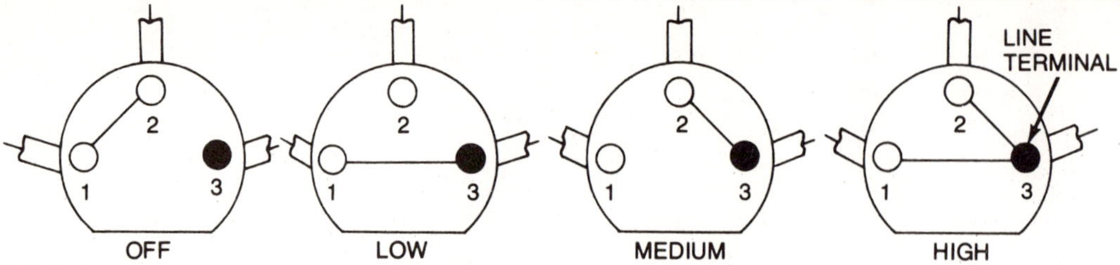

Fig. 2-26 Trilite switch positions for low to high sequence.

lamp with a switch as shown in Figure 2-25 operates in a high, medium, and low sequence. However, if the switch is connected with the line terminal, as shown in Figure 2-26, the lamp will operate in a low, medium, and high sequence.

Trilite bulbs are available in three sizes, or wattage ranges. The largest has a mogul base equipped with 100-W and 200-W filaments. Both filaments operating together provide 300 W of light in the high position. The other two bulb sizes use a medium base, and are available in 50-100-150-W and 30-70-100-W combinations. These units are popular for smaller light fixtures.

Figure 2-27 shows the switch and socket assembly wired as a working unit. Some table lamps and floor lamps are manufacturered with a switch built into the socket assembly. The only connections required are for the live and the neutral wires in the lamp's cord. The switch may be purchased as a separate unit for connection to two independent light sockets, as shown in Figure 2-28. This type of connection may be installed in fixtures using two standard light bulbs, rather than the trilite bulb.

Three-circuit Electrolier. Two types of three-circuit electroliers are available. The *alternate* type, Figure 2-29, puts one bulb on at a time and may be used for pole lamps or other decorative units. The *consecutive* type, Figure 2-30, is often fitted into pole lamps or swag lamps. All three lamps may be on at the same time.

DIMMER SWITCHES

Lighting is used to decorate as well as to illuminate. For example, the dimming or brightening of lights in dining and recreation areas can create a more pleasant atmosphere. Two types of dimmer switches, the *high-low* type and the *infinite-control* type, are commonly used in homes for these purposes. Both types function on electronic circuitry. Other types of dimmers, functioning on electrical resistors alone, produce a great deal of heat which is difficult to dissipate, and are uneconomical to use. These types are not desirable for home use.

The dimmer switch is usually equipped with leads for connection to the electrical circuit. The circuit must be deenergized before installing the switch. Take care to ensure that the total wattage of the circuit load does not *exceed* the rating marked on the switch.

High-low Switch. The high-low switch is a two-stage, flush-mounted, wall switch usual-

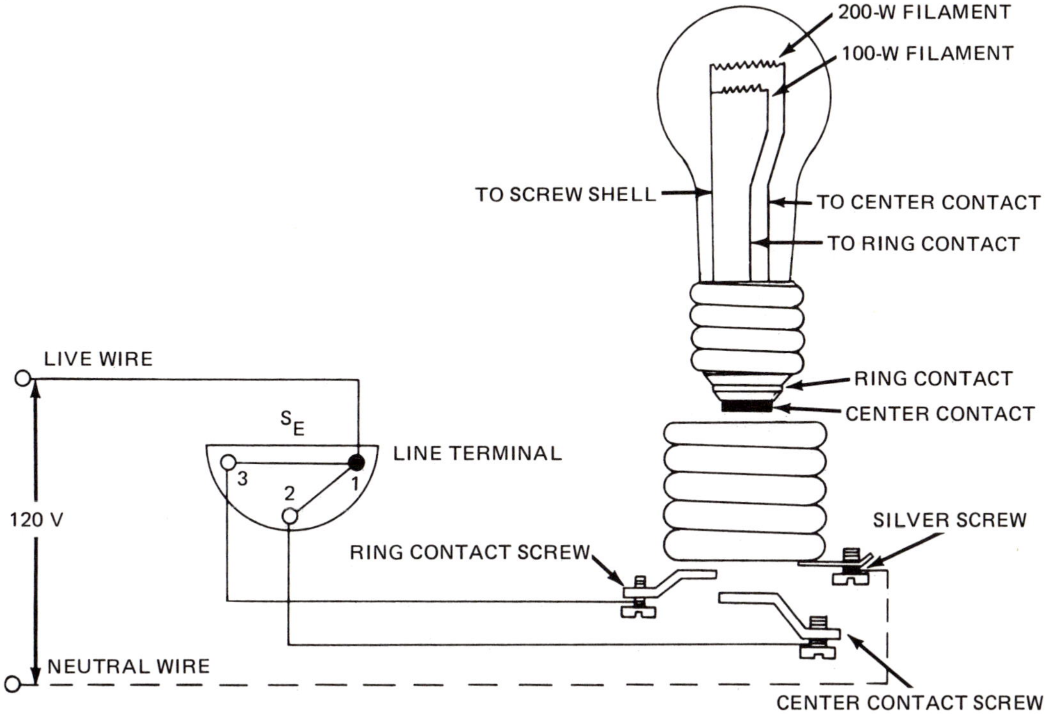

Fig. 2-27 Schematic wiring diagram showing 2-circuit electrolier switch.

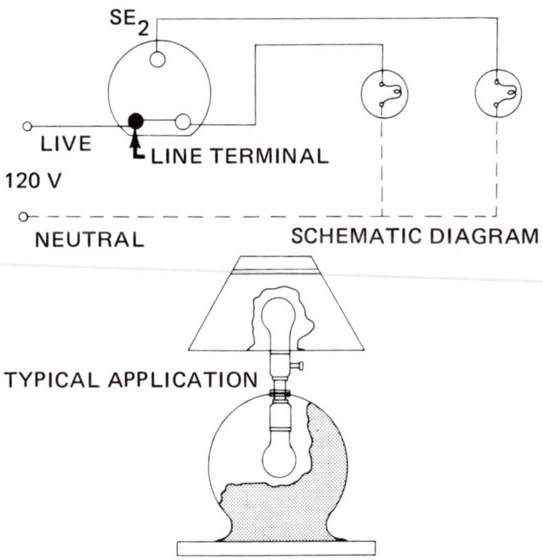

Fig. 2-28 Schematic wiring diagram and typical application showing a 2-circuit electrolier switch using two standard light bulbs.

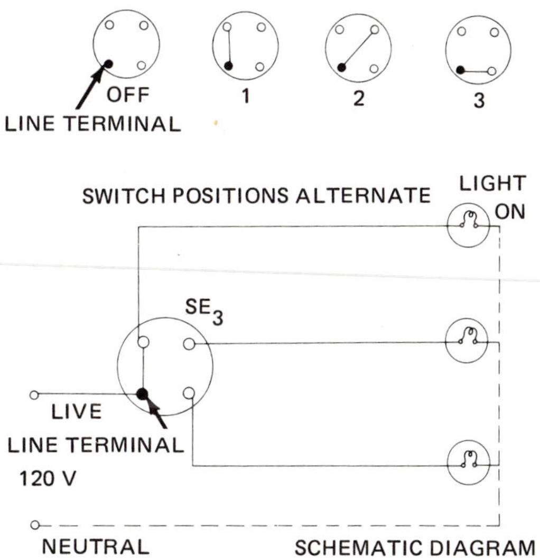

Fig. 2-29 Schematic wiring diagram and alternate switch positions for 3-circuit electrolier.

ly having toggle action, Figure 2-31. When the operating handle is in the *up* position, the standard light bulb in the fixture glows at full brilliancy. When the operating handle is placed in the *down* position, the bulb glows at the lower level of light. The light is turned off by moving the handle of the switch to a *center* position.

The high-low switch is wired in a way similar to any single-pole switch. Take care to install a switch of sufficient wattage rating to handle the lighting load of the circuit.

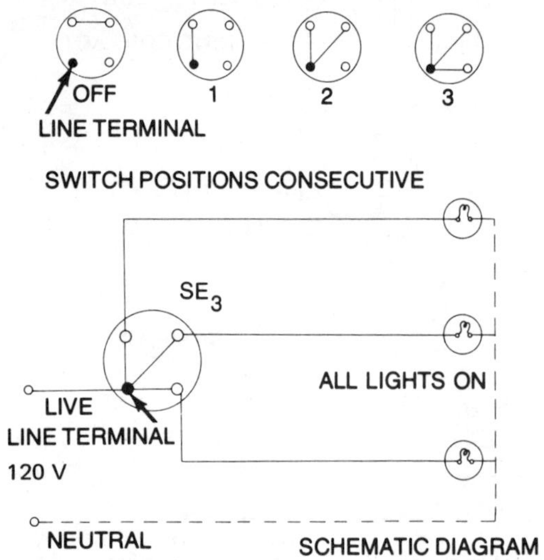

Fig. 2-30 Schematic wiring diagram and consecutive switch positions for a 3-circuit electrolier.

Infinite Dimmer Control. The infinite dimmer control mounts in a standard electrical switch box, and can replace any standard flush-mounted switch. The majority of these controls are wall-mounted; however, units built into light sockets for table lamp and floor lamp applications are available. Both single-pole and three-way units are made. The standard light bulbs used are turned on or off by pressing the control knob, Figures 2-32 and 2-33. Any level of light, between off and full brilliancy, can be obtained by rotating the control knob.

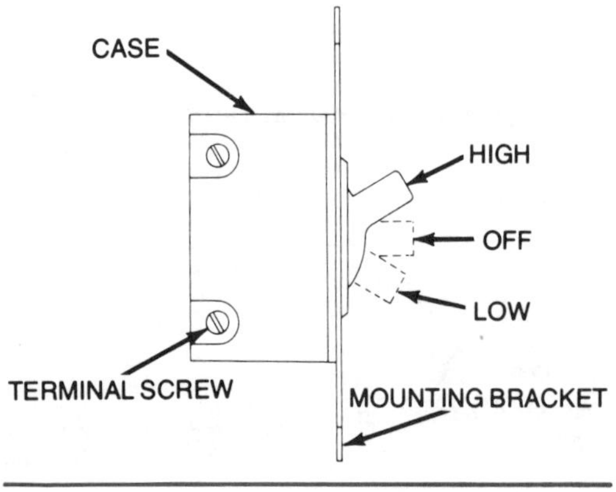

Fig. 2-31 Toggle action of a high-low dimmer switch.

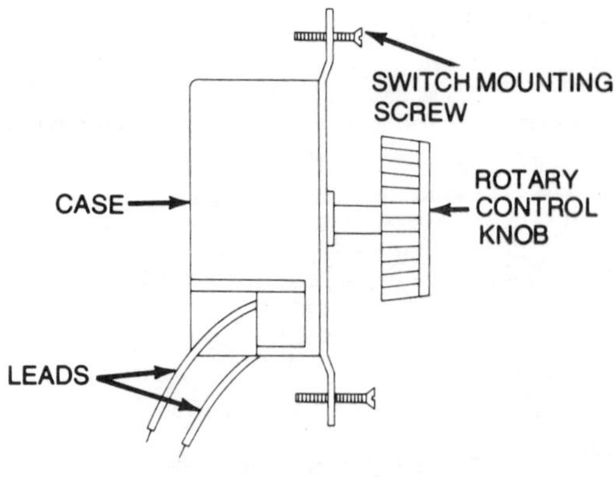

Fig. 2-32 Typical infinite dimmer switch.

Fig. 2-33 Dimmer switch with metal cooling-fin faceplate (wall-mounted).

REVIEW

1. What is the purpose of a switch in a circuit?
2. List and describe the two main types of switches, according to method of installation.
3. List seven types of switches used in lighting control.
4. List the parts of a switch, and state the purpose of each.
5. Define *arcing,* and describe its effect on a switch.
6. List the six electrical ratings that may be marked on switches.
7. Why do manufacturers mark electrical ratings on switches?
8. Explain the function of the Underwriters' Laboratories.
9. An oil burner has a 560-W motor operating on 240-V ac. If the motor draws 6.9 A, what type of switch control is needed? Which electrical ratings should appear on the switch?
10. A room is equipped with twelve 120-V light fixtures, each having a 100-W bulb. What type of switch is required to control these lights? Which electrical ratings should be marked on the switch?
11. Explain how to use a series-lamp tester to locate the internal connections of a switch.

12. Explain why the neutral wire should *never* be used to control a circuit.
13. Explain why a single-pole switch should *never* be used to control a 240-V circuit.
14. May a double-pole switch be used safely to control a 120-V circuit? Explain, with the aid of a diagram.
15. Explain why care must be taken when connecting a double-pole switch to a 240-V circuit.
16. List three places in a home where three-way switch control is useful.
17. By diagram show a three-way switch being used to replace a single-pole switch.
18. What name is given to the two conductors that connect two three-way switches to each other?
19. List the types and quantities of switches required for stairway lighting with control from each floor of a six-story building.
20. List three applications of a two-circuit electrolier (trilite) switch control in the home.
21. Which terminal of the electrolier switch must be located to determine lighting sequence?
22. Explain why you think it would be desirable to be able to change the level of light in a room.
23. Explain how a trilite lamp produces three levels of light.
24. What advantage does an infinite dimmer control switch have over a trilite switch?
25. Explain why care must be taken not to exceed the wattage rating on a dimmer switch.

Chapter 3

Lampholders

SCREW-BASE LAMPHOLDERS

Many types of lampholders are used for residential and commercial lighting installations. This chapter covers only the *screw-base* type of lampholder.

SCREW-BASE SIZES

Five standard screw-base sizes are manufactured, each of which has a particular area of use. As shown in Figure 3-1, these sizes are: mogul, medium, intermediate, candelabra, and miniature.

Mogul. The mogul is the largest of the screw-base sizes. It is used in residential trilite lamps, and for commercial lighting installations in such areas as parking lots, service garages, and warehouses. The larger light bulbs employed in these areas make use of the mogul-size base.

Mogul-base light bulbs range in size from 300 W to 1500 W. The National Electrical Code has set 1500 W as the maximum size for incandescent bulb mogul sockets. The 1500-W bulbs produce a great deal of heat as well as light. For this reason, mogul-base sockets are usually manufactured with a heavy-gauge brass socket, and a porcelain covering to withstand the heat.

Bulbs for the mercury-vapor type of lamp exceed 1500 W. However, they are also equipped with a mogul base.

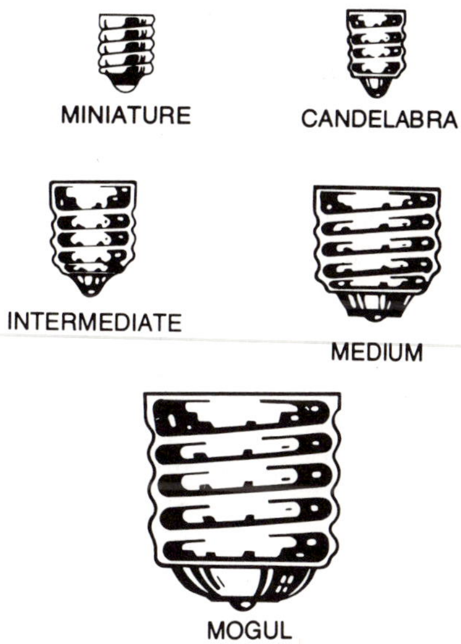

Fig. 3-1 Standard screw-base sizes.

Medium. Medium is the most common residential socket size. Standard incandescent light bulbs from 7.5 W to 300 W are equipped with medium bases, as is the 120-V plug fuse for the distribution panel. Any commercial lighting installation using a bulb size of up to 300 W is also equipped with a medium-base socket.

The National Electrical Code requires that a lamp in excess of 300 W shall not be used in medium-base lampholders. This regulation is designed to protect the user by ensuring safe operation for the range of bulbs available.

Intermediate. The intermediate-size socket is used for such purposes as outdoor Christmas tree lights, showcase and aquarium lighting, sewing-machine lamps, and such appliances as electric stoves. The intermediate socket has an electrical rating of 75 W and 125 V.

Candelabra. The candelabra is the smallest screw-base lampholder that may be connected directly to a 120-V circuit, and its electrical rating is up to 75 W and 125 V. Decorative lighting takes full advantage of this small socket. The light bulbs used in this socket are produced in a wide variety of shapes, often simulating the flames of candles, and are used for indoor Christmas tree lights and crystal chandeliers.

Miniature. The miniature is the smallest screw-base lampholder in the standard group. It is used for such installations as panel lights in radio or television sets to illuminate the dial. One type of Christmas tree light string also uses these sockets in a 120-V series circuit. The voltage is divided among the number of lights in the string, so that, for example, if there are eight lights, each light receives approximately 15 V.

The miniature-size lampholder does not carry a 125-V rating due to the limited space for contact separation inside the socket.

ADAPTERS

An *adapter* is a device that can be used to *reduce* a socket size so that the lampholder will accept a bulb with a smaller base. An adapter has an external thread to fit the socket being reduced, and an internal thread to fit the smaller base of the bulb. Adapters are also made that allow a lampholder to accept two bulbs, rather than one, and/or a plug from an extension cord.

When using adapters, take care *not* to place a load on the lampholder in excess of its electrical rating in amperage, voltage, or wattage. When an extension cord adapter is being used, remember that no effective ground connection is available. This means that the protection offered by a grounded tool or device is lacking. Remember, also, that it is *dangerous* to attempt to *increase* socket size by using adapters, because the heat and current from the larger bulb may damage the smaller socket.

LAMPHOLDER CONSTRUCTION

The basic components of all lampholders are similar, Figure 3-2. The lampholder body, however, may be made of different materials, such as Bakelite, rubber, porcelain, brass, or aluminum, Figure 3-3.

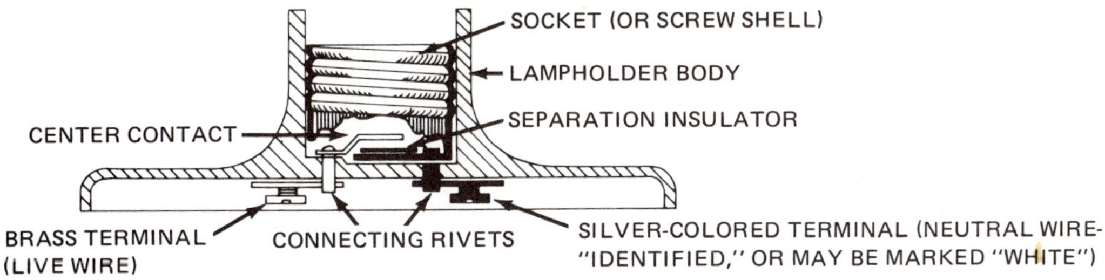

Fig. 3-2 Typical lampholder construction.

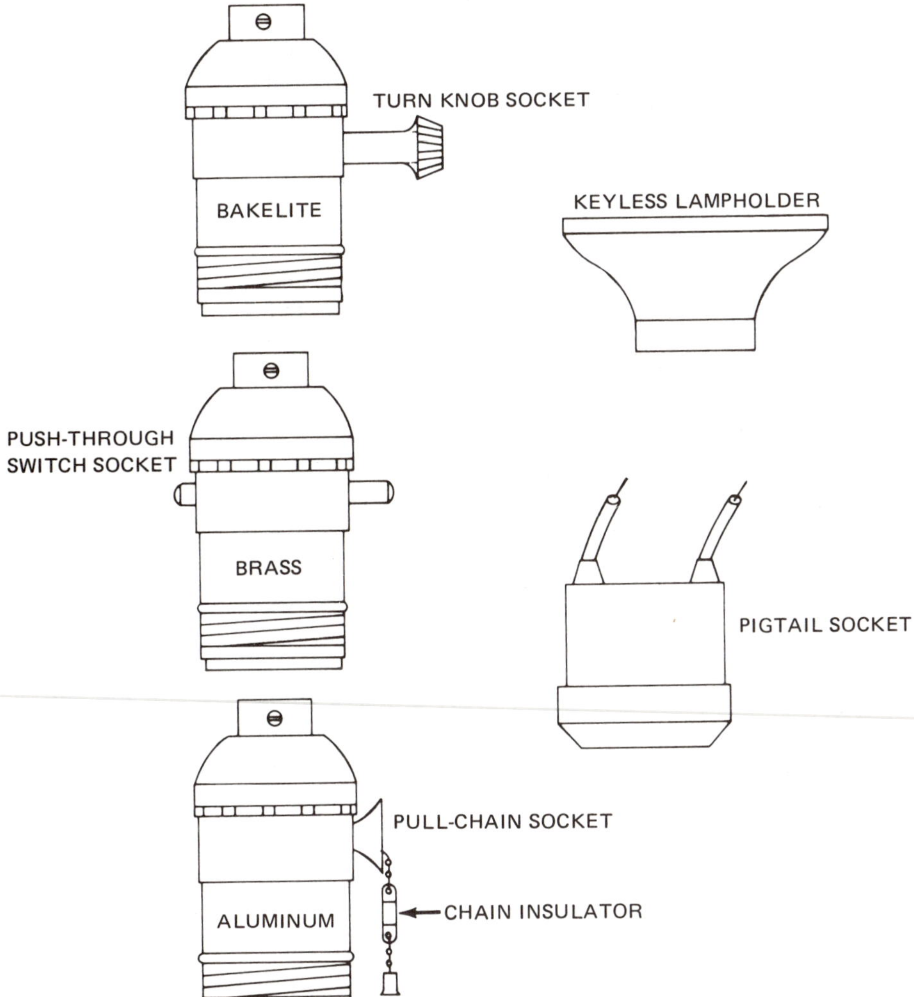

Fig. 3-3 Types of medium-base lampholders.

Fig. 3-4 Key-type push-through switch socket.

Fig. 3-5 Keyless-type lampholder.

LAMPHOLDER SWITCH MECHANISMS

Lampholders with built-in switch mechanisms are called *key* types, Figure 3-4. The key term came into use when lampholders had ornate, rotary switch handles similar to door keys. Lampholders without switches built into the socket assembly are called *keyless* types, Figure 3-5.

CIRCUIT CONNECTION

The terminals on lampholders are color coded for easier connection of the circuit conductors. (Most of the lampholders mentioned in this chapter are designed for 120-V circuits and make use of this color code.)

A 120-V circuit has a live and a neutral wire, which must be connected to the proper terminals at the socket. One terminal is a natural brass color and is designed to receive the live wire (usually black). (Inside, this brass terminal joins with the center contact.) The white neutral wire is fastened to the silver-colored terminal (or the word "white" may be imprinted). The terminal screw is joined to the screw shell of the lampholder, (see Figure 3-2.)

Note: The National Electrical Code requires that a terminal intended solely for the connection of a neutral wire *must* be identified by a tinned finish, a nickel-plated finish, the word "white" imprinted, or by means of some distinguishing mark.

It is important to use the proper method of connection in order to avoid the danger of shock. A person changing light bulbs or cleaning a light fixture with a damp cloth could easily touch the screw shell and become grounded on the bathroom or kitchen sink. If the screw shell is alive, the person could receive a 120-V shock. With proper wiring connections, the *only* live part of the socket is the *center contact,* which is not easily touched by accident.

PULL-CHAIN INSULATORS

The National Electrical Code requires that lampholders with a pull-chain switch mechanism have an insulating link in the chain, Figure 3-6. This is to prevent a person in contact with ground from receiving a 120-V shock if a defective switch unit makes the chain alive.

Fig. 3-6 A lampholder with a pull-chain mechanism.

LOCATION OF LAMPHOLDERS

Article 410 of the National Electrical Code outlines the location of lampholders. These guidelines are subject to frequent change, because new products and materials are constantly being introduced. When planning an installation, refer to the most recent edition of the NEC.

REVIEW

1. List the five standard screw-base lampholder sizes, and give one practical application for each.
2. What is the wattage rating for incandescent lamps for the largest screw-base lampholder?
3. Which electrical ratings for screw-base lampholders are required by the National Electrical Code?
4. Why is the bulb size for the medium-base socket limited to 300 W?
5. List two types of adapters used with light sockets.
6. What precaution is necessary when using an adapter?
7. List the types of lampholder switch mechanisms. What is meant by the term *keyless*?
8. Explain how the conductors from a 120-V circuit must be connected to a lampholder. Why?
9. Why is an insulating link required in the chain of one type of switch mechanism?
10. Where can information be found regarding the installation of lampholders?

Chapter 4

Receptacles

RECEPTACLE USES

The *receptacle,* or outlet, is the point in a circuit at which power may be taken to supply lamps, appliances, or portable plug-in devices. Receptacles are the most widely used electrical devices. A residential receptacle delivers 120 V to any electrical device plugged into it. Every modern home is equipped with a number of duplex receptacles.

DUTY RANGES

Receptacles are available in a variety of qualities and duty ranges. *Light-duty* receptacles (rated 15 A) may be used in living rooms or bedrooms, where a table lamp or radio will be the largest current-demanding device plugged in. *Heavy-duty* receptacles (rated 20 A) should be installed in the kitchen area, however, where such appliances as electric frypans and coffee makers, requiring a current of 12 A to 15 A, will be plugged in. If light-duty receptacles were installed, they would heat up, the contacts would become soft and lose their ability to grip the plug firmly, and *dangerous* overheating would follow. The result would be damage to the receptacles, wiring, and plugs. The danger of an electrical fire also exists as a result of this overheating.

Careful advance planning is necessary in order to prevent overloading of receptacles. For trouble-free service, the duty range of the receptacle should be matched to the type of appliances expected to be using the outlet. The ampere rating of the receptacle must also be matched to the branch-circuit rating and conductor ampacity.

NUMBER OF OUTLETS ON A CIRCUIT

Article 220 of the National Electrical Code outlines requirements for the installation of receptacles. The number of outlets per circuit, the location, and the number of outlets required for a given room or location are discussed in detail. The NEC should be consulted for accurate, up-to-date information.

RECEPTACLE CONSTRUCTION

Receptacles are made in many shapes and styles, but all have basic similarities. Figure 4-1 shows the standard duplex, "U"

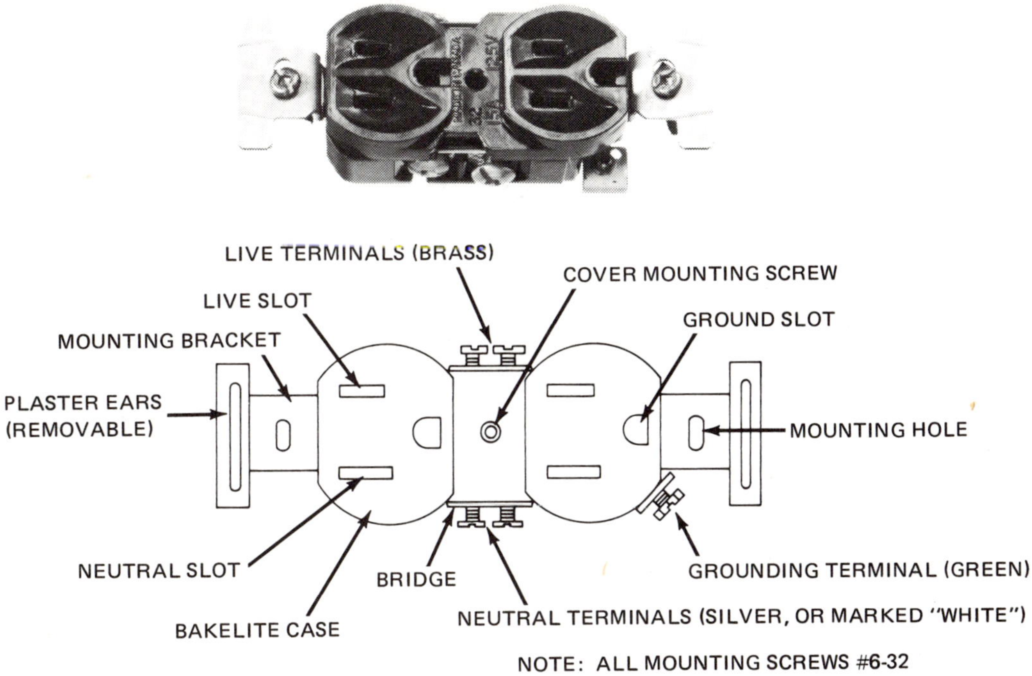

Fig. 4-1 Standard NEMA "U" ground receptacle, and illustration of its construction.

ground receptacle used in the modern home. This receptacle is also widely used in office buildings, stores, and industry. It provides a firm electrical connection for 120-V equipment, together with ground protection for portable equipment.

RECEPTABLE COMBINATIONS

Receptacles are available in single, duplex, and three-plug units. The duplex receptacle, which is designed to receive two electrical plugs, is the most common. Some duplex units have one half as a "U" ground, and the other half in a different shape. Some receptacles are teamed with switches which allow a light to be turned on or off either in the room or at the receptacle itself.

RECEPTACLE AND PLUG SHAPES

Figure 4-2 shows some of the configurations (shapes) of receptacles that are available. Many other shapes are made for specialized use. This chapter discusses only the more common receptacles and plugs.

The National Electrical Manufacturers Association (NEMA) specifications and standards cover the design, construction, and configurations of receptacles.

Old-style Two Prong. Many buildings constructed during the 1940s and 1950s were equipped with two-prong units. The main disadvantage of this type of receptacle is the lack of grounding protection. When this type of receptacle is replaced, a grounding-type receptacle must be used.

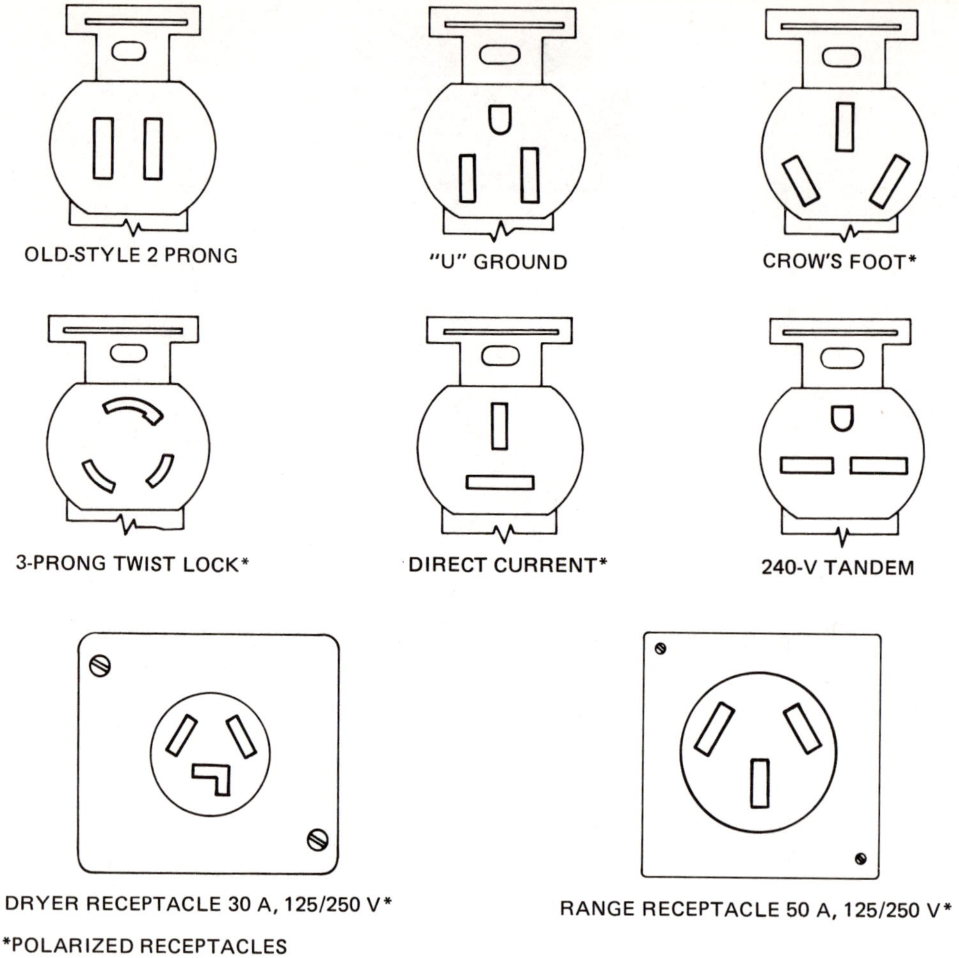

Fig. 4-2 Common receptacle and blade patterns (configurations).

"U" Ground. The "U" ground receptacle accepts both the two-prong plug and the three-prong "U" ground plug, Figure 4-3. If the ground prong is removed from the three-prong plug, it will still fit the receptacle in one direction only. (*Caution:* removing the ground prong violates the Code and renders the plug unsafe.) Careful inspection of the receptacle shows that the live and neutral slots are of different sizes to prevent interchanging of the connections.

Crow's Foot. The crow's-foot plug was an early attempt to provide ground protection in a receptacle. Unfortunately, the shape of the two-prong plug could be modified with pliers, allowing it to be fitted into the receptacle. The crow's-foot plug also could have its ground prong removed and blades reshaped for use in standard two-prong receptacles. Ground protection was lost in both cases, and it became obvious that a change in design was needed.

Fig. 4-3 A specification-grade "U" ground receptacle, 125 V ac.

Twist lock. The twist-lock receptacle is available in two-, three-, and four-prong units, and in several voltage and ampere ratings. Its major advantage is that the plug cannot be pulled out of the receptacle accidentally. The three- and four-prong units provide ground protection.

Tandem. Air conditioners, motors, and heaters of 240 V make use of the tandem device, Figure 4-4. It is equipped with a ground prong, and is used primarily on 240-V circuits where its blade shape prevents 120-V devices from being plugged in.

Direct current. The main use for the direct-current receptacle is to keep the positive and negative conductors from being interchanged in dc systems. Some models have ground protection. Direct-current receptacles are shown in Figures 4-5 and 4-6.

Fig. 4-4 A 240-V tandem "U" ground receptacle.

Fig. 4-5 A 125-V direct-current "U" ground receptacle.

Fig. 4-6 A 250-V direct-current "U" ground receptacle.

Range and Dryer Receptacles. With a wire-spliced installation, each time an electric range or clothes dryer is connected to the cable in a house or apartment building, the cable is shortened slightly. Sometimes the cable is shortened to the point where it can no longer be used. The heavy-duty range and dryer plug and receptacle are designed to prevent this from happening, Figure 4-7. They allow the range and dryer to be pulled out from the wall for cleaning or simple removal.

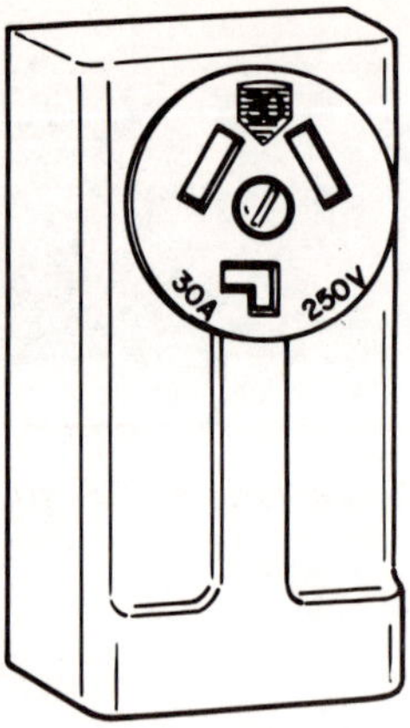

Fig. 4-7 A residential electric dryer receptacle (30 A).

RECEPTACLE GROUNDING

Receptacles equipped with a *grounding slot* are designed to provide safety for the person using the equipment connected to the receptacle. If a fault occurs in an electric drill, for example, the current will travel to the frame of the drill, then back along the grounding conductor to the distribution panel. The circuit fuse in the panel will blow and prevent the drill user from receiving a shock. If there is no grounding protection, however, the user could suffer a 120-V shock as the current flow passes from the tool's frame through the operator on its way to ground.

The grounded-equipped receptacle has a green, hexagonally shaped, terminal screw for connection to the grounding circuit of the electrical box that supports the receptacle. The grounding wire in a nonmetallic wiring system is connected to the electrical box as well as to the grounding terminal on the receptacle. A grounding terminal in the distribution panel receives the grounding wires from the circuit conductors. Metal-conduit or armored-cable systems make use of the metal casing to carry the grounding connection from the distribution panel to the receptacle box. These connections conform to the National Electrical Code requirements.

SPLIT RECEPTACLES

Split receptacles are designed so that the bridge linking the two brass-colored terminal points on the live side can be removed. The neutral side is left intact. The bridge separates electrically the upper and lower portions of the receptacle, Figure 4-8. Dividing the receptacle electrically has two major advantages.

One advantage of a split receptacle is that half of the receptacle placed, for example, in a living room may be controlled

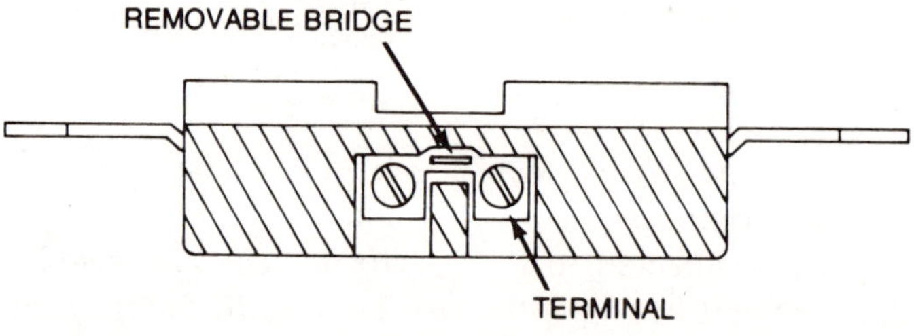

Fig. 4-8 Receptacles can be "split" by removing the bridge.

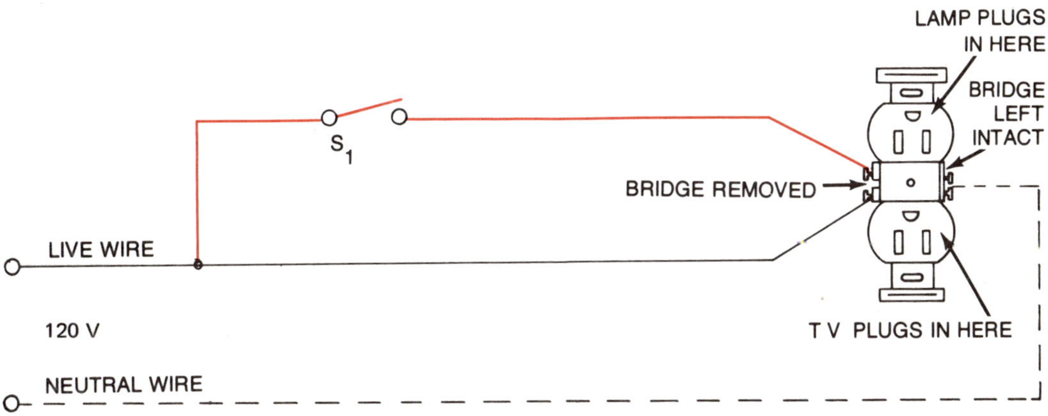

Fig. 4-9 Wiring diagram showing a split-switched receptacle.

by a wall switch. This allows a lamp to be turned on without entering a dark room. The other half of the receptacle may be made to remain alive for use with radios, televisions, or other appliances in the room. Figure 4-9 shows a diagram of such a circuit.

The second advantage of a split receptacle is its use for appliances in the kitchen area. Splitting the receptacle allows two separately protected circuits to be run to one receptacle at the counter. This permits two high-current-consuming devices to be plugged into one receptacle without overloading the circuit. Without a split receptacle, using two small electrical appliances in the same outlet would normally blow a 15-A protective device. Adequate protection for the circuit conductors is not lost when the split receptacle system is used. Figure 4-10 shows a wiring diagram of such a circuit.

The National Electrical Code requires that both live conductors (ungrounded) must be opened simultaneously. This calls for a two-pole, common-trip circuit breaker or, as it is also called, a double-pole fuse pullout.

REPLACEMENT OF RECEPTACLES

While most receptacles deliver years of trouble-free service, replacements are sometimes necessary. When a receptacle no longer holds the plug firmly, or if it heats up during use, a new unit should be installed. Receptacles are available in a variety of styles and colors to suit the room's decor, but attention must be paid to a receptacle's electrical rating to ensure safe operation.

NEMA RECEPTACLES

The National Electrical Manufacturers Association (NEMA) is an organization dedicated to standardizing electrical devices. Through its efforts, a radio of any make can be plugged into a receptacle in any town and be expected to operate satisfactorily. Receptacles have been standardized in mount-

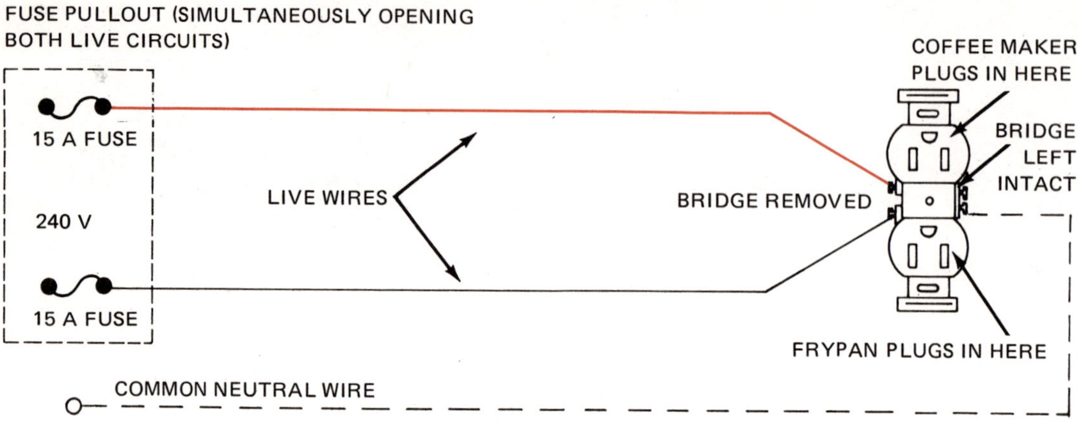

Fig. 4-10 Wiring diagram showing a double-fused split receptacle.

ing technique and slot placement. Receptacles, wall plates, and receptacle boxes from most manufacturers are completely interchangeable.

POLARIZED RECEPTACLES

The crow's-foot, dc, and twist-lock receptacles are examples of polarized units. This simply means that they will accept *only* their own style of plug, and that the plug can be inserted in *only* one way – a necessary feature in circuits where it is *dangerous* to interchange any of the circuit conductors.

PLASTER EARS

The mounting bracket of many receptacles is equipped with small extensions called *plaster ears,* Figure 4-11. When a receptacle is installed in a building with plastered walls or wood panels, plaster ears are very useful. Plaster around the receptacle box often crumbles, leaving a space with little or no support for the receptacle. The plaster ear extensions provide extra length and width to the mounting bracket, making possible a secure, flush mounting.

Receptacle installations in surface wiring boxes, such as the utility box or FS fitting, do not require plaster ears, Figure 4-12. When it is necessary to remove the plaster ears to ensure a correct surface fit, removal is a simple matter. Bend them back and forth several times with a pair of pliers.

GROUND-FAULT CIRCUIT-INTERRUPTER RECEPTACLES (GFCI)

Modern wiring regulations require 15- and 20-ampere outdoor receptacles to be of the ground-fault circuit-interrupter (GFCI) type. GFCI receptacles are also required in bathrooms, and for certain applications in garages. The local electrical code for your

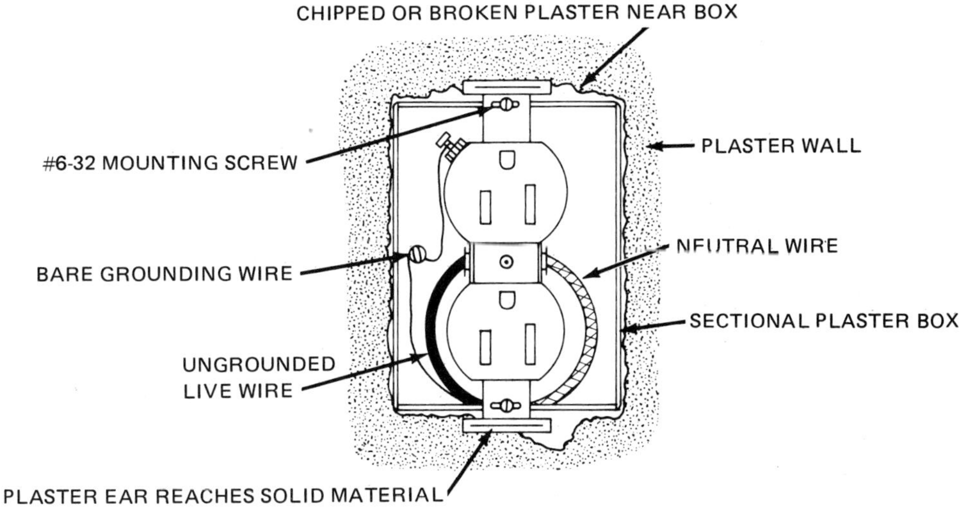

Fig. 4-11 Flush-mounting a receptacle using plaster ears.

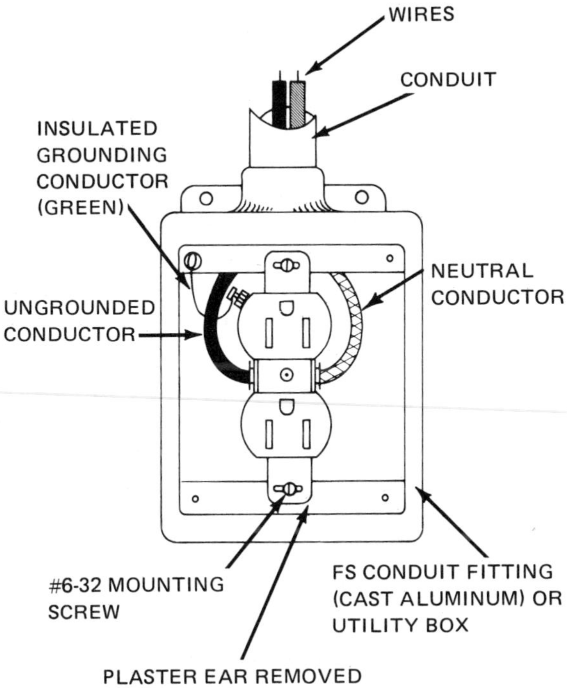

Fig. 4-12 Surface-wiring a receptacle without plaster ears.

area should be consulted to determine exactly where these units must be installed. GFCI receptacles, Figures 4-13 and 4-14, are designed to protect users of portable tools and equipment from electrical shock, which can occur if the tool or equipment becomes faulty. Each year, many people are killed or endangered by portable electric devices that have developed a defect or insulation breakdown within the device. It is not always obvious to the operator that this breakdown has occurred until the device is plugged in and a shock transmitted.

A normal fuse or circuit breaker will not blow or trip unless the current flow to ground (earth) exceeds the ampere rating of the equipment protection device. These devices are designed to protect the wire of the circuit, *not* the person using it. GFCI receptacles are designed to detect small leakage currents when they develop and then to trip open the circuit to protect the person using that circuit. When the GFCI

Fig. 4-13 Front view of a typical GFCI receptacle.

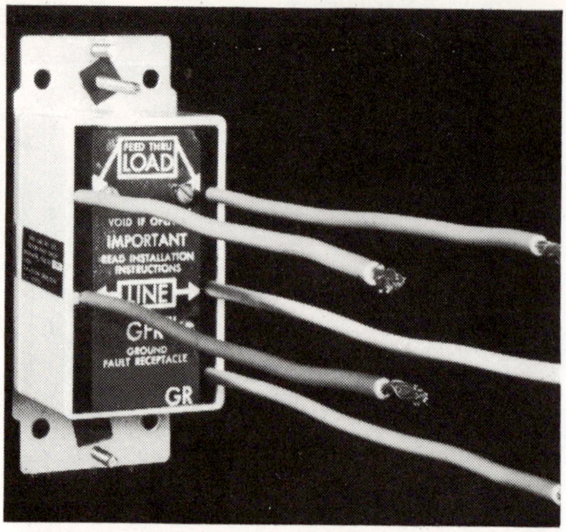

Fig. 4-14 Rear view of a typical GFCI receptacle, showing the connection leads.

is properly installed, currents as low as 5 mA (0.005 A) can be sensed by the unit and it will open the circuit within one-half cycle of current flow.

Dampness in the tool, metal shavings, or rough handling can cause the initial breakdown in the insulation of the tool or equipment. Standing on a concrete floor, outdoors on grass or earth, or on metal plumbing or frame units of a building places the operator in contact with the earth or ground.

An electrical shock can have serious effects on a person who is subjected to it. The human body does not allow a high current to flow, due to its normally high electrical resistance. However, very small amounts of current can cause pain or upset of bodily functions, such as breathing and heartbeat.

The unit of measurement for this discussion will be the milliampere (mA). One mA of current is equal to 0.001 A. A person will feel a slight shock if 5 mA of current flows through the body; not enough to harm, but sufficient to know that a shock hazard exists. Between 10 mA and 15 mA of current flow, *muscular freeze* may occur, preventing the operator from letting go of the tool being used. **At the 50 mA to 100 mA level, heart fibrillation and death occur.** As can be seen by these figures, a normal 15-A circuit fuse would be of no value to protect a person using the circuit. Chapter 16 in this text explains the functions of ground-fault devices in more detail.

One feature of ground-fault circuit-interrupter receptacles is that, once connected to a circuit, any receptacle connected *after* the GFCI will also offer ground-fault circuit-protection to the circuit. Pool areas, workshops, laundry rooms, and numerous other locations can well take advantage of the protection offered by ground-fault circuit-interrupters. Great care must be taken to connect the leads of the receptacle according to the manufacturer's instructions. If connected incorrect-

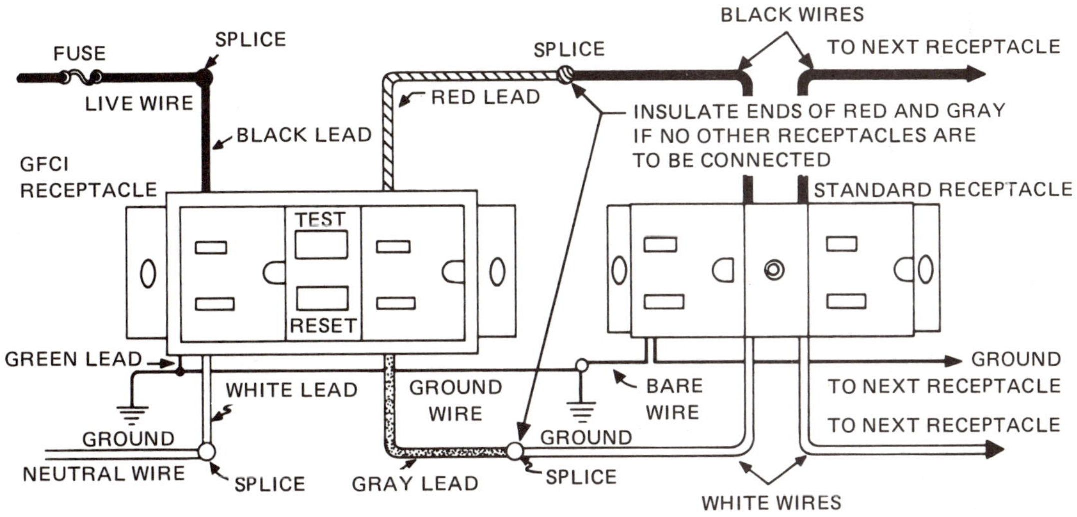

Fig. 4-15 Connection diagram for a typical GFCI receptacle installation. (Protection to other receptacles from one GFCI receptacle.)

ly, the GFCI receptacle will not provide the protection desired. Figure 4-15 illustrates a typical wiring diagram for the correct installation of a GFCI receptacle.

GFCI receptacles are equipped with test and reset buttons. The unit should be tested at least once a month by pressing the test button on the face of the receptacle. This simulates a ground fault in the circuit and the relay within the receptacle should react and open the circuit. At this time, the reset button will protrude from the face of the receptacle, indicating successful operation of the unit. Pressing the reset button back into place will reactivate the relay in the receptacle for further use. This simple test assures the owner that the unit is in proper working order. Circuit-breaker type GFCIs are also available to protect the complete branch circuit.

Prior to recent GFCI developments, electric shaver receptacles that isolated the supply voltage from the shaver were available. This eliminated the shock hazard present when electric shaving equipment was used near plumbing fixtures. However, these receptacles were too small, electrically, for combined use with modern hair dryers.

In newer homes, such receptacles have been replaced with a grounded-type GFCI-protected receptacle in the bathroom. The NEC now requires that at least one wall receptacle outlet shall be installed in the bathroom adjacent to the basin location. This receptacle shall be of the grounding type, and have GFCI protection for personnel. This subject is covered in NEC *Sections 210-7(a), 210-8(a)(1),* and *210-52(c).*

REVIEW

1. What duty range of receptacle should be selected for kitchen use? Why?
2. Where in the home may light-duty receptacles be installed safely?
3. Where can information be obtained regarding the number of receptacles required for a room?
4. Explain where the live, neutral, and grounding wires are connected on a duplex "U" ground receptacle.
5. List two uses for a switch and receptacle combination unit.
6. What is the main disadvantage of the old-style two-prong receptacle?
7. Would an extension cord fitted with twist-lock units be an advantage on a construction job site? Why?
8. Explain the purpose of the grounding slot on a receptacle.
9. Where does the ground in a receptacle originate?
10. How is the grounding circuit carried in a conduit system?
11. What is the danger in using a portable electric tool that has had the ground prong removed from its plug? Explain.
12. Name two areas in the home for the use of split receptacles.
13. What are the advantages to using split receptacles?
14. What is a NEMA receptacle? Explain.
15. List three examples of polarized receptacles.
16. What are plaster ears? Where and why are they used?
17. What type of receptacle is recommended by the National Electrical Code for use in outdoor areas?
18. What advantage does a GFCI receptacle have over other types of receptacles?
19. Why will a normal fuse or circuit breaker not protect a person using portable tools on a circuit?
20. Why is it important for a GFCI to detect leakage currents as low as 5 mA?

Chapter 5

Conductors

CONDUCTOR IMPORTANCE

The conductor that makes up the wiring circuit is one of the most important parts of any electrical system. Most conductors are rarely checked after the installation has been approved. For this reason, and others, the choice of conductor is a very important feature of circuit design.

CONDUCTOR MATERIALS

NEC *Article 310* covers conductors for general wiring. Conductors used for residential and industrial wiring circuits are usually made from copper, aluminum, or steel. If exposed to the air, these metals combine with the oxygen. This *oxidation* produces a layer of oxide on the surface of the conductor. If a conductor is allowed to oxidize at the terminal point or splice, the current-carrying ability of the terminal will be seriously reduced, because the oxide layer does not conduct electricity.

Copper. Conductors are often made of copper, because it is an excellent conductor, easy to work with and handle, and does not oxidize as much as aluminum or steel. Copper is also used in electronic circuits because it solders readily, ensuring a secure electrical connection. Copper has become very expensive, however, and a substitute is being sought.

Copper wire splices must be made secure to prevent the oxide coating from reducing the current-carrying ability of the splice.

Aluminum. Aluminum is lighter in weight than copper but not as good a conductor. To obtain the same current-carrying capacity, therefore, an aluminum conductor must be slightly larger than one made of copper. Aluminum oxidizes rapidly. The formation of aluminum oxide acts as an insulator, and reduces the flow of electrical current.

The use of aluminum poses two other major problems. One is *electrolysis,* which is the chemical breakdown of two metals reacting with each other. Aluminum will react with some metals in this way. The process is accelerated by moisture and the flow of current in the conductors. Approved connection devices resist this process, so approved terminal splice connectors must be used. Antioxidant chemicals to prevent oxidation of the wire are available. These chemicals are usually applied with a brush, forcing the chemical into the strands of the cable. This must be done before installing the wire under the mechanical termination.

Another major problem with aluminum conductors is that the connection points lose their grip. A terminal screw can be tightened firmly but, in a few days, the aluminum will accept the new shape into which it has been forced. This reduces the pressure of the terminal screw on the wire, and the aluminum flows into the new shape. This flow – called *cold flow* – causes a loosened connection. As a result, the terminal connection will overheat, causing damage to the connection and insulation. An inspection should be made periodically, and connection points retightened as needed.

If handled properly, aluminum is an effective conductor that provides many years of service. However, aluminum is not effective for circuits where solder connections are necessary, because it cannot be soldered simply.

Steel. Steel is the strongest of the three metals, and is used mainly as a supporting material. For example, aluminum and copper cables are often wound around a steel center core for outdoor wiring. Conductors made this way can withstand a great deal of stress. Electric railway or subway systems often use steel rails as the conductors in their supply system. The rails are sometimes made of special steels containing a small amount of copper to improve the current-carrying ability of the rails.

While the metal conduit and boxes are not used to carry current for the operation of electrical equipment, steel enclosures are used to complete the grounding circuit. In rural service installations, for example, two steel rods are often driven into the ground to provide a grounding connection where there is no metallic water-supply system to the building. *Article 250* of the NEC covers grounding requirements.

CONDUCTOR FORMS

Conductors used for residential and industrial wiring circuits are made in the form of wire, cable, and cord.

Wire. Wire that is made in a single, solid strand, is the least flexible form of conductor, when compared with others of the same size.

Copper wire is usually made by drawing (pulling) a soft copper rod, which is between 1/4 inch and 1/2 inch (6 mm and 13 mm) in diameter, through a series of doughnut-shaped metal blocks called *dies.* The hard center of each die has a hole slightly smaller than the hole of the die through which the rod has just been drawn. As the rod passes through the dies, it is reduced in diameter and lengthened.

The wire tends to work-harden after passing through several dies. This resultant *hard-drawn* copper is used for some outdoor wires because it is quite strong.

If the diameter must be made even smaller, the wire is first passed through an *annealing oven,* which heats and softens it. The wire can then be further reduced to any required size. The annealing process may have to be repeated several times. The wire may also be insulated with a variety of materials, Figure 5-1. Annealing can also

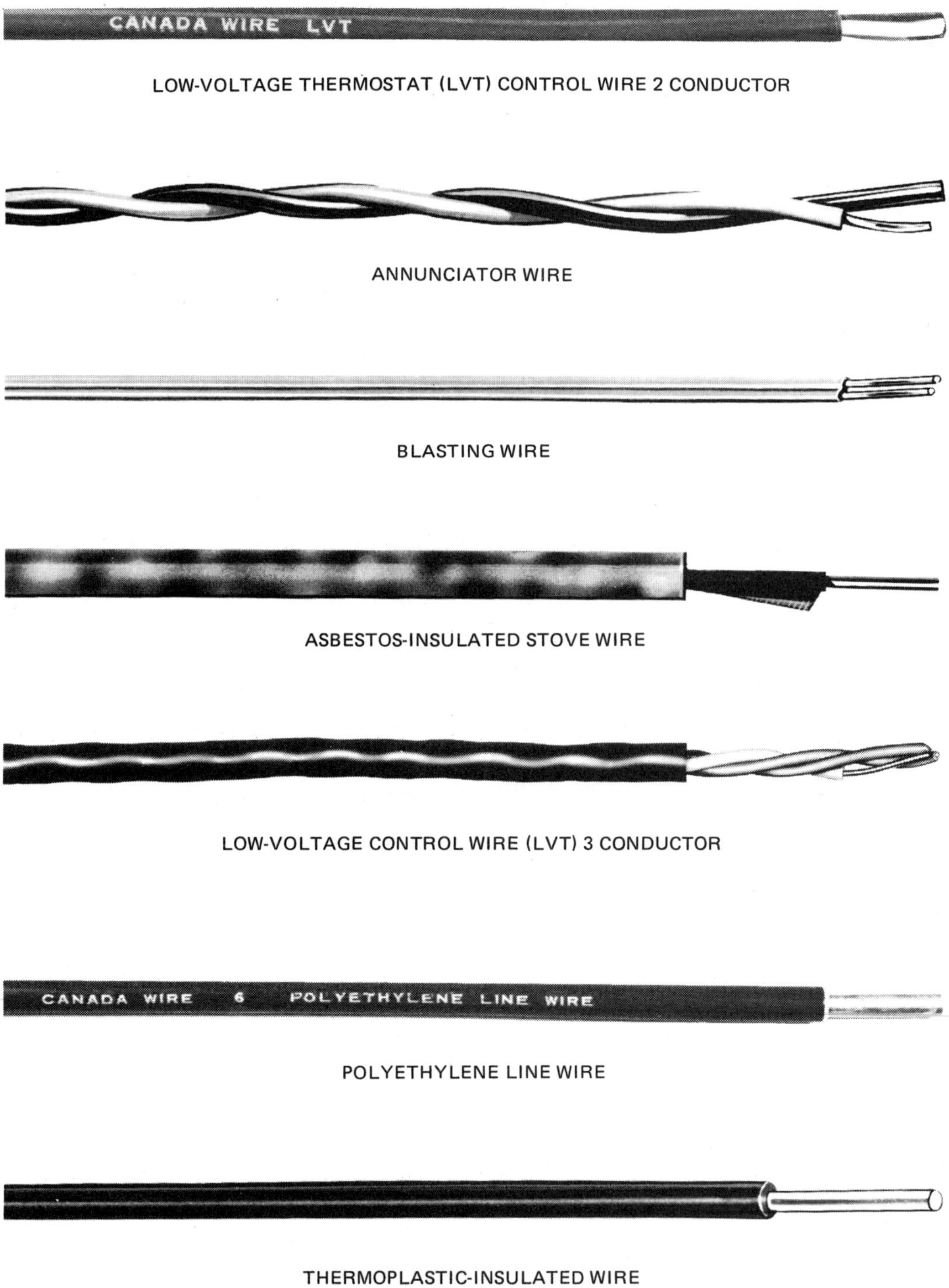

Fig. 5-1 Common types of insulated wires.

be used to produce *soft-drawn* wire for interior wiring use.

Aluminum wire is produced in the same way.

Cable. A cable is a *compound* conductor made of a number of wire strands. Cable is more flexible than a wire of the same size. Some cables are composed of wires twisted together to form a large conductor *before* being insulated. This type is usually used in circuits where there is a heavy flow of current. Other cables are assembled and placed under a common cover *after* each wire has been insulated, Figure 5-2. This type is used extensively for the wiring of buildings.

Cord. A cord's conductors are made of many strands of fine wire twisted together. The cord is made of two or more separately insulated conductors assembled within an insulating jacket, Figure 5-3. Cord is the most flexible form of conductor, and it is used to supply current to hand-held appliances and portable tools, where freedom of movement is important. Appliance and tool cords sometimes have a third conductor (with green insulation) which is used for grounding.

CONDUCTOR SIZES

Conductor sizes are measured and listed in two ways. One method is based upon the American Wire Gauge. The other is based upon the cross-sectional area of the wire.

American Wire Gauge (AWG). The American Wire Gauge is used to measure only *solid* conductors (wire). The outer edge of the gauge has numbered slots. The *smallest slot* into which the wire will fit is the *gauge number* of the wire, (see Figures 5-4 and 5-5). The AWG measures wire sizes from No. 36 (the smallest size) to No. 0 (the largest size). *Stranded* conductor (wire) size is determined by the diameter of each solid wire, and the number of individual wires making up the conductor.

Wire produced for special applications can be as small as No. 44, and as large as No. 0000 (4/0). The small, hairlike wires are used in the windings of electric motors and similar equipment.

Cross-Sectional Area. The size of large conductors is determined by calculating the cross-sectional area of the conductor. When it is necessary to measure a compound conductor, such as a cable, first calculate the area of one strand. Then, multiply the area of that strand by the number of strands in the cable. The area of cross-section is expressed as the *circular mil area.* This term will be phased out, however, as metric conversion proceeds, and the cross-sectional area will be measured in square millimetres (mm^2).

The cross-sectional area is obtained by using the formula $A = \pi \times r^2$, where A is area, r is the radius of the conductor in inches and π is a constant (3.14). *Note:* 1 inch = 25.40 millimetres.

The term MCM is still being used as a name for particular sizes of wires, as is AWG. Therefore, references to MCM are made in this text. Wire size MCM is used for large conductors (over 4/0 in size) and refers to thousands of circular mils. For example, a 500-MCM conductor has a cross-sectional area of 500 000 circular mils.

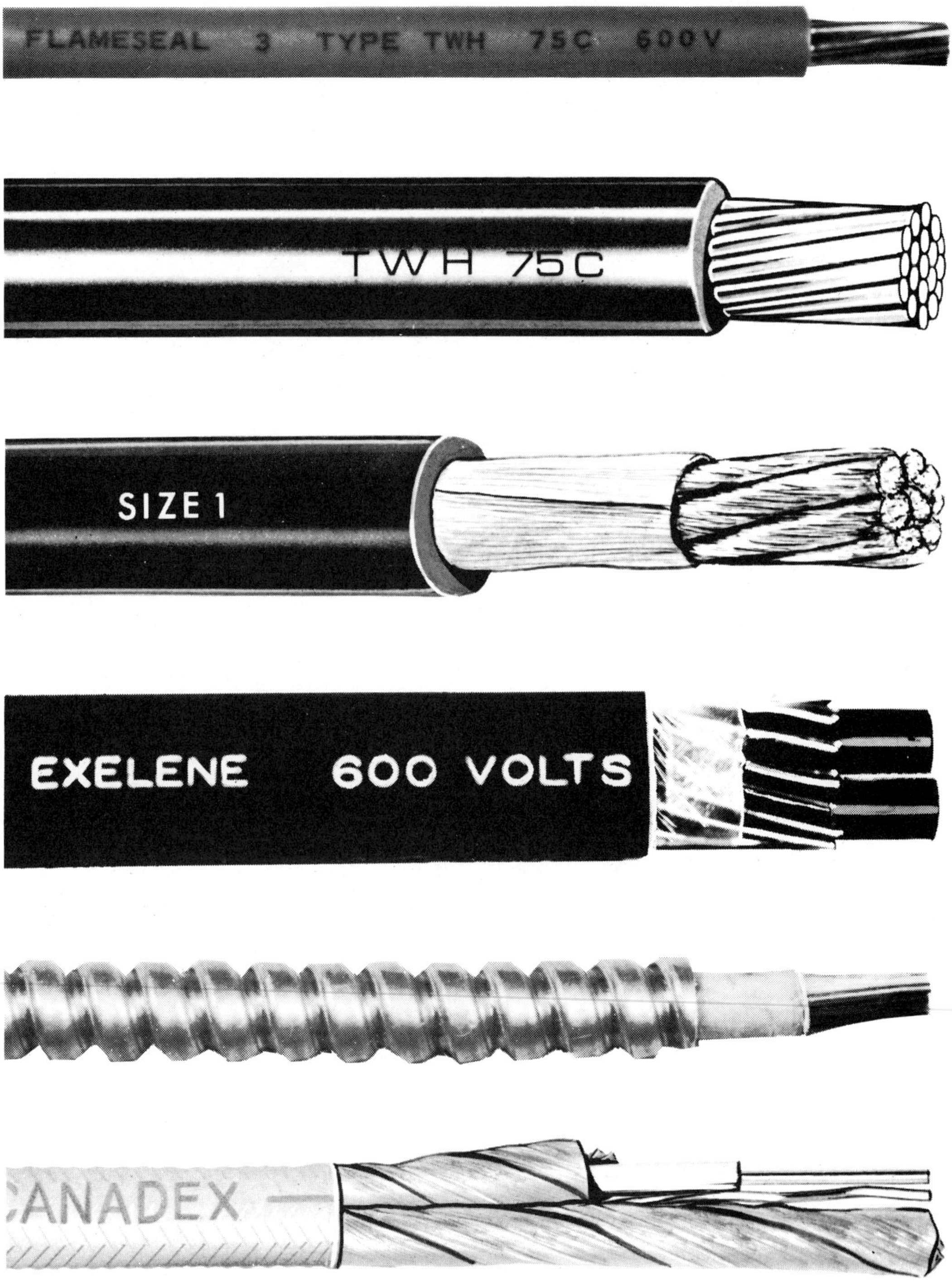

Fig. 5-2 Typical insulated cables.

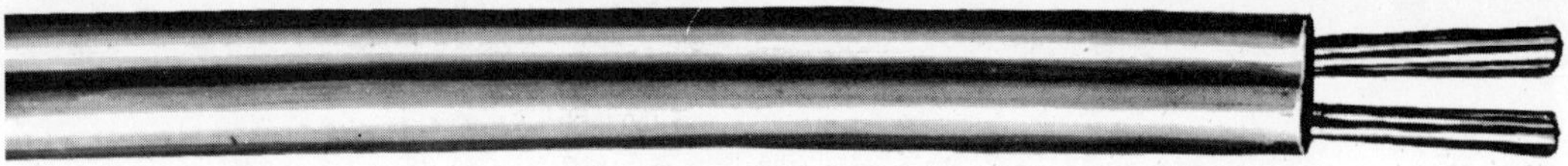

LAMP CORD $^{18}/_{2}$

HEAT-PROOF NEOPRENE (HPN) CORD 16/2

HEATER CORD $^{16}/_{2}$

Fig. 5-3 Insulated cords for lighting fixtures and appliances.

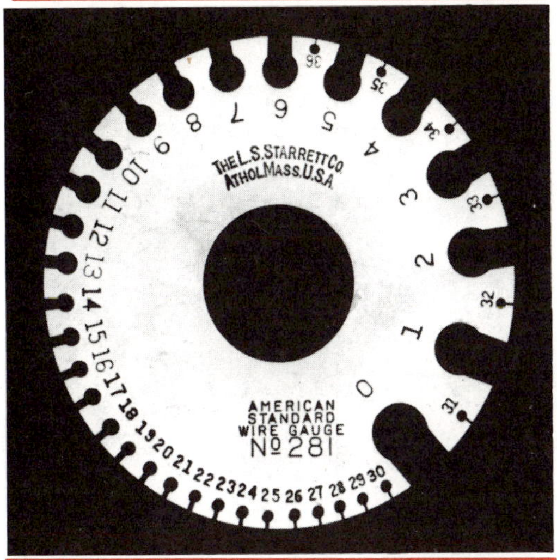

Fig. 5-4 American Wire Gauge used to measure size of solid conductors (wires).

NEC *Table 5,* shown in Table 5-1, lists the sizes of various types of conductors. Always check the most recent edition of the National Electrical Code for up-to-date information on the standards to be used for measuring conductors.

WIRE SIZE USES

Wire and cable for buildings is made in even-numbered gauge sizes, such as Nos. 14, 12, 10, 8, and so forth. Odd-numbered gauge sizes, such as Nos. 15, 13, 11, 9, and so forth, are not normally used for building wire and cable because there is not enough difference in current-carrying capacity (ampacity) to make production of these odd-numbered sizes worthwhile. However, these sizes are available from some manufacturers.

The National Electrical Code specifies that No. 10 gauge wire shall be the largest single-strand conductor allowed under a terminal screw. Larger conductors *must* be inserted into a compression-type fitting, called a *lug*, Figure 5-6. Since wire and cable for buildings must be flexible, conductors for this use are usually stranded when made in sizes larger than No. 10 gauge.

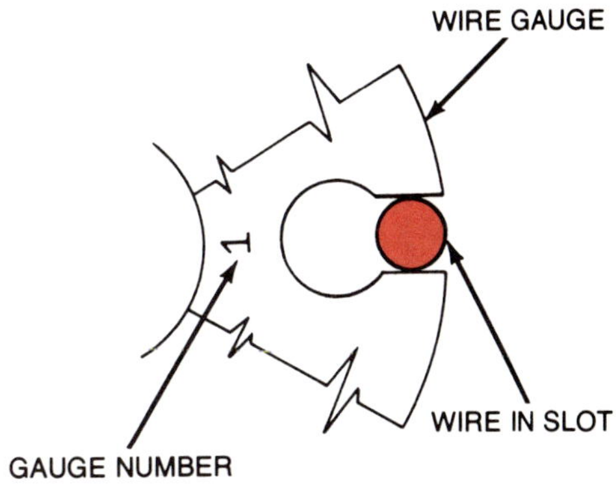

Fig. 5-5 Using the American Wire Gauge.

Wire for motors, transformers, and other magnetic equipment is made in *both* odd and even gauge numbers, because wire size for these uses is more critical than it is for other purposes. Conductors with an area-of-cross-section rating are used for industrial applications where a large amount of current must be carried.

CONDUCTOR INSULATION

Conductor insulators are made for a wide variety of uses and situations. The more common types of conductor insulation materials are thermoplastic, rubber and cotton, neoprene, asbestos, varnish, glass over thermoplastic, and cross link.

Thermoplastic. Thermoplastic (Type T) insulation is one of the most common insulators used for residential and industrial wir-

Table 5. Dimensions of Rubber-Covered and Thermoplastic-Covered Conductors

Size AWG MCM	Types RFH-2, RH, RHH,*** RHW,*** SF-2		Types TF, T, THW,† TW, RUH,** RUW**		Types TFN, THHN, THWN		Types**** FEP, FEPB, FEPW, TFE, PF, PFA, PFAH, PGF, PTF, Z, ZF, ZFF		Type XHHW, ZW††		Types KF-1, KF-2, KFF-1, KFF-2	
	Approx. Diam. Inches	Approx. Area Sq. In.	Approx. Diam. Inches	Approx. Area Sq. In.	Approx. Diam. Inches	Approx. Area Sq. In.	Approx. Diam. Inches	Approx. Area Sq. Inches	Approx. Diam. Inches	Approx. Area Sq. In.	Approx. Diam. Sq. In.	Approx. Area Sq. In.
Col. 1	Col. 2	Col. 3	Col. 4	Col. 5	Col. 6	Col. 7	Col. 8	Col. 9	Col. 10	Col. 11	Col. 12	Col. 13
18	.146	.0167	.106	.0088	.089	.0064	.081	.0052	...	....	.065	.0033
16	.158	.0196	.118	.0109	.100	.0079	.092	.0066	...	....	.070	.0038
14	30 mils .171	.0230	.131	.0135	.105	.0087	.105 .105	.0087 .0087	...	....	.083	.0054
14	45 mils .204*	.0327*	...	....	...	....	... ...	....	...	....	...	....
14	....	....	.162†	.0206†	...	....	... ...	....	.129	.0131	...	....
12	30 mils .188	.0278	.148	.0172	.122	.0117	.121 .121	.0115 .0115	...	....	.102	.0082
12	45 mils .221*	.0384*	...	....	...	....	... ...	....	...	....	...	....
12	....	....	.179†	.0251†	...	....	... ...	....	.146	.0167	...	....
10	...... .242	.0460	.168	.0224	.153	.0184	.142 .142	.0159 .0159	...	....	.124	.0121
10	....	....	.199†	.0311†	...	....	... ...	....	.166	.0216	...	....
8	...... .328	.0854	.245	.0471	.218	.0373	.206 .186	.0333 .0272	...	....	...	....
8	....	....	.276†	.0598†	...	....	... ...	....	.241	.0456	...	....
6	.397	.1238	.323	.0819	.257	.0519	.244 .302	.0467 .0716	.282	.0625	...	....
4	.452	.1605	.372	.1087	.328	.0845	.292 .350	.0669 .0962	.328	.0845	...	....
3	.481	.1817	.401	.1263	.356	.0995	.320 .378	.0803 .1122	.356	.0995	...	....
2	.513	.2067	.433	.1473	.388	.1182	.352 .410	.0973 .1316	.388	.1182	...	....
1	.588	.2715	.508	.2027	.450	.1590	.420 ...	.1385	.450	.1590	...	....
0	.629	.3107	.549	.2367	.491	.1893	.462 ...	.1676	.491	.1893	...	....
00	.675	.3578	.595	.2781	.537	.2265	.498 ...	.1974	.537	.2265	...	....
000	.727	.4151	.647	.3288	.588	.2715	.560 ...	.2463	.588	.2715	...	....
0000	.785	.4840	.705	.3904	.646	.3278	.618 ...	.2999	.646	.3278	...	....

Table 5-1

Table 5

Size AWG MCM	Types RFH-2, RH, RHH,*** RHW,*** SF-2		Types TF, T, THW,† TW, RUH,** RUW**		Types TFN, THHN, THWN		Types**** FEP, FEPB, FEPW, TFE, PF, PFA, PFAH, PGF, PTF, Z, ZF, ZFF		Type XHHW, ZW††	
	Approx. Diam. Inches	Approx. Area Sq. In.	Approx. Diam. Inches	Approx. Area Sq. In.	Approx. Diam. Inches	Approx. Area Sq. In.	Approx. Diam. Inches	Approx. Area Sq. Inches	Approx. Diam. Inches	Approx. Area Sq. In.
Col. 1	Col. 2	Col. 3	Col. 4	Col. 5	Col. 6	Col. 7	Col. 8	Col. 9	Col. 10	Col. 11
250	.868	.5917	.788	.4877	.716	.4026	...	....	.716	.4026
300	.933	.6837	.843	.5581	.771	.4669	...	....	.771	.4669
350	.985	.7620	.895	.6291	.822	.5307	...	....	.822	.5307
400	1.032	.8365	.942	.6969	.869	.5931	...	....	.869	.5931
500	1.119	.9834	1.029	.8316	.955	.7163	...	....	.955	.7163
600	1.233	1.1940	1.143	1.0261	1.058	.8792	...	....	1.073	.9043
700	1.304	1.3355	1.214	1.1575	1.129	1.0011	...	....	1.145	1.0297
750	1.339	1.4082	1.249	1.2252	1.163	1.0623	...	....	1.180	1.0936
800	1.372	1.4784	1.282	1.2908	1.196	1.1234	...	....	1.210	1.1499
900	1.435	1.6173	1.345	1.4208	1.259	1.2449	...	....	1.270	1.2668
1000	1.494	1.7531	1.404	1.5482	1.317	1.3623	...	....	1.330	1.3893
1250	1.676	2.2062	1.577	1.9532	...	....	...	....	1.500	1.7672
1500	1.801	2.5475	1.702	2.2748	...	....	...	....	1.620	2.0612
1750	1.916	2.8895	1.817	2.5930	...	....	...	....	1.740	2.3779
2000	2.021	3.2079	1.922	2.9013	...	....	...	....	1.840	2.6590

* The dimensions of Types RHH and RHW.
** No. 14 to No. 2.
† Dimensions of THW in sizes No. 14 to No. 8. No. 6 THW and larger is the same dimension as T.
*** Dimensions of RHH and RHW without outer covering are the same as THW No. 18 to No. 10, solid; No. 8 and larger, stranded.
**** In Columns 8 and 9 the values shown for sizes No. 1 thru 0000 are for TFE and Z only. The right-hand values in Columns 8 and 9 are for FEPB, Z, ZF, and ZFF only.
†† No. 14 to No. 2.

Table 5-1

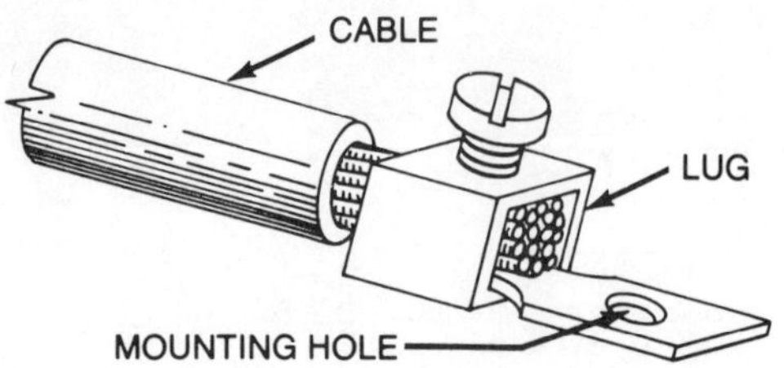

Fig. 5-6 Method for installing cable in a lug.

ing. Thermoplastic is an excellent insulator, but it is sensitive to extremes of temperature. At high temperatures, it melts; at low temperatures, it becomes brittle, and cracks if handled roughly. Weatherproof (Type TW) and heat-resistant (Type TH) thermoplastics are available, however. Thermoplastic weatherproof insulation is called Type TW. Thermoplastic which is weatherproof and heat resistant is called Type THW. Some of the heat produced by equipment such as electric heaters, stoves, and light fixtures travels back through the conductor, damaging the insulation. THW insulation is capable of withstanding such heat.

Two other insulators commonly used are Types THHN and THWN, where the letter "N" represents an additional nylon covering. See NEC *Article 310,* "Conductors for General Wiring."

Rubber and Cotton. Homes wired with the knob and tube system have conductors covered first with a layer of rubber and, then, an outer braid of cotton for extra protection. The copper wire is also coated with tin to prevent premature oxidation of the copper caused by the sulphur content of the rubber. This form of insulation is rarely used for modern wiring systems.

Neoprene. Neoprene is a special type of rubber insulation used widely on heatproof line cords for such heat-producing appliances as coffee makers, frypans, and soldering irons. Since neoprene is also oil resistant, it can be used on extension cords for service stations. The abbreviation for heatproof neoprene insulation is HPN.

Asbestos. Major appliances, such as electric stoves, need heatproof insulation for internal circuits. Asbestos-coated wire is used, for example, to connect the elements of stoves to the control switches. Asbestos is also used in the cords of cloth-pressing irons. A cotton-braid jacket is wrapped around the cord for extra strength.

Varnish. Copper wire with a baked-on varnish insulation is used extensively for motor windings. The high quality of the insulating varnish means that the insulation can be very thin, allowing space for the many windings required. This type of insulation has a temperature rating in excess of 200°C (392°F). The abbreviation for varnish insulation is V.

Glass over Thermoplastic. Conductors supplying recessed fixtures require a heat-resistant insulation because there is little, if any, air circulation to cool the conductors. PGF (glass and thermoplastic for fixtures) is used for this purpose.

Cross Link. Cross link material is thermosetting and will not melt. It has a higher temperature rating than TW, and is now widely used for building wire instead of rubber and cotton.

AMPACITY

The purpose of a conductor is to carry current from one place in a circuit to another. *Ampacity* refers to the ability of a conductor to carry current. (See NEC *Table 310-16* in Table 5-2.)

The ampacity rating of a conductor is determined by its material, size, insulation, and operating temperature.

Table 310-16. Allowable Ampacities of Insulated Conductors Rated 0-2000 Volts, 60° to 90°C

Not More Than Three Conductors in Raceway or Cable or Earth (Directly Buried), Based on Ambient Temperature of 30°C (86°F)

Size	Temperature Rating of Conductor, See Table 310-13								Size
	60°C (140°F)	75°C (167°F)	85°C (185°F)	90°C (194°F)	60°C (140°F)	75°C (167°F)	85°C (185°F)	90°C (194°F)	
AWG MCM	TYPES †RUW, †T, †TW, †UF	TYPES †FEPW, †RH, †RHW, †RUH, †THW, †THWN, †XHHW, †USE, †ZW	TYPES V, MI	TYPES TA, TBS, SA, AVB, SIS, †FEP, †FEPB, †RHH †THHN, †XHHW*	TYPES †RUW, †T, †TW, †UF	TYPES †RH, †RHW, †RUH, †THW †THWN, †XHHW, †USE	TYPES V, MI	TYPES TA, TBS, SA, AVB, SIS, †RHH, †THHN, †XHHW*	AWG MCM
	COPPER				ALUMINUM OR COPPER-CLAD ALUMINUM				
18	....	....	....	14	....	....	....	....	....
16	....	....	18	18	....	....	....	....	....
14	20†	20†	25	25†	....	....	....	....	....
12	25†	25†	30	30†	20†	20†	25	25†	12
10	30†	35†	40	40†	25†	30†	30	35†	10
8	40	50	55	55	30	40	40	45	8
6	55	65	70	75	40	50	55	60	6
4	70	85	95	95	55	65	75	75	4
3	85	100	110	110	65	75	85	85	3
2	95	115	125	130	75	90	100	100	2
1	110	130	145	150	85	100	110	115	1
0	125	150	165	170	100	120	130	135	0
00	145	175	190	195	115	135	145	150	00
000	165	200	215	225	130	155	170	175	000
0000	195	230	250	260	150	180	195	205	0000
250	215	255	275	290	170	205	220	230	250
300	240	285	310	320	190	230	250	255	300
350	260	310	340	350	210	250	270	280	350
400	280	335	365	380	225	270	295	305	400
500	320	380	415	430	260	310	335	350	500
600	355	420	460	475	285	340	370	385	600
700	385	460	500	520	310	375	405	420	700
750	400	475	515	535	320	385	420	435	750
800	410	490	535	555	330	395	430	450	800
900	435	520	565	585	355	425	465	480	900
1000	455	545	590	615	375	445	485	500	1000
1250	495	590	640	665	405	485	525	545	1250
1500	520	625	680	705	435	520	565	585	1500
1750	545	650	705	735	455	545	595	615	1750
2000	560	665	725	750	470	560	610	630	2000
	CORRECTION FACTORS								
Ambient Temp. °C	For ambient temperatures over 30°C, multiply the ampacities shown above by the appropriate correction factor to determine the maximum allowable load current.								Ambient Temp. °F
31-40	.82	.88	.90	.91	.82	.88	.90	.91	86-104
41-45	.71	.82	.85	.87	.71	.82	.85	.87	105-113
46-50	.58	.75	.80	.82	.58	.75	.80	.82	114-122
51-60	....	.58	.67	.71	....	.58	.67	.71	123-141
61-70	....	.35	.52	.58	....	.35	.52	.58	142-158
71-80	....	....	.30	.41	....	....	.30	.41	159-176

† The load current rating and the overcurrent protection for conductor types marked with an obelisk (†) shall not exceed 15 amperes for 14 AWG, 20 amperes for 12 AWG, and 30 amperes for 10 AWG copper; or 15 amperes for 12 AWG and 25 amperes for 10 AWG aluminum and copper-clad aluminum.

* For dry locations only. See 75°C column for wet locations.

Table 5-2

Material. The material of which the conductor is made determines how easily it will carry current. For example, copper is a better conductor than aluminum and it will, therefore, carry more current.

Size. The larger the conductor, the more current it will carry without heating. Since conductors are often enclosed in conduits and boxes, take care to use conduits and boxes large enough to give the necessary cubic-inch area for the number of conductors. Sufficient space for air circulation must be provided in order to prevent damage to the insulation from overheating. The temperature of the air surrounding the conductors (ambient temperature) will also affect their ampacity. See *Article 370* and *Chapter 9 Tables* of the National Electrical Code.

Insulation. A conductor with insulation capable of withstanding heat has a higher ampacity rating than a conductor of the same size with a lower insulator temperature rating. Wires with heat-resistant covering can be contained in a more confined area than normal conductors, without heat damage to the insulation.

The voltage of the circuit also affects the choice of insulator. The conductors *must* be insulated from one another as well as from ground, and the insulation must be capable of withstanding the electrical pressure of the circuit.

Under high-current conditions, the current flows mainly along the surface of the conductor. This condition is known as the *skin effect,* and steel-core conductors can be used in such situations to combine strength and ampacity, because the outer section of the conductor carries the bulk of the current.

Conductors must be handled carefully (particularly at the time of installation) to prevent surface nicks and scratches. The conductor will heat at the point of a surface nick, because the nick will have reduced the cross-sectional area. If the nick is deep enough, there may be serious heat damage to the surrounding insulation, resulting in a possible ground-fault short circuit.

ELECTRICAL RESISTANCE

Current does not flow through a conductor by itself. Electrical pressure (voltage) must be applied to force the current along the conductor. Some form of resistance to current flow is always present in the conductor. Resistance is measured in *ohms* (Ω).

Resistance is determined by the conductor's diameter, length, temperature, and material. The smaller the diameter, the longer the length; the higher the temperature, the greater the resistance.

Resistance has little, if any, effect on electrical performance in the average residential circuit. However, resistance is an important factor in tall buildings, sprawling industrial complexes, or street lighting, where conductors must travel long distances. Considerable voltage may be lost (dropped) in delivering the current to the end of the circuit. The result is dimming lights and less efficient electrical motors. Voltage drop is equal to $I^2 \times R$. That is, amperes squared, multiplied by the resistance of the wire. Voltage drop is reduced by increasing the diameter of the wire used in the circuit.

REVIEW

1. What two materials are most commonly used for electrical conductors? Why?
2. Define *oxidation,* and describe its effect on metal conductors.
3. Describe the process for producing wire.
4. Why are cables and cords made from many strands of wire?
5. What is the name of the gauge used to measure solid conductors? What sizes of wire can this gauge measure?
6. What method is used to measure the size of large cables?
7. List five or more materials commonly used for insulating conductors.
8. Define *ampacity,* and list and explain the factors that affect it.
9. Why should circuit voltage be considered when selecting a conductor?
10. Explain why cuts or other damage to conductors should be avoided during installation.
11. List and explain the factors that determine a conductor's resistance.
12. What is the effect of conductor resistance on circuit voltage? Give examples.

Chapter 6

Cord Fittings

LINE CORD

Any electrical device that is designed to receive its electrical supply from a receptacle is fitted with a line cord. Cord fittings attached to the line cord are used as a convenient method for connecting the device to the power source.

MALE AND FEMALE ATTACHMENT PLUGS (CAPS)

Two main types of cord fittings are the male and the female attachment plugs (caps). Both are produced with the same blade shapes, as described in Chapter 4, and in Figure 4-2.

The male attachment cap is designed to be inserted into the slots of a receptacle, Figure 6-1. It is usually attached to the end of line cords on electrical appliances, lamps, and power tools.

The male attachment plug is never permitted to have a *live* electric supply to the exposed prongs. An extension cord made with a male attachment cap on both ends is a violation of the National Electrical Code.

The female attachment plug cap is designed to receive the male plug cap, Figure 6-2. It is attached to one end of an extension cord which has a male plug cap on the other end. Usually, both have the same blade shape.

USES FOR HEAVY-DUTY CORD CAPS

Mobile homes (trailers), electric stoves, and clothes dryers require heavy-duty cord caps for the power supply. In industry, cord fittings with current and voltage ratings other than the usual residential ratings of

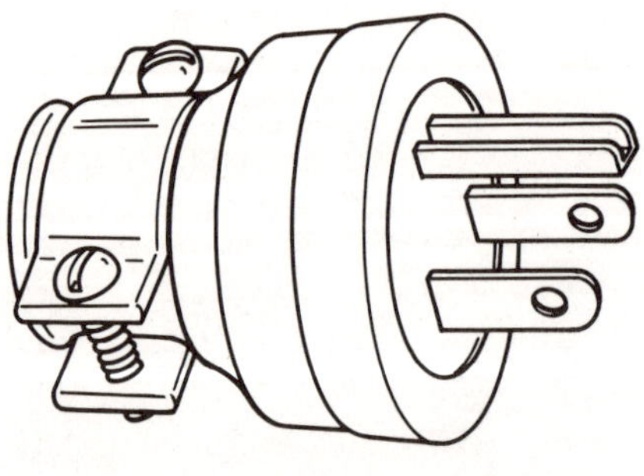

Fig. 6-1 Male plug cap.

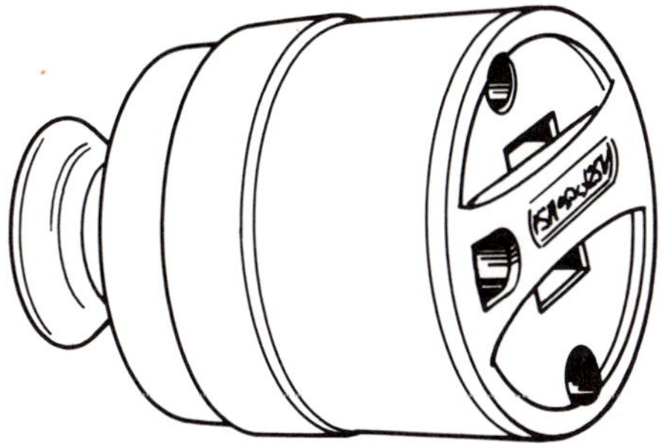

Fig. 6-2 Female plug cap (connector body).

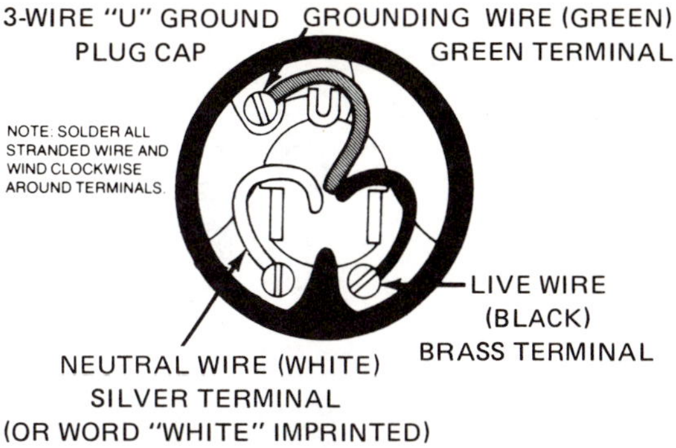

Fig. 6-3 Method for connecting cord to plug cap.

10 A to 15 A at 115 V are often required. Some examples are cord caps for welding machines, floor finishers, battery chargers, and marine shore lines.

PLUG CAP CONNECTION

Plug caps are often damaged by carelessness. For example, many people remove plugs from the receptacles by tugging on the cord. Figure 6-3 shows how line cords and plug caps are connected so that the strain from pulling on the cord will be absorbed by the *blades* rather than by the terminals. To prevent loose strands of the flexible cord from slipping out of the terminal connection, the strands are twisted together and then wound clockwise (the direction in which the screw tightens) around the terminal screw. Soldering the groups of twisted strands before assembly also helps make the electrical connection secure.

An *underwriter's knot* tied in the cord was once used to prevent it from being pulled out of the cord cap. But such a knot places stress on the cord's insulation, so it is seldom used with modern plug caps in which conductor space is limited. Instead, industrial or heavy-duty plug caps are now equipped with metal *clamping* devices to hold the cord in the plug cap, Figure 6-4. These devices are particularly useful on construction sites, where extension cords and caps are walked on, driven over, and treated roughly.

DEAD-FRONT ATTACHMENT PLUG CAPS

The *insulating disc* of the male attachment plug cap is a separate part fitted over the prongs where the cord has been connected. The purpose of the disc is to prevent loose strands or terminals from coming in contact with the cover plate of the receptacle. If the cover plate is made of metal and there is no insulating disc, a short-circuit flash can occur as the plug is inserted into the receptacle, burning the hand of the person plugging in the cap.

Dead-front attachment plug caps, with the prongs and terminals assembled as a removable unit, are available, Figure 6-5. The cord connects to the rear of the unit. The front, which is the exposed portion of the plug cap, is free of terminals, conductors, and an insulating disc.

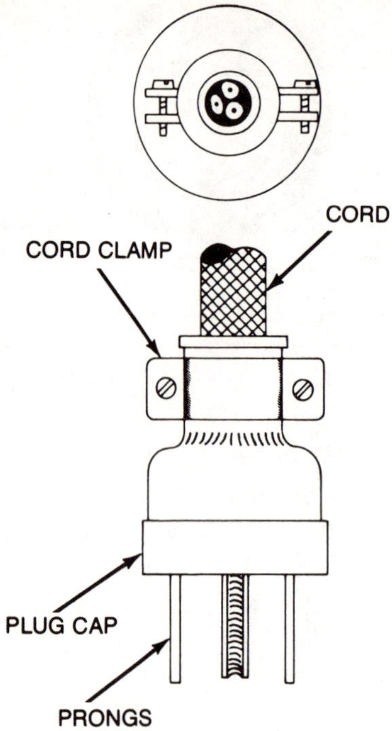

Fig. 6-4 Industrial (heavy-duty) plug cap equipped with cord clamp.

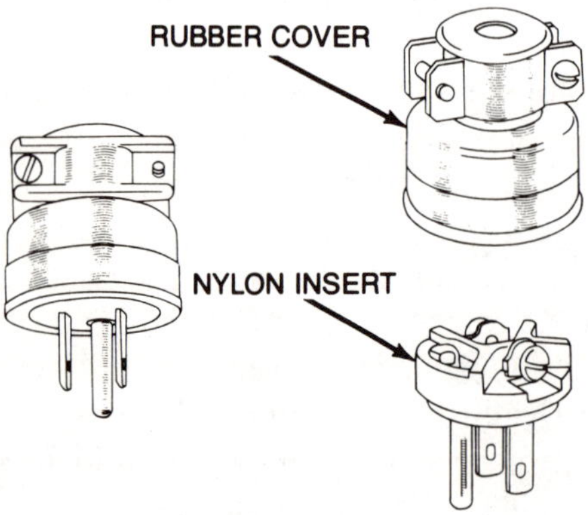

Fig. 6-5 Dead-front attachment plug cap.

APPLIANCE ATTACHMENT PLUG CAPS

Small portable appliances, such as fry-pans and coffee makers, are often made with removable cord sets. These sets must have a plug that is easy to grip, provides strain relief for the cord, and resists heat. Figures 6-6 and 6-7 show a typical appliance attachment plug and its cord connections.

Some large stationary appliances, such as electric stoves and clothes dryers, are now being made with cord sets. Cord sets allow larger units to be pulled away from the wall for cleaning and to be quickly disconnected from the power source for servicing. Figures 6-8 and 6-9 show typical cord and cap sets for use with large appliances.

ELECTRICAL RATINGS

Attachment plug caps are rated in volts and amperes. The ratings of the unit *must* be matched to the requirements of the circuit. If a light-duty unit is installed where a heavy-duty unit is necessary, the plug will overheat. This can cause damage to the plug and/or receptacle and the conductor insulation. Look for the UL stamp on the plug cap; this is a guarantee that the manufacturer's ratings are correct.

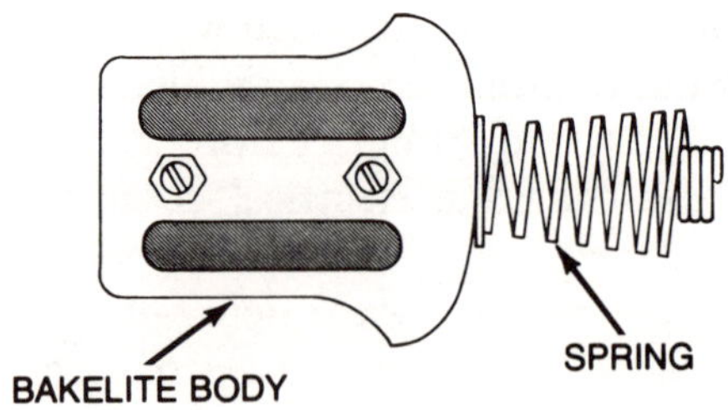

Fig. 6-6 Typical appliance plug.

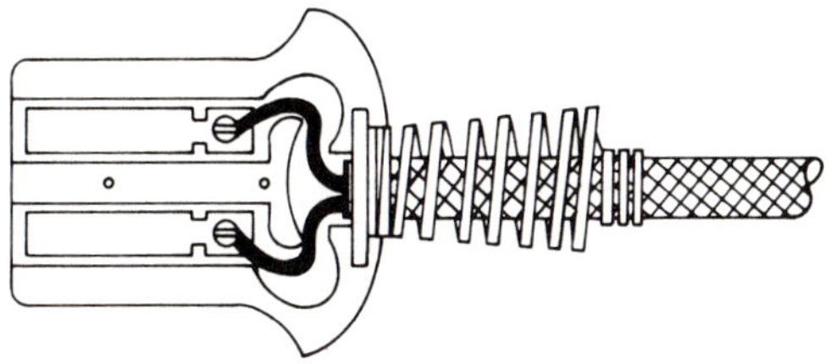

Fig. 6-7 Typical applicance plug cord connection.

GROUNDED CORD CAPS

As explained in Chapter 4, ("Receptacle and Plug Shapes") it is necessary to have a ground prong on a male plug cap and a corresponding slot in the matching female plug cap. These units are placed on any tool or device that could transmit a shock to the operator if the device is used in an area where the operator could become grounded. Faulty equipment can often allow current to flow to the body or frame of the tool, and then through the opèrator into the ground. This current flow produces a *serious* electrical shock. The ground prong and grounding wire can carry this stray or leakage current into the ground rather than have it go through the operator. This can prevent electrical shock to the operator. Intentional removal of the ground prong is a violation of the NEC and *all* safety requirements.

Figure 6-10 illustrates a typical male plug cap with its rugged U-shaped ground prong, recommended for all portable tools and equipment.

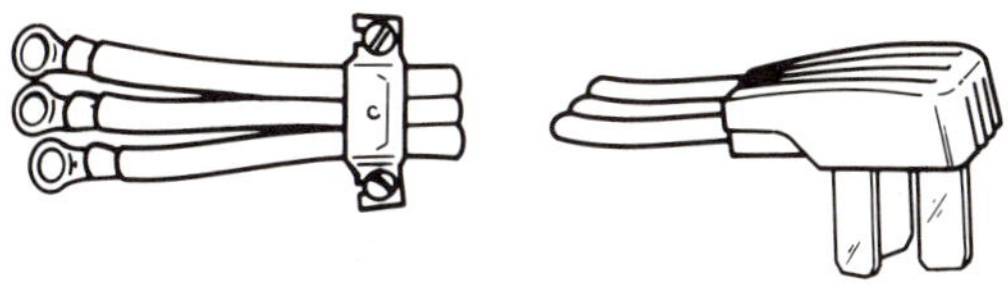

Fig. 6-8 Typical 50-A range cord and plug cap.

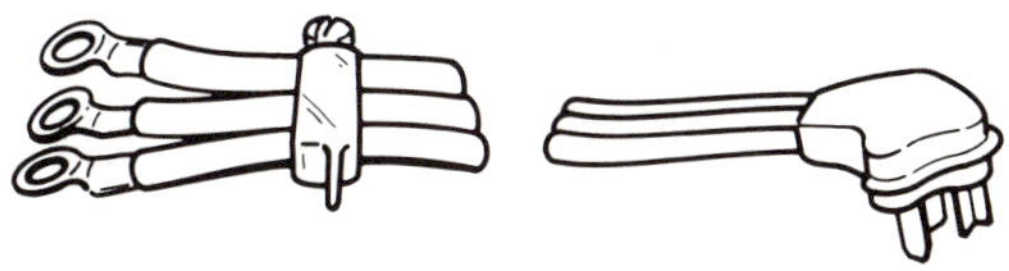

Fig. 6-9 Typical 30-A dryer cord and plug cap.

Fig. 6-10 Typical male attachment plug cap with ground prong in dead-front plug design.

REVIEW

1. Describe the two main types of cord fittings.
2. List five common cord cap blade shapes.
3. What may happen if a plug is removed from a receptacle by pulling on the cord?
4. Why are the conductors taken around the blade of the plug cap before securing them to the terminal screw?
5. What is the purpose of soldering the cord strands before winding them around the terminal screw?
6. Why are heavy-duty plug caps equipped with clamping devices for the cords?
7. Describe the insulating disc, and explain its purpose.
8. What is the purpose of the dead-front plug cap?
9. Give two reasons why some large appliances are fitted with cord sets.
10. List the electrical ratings that must be marked on plug caps.
11. Why must the ground prong never be removed either intentionally or accidentally?

Chapter 7

Electrical Outlet Boxes

OUTLET BOX REQUIREMENTS

Modern wiring systems require an electrical outlet box at each point in the circuit where a switch, lampholder, receptacle, or splice is located. *Article 370* of the National Electrical Code provides an up-to-date summary of installation procedures. NEC *Article 90 (Introduction)* states that dual designation SI units are not given for box sizes; these sizes are given in English measurements only.

Most electrical boxes are made of galvanized or cadmium-plated steel. Nonmetallic wiring systems may use boxes made of Bakelite. This heat-resistant material cannot be used for metallic wiring systems because it is quite fragile and unable to withstand the strain of metal box connectors.

Boxes made of brass or Everdur may be used where there is a great deal of moisture. These nonrusting, highly corrosion-resistant materials prevent box damage from chemicals or moisture in the surrounding air.

Several types of boxes are available for various uses: round, octagon, pancake, square, sectional plaster, utility, and concrete-masonry-tile.

ROUND BOX

The NEC does not permit round boxes to be used where conduits or connectors requiring the use of locknuts or bushings are to be connected to the side of the box. Round boxes with conduit hubs are used.

OCTAGON BOX

The Octagon box is usually used to support light fixtures or to serve as a junction point for wire splices. It can also be used with special covers as a supporting box for a switch or receptacle.

Dimensions. Octagon boxes are available in three diameters. The 4-inch diameter box is the most common, Figure 7-1. Boxes with diameters from 3 1/4 inches to 3 1/2 inches are available for applications where the box size is limited. *Note:* Check the local electrical code requirements for your area.

Depth is another important box dimension. The depth measurement determines the amount of conductor space within the box. The most common depth is 1 1/2 inches. A box 2 1/8 inches in depth is available for installation where more conductor space is required.

Box extensions are designed to mount directly on the top of an existing octagon box. *Box extensions* are simply octagon boxes without a bottom surface, and they are used to provide an increase in conductor space when required, Figure 7-2.

Covers. A variety of octagon box covers allows the box to serve many purposes, Figure 7-3. Box covers and extension boxes are fastened with two No. 8-32 round-head machine screws, which are provided with each box.

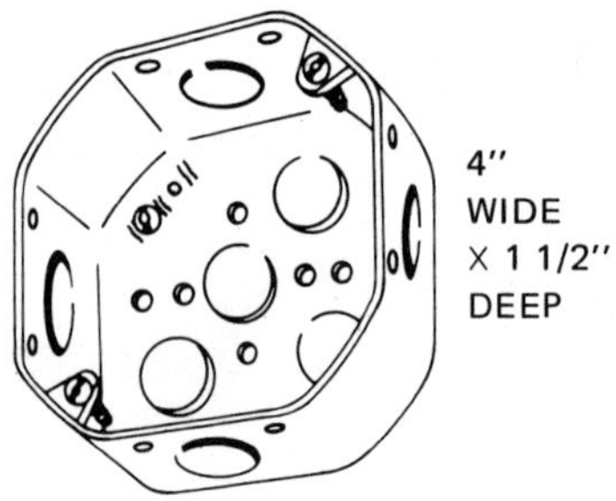

Fig. 7-1 Octagon box.

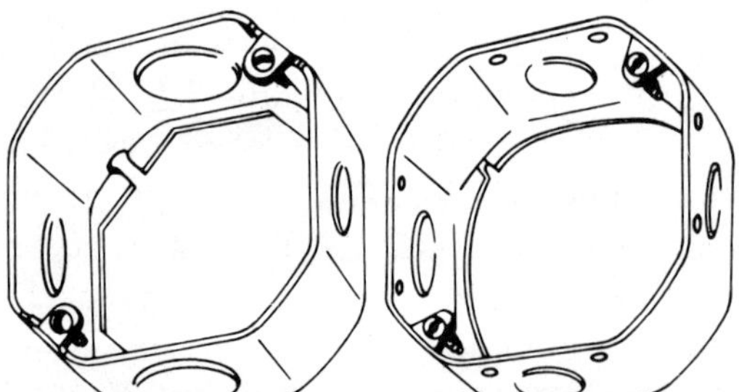

Fig. 7-2 Octagon box extensions.

Octagon Concrete Rings. Concrete rings are used in buildings where the boxes and wiring system are installed before the concrete is poured, Figure 7-4. They are available in depths of 1 1/2 inches to 6 inches. Either end of the ring may have a cover. Figure 7-5 shows a concrete-ring ceiling installation.

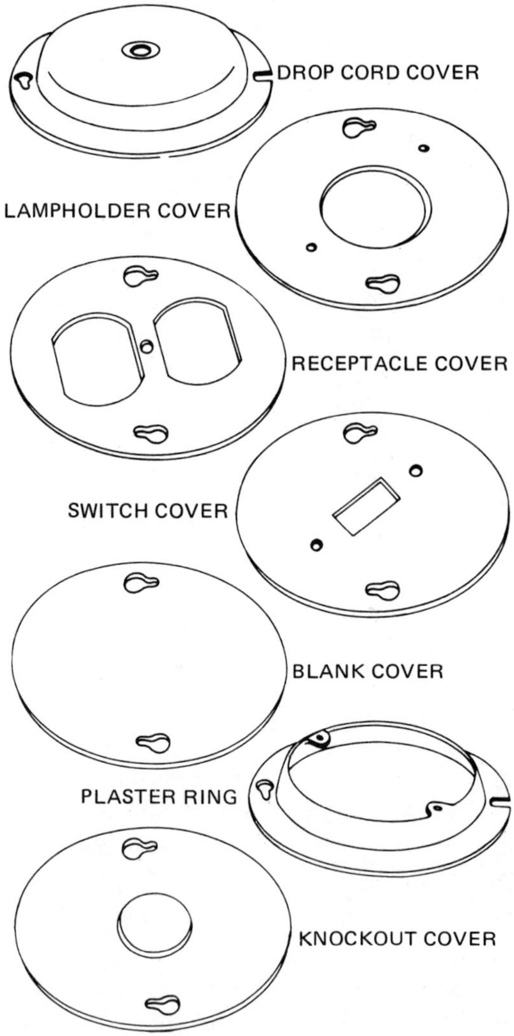

Fig. 7-3 Octagon box covers.

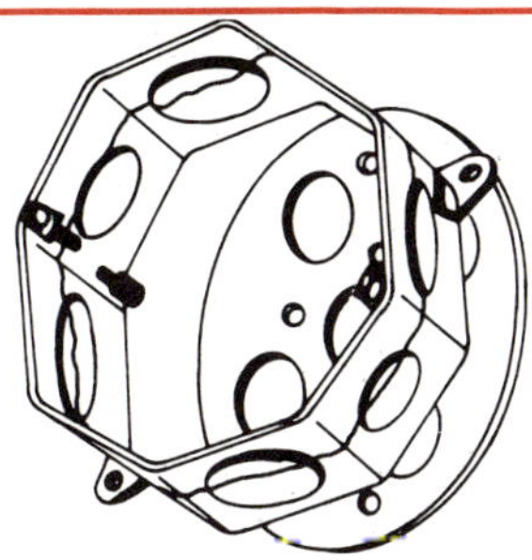

Fig. 7-4 Typical 4-inch octagon concrete ring. (Cover screw may be wax protected.)

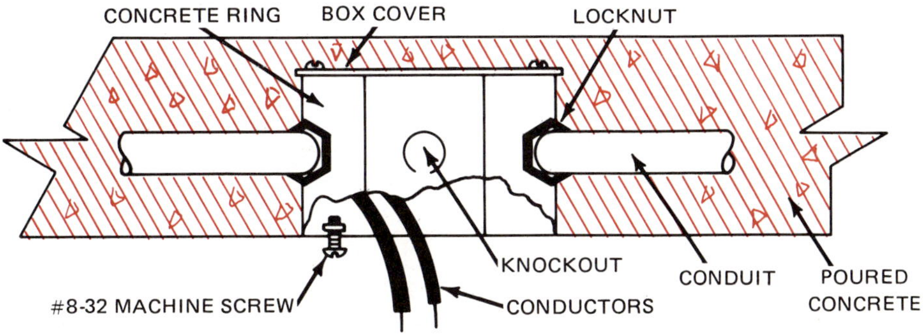

Fig. 7-5 Concrete-ring ceiling installation.

Knockouts. Conductor access to the box is made through removable portions called *knockouts*. Knockouts provide for the entrance of wire, cable, or conduit to the box. They are made in common trade sizes; for example, 1/2 inch, 3/4 inch, or 1 inch. A 1/2-inch knockout is designed to accept a conduit with an internal diameter of 1/2 inch. The actual diameter of the knockout is approximately 7/8 inch.

Methods of Mounting. Octagon boxes can be used with either a surface- or a concealed-wiring system. Screw holes are provided in the bottom of the box for fastening the box on the surface of a wall or ceiling. Concealed wiring methods allow for several mounting techniques. Figure 7-6 shows the mounting-bracket and hanger-bar assemblies. Figure 7-7 shows three ways of supporting the boxes.

PANCAKE BOX

The pancake box is a 4-inch round box used primarily with an outdoor porch light, Figure 7-8. Its 1/2-inch depth allows a fixture to be mounted over it without the

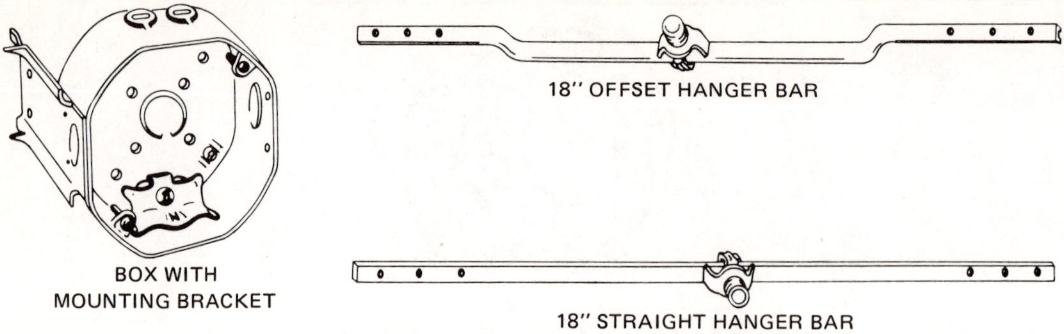

Fig. 7-6 Mounting bracket and hanger bars.

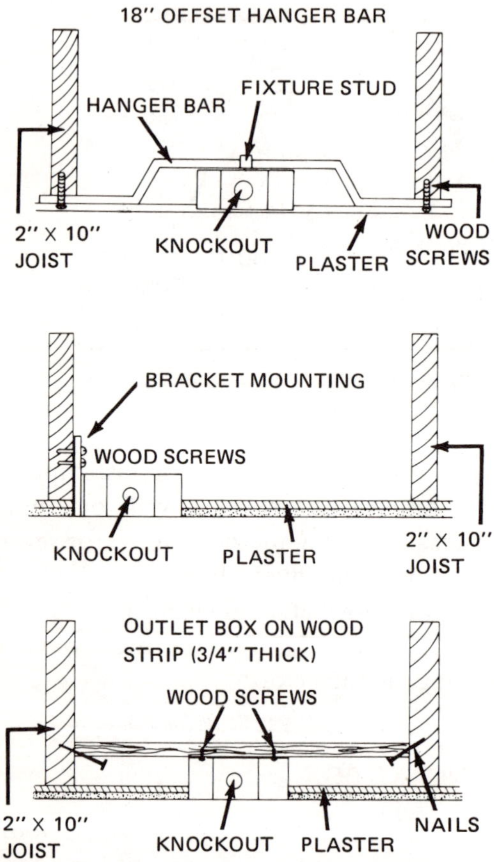

Fig. 7-7 Methods for supporting octagon boxes.

box being visible. When it is necessary to install a fixture on a finished wall, using a pancake box eliminates the need to make a hole in a wooden or brick wall, Figure 7-9. Conductor space in the box is limited to two wires, and access is through knockouts in the back of the box. (*Caution:* Some local electrical codes prohibit the use of this box. The local code should be consulted.)

SQUARE BOX

The square box is used primarily as a junction box for surface- and concealed-wiring systems, Figure 7-10. Special-purpose covers permit this box to support switches,

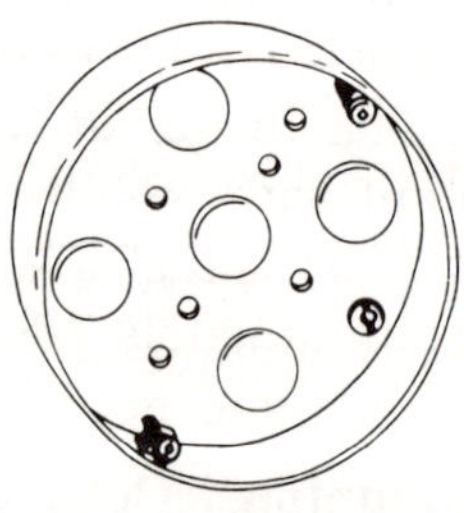

Fig. 7-8 Typical 4-inch round pancake box.

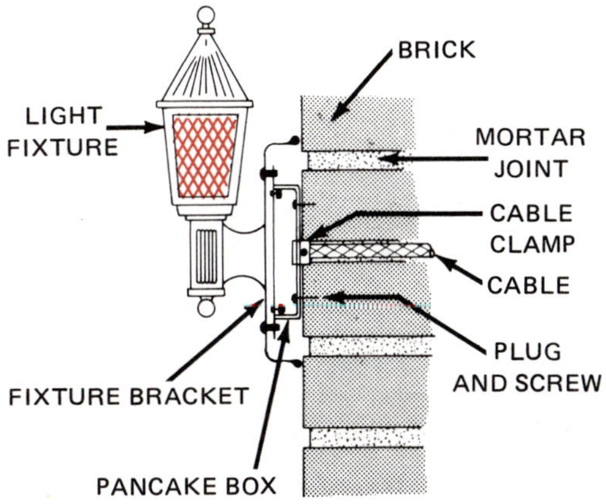

Fig. 7-9 Installation of fixture on pancake box.

receptacles, and pilot lights and to convert from square to round covers, Figure 7-11 and 7-12. Square boxes are very useful in block-wall, metallic and PVC conduit construction. Raised plaster rings are used to extend the square box flush with the finished wall surface.

Dimensions. Square boxes are made in two widths. The 4-inch width is the most common. A box 4 11/16 inches wide is also available.

The standard square box depth is 1 1/2 inches. Box depths of 1 1/4 inches and 2 1/8 inches are also available.

When extra conductor space is needed, extension rings made for the square box are used. Box covers and extension rings are fastened with two No. 8-32 round-head machine screws. Knockouts in a combination of 1/2-inch, 3/4-inch and 1-inch sizes are available. Figure 7-13 shows a square box extension.

Methods of Mounting. The square box is mounted in much the same way as the octagon box. (See Figure 7-7.)

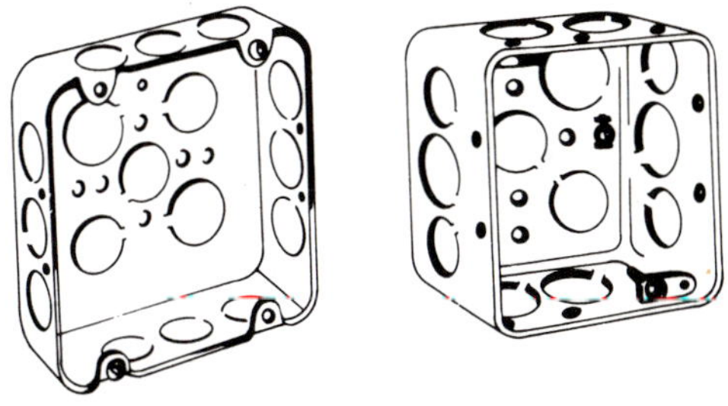

Fig. 7-10 Typical square boxes.

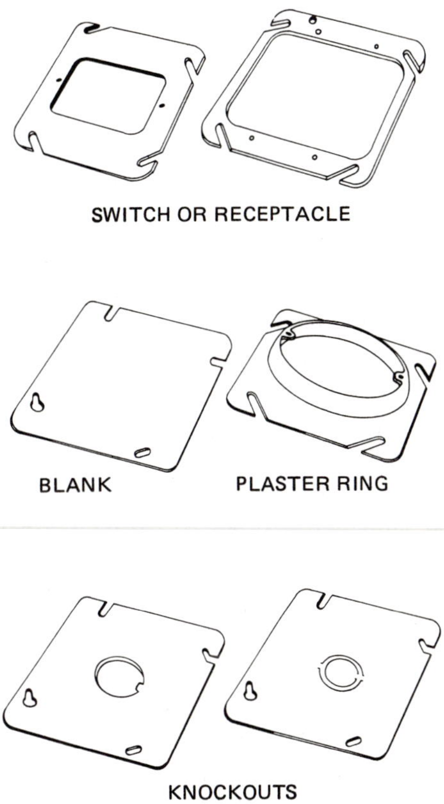

Fig. 7-11 Types of square box covers.

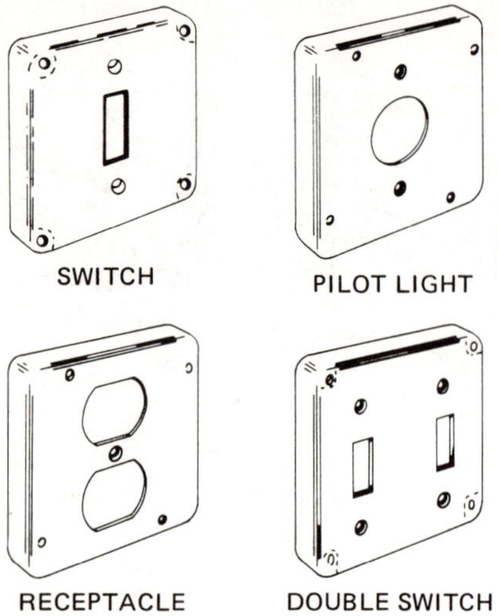

Fig. 7-12 Special-purpose surface covers for square boxes.

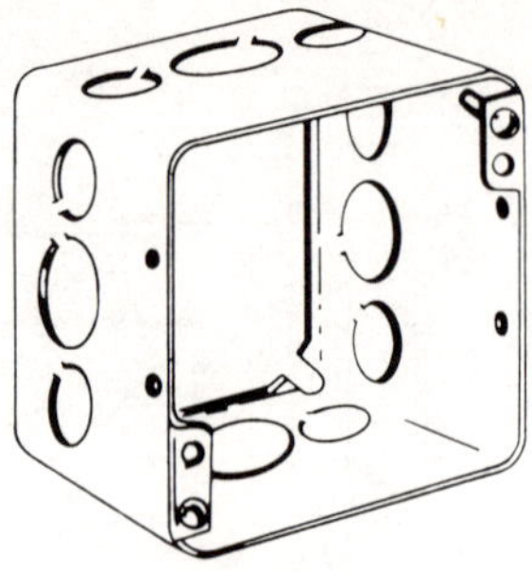

Fig. 7-13 Typical square box extension.

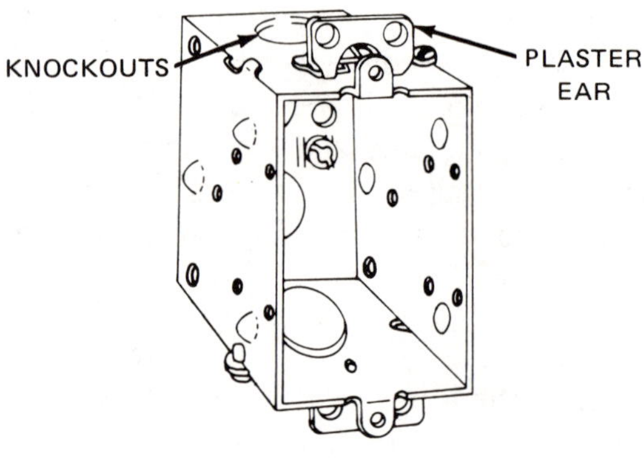

Fig. 7-14 A sectional plaster box.

SECTIONAL PLASTER BOX

The sectional plaster box is used to support switches and receptacles in many concealed wiring systems, Figure 7-14. It is not suitable for surface wiring because of its irregular outer surface.

The sectional plaster box is so named because it was designed for use in walls finished in plaster or masonry. However, it is used with many types of drywall construction.

The National Electrical Code requires that the front of the box be flush with the surface of walls finished in *combustible* materials, such as wood paneling. This is a precaution to prevent a flash fire in the box from spreading to the surrounding material. The box may be recessed to a depth of up to 1/4 inch in walls of such noncombustible materials as plaster or masonry.

Construction. Most sectional plaster boxes are made of galvanized steel. Bakelite, or phenolic, boxes are available for use with *nonmetallic* wiring systems.

The sides of the *metal* boxes are easily removed for grouping, or ganging together, a series of boxes, Figure 7-15. This allows an installer to assemble quickly a box capable of supporting any number of switches or receptacles. Sectional boxes made by one manufacturer may not link up with those

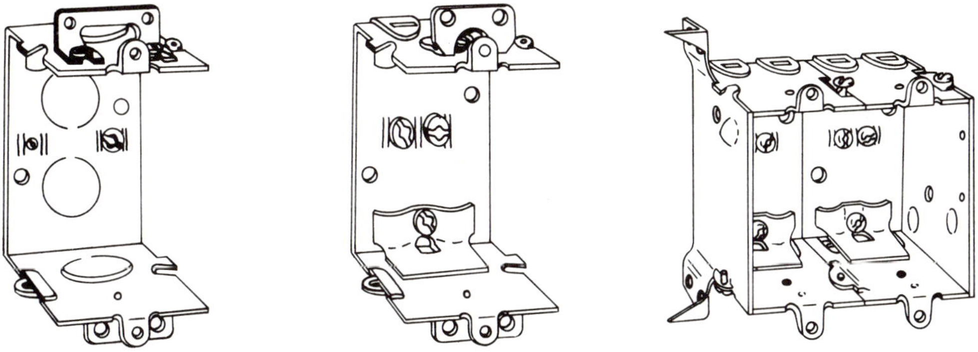

Fig. 7-15 Sectional plaster boxes for ganging.

made by another. When assembling a gang unit, therefore, take care to select boxes of the same type.

Dimensions. The standard sectional box is 2 inches wide and 3 inches in height. The depth of the box varies with the amount of conductor space required. The standard depth is 2 1/2 inches. Units in depths of 3 inches, 2 inches, and 1 1/2 inches are also available.

Covers. Covers for sectional plaster boxes are made primarily for switches or receptacles. A *blank* cover is used when the box is to be installed as a junction point for splices in the wires. Plaster boxes have two No. 6-32 threaded mounting lugs spaced 3 1/4 inches (80 mm) apart, which accept any manufacturer's switch or receptacle. The covers are usually fastened to the switch or receptacle with No. 6-32 machine screws, Figure 7-16.

Gang covers are used for multiple-switch or receptacle units mounted in grouped boxes. Sectional boxes are used, sometimes, to enclose pilot lights which indicate when a piece of electrical equipment is operating. Covers for this purpose are also available.

Fig. 7-16 Sectional plaster box covers.

Bakelite, or phenolic, is usually used in the manufacture of covers. It is a good insulator, is heat resistant and easily molded, and it is generally low in cost. Covers of brass, aluminum, stainless steel, and galvanized steel are produced where a stronger or more decorative cover is needed. (These metal covers may require special grounding.)

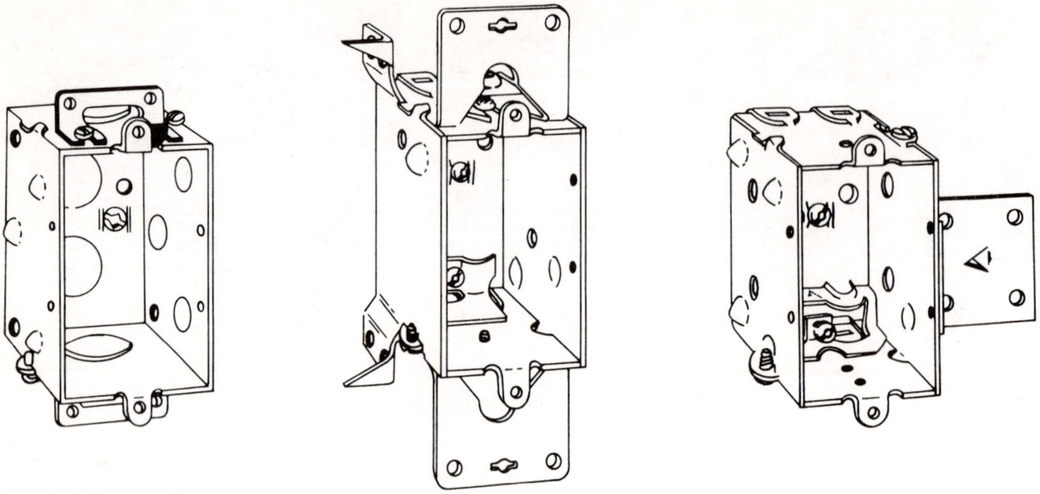

Fig. 7-17 Sectional plaster boxes with mounting brackets.

Methods of Mounting. Sectional plaster boxes can be supported in a number of ways. Figure 7-17 shows a group of boxes equipped with mounting brackets. Figure 7-18 shows how single or ganged units can be supported.

When it is necessary to install a box in a finished wall, a special-purpose unit can be used. This type of sectional box is equipped with an expanding bracket that passes through a precut hole in the wall, then expands and grips the plaster when a tension bolt is tightened, Figure 7-19. The cable *must* be fastened securely to the box *before* inserting the box into the hole. Once the bracket has expanded, it cannot be removed easily. Special clamps, referred to as "jiffy clamps," are available to mount standard sectional boxes into a finished wall.

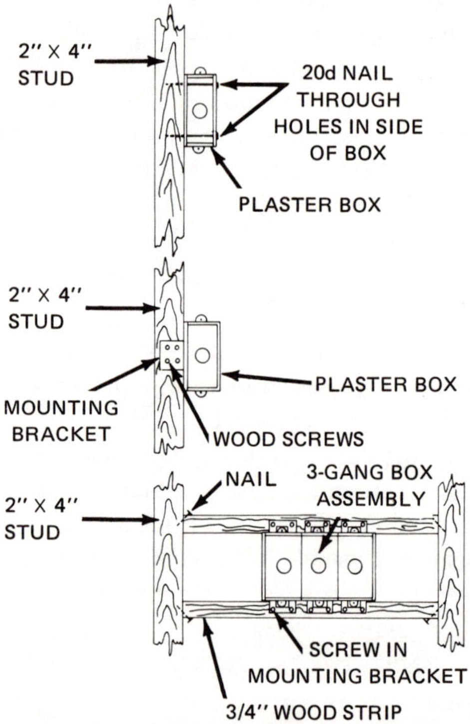

Fig. 7-18 Methods for mounting sectional plaster boxes.

UTILITY BOXES

The utility box (or handy box) is used to support receptacles, switches, and pilot lights in *surface* wiring systems. Its rounded corners and smooth exterior design make this versatile box an ideal unit for surface-wiring systems with cable or conduit.

Construction. The typical utility box is usually made of galvanized or cadmium-plated steel, Figure 7-20.

Dimensions. Standard utility box widths vary between 2 1/4 inches and 2 1/2 inches, depending upon the manufacturer. The length of the box is a standard 4 inches. The depth is usually 1 1/2 inches. A unit 2 1/4 inches in depth is available.

Covers. Utility box covers are usually made of plated steel, Figure 7-21. Bakelite covers should not be used because the sharp overlapping corners break easily when used with surface wiring materials.

Most utility box covers are made for receptacles and switches. However, pilot light and blank covers are also available.

Methods of Mounting. Some utility boxes are made with mounting brackets attached, Figure 7-22. Holes for bolts or screws are provided in the back of the box for mounting it directly on a surface. The devices are fastened to the box with No. 6-32 machine screws.

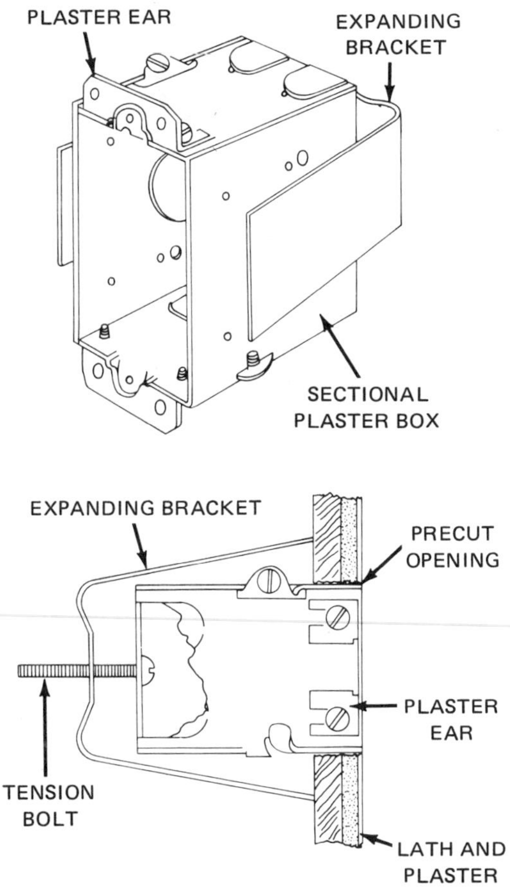

Fig. 7-19 Methods for mounting sectional plaster box in finished wall.

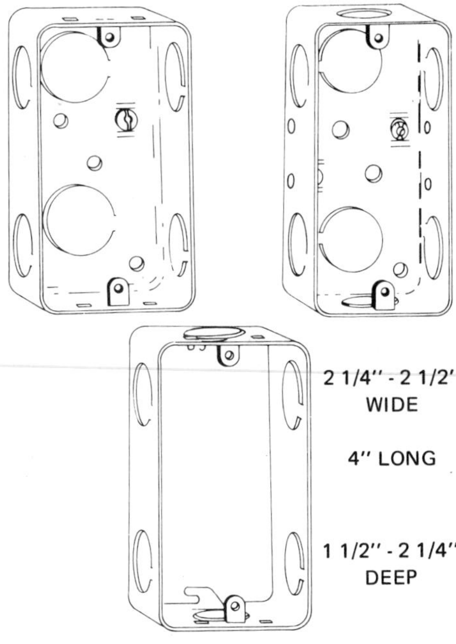

Fig. 7-20 Typical utility box and extension.

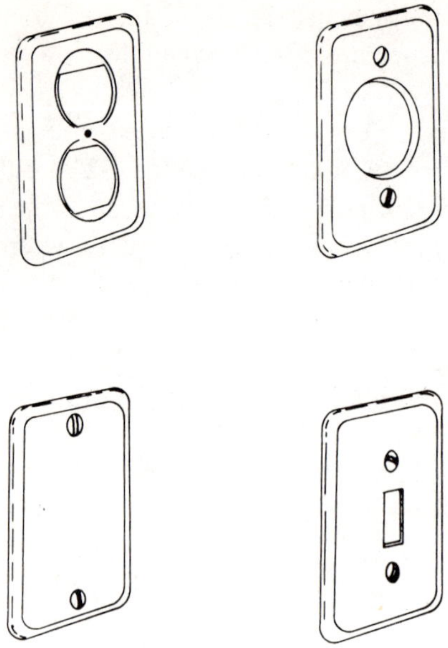

Fig. 7-21 Utility box covers.

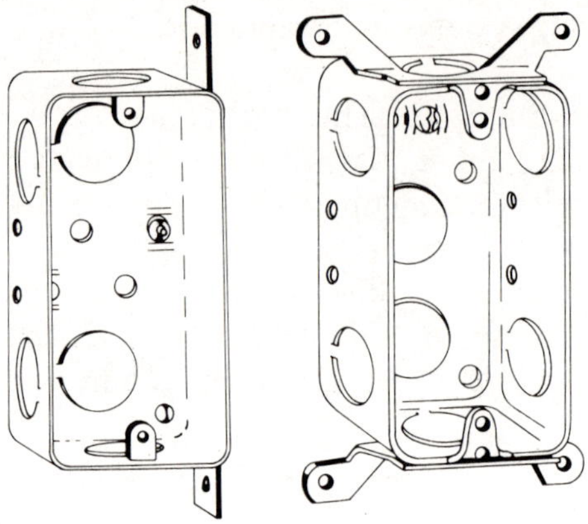

Fig. 7-22 Utility boxes with mounting brackets.

CONCRETE-MASONRY-TILE BOXES

Industrial and commercial buildings made with poured concrete, or other masonry materials, require a box that can be set directly into the material *during* construction. Concrete-masonry-tile boxes are intended for use with *concealed* metallic wiring systems, Figure 7-23.

Construction. Concrete-masonry-tile boxes are usually made of galvanized steel.

Dimensions. A single-gang unit is 2 inches wide and 3 3/4 inches in height. A multi-switch unit capable of supporting up to five switches or receptacles is also available.

Covers. Covers are made for single or group installation of switches, receptacles, and pilot lights, or combinations of all three.

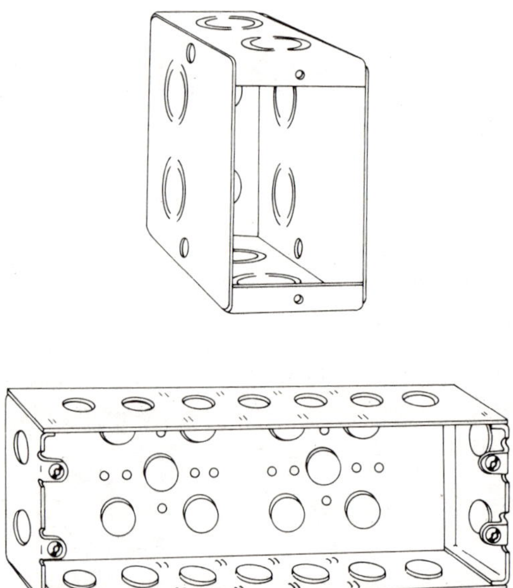

Fig. 7-23 Concrete-masonry-tile boxes.

Fig. 7-24 Typical knockout and pryout.

Methods of Mounting. Concrete-masonry-tile boxes are held in place by the concrete or mortar of the masonry wall. When the building is made with poured concrete walls and floors, the boxes (supplied with the conduit system) may be wired into position to hold them securely while the concrete is being poured.

CABLE CLAMPS IN BOXES

Electrical boxes provide access for conductors in two ways. The removable disc (the knockout) allows conduit and/or cable connectors in trade sizes of 1/2-inch, 3/4-inch, and 1-inch internal diameter to be fitted to the box.

Boxes designed for use with nonmetallic or armored-cable systems are available with built-in clamps. These cable clamps eliminate the need for separate connecting devices, and also shorten the time required for this type of installation.

The removable disc designed for boxes equipped with cable clamps is called the pryout. A *pryout* is a knockout having a slot for the insertion of a screwdriver which is used to remove the knockout, Figure 7-24.

Several types of built-in cable clamps are available, Figure 7-25. Built-in cable clamps are usually found in octagon and sectional plaster boxes, Figure 7-26.

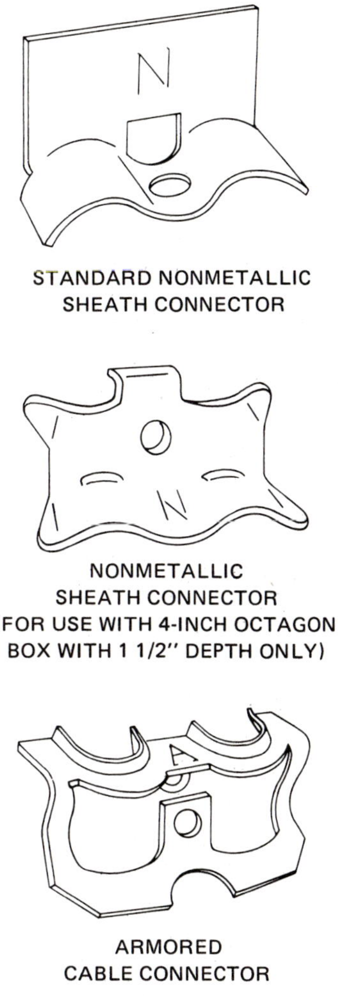

Fig. 7-25 Cable clamps.

BOX GROUNDING

Modern electrical wiring boxes provide one or two machine screws in the back of the box for grounding purposes. *Nonmetallic* cable systems have a bare grounding wire within the cable. The grounding wire must be connected to one of the screws provided in the box. A receptacle should have a grounding conductor (usually "green"

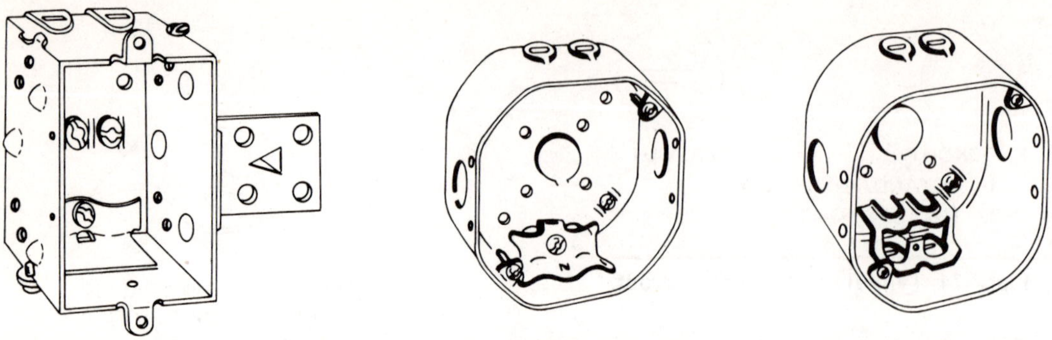

Fig. 7-26 Boxes with built-in cable clamps.

or bare) joining its grounding terminal with the grounding screw in the box.

Metallic wiring systems normally rely on a secured metal-to-metal connection with the box to complete their equipment grounding circuits.

CONDUCTOR CAPACITY OF A BOX

The *number* of current-carrying conductors contained in a box is necessarily limited. When too many conductors are contained in a box, the conductors might be forced into a sharp edge or mounting screw within the box. The sharp edge might penetrate the insulation on the conductor, allowing the current to take other than its intended path. If the *damaged* conductor is a live wire, a short-circuit condition will occur and the circuit fuse will blow. If, for some reason, the box is not grounded, the box and any metal object in contact with it will become *alive and dangerous.*

A second, and equally important, reason for limiting the number of conductors in a box is *overheating.* Any conductor carrying current produces some heat as a side effect. The more current passing through the conductor, the more heat that is produced. The more conductors in a box, the more heat that accumulates within the box. Once the cover is placed on the box, there is little or no air circulation to cool the conductors. Modern conductor insulation is designed to withstand some heat, but it will become hard and brittle if overheated for an extended period of time.

Table 370-6(a) of the National Electrical Code lists the number of conductors in standard electrical boxes, (see Table 7-1). Mathematical calculation of the volume of a box (length × width × depth) will *not* give the same results as listed in Table 7-1. Because of the differences in tolerance during manufacture of boxes, the National Electrical Code-Making Committee decided on standard capacities. The actual internal dimensions of a sectional plaster box, for example, are slightly smaller than those listed in catalogs and tables.

Each size of conductor requires a certain amount of free air space for cooling. *Table 370-6(b)* of the National Electrical Code lists the volume of air space required by some of the common conductor sizes, (see Table 7-2).

Table 370-6(a). Metal Boxes

Box Dimension, Inches Trade Size or Type	Min. Cu. In. Cap.	Maximum Number of Conductors No. 14	No. 12	No. 10	No. 8	No. 6
4 X 1 1/4 Round or Octagonal	12.5	6	5	5	4	0
4 X 1 1/2 Round or Octagonal	15.5	7	6	6	5	0
4 X 2 1/8 Round or Octagonal	21.5	10	9	8	7	0
4 X 1 1/4 Square	18.0	9	8	7	6	0
4 X 1 1/2 Square	21.0	10	9	8	7	0
4 X 2 1/8 Square	30.3	15	13	12	10	6*
4 11/16 X 1 1/4 Square	25.5	12	11	10	8	0
4 11/16 X 1 1/2 Square	29.5	14	13	11	9	0
4 11/16 X 2 1/8 Square	42.0	21	18	16	14	6
3 X 2 X 1 1/2 Device	7.5	3	3	3	2	0
3 X 2 X 2 Device	10.0	5	4	4	3	0
3 X 2 X 2 1/4 Device	10.5	5	4	4	3	0
3 X 2 X 2 1/2 Device	12.5	6	5	5	4	0
3 X 2 X 2 3/4 Device	14.0	7	6	5	4	0
3 X 2 X 3 1/2 Device	18.0	9	8	7	6	0
4 X 2 1/8 X 1 1/2 Device	10.3	5	4	4	3	0
4 X 2 1/8 X 1 7/8 Device	13.0	6	5	5	4	0
4 X 2 1/8 X 2 1/8 Device	14.5	7	6	5	4	0
3 3/4 X 2 X 2 1/2 Masonry Box/Gang	14.0	7	6	5	4	0
3 3/4 X 2 X 3 1/2 Masonry Box/Gang	21.0	10	9	8	7	0
FS — Minimum Internal Depth 1 3/4 Single Cover/Gang	13.5	6	6	5	4	0
FD — Minimum Internal Depth 2 3/8 Single Cover/Gang	18.0	9	8	7	6	3
FS — Minimum Internal Depth 1 3/4 Multiple Cover/Gang	18.0	9	8	7	6	0
FD — Minimum Internal Depth 2 3/8 Multiple Cover/Gang	24.0	12	10	9	8	4

*Not to be used as a pull box. For termination only.

Table 7-1

Table 370-6(b). Volume Required per Conductor.

Size of Conductor	Free Space Within Box for Each Conductor
No. 14	2. cubic inches
No. 12	2.25 cubic inches
No. 10	2.5 cubic inches
No. 8	3. cubic inches
No. 6	5. cubic inches

Table 7-2

CALCULATION OF BOX CAPACITY

Table 7-1 lists a device box 3" X 2" X 2 3/4" as having 14 cu. in. of air space. Table 7-2 lists a No. 14 gauge conductor as requiring 2 cu. in. of air space. Therefore, the number of No. 14 gauge conductors allowed in this box will be:

$$\frac{\text{Air space in the box}}{\text{Air space of one conductor}} = \frac{14}{2} = 7$$

Accordingly, only 7 conductors can be put in the box.

Clamps, receptacles, switches, or other devices inside the box take up some of the free air space. *Section 370-6* of the National Electrical Code states that for each device in the box, one conductor must be *subtracted* from the total listed in Table 7-1. For example, the device box listed as having a capacity of 7 conductors would be limited to 5 conductors, if it contained a switch and built-in cable clamp. A grounding conductor (one or more) also reduces the number of conductors by one.

NEC *Section 370-6* explains in detail how conductors entering and/or leaving a box must be counted to determine the total number of conductors allowed. Always use the most recent edition of the Code, which is revised regularly.

REVIEW

1. What materials are used for the construction of electrical boxes?
2. Where and why are brass boxes used?
3. What type of box is used for indoor light fixtures?
4. List the sizes in which octagon boxes are made.
5. Explain how switches or receptacles are fastened to the octagon box.
6. When are octagon concrete rings used?
7. Where are pancake boxes usually used? Why? Are they prohibited by some local codes?
8. Explain where and for what purposes square boxes are used.
9. How did the sectional plaster box get its name?
10. What is a gang box? What is it used for?
11. Explain where and how the sectional plaster box special-purpose unit is used.
12. Why is Bakelite not recommended for utility box covers?
13. Which boxes are used for buildings made of poured concrete and block?
14. State two reasons for using boxes with built-in cable clamps.
15. What is a *pryout?* How does it differ from a *knockout*?
16. How are boxes grounded in a conduit system? How are they grounded in a nonmetallic cable system?
17. State two reasons for limiting the number of conductors in a box.
18. Explain how the volume and the conductor capacity of a box are calculated.
19. Calculate the number of conductors allowed in a 4-inch octagon box, 1 1/2-inch in depth, to be used in a circuit with No. 12 gauge wires. Allow for a grounding conductor, a cable clamp, and one device.
20. If the octagon box described in question 19 contains two built-in clamps and a fixture stud, how many wires are allowed, according to the National Electrical Code?

Chapter 8

Nonmetallic-sheathed Cable Wiring

DESCRIPTION

A *nonmetallic-sheathed cable* is an assembly of two or more conductors having an outer sheath of moisture-resistant, flame-retardant, nonmetallic material. Type NM cable is the type most commonly used, especially in residential wiring. Another type of cable, Type NMC, not only has an overall covering that is flame retardant and moisture resistant, but it is also fungus resistant and corrosion resistant.

Nonmetallic-sheathed cable must be of an approved type. Occasionally, some nonmetallic-sheathed cable appears on the market that does not meet the UL listing. Cable of this kind should not be used.

CABLE CONSTRUCTION

Several basic types of nonmetallic-sheathed cable are produced, Figures 8-1 through 8-4. The exterior of the cable must be clearly marked, throughout its entire length, specifying the cable type and the name of the manufacturer. The box in which the cable is packaged will bear the UL listing, if any.

Types NM and NMC cable are constructed of conductors having an ampacity rating of the 60°C (140°F) conductors in NEC *Table 310-16*. These cables are available in sizes No. 14 through No. 2 with cop-

Fig. 8-1 Typical 2-wire Type NM cable with ground. Copper conductors.

Fig. 8-2 Type NM cable, 3-wire, with ground.

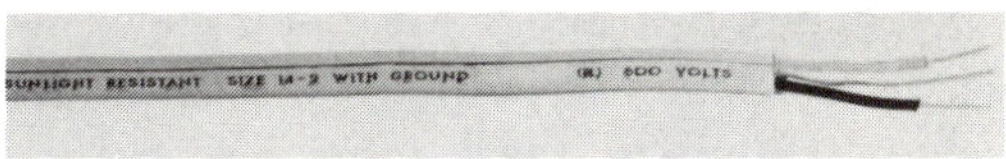

Fig. 8-3 Type UF moistureproof 2-wire cable, with ground.

Fig. 8-4 Type SE cable used for service entrance, ranges, or dryers.

per conductors, and in sizes No. 12 through No. 2 with aluminum or copper-clad aluminum conductors. In addition to the current-carrying conductors, these cables may have an uninsulated (bare) conductor (the same size as the current-carrying conductors), or a green insulated conductor for equipment grounding purposes only.

Regardless of the type of insulation used to cover the conductors, NM and NMC cables are rated to operate under the *maximum allowable temperature* of 60°C (140°F).

USES PERMITTED

Type NM cable *may* be used for concealed or exposed installation, installed or fished in the hollow voids of masonry or tile walls that are not exposed to excessive moisture or dampness.

Moisture- and corrosion-resistant Type NMC cable *may* be used for both concealed and exposed work in dry, moist, damp, or corrosive locations; in outside and inside walls of masonry block or tile; and where embedded in plaster walls and covered with plaster within 2 inches (50.8 mm) of the finished surface.

RESTRICTIONS

Types NM and NMC cable have been used for residential wiring for a number of years. However, these cables are *not* suitable for use with modern light fixtures. Heat from the fixture bulbs often dries out the cable, causing the insulation to become brittle and of little value.

Type NM cable may *not* be used where exposed to corrosive vapors or fumes; shall *not* be embedded in masonry, concrete, adobe, fill, or plaster; and shall *not* be run in a shallow chase in masonry, concrete, or adobe and covered with plaster, adobe, or similar finish.

Type NMC or NM cable is *not* permitted to be used as service-entrance cable; may *not* be used in commercial garages, theaters, and assembly halls, except as permitted in NEC *Article 518;* and may *not* be installed in storage-battery rooms, hoistways, or hazardous locations. These cables may *not* be embedded in poured cement, concrete, or aggregate. Local codes may impose other restrictive uses.

Trade Names. Type NM cable was first produced by the Rome Wire and Cable Company, which named its new product *Romex.* This name is still used in the trades, and is also commonly referred to as *loom wire.* Each company producing NM and NMC however, has its own product name ending in "ex"; for example, *Canadex* from Canada Wire and Cable Limited, and *Philex* from Phillips Cables Limited.

CABLE INSTALLATION

The National Electrical Code requires that NM and NMC cable be installed in a *loop system,* meaning that all cables are run in continuous lengths between the electrical boxes, Figure 8-5. Joints or splices in the cable *must* be made in a box or wiring device identified for the use. All electrical boxes in the system *must* be accessible for inspection or circuit repair after the building is completed.

In the past, the *knob* and *tube* wiring systems, used before the loop system was developed, allowed splices at almost any point in the circuit. This, and the

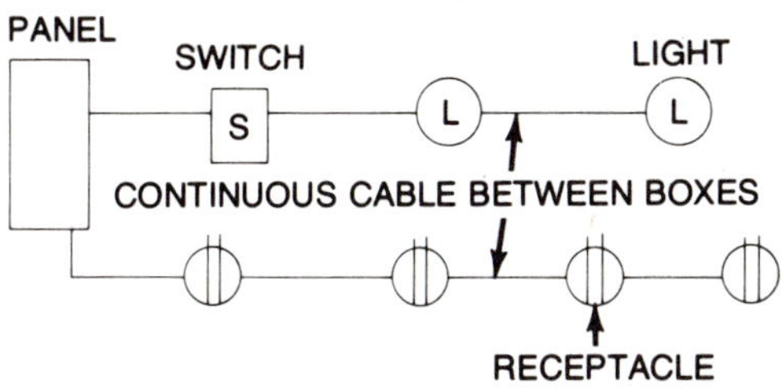

Fig. 8-5 Loop system.

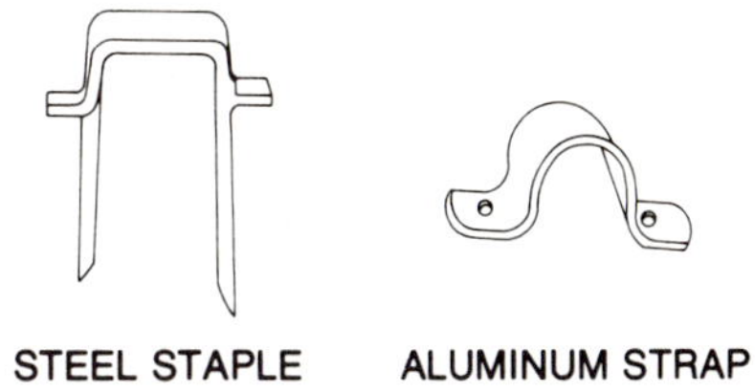

Fig. 8-6 Cable supports.

fact that the live and the neutral wire of the same circuit did not always travel side by side, made troubleshooting difficult, even for experts.

CABLE SUPPORTS

Nonmetallic-sheathed cable *must* be fastened to the *wooden* members of a building by means of straps, staples, or other devices permitted by the NEC and identified for the purpose, Figure 8-6. Staples take the least time to fasten, but the cable may be damaged if the staples are driven too deeply into the wood, Figure 8-7. Too often, it is the last blow of the hammer that does the damage. Sometimes there is an easy tendency to use a nail bent over to secure a cable. This practice should *never* be tolerated. Straps may be secured with screws or nails.

The National Electrical Code requires that a strap or staple be placed within 12 inches (300 mm) of *every* box. This is to prevent any undue strain on the cable from pulling the conductors out of the box. Cable supports may be placed as far apart as 4 1/2 feet (1.4 m) on the runs between the boxes, but it is often a good idea to place them closer together. Cable installations are usually in service for many years, so neat and secure installations are important.

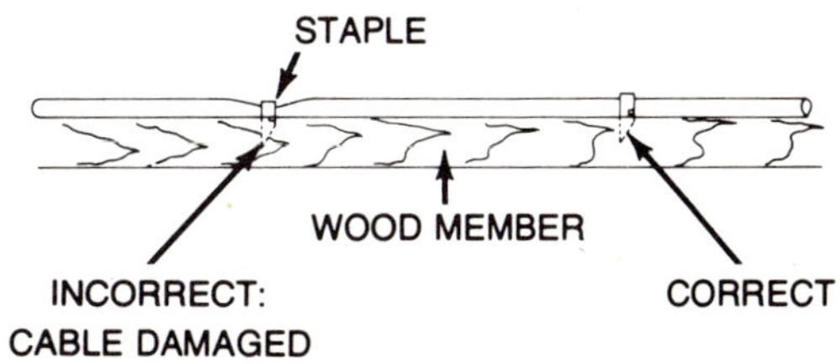

Fig. 8-7 Fastening cable with staples.

Cable assemblies *not* smaller than two No. 6 conductors or three No. 8 conductors may be fastened directly to the lower edge of the joists in an unfinished basement. In other locations, cables may be run either through bored holes or on running boards.

FASTENING A CABLE TO A BOX

Nonmetallic-sheathed cable must be held securely at the box by the clamp or connector. Do not overtighten the clamp as this might crush the cable and create a short circuit.

A maximum of approximately 1/4 inch (6 mm) of outer sheath should extend beyond the clamp to protect the TW insu-

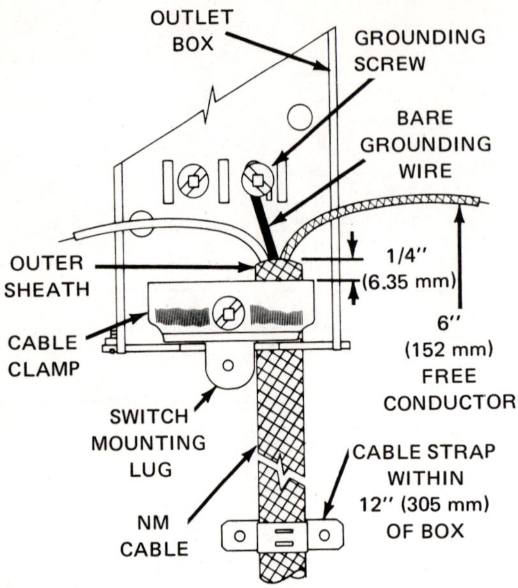

Fig. 8-8 Method for connecting cable to outlet box.

lated wire from the clamp, Figure 8-8. A minimum of 6 inches (152 mm) of free, insulated conductor must be available for connection to devices in the box. The bare ground wire may be trimmed after connecting it to the box ground screw or, in the case of a receptacle, left long enough to connect to the receptacle grounding screw as well. Carefully trim excess kraft paper away from the cable in the box to reduce any fire hazard.

UF CABLE BURIED IN THE EARTH

Cable used for underground *feeders* and *branch circuits* is called Type UF cable, Figure 8-9. It is not to be confused with USE cable (Underground Service Entrance) which is direct-burial cable for *services.* UF is *not* to be used for services.

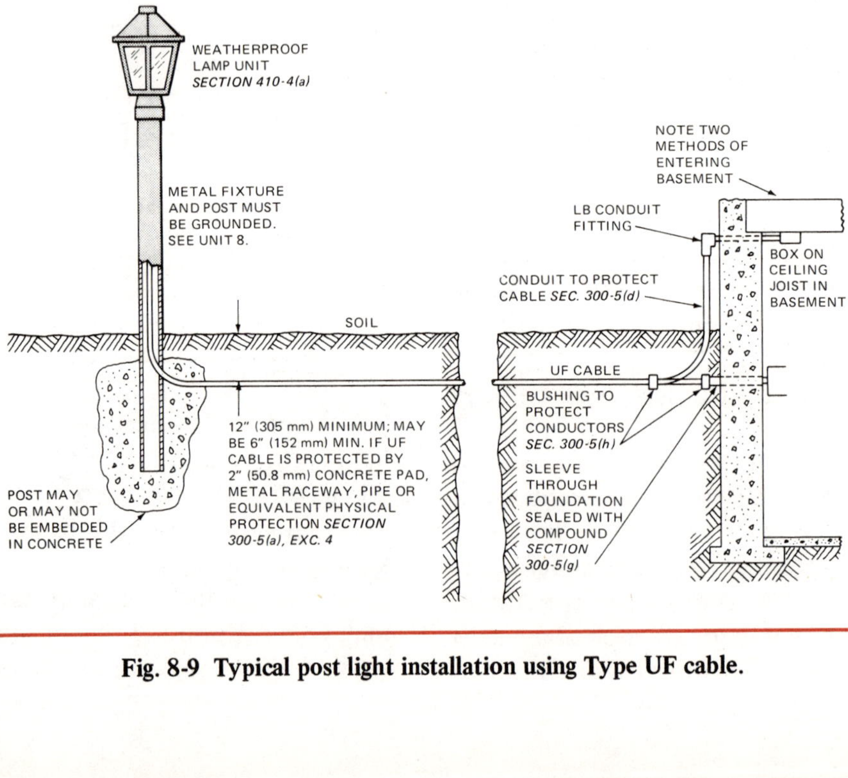

Fig. 8-9 Typical post light installation using Type UF cable.

Type UF cable listed by the Underwriters' Laboratories comes in conductor sizes from No. 14 through 4/0 (0000) AWG. The conductor may be TW or RHW, for example, as identified for the purpose. An outer covering on the insulated conductors is flame retardant, moisture resistant, fungus resistant, and corrosion resistant. This outer covering is identified for direct burial in the earth. UF cable may also carry a bare or insulated conductor to be used for grounding purposes only.

CABLE PROTECTION FOR CONCEALED INSTALLATIONS

The National Electrical Code requires that NM and NMC cables be kept at least 1 1/4 inches (31.8 mm) from the *outer* edge of any wooden member. This is to prevent driven nails or screws supporting baseboard, plaster, wood panels, or other wall products from piercing the cable. When a cable has been pierced with a nail, it is usually not discovered until the building is completed and the circuit made alive. With a finished wall concealing the damaged cable, it is both difficult and costly to locate and repair the fault.

Normal procedure is to drill holes in the *center* of the wooden members that the cable must pass through. Water and air circulation systems in the walls will often make it necessary to run the cable closer to the edge of a wooden member than the required 1 1/4 inches (31.8 mm). In these cases, a *steel plate* is fastened to the wooden member in *front* of the cable to protect it, Figure 8-10. Often, the plate is a side of a sectional plaster box left over from a gang box assembly. Take care not to damage the cable by locating it too close to a hot-water pipe or hot-air duct.

RESIDENTIAL CABLE APPLICATIONS

Types NM and NMC cable have a maximum rating of 300 V and will readily accept the 120/240 V supplied to a residence.

Most circuits in houses consist of No. 14 gauge NM cable and should be fused at a maximum of 15 A. Kitchen receptacles supplying current to frypans, coffee makers, or similar appliances use No. 12 gauge NM fused at 20 A. This increase in cable size and ampacity provides a margin of safety to a circuit often operating very close to its capacity. Home electric heating systems often have circuits consisting of No. 12 gauge NM. Type S safety fuses should be used to protect these circuits.

Clothes dryers are supplied with three-conductor, No. 10 gauge NM, commonly called *dryer cable.* Fuse protection for this cable should not exceed 30 A. Cartridge-type fuses or circuit-breakers are normally used. Electric stoves use a three-conductor, No. 8 or No. 6 gauge range cable fused at 40 A to 50 A, depending upon the size of stove, Figure 8-4. Refer to NEC *Article 220 (Branch-Circuit, Feeder Calculations).*

CABLE WIRING DIAGRAMS

The preparation of a circuit diagram is an important part of any wiring system. This diagram helps determine the *sequence* in which the devices are to be connected, and the *number* of wires required in the cable between the boxes.

Graphic symbols from the American Standards Association (ASA) are used to simplify the drawing of electrical devices in a circuit, Figure 8-11. These symbols are universal. Cables are represented by a single, solid line, with the number of insulated wires in the cable shown by short dashes

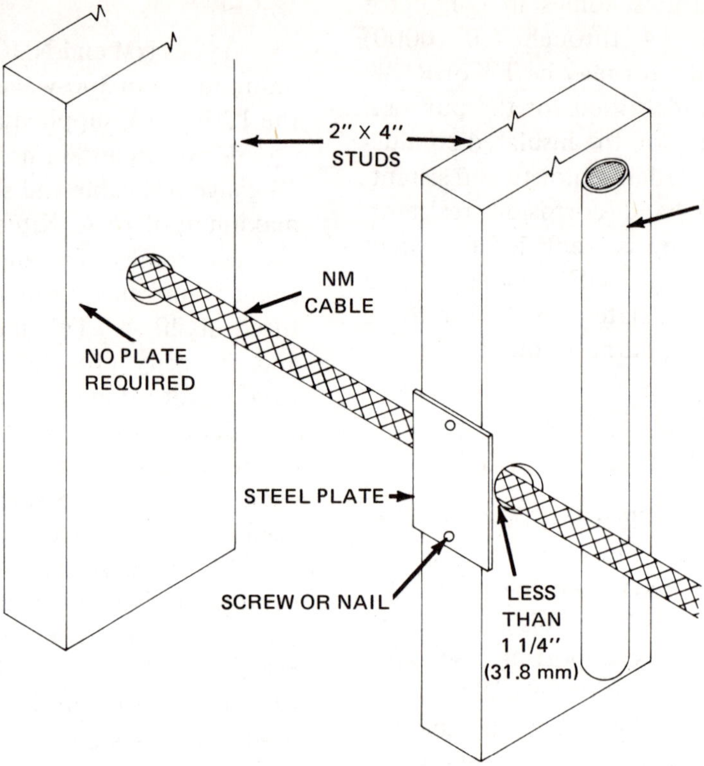

Fig. 8-10 Steel plate for cable protection.

across the cable line. For example, a two-wire cable is shown as —//—, and a three-wire cable is shown as —///—.

Remember that nonmetallic-sheathed cable is available in two- and three-wire combinations. Any wiring circuit must be completed using *only* the white, black, or red NMC wires available.

Figures 8-12, 8-13, and 8-14 compare architectural symbols with wiring diagrams of basic lighting circuits. When a cable system of wiring is used, *all* splices and connections *must* be in a box. This makes it unnecessary to show the individual conductors between the boxes. A single line is used to represent the cable in these circuits, Figure 8-12.

Figure 8-13 shows simple lighting circuits with a duplex receptacle added. Receptacles are considered to be alive at all times, unless otherwise marked on the diagrams. Three-conductor cable is required for some of these circuits.

Three- and four-way switch-control circuits are shown in Figure 8-14. Three-conductor cable is often used in this type of circuit.

NMC ACCESSORIES

Not all electrical boxes are equipped with built-in clamps. Distribution panels and outlet boxes with 1/2-inch (13 mm)

GRAPHIC ELECTRICAL SYMBOLS FOR ARCHITECTURAL PLANS

Ceiling	Wall	GENERAL OUTLETS
○	-○	Outlet
Ⓑ	Ⓑ	Blanked Outlet
Ⓓ		Drop Cord
Ⓔ	-Ⓔ	Electric Outlet; for use only when circle used alone might be confused with columns, plumbing symbols, etc.
Ⓕ	-Ⓕ	Fan Outlet
Ⓙ	-Ⓙ	Junction Box
Ⓛ	-Ⓛ	Lampholder
Ⓛ PS	-Ⓛ PS	Lampholder with Pull Switch
Ⓢ	-Ⓢ	Pull Switch
Ⓥ	-Ⓥ	Outlet for Vapor Discharge Lamp
⊗	-⊗	Exit Light Outlet
Ⓒ	-Ⓒ	Clock Outlet (Specify Voltage)

RECEPTACLES

	Duplex Receptacle
1, 3	Other than Duplex Receptacle 1 = Single, 3 = Triplex, etc.
	Split-Switched Duplex Receptacle
	Three-Conductor Spit-Duplex Receptacle
	Three-Conductor Split-Switched-Duplex Receptacle
WP	Weatherproof Receptacle
R	Range Receptacle
S	Switch and Receptacle
R	Radio and Receptacle
	Special Purpose Receptacle (Described in Specification)
⊙	Floor Receptacle

SWITCHES

S	Single-Pole Switch
S_2	Double-Pole Switch
S_3	Three-Way Switch
S_4	Four-Way Switch
S_D	Automatic Door Switch
S_E	Electrolier Switch
S_K	Key Operated Switch
S_P	Switch and Pilot Lamp
S_{CB}	Circuit Breaker
S_{WCB}	Weatherproof Circuit Breaker
S_{MC}	Momentary Contact Switch
S_{RC}	Remote-Control Switch
S_{WP}	Weatherproof Switch
S_F	Fused Switch
S_{WF}	Weatherproof Fused Switch

SPECIAL OUTLETS

○ a, b, c, etc.	Any standard symbol as given above with the addition of a lower case subscript letter may be used to designate some special variation of standard equipment of particular interest in a specific set of architectural plans.
a, b, c, etc.	When used they must be listed in the Key of Symbols on each drawing and, if necessary, further described in the specifications.
S a, b, c, etc.	

PANELS, CIRCUITS, AND MISCELLANEOUS

▬	Lightning Panel
▨	Power Panel
——	Branch Circuit; Concealed in Ceiling or Wall
— · —	Branch Circuit; Concealed in Floor
- - - -	Branch Circuit; Exposed
→→	Home Run to Panelboard. Indicate number of circuits by number of arrows.

NOTE: Any circuit without further designation indicates a two-wire circuit. For a greater number of wires indicate as follows:

—/// —	*(3 wires)*
—//// —	*(4 wires), etc.*

Fig. 8-11 Electrical symbols for architectural plans.

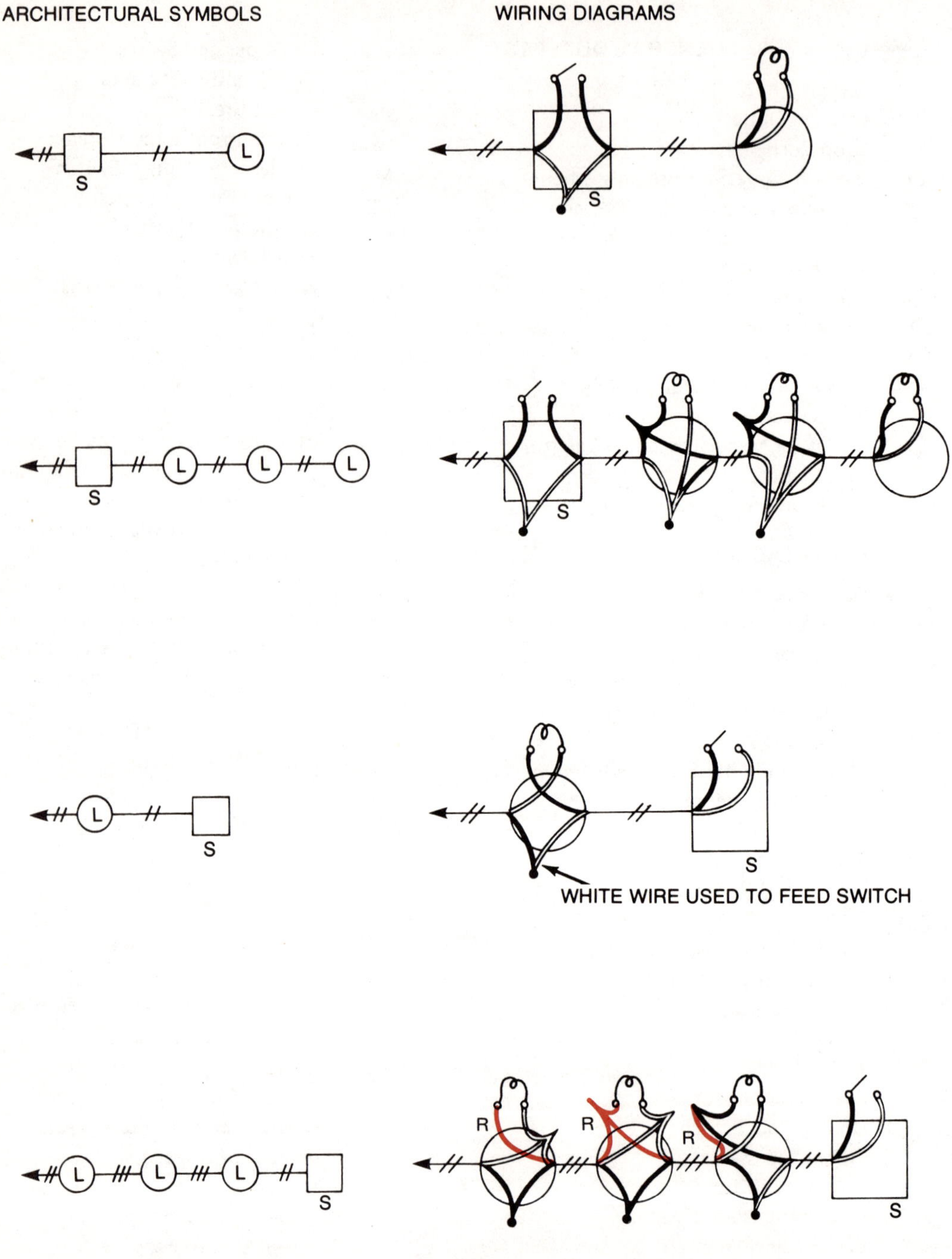

Fig. 8-12 Cable wiring diagrams for single-pole switch.

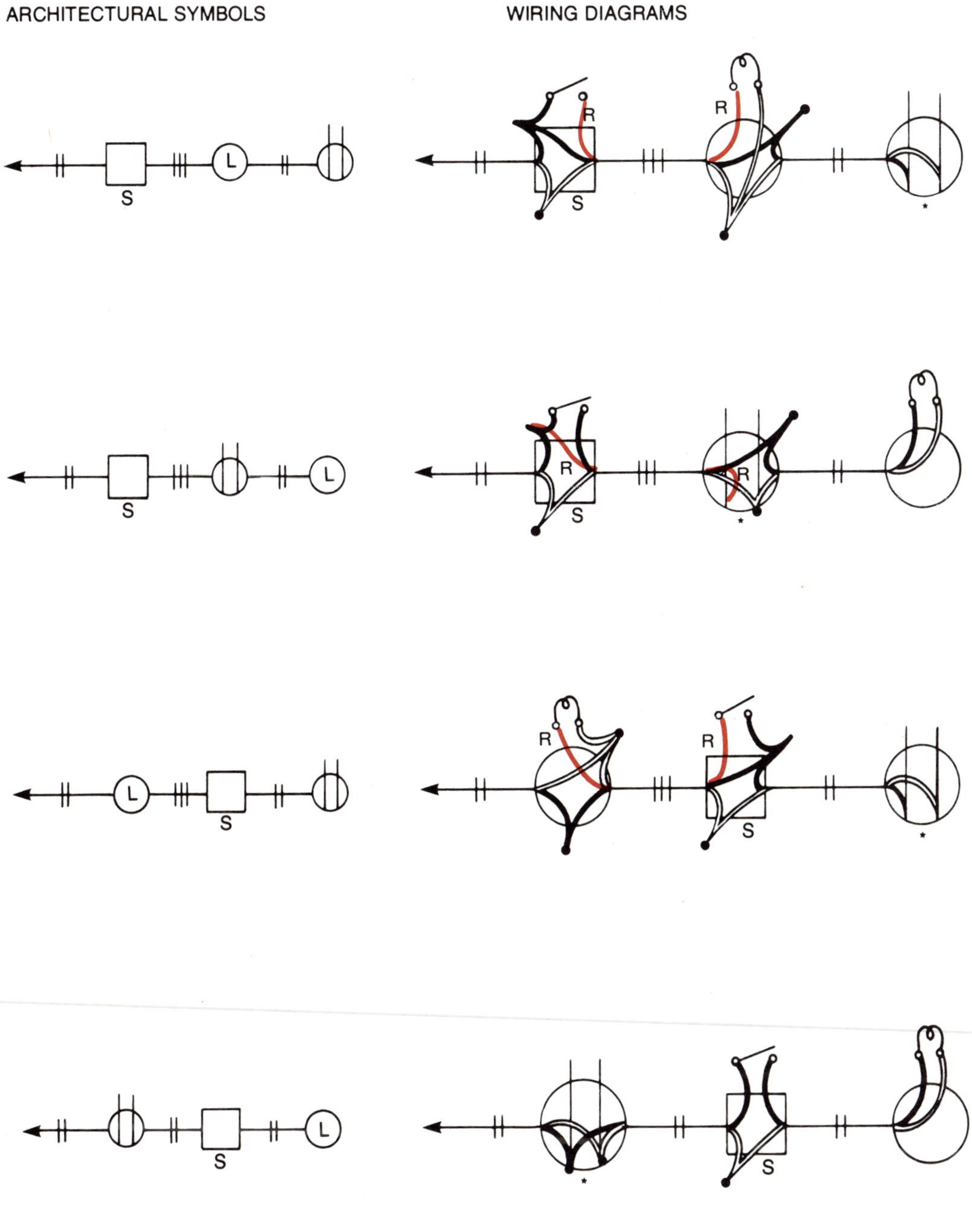

*NOTE: RECEPTACLES ALIVE AT ALL TIMES. SWITCH CONTROLS ONLY THE LIGHT.

Fig. 8-13 Wiring diagrams with receptacles.

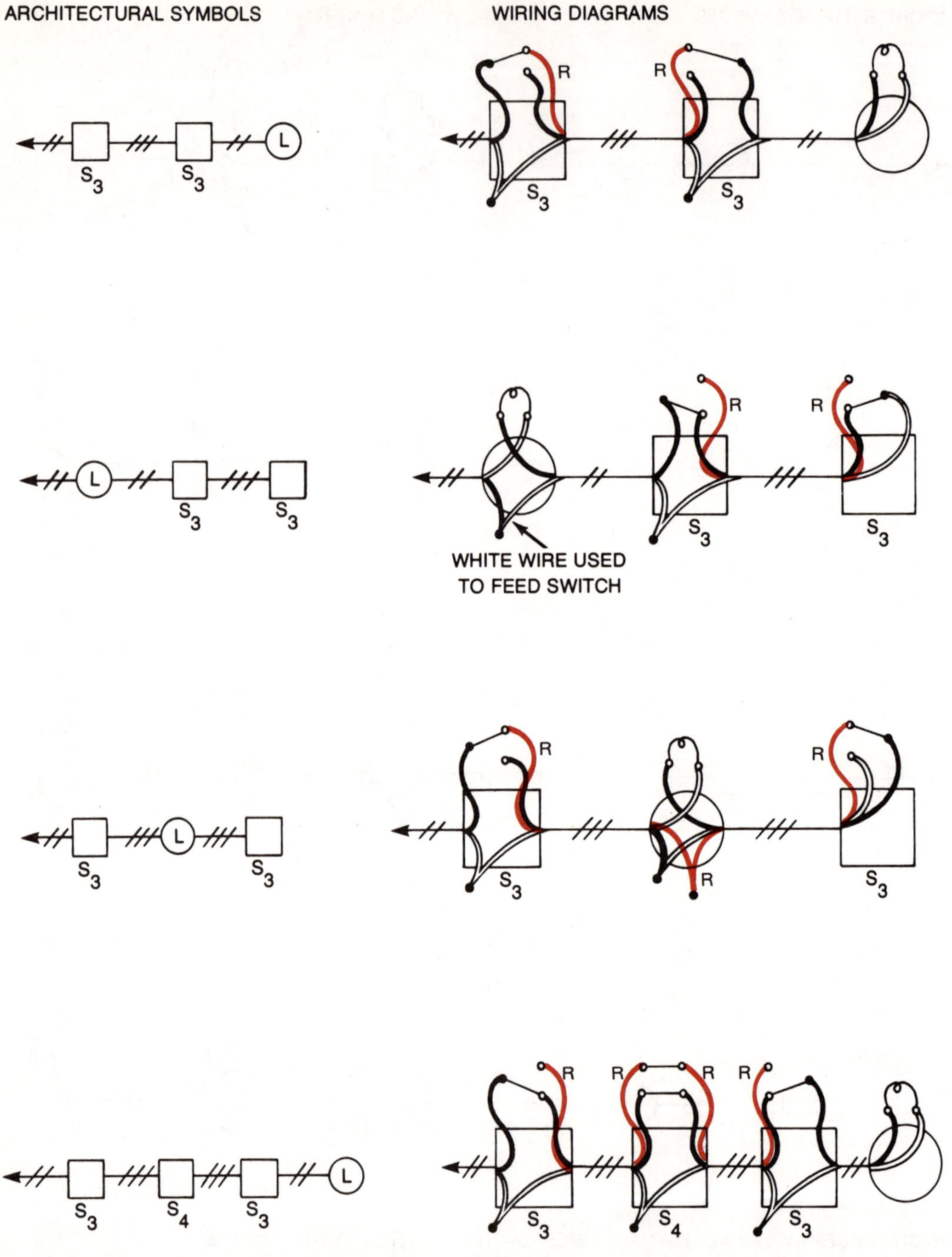

Fig. 8-14 Wiring diagrams showing 3- and 4-way switch control.

knockouts require a cable *connector*, Figure 8-15.

Some 6 inches (152 mm) of the cable's outer sheath is easily split with a *cable ripper*, Figure 8-16. This simple metal tool saves time, and prevents damage to the cable during the ripping process. *Cutting pliers* (diagonal cutters) are used to trim off the loose ends of kraft paper and outer sheath.

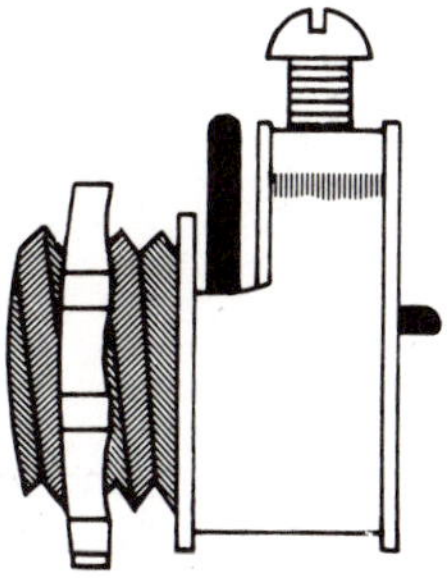

Fig. 8-15 An aluminum die-cast connector (with locknut).

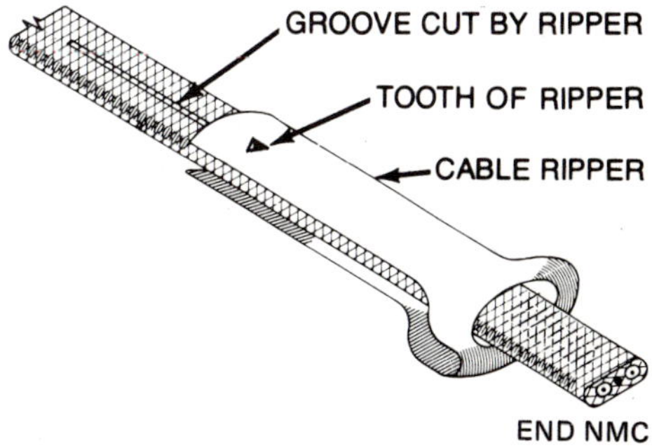

Fig. 8-16 Cable ripper for preparation of NMC for box.

REVIEW

1. With which types of building materials may NM be used?
2. What are the two types of NMC, and where are they used?
3. Which two materials are used for conductors in NMC?
4. What is the purpose of the bare wire in NMC? What size must it be?
5. Describe the *loop* system used with NM or NMC wiring.
6. Why does the National Electrical Code require that all joints or splices be made in a box?
7. List the devices that may be used to support NMC. How should these devices be spaced, and why?
8. When fastening a cable to a box, what length of conductor should be left free? Why? How much of the outer sheath is allowed inside the box past the cable clamp?
9. Describe how Type UF cable is run underground and connected to an outdoor post light.
10. When running NM or NMC through wooden members, what precautions should be taken? Why?

Chapter 9

Solderless Connectors

THE DEVELOPMENT OF SOLDERLESS CONNECTORS

The National Electrical Code once required that joints and splices in insulated conductors be soldered, and then covered with an insulating tape equivalent to the insulation on the conductor. Soldering the splice prevented the weakening of the electrical connection by the action of oxidation. But it was a time-consuming method, and when large cables were connected they were heated to a temperature that often damaged the insulation near the splice. In addition, to disconnect the conductors, more heat was needed to melt the solder. The number of tools required and the cost of materials that could be used only once made this system unsatisfactory. The need for a convenient method of making electrical splices resulted in the development and subsequent UL listing of solderless connectors identified for this use.

Soldering of splices or joints at conductor termination is accepted by the NEC, provided that such materials as solder, fluxes, inhibitors and compounds, where employed, are suitable for the use. The materials must be of a type which will not adversely affect the conductors, installation, or equipment. Before soldering, soldered splices must first be spliced or joined so as to be mechanically and electrically secure without solder.

SOLDERLESS WIRE CONNECTORS

One of the first solderless wire connectors consisted of a tapered porcelain cap with an internal screw thread. While porcelain is still used in the production of some heat-proof units, Bakelite and/or nylon are now most often used in the construction of wire connectors.

Three main types of solderless wire connectors are: twist-on, set-screw, and compression. The names for each refer to the system used to apply the unit.

Twist-on Connector. The twist-on connector has a cone-shaped metal spring that threads itself around the conductors as the connector is rotated, Figure 9-1. The internal spring design takes advantage of leverage and vise action to multiply the strength of a person's hand. Thus, the conductors are forced into a solid, effective splice, Figure 9-2.

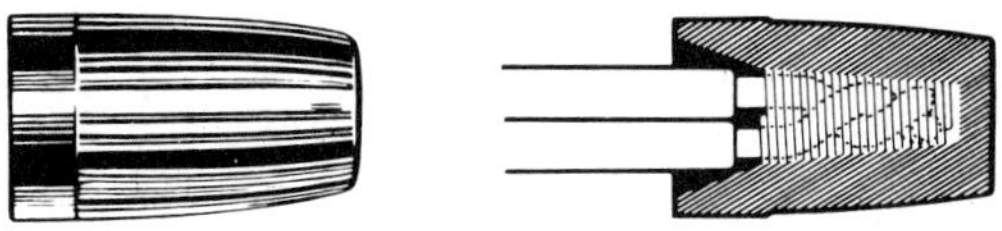

Fig. 9-1 A twist-on connector.

Using the twist-on connector is one of the quickest ways of splicing and insulating electrical conductors. This connector is suitable for use with solid and/or stranded conductors operating at 600 V or less. Some models have been approved for use at 1000 V.

When using twist-on connectors on combinations of solid and stranded wire, the stranded wire should extend approximately 1/4 inch (6.35 mm) beyond the solid conductor before twisting on the connector.

The twist-on connector is made in a range of sizes for splicing conductors from No. 18 gauge up to No. 8 gauge.

Set-screw Connector. The set-screw connector is a two-piece connector widely used in circuits where equipment (such as motors) must be changed frequently for maintenance purposes, Figures 9-3 and 9-4. The simple set-screw design allows solid and/or stranded conductors to be interchanged easily. For use on dissimilar types of conductor material, such as an intermix of aluminum and copper, the connector must be identified as suitable for the purpose.

Set-screw connectors are made for use on circuits up to 600 V, with approval for use on some lighting circuits of 1000 V. They are designed to splice conductors from No. 18 gauge up to No. 10 gauge in size.

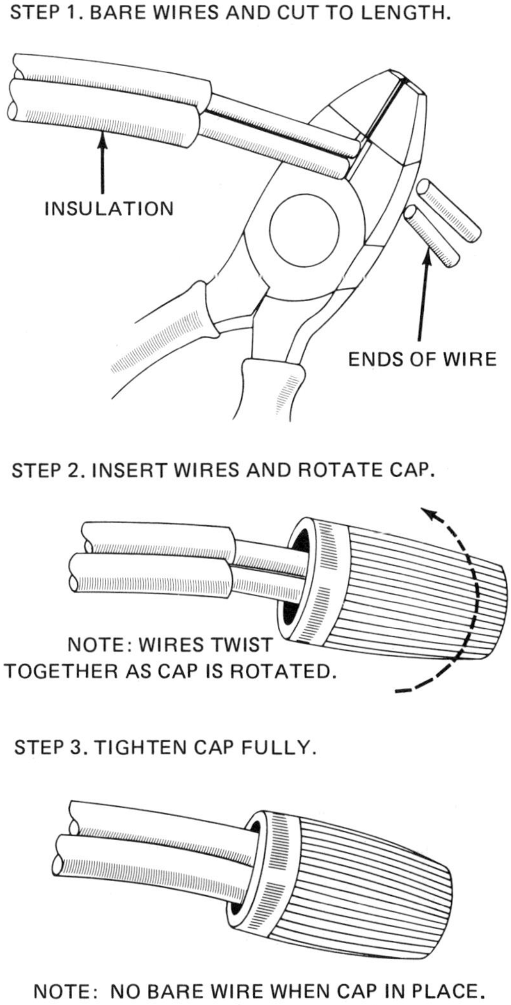

Fig. 9-2 Method for installing a twist-on connector.

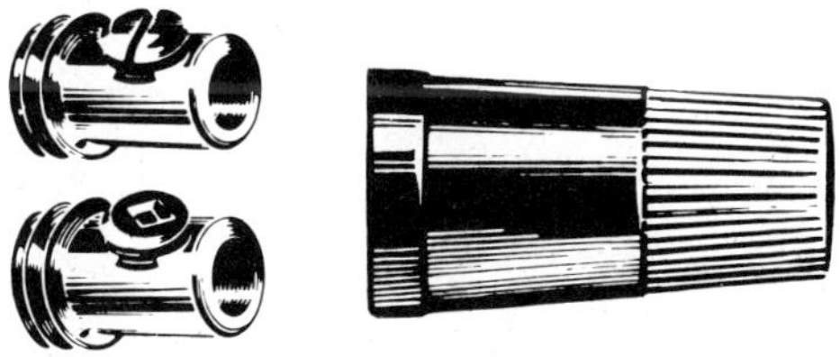

Fig. 9-3 A set-screw connector.

Figure 9-5 shows typical applications of both the twist-on and set-screw connectors. Both types are for use in electrical boxes or enclosures only.

Compression Connector. Unlike the mechanical (set-screw) connector, the two-piece compression connector requires a special crimping (compression) tool to install the conductor-retaining sleeve. The retaining sleeves are made of copper and/or zinc-plated steel. An insulating cap of plastic or nylon is fastened over the sleeve after the conductors have been crimped firmly, Figure 9-6. This type of connector is for permanent installation, because the con-

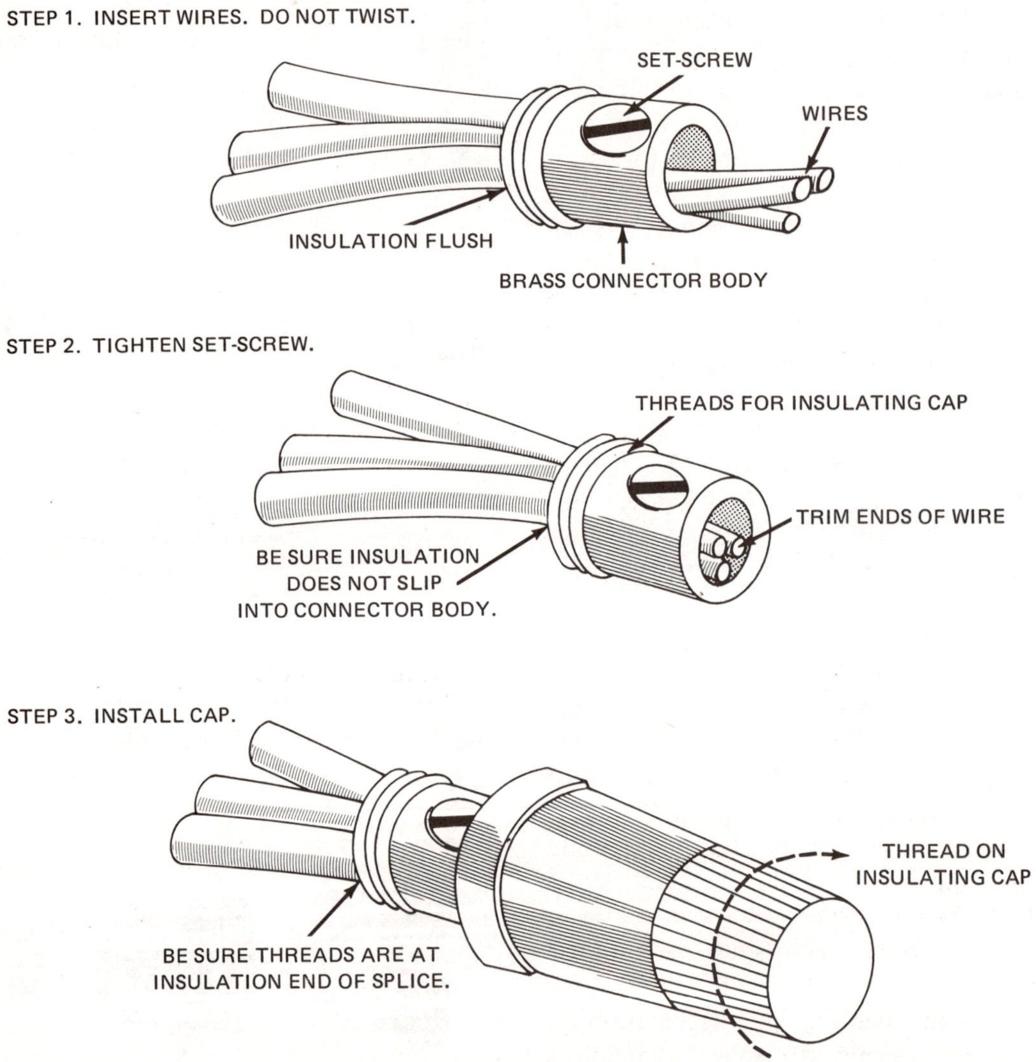

Fig. 9-4 Method for installing a set-screw connector.

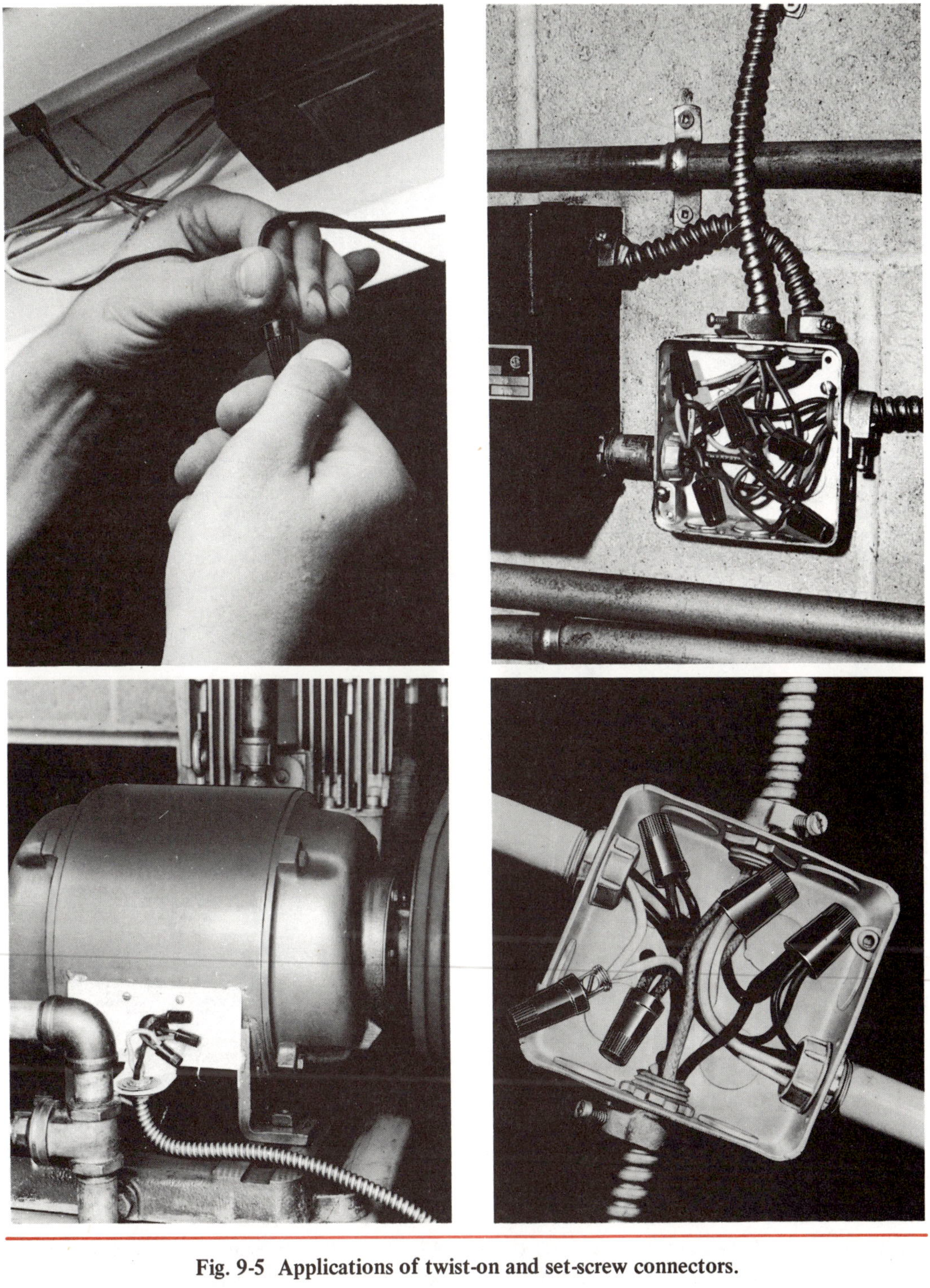

Fig. 9-5 Applications of twist-on and set-screw connectors.

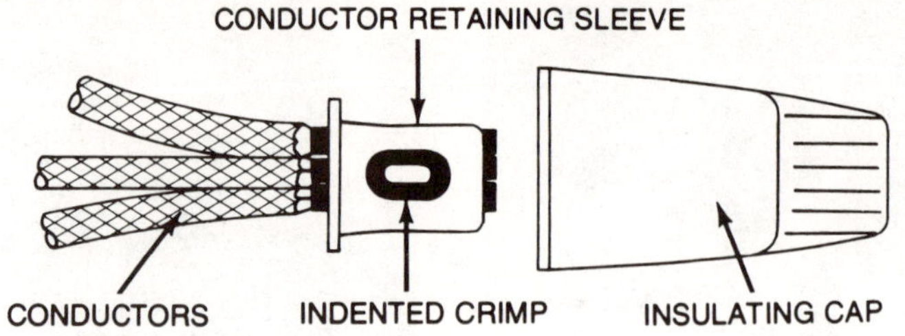

Fig. 9-6 Method for installing a compression connector.

ductor cannot be easily removed from the retaining sleeve.

Two-piece compression connectors are made for use on circuits of 600 V and may be used on lighting fixtures up to 1000 V. The retaining sleeves are made for splicing conductors from No. 18 gauge up to No. 6 gauge. The zinc-plated steel sleeves can be used only on copper conductors, due to the possibility of electrolysis acting upon these sleeves.

Some compression connectors are used to simplify the connection of stranded conductors to a terminal screw. These connectors prevent loose strands from slipping out from under the terminal screw and reducing the current-carrying capacity of the connection. (See Figures 9-7, 9-8, and 9-9.) Special crimping pliers are needed to install these connectors on conductors ranging from No. 18 gauge to No. 10 gauge.

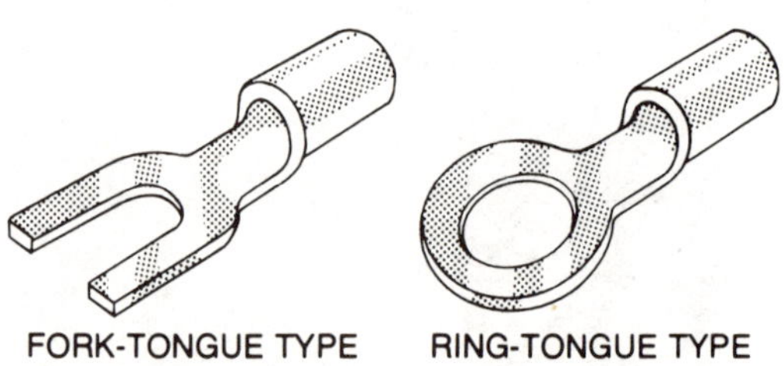

Fig. 9-7 Typical crimp-on terminal lugs.

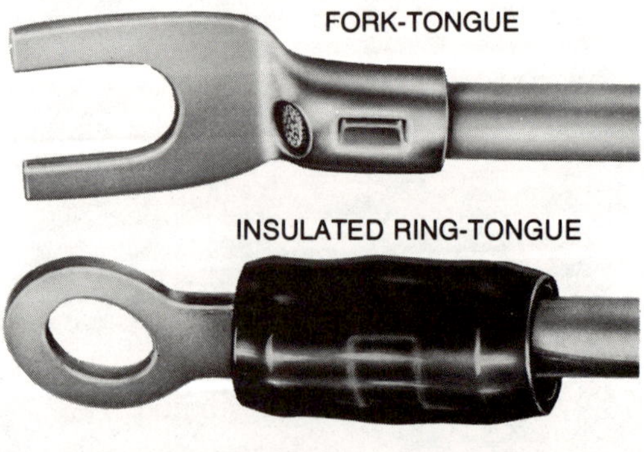

Fig. 9-8 Fork-tongue and insulated ring-tongue compression lug installations.

MECHANICAL CABLE CONNECTORS

The connecting of cables poses several difficulties. All strands must be held securely without damage to any strand, because damage will reduce the conductivity of the connection. The connection also must main-

Fig. 9-9 Typical compression tool and crimp-on terminal lugs.

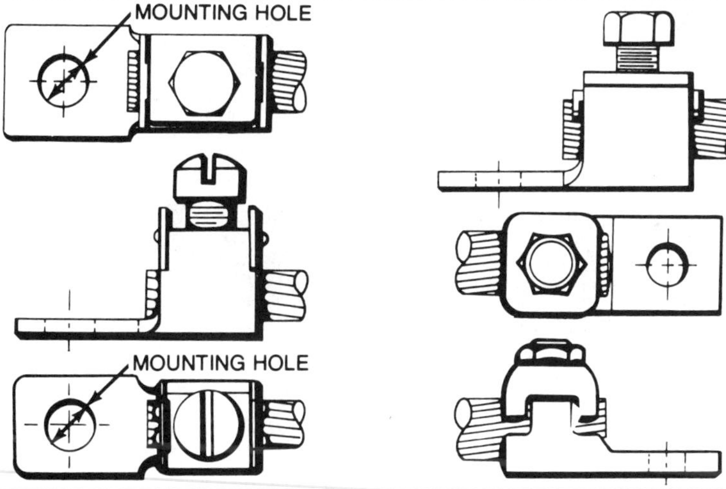

Fig. 9-10 Types of mechanical lugs.

tain a firm grip on the cable, compressing the strands into a solid group that will not loosen after a time. It is also necessary that the connector be made of a metal that will not encourage electrolysis between itself and the cable. Mechanical connectors, called *lugs,* meet these requirements, Figure 9-10.

Mechanical cable connectors are available in several types, Figures 9-11, 9-12, and 9-13. Service-entrance equipment for buildings makes extensive use of these connectors, particularly for meter sockets. Figure 9-14 shows a neutral block assembly from a distribution panel. This neutral block is capable of connecting the main

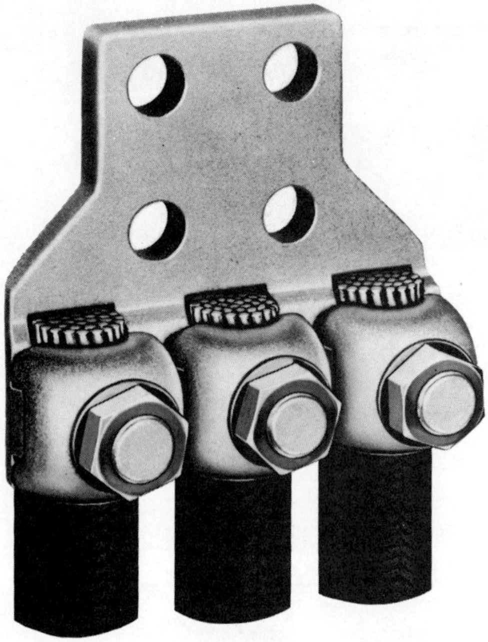

Fig. 9-11 **Typical 3-conductor mechanical connector.**

neutral cable to the neutral wires of every circuit within the building.

Cable connectors are made in sizes for conductors from No. 14 gauge up to 1000 MCM. Connectors for cables larger than 1000 MCM may be obtained from some manufacturers by special order.

Preparation of the Cable. Use the following procedure for installing copper and aluminum conductors in lugs.

Step 1. With a knife, remove the insulation from the conductor to the proper length for insertion into the lug. Whan scraping, take care not to nick any of the strands. Do *not* circle the cable with the knife, since this usually nicks the wire strands, which reduces their strength and conducting capacity.

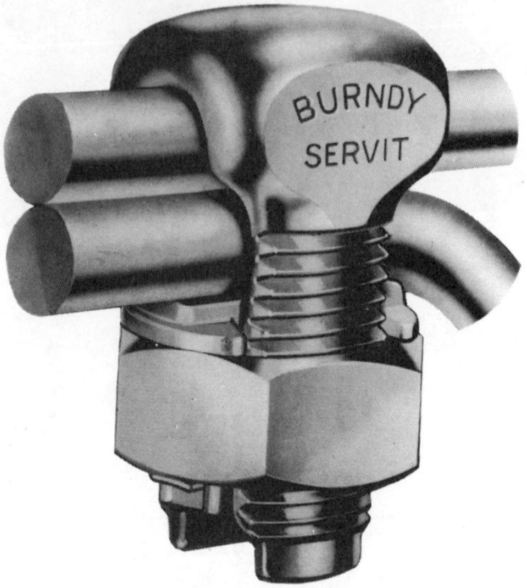

Fig. 9-12 A two-piece, split-bolt connector.

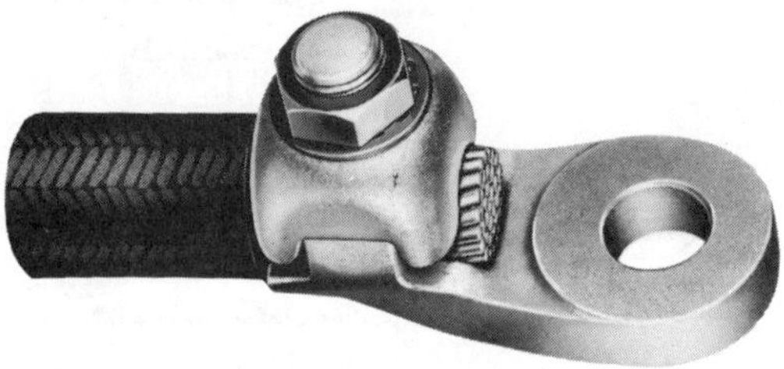

Fig. 9-13 A single-conductor mechanical connector.

Step 2. Remove any oxide coating (a dark dull coating) that is visible on the bared portion of the cable. This is an important procedure when using aluminum cable, because it oxidizes rapidly. Use a wire brush to apply the chemical for removing oxidation. At the same time, apply antioxidant chemicals to aluminum conductors.

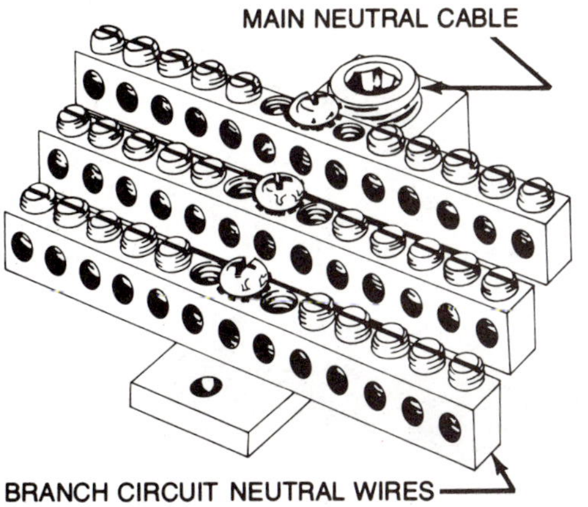

Fig. 9-14 Neutral block assembly for distribution panels.

Step 3. Tighten the holding screw, once the cable has been fully inserted into the lug. Allow several minutes to elapse, and then retighten the holding screw. (The strands will have settled into place, making a second tightening necessary for a secure connection.)

COMPRESSION CABLE CONNECTORS (TUBULAR)

Some solderless connectors are made from a one-piece tubular form for installation with a hand- or hydraulic-powered compression tool, Figure 9-15. The tubular forms are made from a high-conductivity, electrolytic copper. One end (called the *tongue)* is flattened and drilled for fastening to a terminal block. These connectors are often electro-tinplated to minimize corrosion.

The strands of the cable are compressed within the copper tube by the compression tool until they form a solid mass of copper. This process ensures long life and maximum

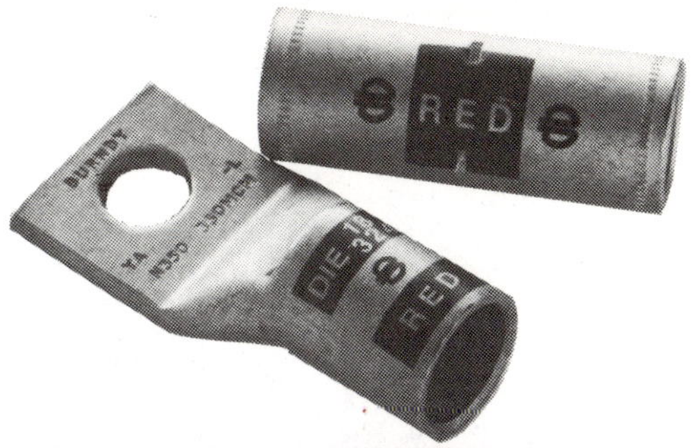

Fig. 9-15 Tubular compression-type solderless terminal and splice.

current-carrying capacity for the terminal connection. Three types of compression tools are: the hand-operated mechanical type, the hand-operated hydraulic type, and the motor-driven hydraulic type. (See Figures 9-16, 9-17, and 9-18.)

Aluminum cables require special attention when terminated with a compression-type solderless lug. Figure 9-19 shows the correct procedure for terminating aluminum cables. Figure 9-20 shows the effect of a compression tool on a cable and lug.

Figure 9-21 shows two outdoor applications for solderless compression connectors. In both cases, it would be very difficult and inconvenient to install solder-type connectors.

INSULATING TAPES

Once installed, solderless cable connectors require an insulating tape or other covering to replace the insulation removed during the making of the splice, Figure 9-22. This tape or covering material must be capable of withstanding both the circuit voltage and the normal wear and tear on the cable. In other words, the covering material must have the same voltage rating

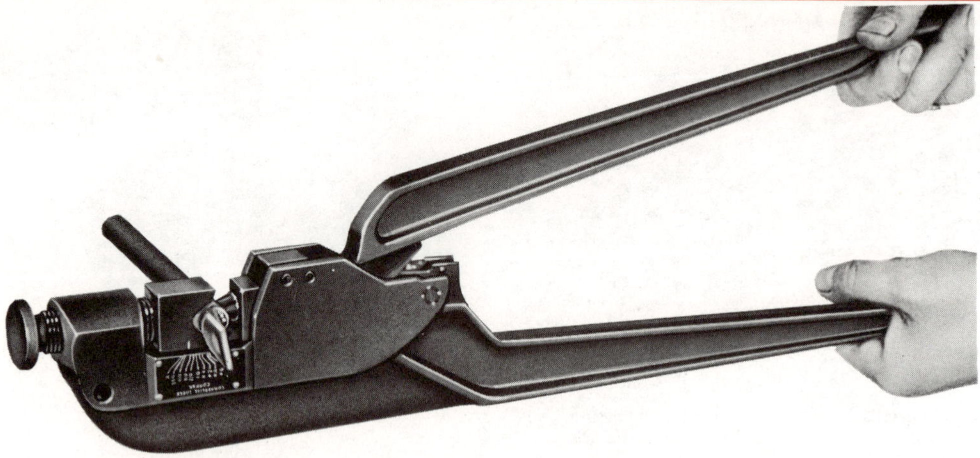

Fig. 9-16 A hand-operated, mechanical-type compression tool.

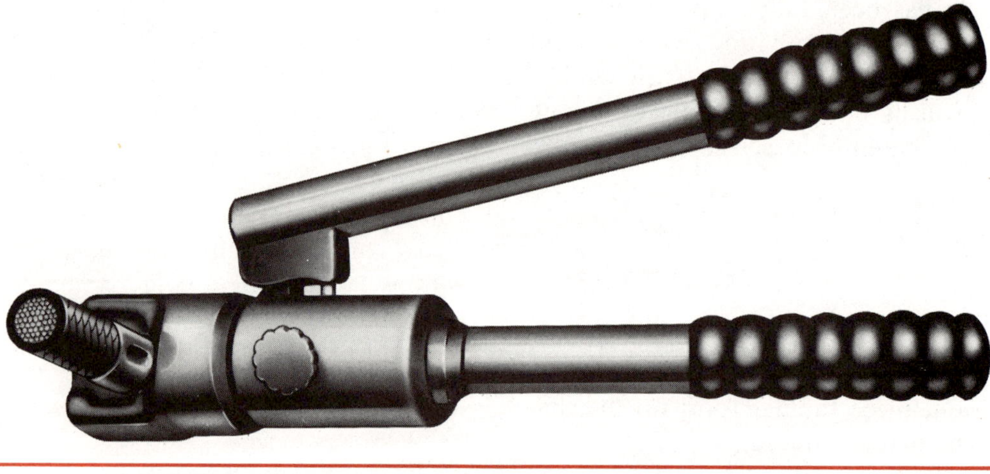

Fig. 9-17 A hand-operated, hydraulic-type compression tool.

and physical properties as the cable's original insulation.

Some splices require several types of insulating material to provide adequate electrical protection. Common coverings are friction tape, vinyl plastic, fiberglass, and insulating putty.

Friction Tape. Friction tape is basically a cotton tape with an insulating compound impregnated into the weave. It is not a quality insulator and should not be used on circuits higher than 115 V.

Vinyl Plastic. Vinyl plastic is an excellent insulating product which is available in as-

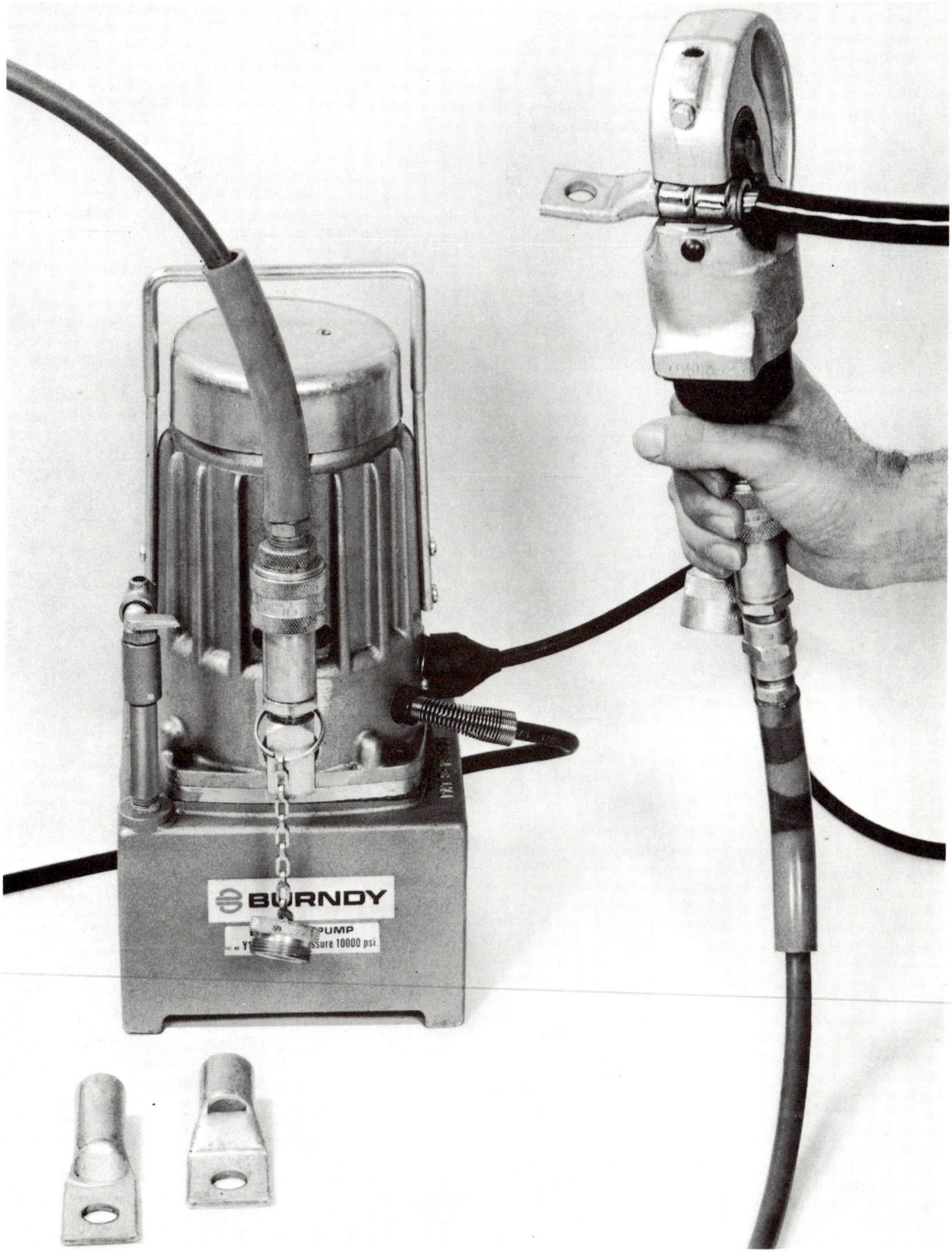

Fig. 9-18 A motor-driven, hydraulic-type compression tool.

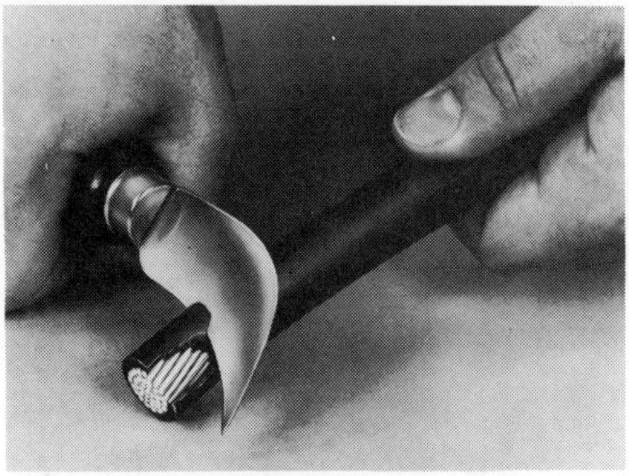

STEP 1. CAREFULLY REMOVE INSULATION WITHOUT NICKING CONDUCTOR.

STEP 2. WIRE BRUSH CONDUCTOR TO REMOVE ANY OXIDE.

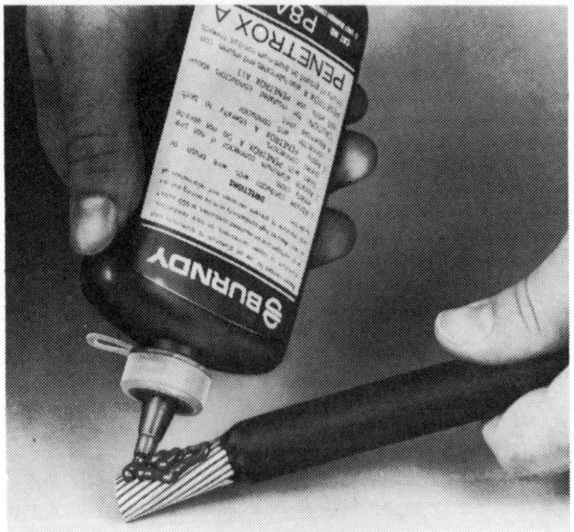

STEP 3. APPLY ANTIOXIDANT TO PREVENT FORMATION OF SURFACE OXIDE.

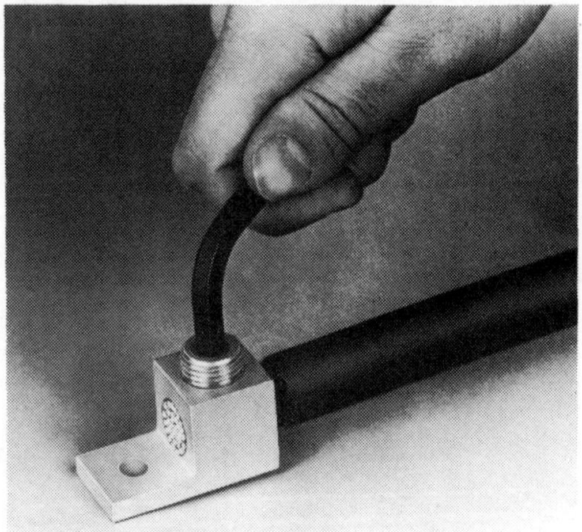

STEP 4. TIGHTEN MECHANICAL CONNECTORS SECURELY.

STEP 5. CRIMP COMPRESSION TYPE CONNECTORS WITH PROPER DIE AND TOOL THE RECOMMENDED NUMBER OF TIMES.

Fig. 9-19 Procedure for terminating aluminum cables.

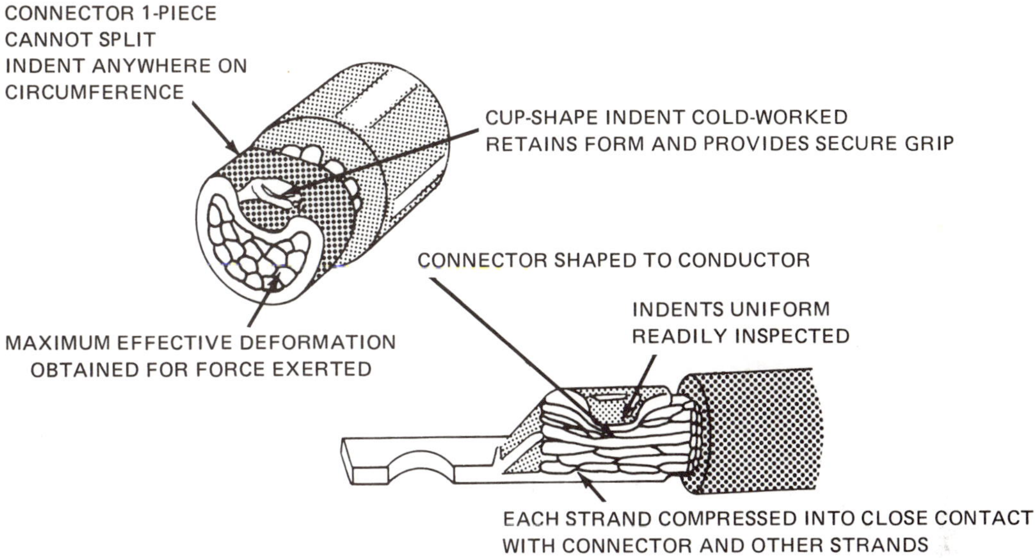

Fig. 9-20 Effect of compression tool on cable and lug.

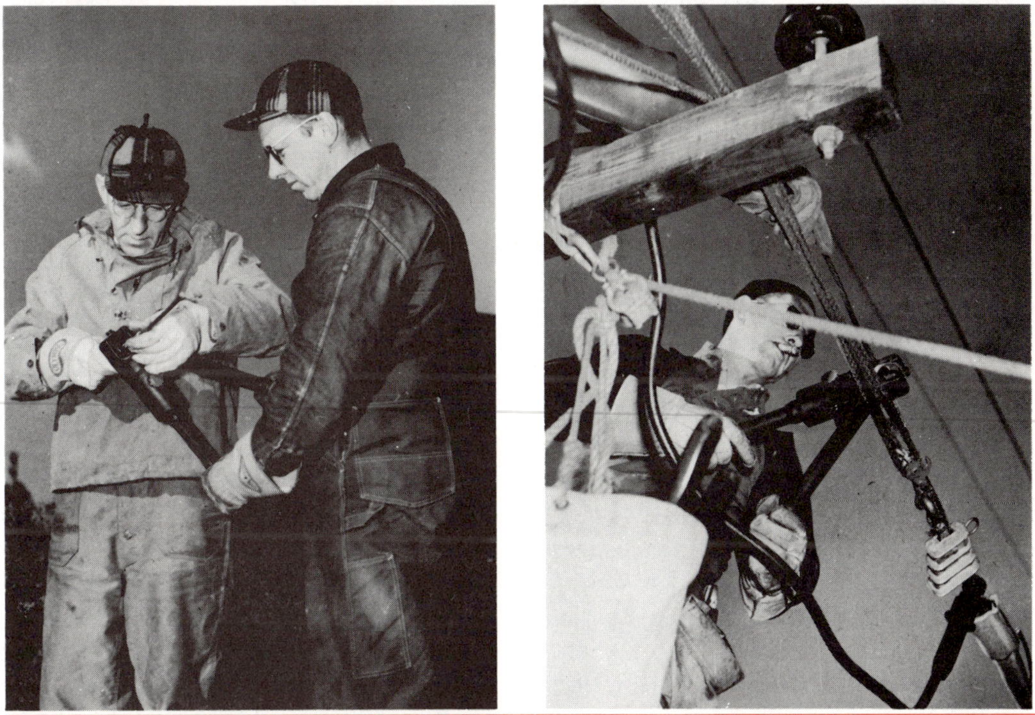

Fig. 9-21 Typical outdoor uses for hand-operated, hydraulic-type compression tool.

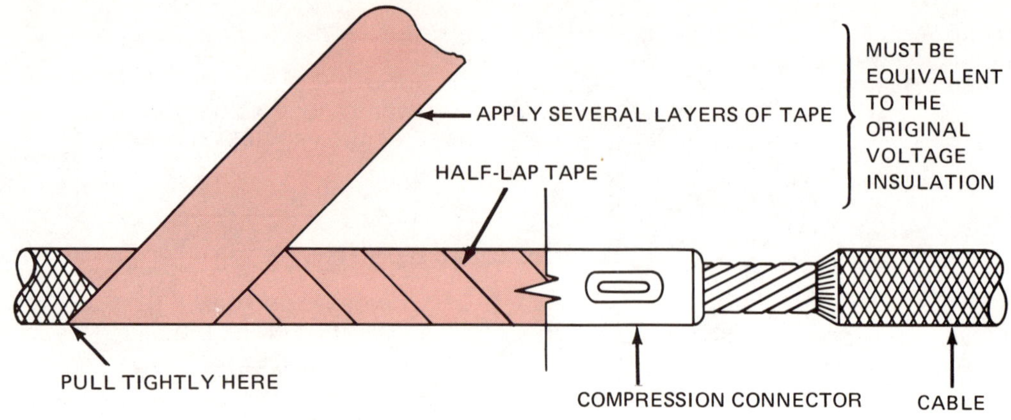

Fig. 9-22 Taping a splice.

sorted types and colors. This tape is available also in several thicknesses and widths, and it adheres readily to most surfaces.

A general-purpose vinyl tape, as seen in Figure 9-23, is approximately 0.00787 inch (0.2 mm) in thickness, and resists abrasion, sunlight, moisture, alkalis, and many acids. It has UL listing for 5000 V at 90°C (194°F). A more specialized type of vinyl tape is produced for use where cold temperatures are experienced. It is slightly thicker than the general-purpose tape, and has extra pliability for application at extremely low temperatures, yet remains easy to handle at normal temperatures. Figure 9-24 illustrates an application of this tape.

More rugged vinyl tapes, 0.01181 inch (0.3 mm) in thickness, are available where more abrasion resistance and mechanical strength are required. Wider widths are produced for use in speeding up the insulating of larger splice areas. Where temperatures up to 105°C (221°F) are encountered, a more heat-resistant tape is available. This tape can be used in and

Fig. 9-23 One type of general-purpose vinyl plastic electrical all-weather tape.

around electric motors, and is equipped with a special oil-resistant adhesive and vinyl backing material.

EPR Tape. Ethylene propylene rubber (EPR) tape is a 0.03150-inch (0.8 mm), nonvulcanizing material that can be used for low-voltage as well as high-voltage applications up to 69 000 V. It can be stretched upon application to 10 times its normal length, thus forming a moisturetight layer over the splice, Figure 9-25.

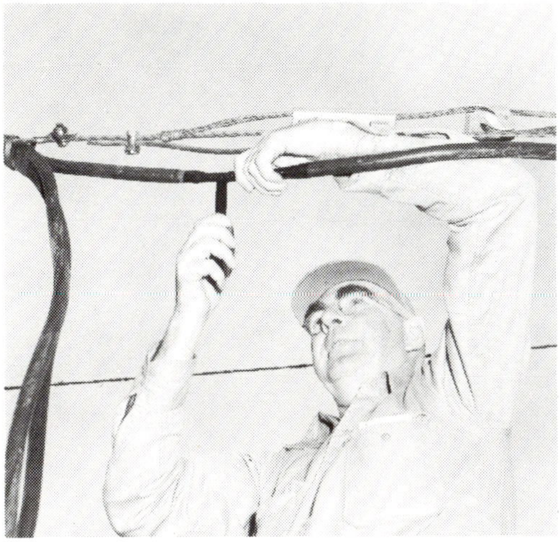

Fig. 9-24 Application of vinyl plastic electrical tape suitable for cold weather.

Fiberglass. Fiberglass tape is used on splices where the temperature may reach 130°C (266°F). It has a pressure-sensitive, thermosetting adhesive that makes it suitable for use in such high-temperature applications as furnace connections and water heaters, among others.

Insulating Putty. Large connectors require a quality insulation that will fill voids, and pad any irregular shapes produced by the connector. An electrical-grade, rubber-based, self-fusing elastic-type putty is available in tape form for the insulating of large connectors. It should be used on low-voltage circuits (600 V or less) where it will resist aging and not dry out. Figure 9-26 illustrates the proper method of insulating a large connector.

Another form of insulating putty is the vinyl mastic pad, Figure 9-27. These pads are made of a self-fusing, rubber-based compound with a very strong adhesive. They

Fig. 9-25 Typical high-voltage, EPR rubber tape.

mold around difficult shapes and have excellent resistance to alkalis, acids, moisture, and varying weather conditions. Figure 9-28 illustrates the proper method of using these pads.

Resin Splicing Kits. A resin splicing kit can be obtained with sufficient materials included to complete one splice. A plastic mold, funnels, insulating and sealing compound, and pouring resin are included in the kit. Figure 9-29 illustrates the pouring of the resin into the mold. These unique "field splicing kits" can be used for overhead, underground, or direct-burial applications up to 5000 V. Figure 9-30 illustrates a cutaway view of a resin splice after the mold has been removed.

Resin splicing kits are available in a variety of forms to handle numerous shapes and types of splices. They are produced in a variety of voltage ratings and care should be taken to select the proper kit and follow all instructions for the splice or connection to be insulated. (An electrical distributor can furnish this information.) When properly installed, the kit forms a moisturetight

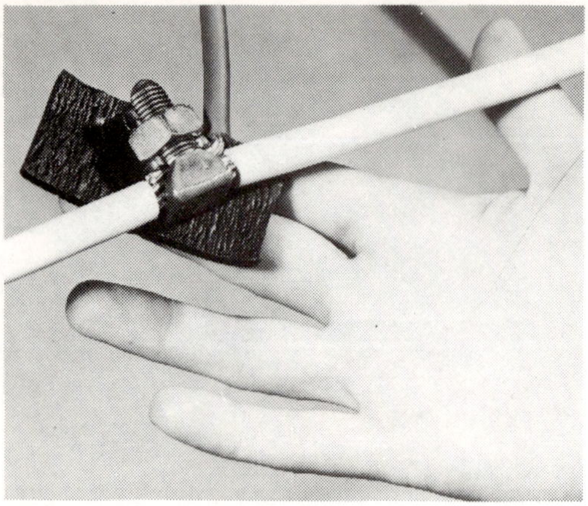

STEP 1. APPLY FIRST PIECE OF INSULATING PUTTY.

STEP 2. OVERLAP A SECOND PIECE OF INSULATING PUTTY.

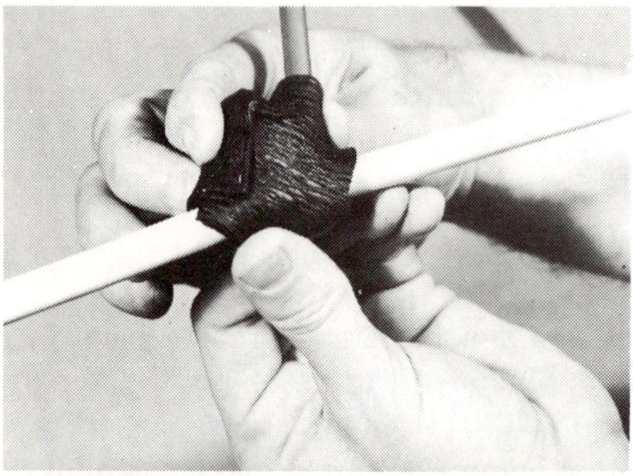

STEP 3. PRESS AND FORM PUTTY TO SHAPE OF CONNECTION.

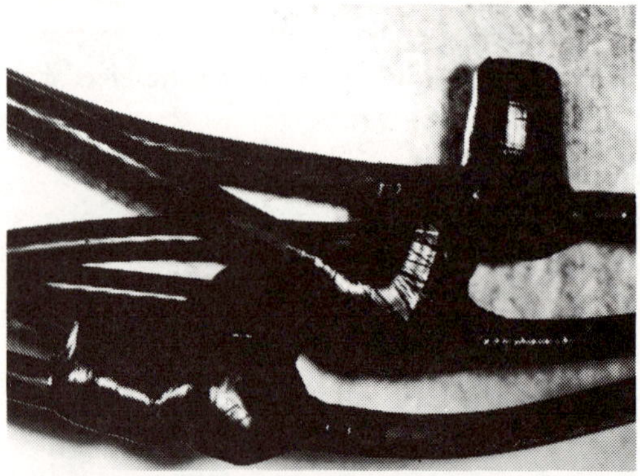

STEP 4. COMPLETE INSULATING PROCESS WITH LAYER OF VINYL PLASTIC TAPE.

Fig. 9-26 Insulating a splice using self-fusing elastic putty.

Fig. 9-27 Typical vinyl mastic pad insulating putty.

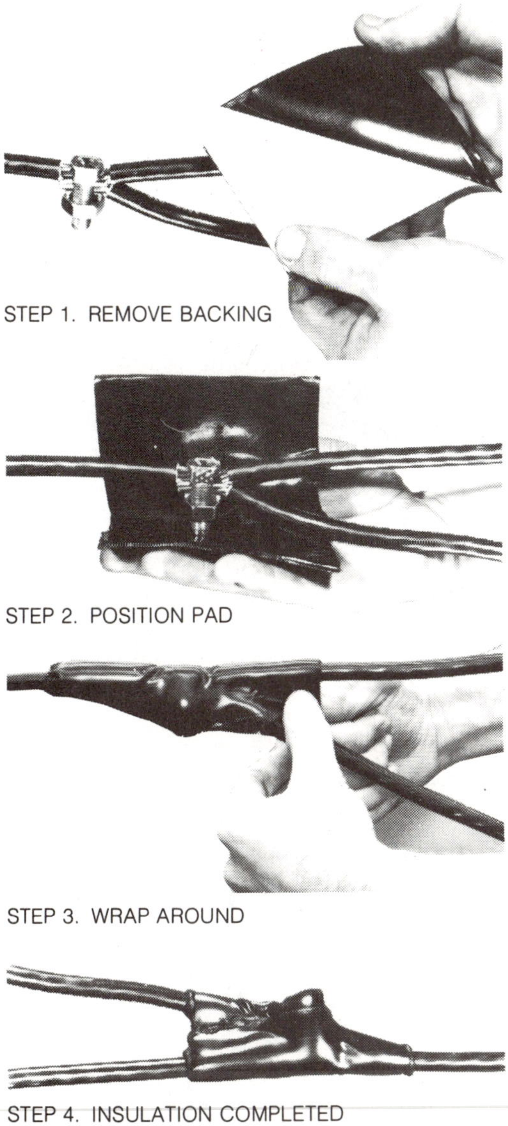

Fig. 9-28 Proper technique for insulating with vinyl mastic pads.

and watertight cover over a splice that otherwise could take considerable time and material to insulate.

A resin-pressure system of insulating splices is produced which insulates cables up to a capacity of 8000 V. The first step is to apply an open-weave spacer tape around the splice. Next, an injection fitting is taped into place, and a final taping of vinyl plastic is applied, forming a liquidtight mold. A resin-pressure gun is then used to pump the insulating resin into the wrapped splice. The result is a tough, moistureproof insulation on the splice. Figure 9-31 illustrates the steps in making such a splice.

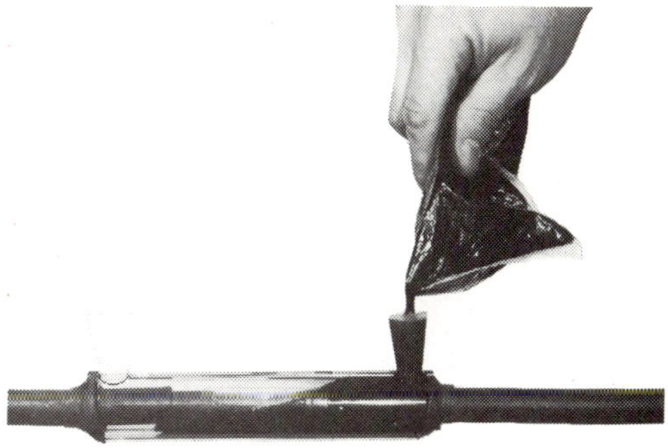

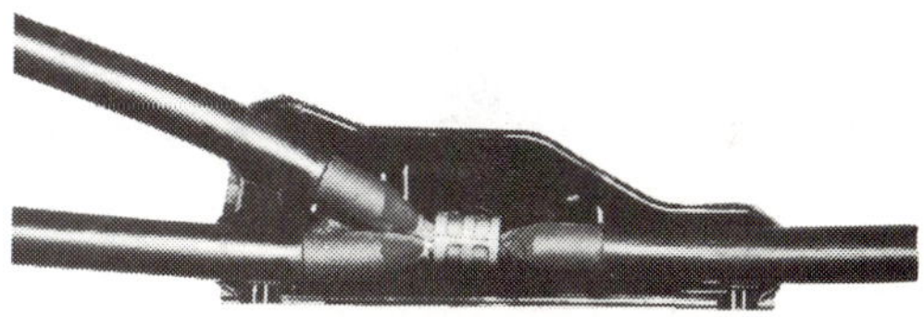

Fig. 9-30 Cutaway view of a completed resin-insulated splice.

Electrical Coating. Occasionally, a splice that has been taped, or one that has been in service for awhile, may require a little extra moistureproofing or insulating. A liquid, brush-on coating is available to provide this extra protection when required. Figure 9-32 illustrates a typical product being applied to a taped splice.

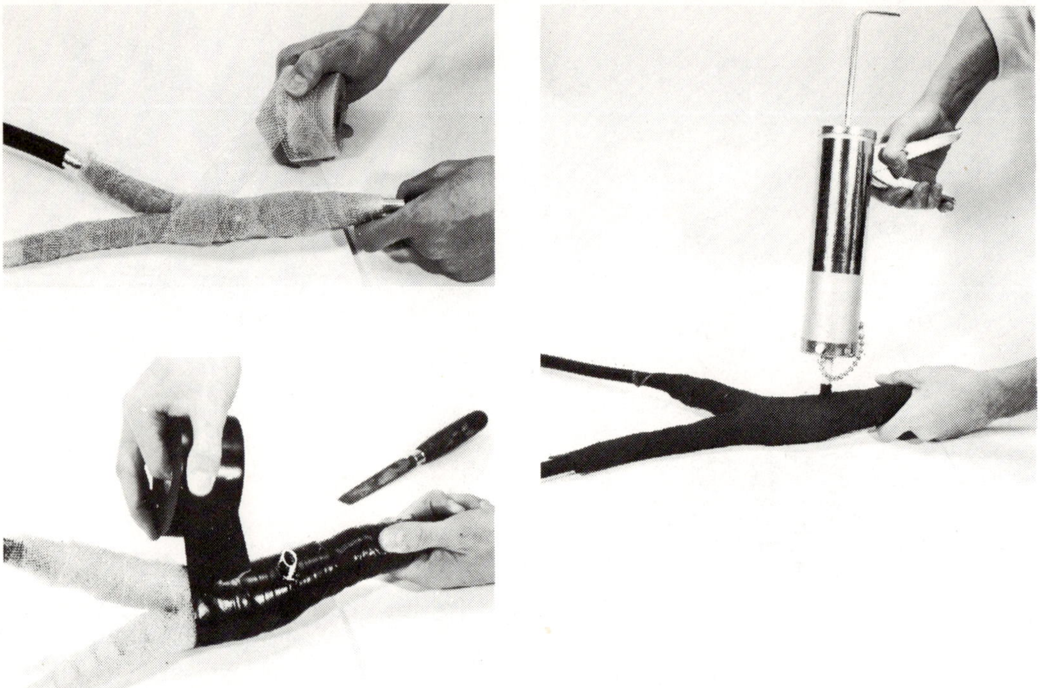

Fig. 9-31 Insulating a splice using the resin-pressure system.

Colored Tapes. Vinyl tape is available in eight fade-resistant colors: red, yellow, blue, green, white, orange, brown, and gray. These tapes can be used to replace colored insulation removed from a conductor before splicing, or they can be used to identify and mark various circuit conductors. They are approved for use on 600 V at 90°C (194°F), and have similar qualities to the other vinyl tapes available.

TAPING A SPLICE

Refer to Figure 9-22, which shows how to tape a splice. Manufacturers provide instruction sheets covering application methods for more specialized materials.

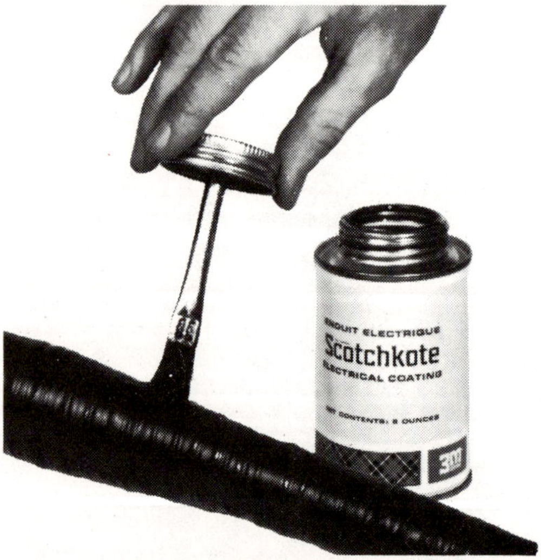

Fig. 9-32 Application of liquid insulating coating.

When removing insulation, take care not to nick the cable, thus producing sharp edges or burrs in the strands. Damaged strands are potential weak areas that might puncture the tape or establish electrical stress in the splice. The higher the circuit current and voltage, the greater the danger of splice damage from electrical stress.

REVIEW

1. List two disadvantages of soldering and taping wire splices.
2. List three types of solderless wire connectors. What is the maximum circuit voltage for each?
3. Why are crimp-on connectors installed on small, stranded conductors fastened under terminal screws?
4. What are the disadvantages of compression connectors?
5. Give three reasons why lugs are useful.
6. List the steps in fastening a copper cable to a terminal lug.
7. List the steps in fastening an aluminum cable to a terminal lug.
8. Why is it necessary to avoid cutting nicks on the strands of a cable when removing insulation?
9. Why must a mechanical connector be retightened several minutes after the first tightening?
10. Describe the methods for crimping copper and aluminum compression connectors to a cable.
11. Which two types of protection must electrical tape provide?
12. Why is it necessary to observe the temperature ratings of electrical tape?
13. List three common types of electrical tape.
14. Describe how large, irregularly shaped cable connectors are insulated.
15. What are the advantages of insulating putty when covering large or irregularly shaped splices?
16. What advantages do resin splicing kits have over more conventional methods of insulating a splice?
17. Why must care be taken to select an insulating product that has the proper voltage rating?
18. What are the advantages of using colored insulating tapes?
19. Why is it desirable to pull or stretch the tapes as they are being applied to the splice area?
20. Under what conditions would a brush-on, liquid insulating material be used?

Chapter 10

Heat-control Switches

SERIES-PARALLEL CONNECTIONS

Electrical cooking appliances, such as ovens, ranges, hot plates, and commercial coffee makers use a variety of heavy-duty switches to control their heating elements. One common method for providing different levels of heat is to use two elements and connect them in series-parallel combinations across 120 V and 240 V. The switches discussed in this chapter are capable of providing the series-parallel connections required for this method of heat control.

THREE-HEAT, SINGLE-POLE SWITCH

The simple three-heat, single-pole, heat-control switch is used primarily for small ranges and hot plates operating at 120 V. Variations of heat are provided by connecting two elements of the same size (wattage) as follows:

Low: two elements in series on 120 V.

Medium: one element on 120 V.

High: two elements in parallel on 120 V.

Figures 10-1 and 10-2 show the three-heat, single-pole switch, its internal connections, and wiring by schematic diagram. The switch connections or the heating elements can be tested with test instruments or a series-lamp tester (described in Chapter 2, "Equipment for Testing Switches").

THREE-HEAT, DOUBLE-POLE SWITCH

The three-heat, double-pole switch uses the same series-parallel combinations of the two elements as does the single-pole version. Switches controlling 240 V (two live wires) must be capable of opening simultaneously both live wires at the switch. When the switch is turned to *off,* there must be *no* voltage present at the elements.

Figures 10-3 and 10-4 show the three-heat, double-pole switch, its internal connections, and wiring by schematic diagram.

FIVE-HEAT, DOUBLE-POLE SWITCH

More variations in heat are available by using the five-heat, double-pole switch to control two elements of the same size.

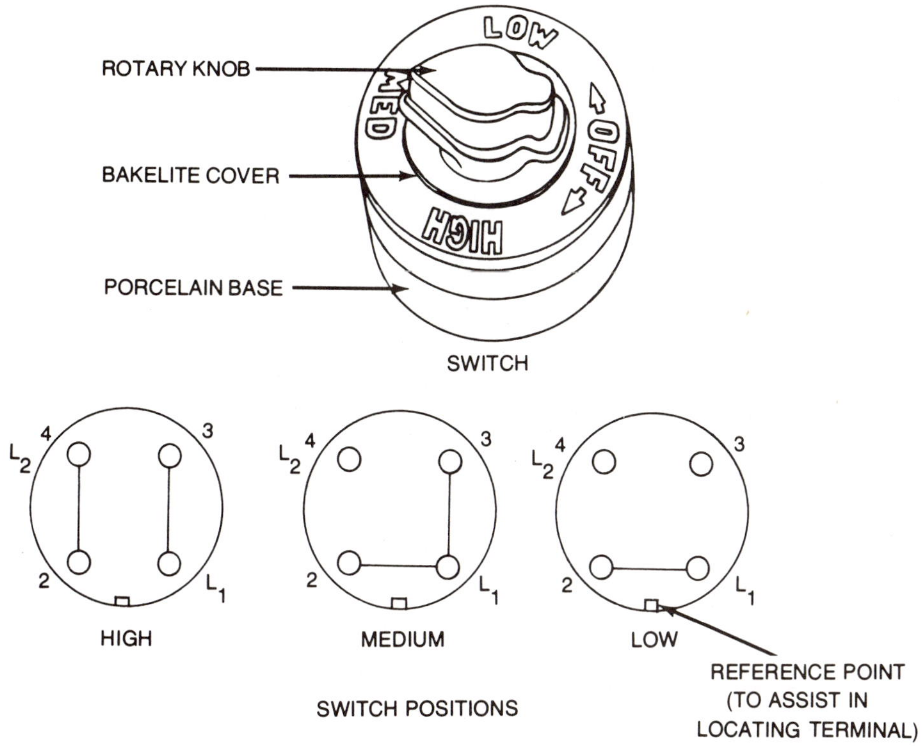

Fig. 10-1 Typical 3-heat, single-pole switch and switch positions.

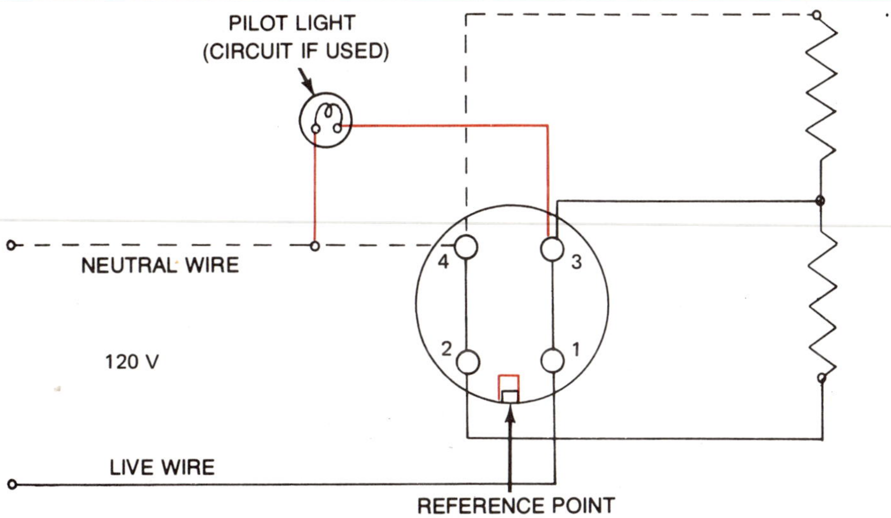

Fig. 10-2 Schematic wiring diagram for a 3-heat, single-pole switch (shown in the "high" heat position).

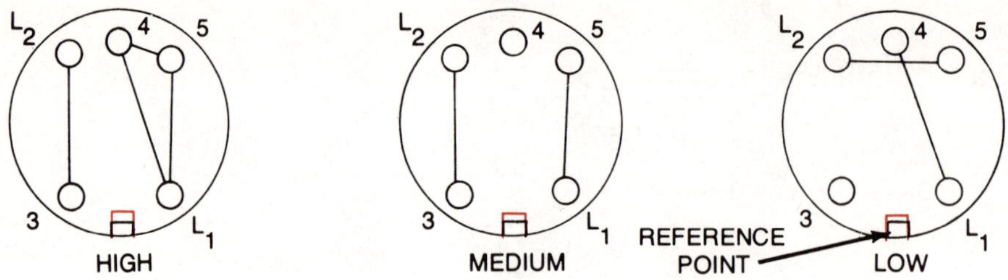

Fig. 10-3 Switch positions for a 3-heat double-pole switch.

Two elements, each rated at 600 W on 240 V, provide variations in heat as follows:

Low: two elements in series on 120 V provide 75 W.

Low-medium: one element on 120 V provides 150 W.

Medium: two elements in parallel on 120 V provide 300 W.

Medium-high: one element on 240 V provides 600 W.

High: two elements in parallel on 240 V provide 1200 W.

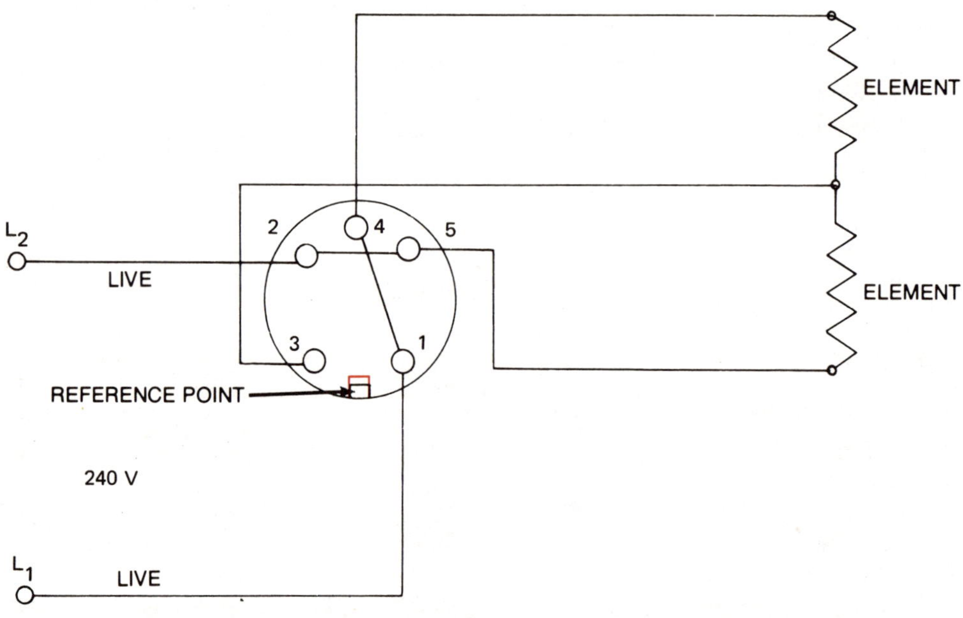

Fig. 10-4 Schematic wiring diagram for a 3-heat, double-pole switch (shown in the "low" heat position).

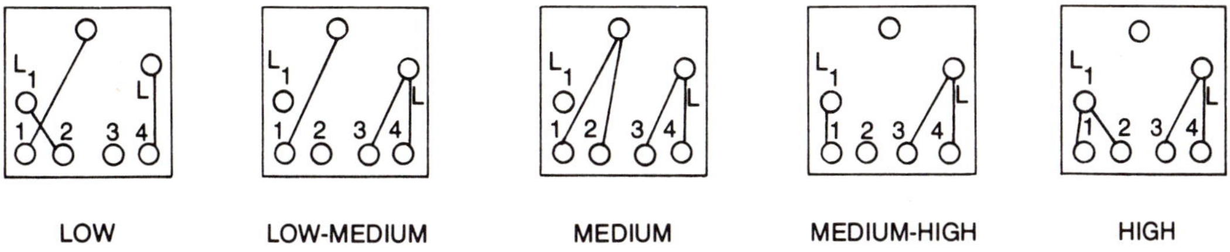

Fig. 10-5 Internal switch connections (rear view) for a 5-heat switch.

Figure 10-5 shows the internal connections for the five-heat switch.

Heat variations other than those listed previously can be obtained by using elements of two different sizes. The connections for a pilot light which indicates the *on* positions of the five-heat switch is shown by the schematic wiring diagram in Figure 10-6. Take care to use a 120-V lamp for the pilot light. Lamps with lower voltage ratings will burn out if connected to this circuit. Lamps with higher voltage ratings may be too dim for effective use.

SEVEN-HEAT SWITCH

The seven-heat switch looks much like the five-heat switch, but it has the advantage of two extra heat levels. Because of this

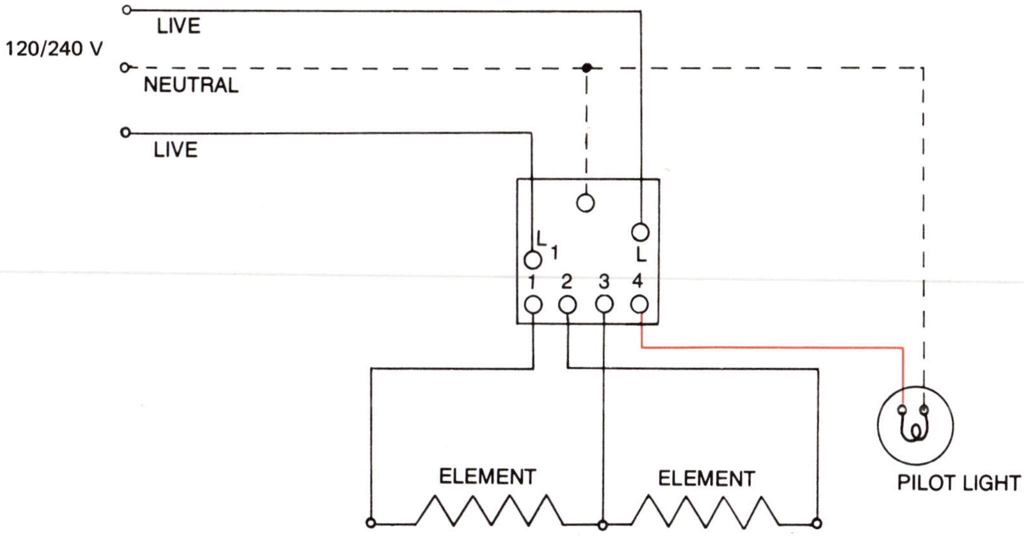

Fig. 10-6 Schematic wiring diagram for a 5-heat switch (shown in the "off" position).

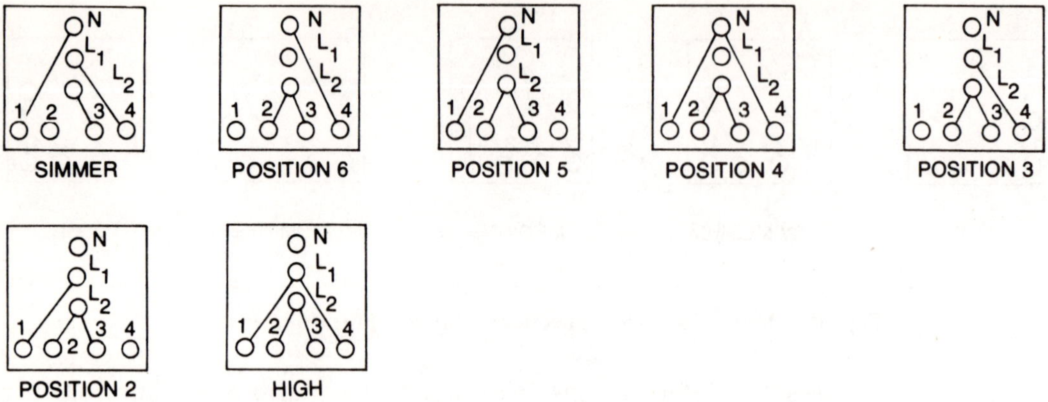

Fig. 10-7 Internal switch connections (rear view) for a typical 7-heat switch.

greater range of heat selection, the seven-heat switch has gradually replaced the older five-heat switch design.

The seven heat levels are obtained by using two different elements, each rated at 240 V. A 600-W element and an 800-W element, in combination, provide heat as follows:

Simmer: both elements in series on 120 V provide 82.5 W.

Position 6: one 600-W element on 120 V provides 150 W.

Position 5: one 800-W element on 120 V provides 200 W.

Position 4: both elements in parallel on 120 V provide 350 W.

Position 3: one 600-W element on 240 V provides 600 W.

Position 2: one 800-W element on 240 V provides 800 W.

High: both elements in parallel on 240 V provide 1400 W.

Figures 10-7 and 10-8 show the internal connections of the seven-heat switch, and its wiring by schematic diagram.

Most modern ranges have a pilot light terminal at the rear of the switch to show the *on* position of the switch. An element that has been left in the *on* position may not glow red, but it will still be hot enough to burn a person touching it.

INFINITE-HEAT SWITCH

Improvements in the design of electric ranges created a need for a greater variety in heat levels. The infinite-heat switch, as the name implies, offers an *unlimited* variation in heat level. Although this switch is often more expensive than other heat-control switches, a single element requiring less wiring offsets this initial cost.

Figure 10-9 shows by schematic wiring diagram the internal connections of a typical infinite-heat switch.

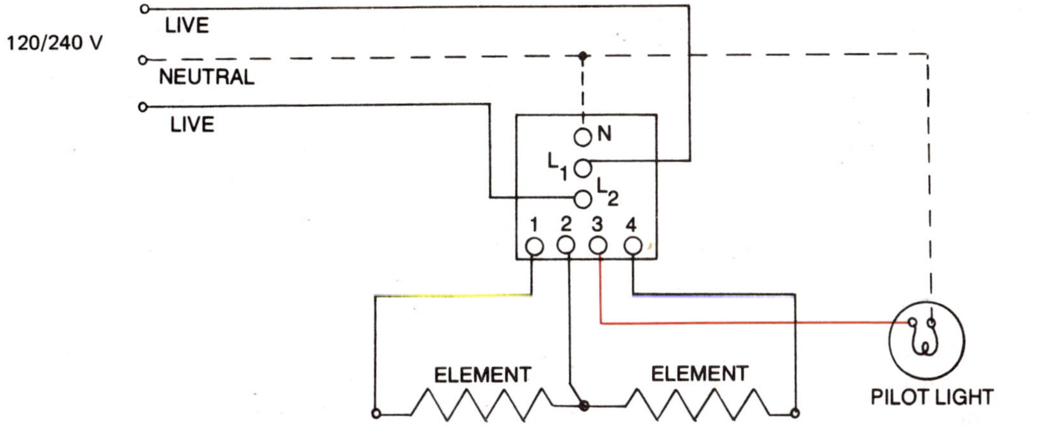

Fig. 10-8 Schematic wiring diagram for a 7-heat switch (shown in the "off" position).

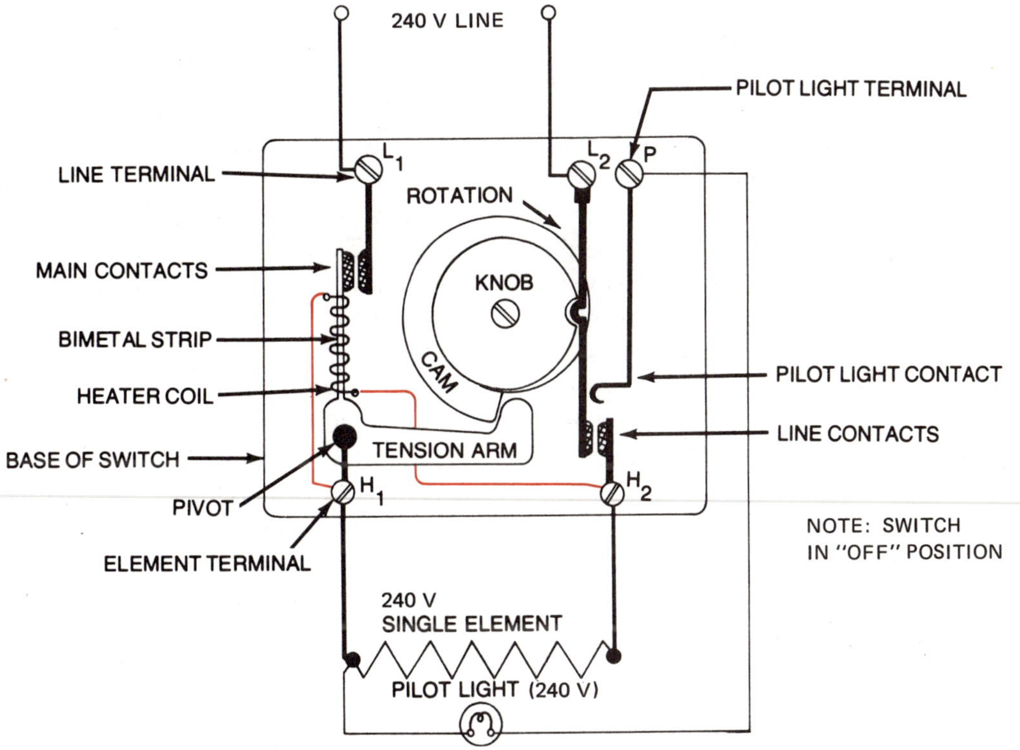

Fig. 10-9 Schematic wiring diagram showing the parts and circuit of an infinite-heat switch.

Operation of an Infinite-heat Switch. A cam is fastened to the knob of the switch. One of the main contacts is on a bimetal strip fitted into a tension arm. As the knob is turned slightly, the cam actuates the tension arm. The arm rotates on the pivot and closes the main contacts, thus heating the element.

A heater coil is wound around the bimetal strip and connected in parallel with the element. The heat from the heater coil warms the bimetal strip and causes the strip to bend outwards. This opens the circuit, and the heater cools. The strip contracts and closes the contacts. This activates the element and heater again.

The more the knob is turned, the more it causes the tension arm to exert pressure on the contacts. Therefore, it takes longer for the heater to become hot enough to actuate the bimetal strip and open the contacts.

The single element operates at full strength each time the contacts are closed, and the heat level is regulated by the speed at which the main contacts open and close. The number of operating cycles in a given period of time and the length of time the element is *on* in each cycle, is controlled by the extent to which the knob on the switch is rotated.

A pilot light terminal simplifies the connection of a light to indicate the *on* position. Figure 10-10 illustrates the internal layout of a typical infinite-heat switch.

Fig. 10-10 Internal layout of an infinite-heat switch.

OVEN CONTROL SWITCH

The switch controlling the heat level in a modern oven also provides infinite-heat control. A liquid expansion system is used to regulate the *on* and *off* cycles of the switch, Figure 10-11. Cycling of the switch is further regulated by an adjustable threaded shaft that operates like a cam. The timing of the heating cycle determines the amount of heat in the oven at a given time. The liquid expansion system replaces the heater coil used in the infinite-heat switch.

The oil-like material used in the bulb of the liquid expansion system senses the oven temperature, expands accordingly, and operates a diaphragm device inside the switch. The expansion and contraction of the liquid, and the movement back and forth of the diaphragm, open and close a set of contacts inside the switch. These contacts switch the electrical power of the oven *on* and *off* at the proper rate of cycling to obtain the temperature selected by the person using the oven.

The oven switch controls two elements inside the oven. The upper broil element is turned on by rotating the knob of the switch to the broil position. Oven manufacturers design the oven door so that it will remain in a slightly open position for broiling. Manufacturers recommend that broiling be done with the oven door partly open so that the oven temperature does not get hot enough to cycle the contacts. The broil element will then re-

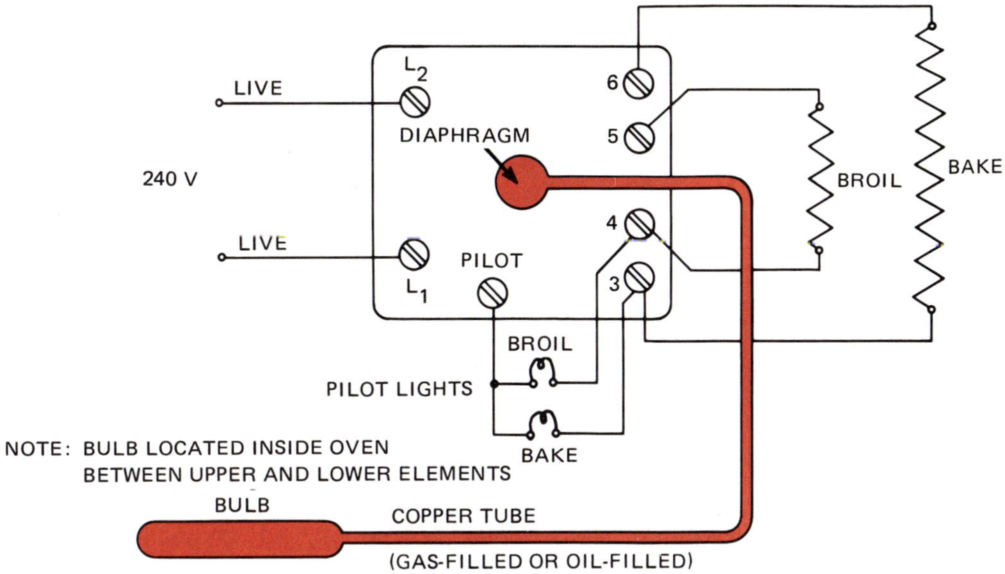

Fig. 10-11 Typical liquid expansion system oven control switch.

main *on*, and food being broiled must be watched carefully to prevent burning. It is also common to have an oven with an automatic preheat process. When the oven is being preheated, the broil element will come on with the bake element when the unit is first turned on. The broil element will be automatically turned *off* when the oven temperature is approximately 40°C (104°F) lower than the knob setting of the switch. The oven temperature will then be brought to its selected heat level by the bake element and kept there as the switch cycles *on* and *off* during the cooking process. Having both elements *on* together produces high heat and gets the oven up to temperature quickly. The purpose of the broil element then is to preheat the oven and/or to brown food quickly. The lower bake element is used for normal heating of the oven and is under the control of the liquid expansion system at all times.

Pilot lamps are normally included in the oven circuit to indicate when the elements are *on* and producing heat in the oven. Figure 10-12 illustrates the external wire connections and internal layout of a typical oven control switch.

SELF-CLEANING OVENS

Many residential ovens now have the self-cleaning feature. This self-cleaning action eliminates the tedious and often messy process of using strong cleaning materials to remove spilt or burnt food from the walls and floor of the oven. Self-cleaning ovens are available in two basic types: continuous self-cleaning, and pyrolytic self-cleaning.

Continuous Self-cleaning Type. The continuous self-cleaning oven uses the normal cooking temperature of the oven to reduce

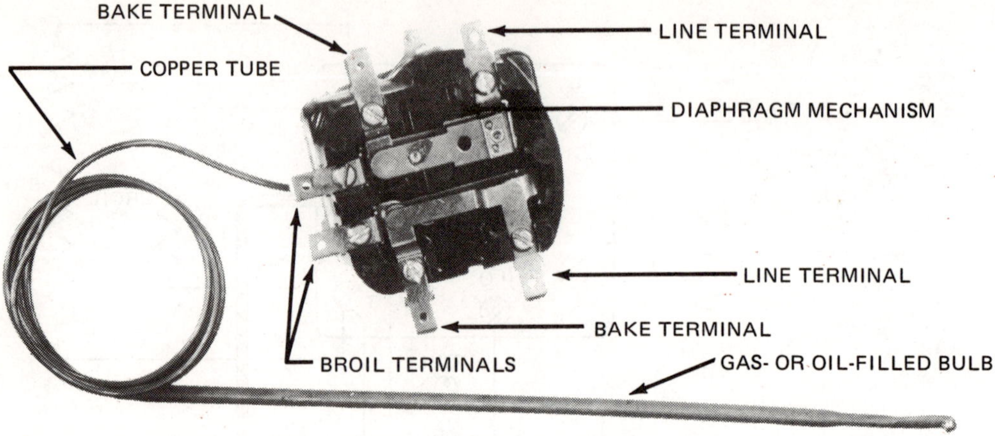

Fig. 10-12 External wire connections and internal layout of an oven control switch.

gradually to ash the burnt food particles on the inner surfaces of the oven. Compared to the pyrolytic process, this method takes longer and the oven will appear to be not as clean.

Pyrolytic Self-cleaning Type. The pyrolytic system of self-cleaning is highly efficient. The oven temperature reaches approximately 480°C (896°F) during its cleaning cycle, and reduces food spills to a small amount of ash in an hour or two. This ash can then be removed easily, leaving the oven clean and ready for use.

The type of control switch and circuit used in the pyrolytic type of oven operates in much the same manner as the temperature-sensitive expansion system used in regular oven switches. However, this system performs several other functions, as well.

Due to the much higher temperatures in the pyrolytic-type of oven, helium gas is used to replace the oil-like material in the expansion system. The bulb is larger in appearance to accommodate the amount of helium required to operate the diaphragm in the switch.

The pyrolytic-type of control switch may have up to four different sets of contacts. All of these contacts are set to operate at different temperatures and to open and close the various circuits as the gas-filled expansion system dictates. One set of contacts can be used for baking and broiling, while a second set, calibrated at the factory, can be used to operate the oven at high temperature during the cleaning cycle. It should be pointed out at this time that, during the cleaning cycle, oven temperatures will be high, but not a great deal of electricity is used. The elements are connected in such a manner that they operate at 120 V, producing less than their full wattage, but they remain *on* for the full length of time in order that the oven can reach the desired high temperature. This simply means that the elements are not cycled *on* and *off* as frequently as they are during the cooking operation, but are allowed to produce heat for a longer period of time. Extra insulation is usually put into a self-cleaning oven to assist the elements in obtaining the proper temperature, and to prevent excessive heat from passing through to the outer surfaces of the stove itself.

A third set of contacts can be used to engage a latching mechanism in the oven door. This prevents the door from being opened during the high-temperature cleaning cycle. For two safety reasons, the oven door latching mechanism is set to keep the door closed whenever the oven temperature exceeds 320°C (608°F). First, burns to the hands or face can be severe when oven surface and air temperatures reach this level. Second, if the oven door were to be opened, a sudden inflow of oxygen from the air could cause food particles in the oven to burst into flames when the temperature is in the high range used for cleaning.

A fourth set of contacts on the switch can be used to hold the cooked food at a warm temperature until serving time. The temperature used is approximately 80°C (176°F).

Most modern ovens have a timer circuit built into their control panel that allows a variety of operations. Ovens can be set to come *on* at a predetermined hour of day or night, cook for a preset length of time, and shut off or keep food warm until needed. Elements, receptacles, and minute minders can all be timed and/or controlled by these units.

Figure 10-13 illustrates a typical circuit diagram for a self-cleaning oven. All switches contained within the dotted lines are controlled by the gas-filled expansion system. The door lock switch operates the door lock solenoid, once the cleaning cycle has been selected, and prevents the door from being opened during the cleaning operation. The door electrical interlock (5 and 6 on the diagram) prevents the solenoid from releasing the door once the oven temperature reaches cleaning level.

The manual door interlock switch bypasses the normal cycling contacts, thus removing the expansion system from the circuit for the cleaning operation and allowing the oven to reach 480°C (896°F). A *hi-temp clean switch* prevents the oven temperature from rising above the desired cleaning temperature.

The function switch, located on the control panel of the stove/oven unit, is used to select the bake, broil, preheat, and other functions of the oven. The bake and broil elements are both further controlled by a thermal relay (part of the gas expansion system) which controls their cycling to obtain the desired oven temperature. Figure 10-14 illustrates the internal and external layouts of a typical pyrolytic-type self-cleaning oven switch.

ROTISSERIE

Many modern ovens are equipped with a motor-driven unit, called a rotisserie, which rotates foods within the oven while cooking. Juices from the food drip from the rotating unit and often land on the lower bake element. To prevent smoking of these greases, the broil element is used by many manufacturers for the rotisserie cooking cycle. The location of the broil element in the upper part of the oven prevents dripping particles from landing on it, thus reducing the smoke problem.

When selecting the cooking temperature for rotisserie cooking, the switch knob is rotated clockwise to the broil position, and then adjusted to the desired cooking temperature. This engages the broil element only, through a mechanical selection system built into the front of the switch, and allows the broil element to heat the oven while being controlled by the gas expansion system.

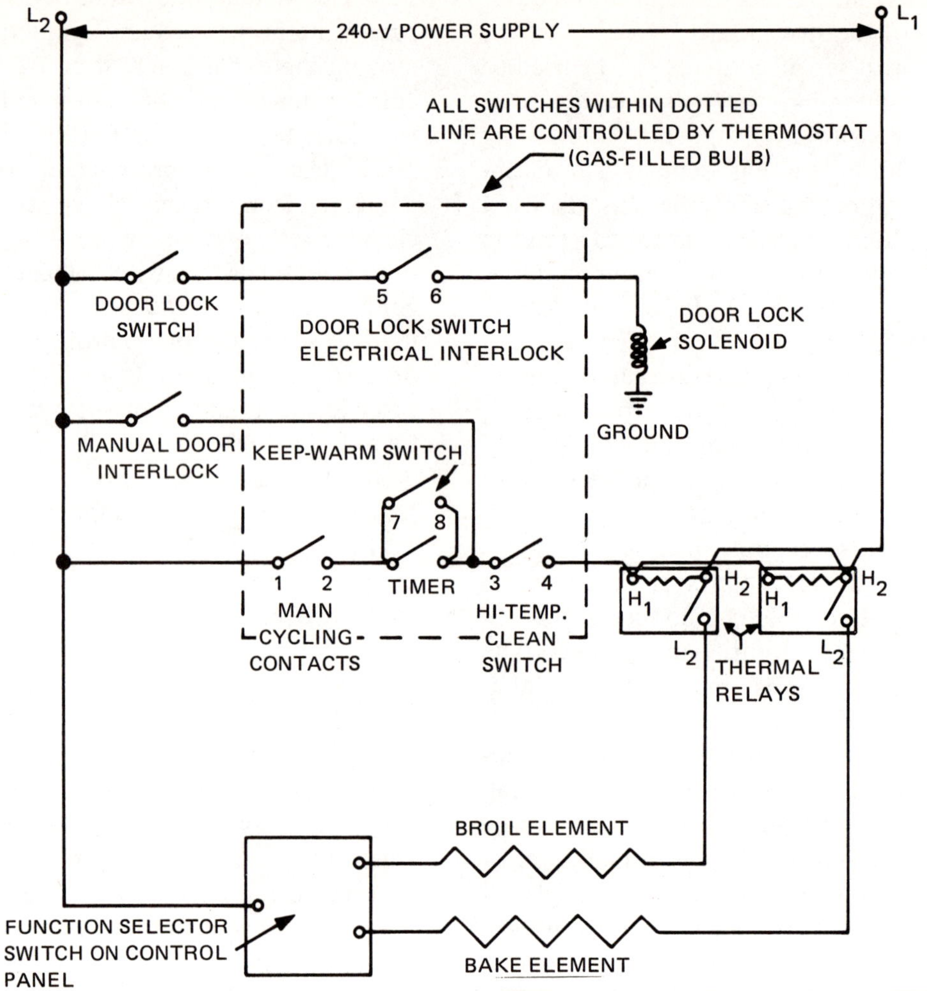

Fig. 10-13 A typical self-cleaning oven circuit.

REVIEW

1. List four appliances that use heat-control switches.
2. Explain how three levels of heat are obtained with the three-heat switch.
3. Of what use is the reference point on the base of a three-heat switch?
4. Why must heat-control switches operating on 240 V be capable of opening both live wires?
5. Why has the seven-heat switch gradually replaced the older five-heat switch?

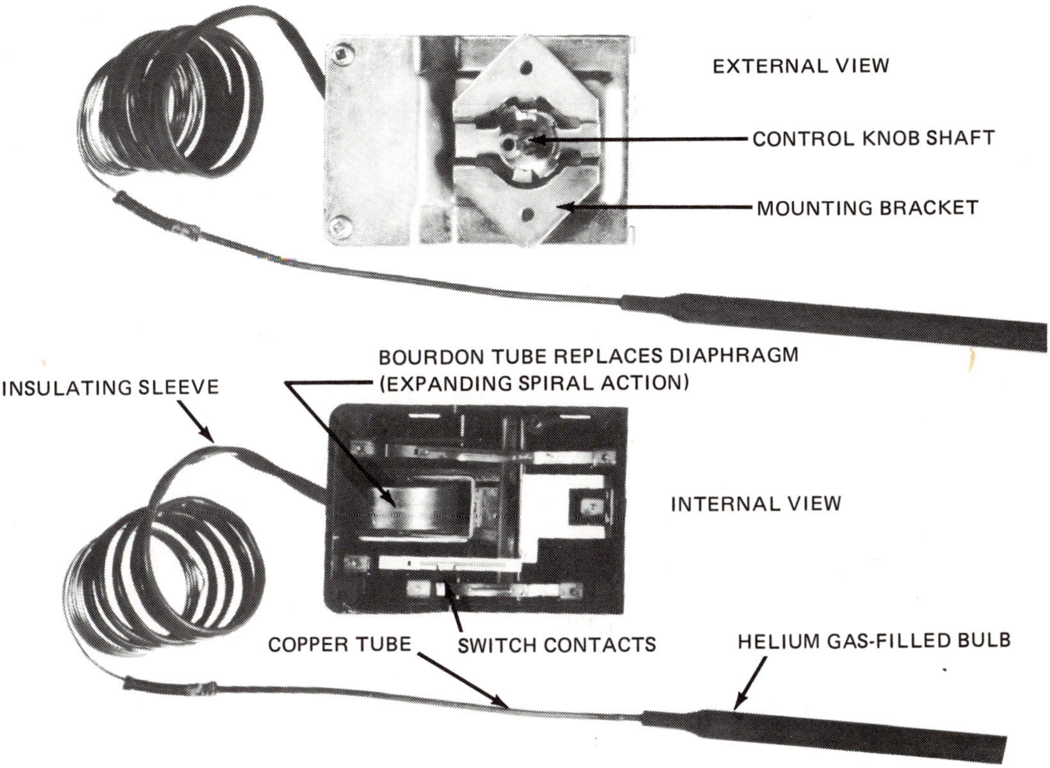

Fig. 10-14 A self-cleaning, pyrolytic oven control switch.

6. Why is a pilot light an important feature on heat-control switches?
7. List two advantages the infinite-heat switch has over other heat-control switches.
8. Explain in your own words the operation of the infinite-heat switch.
9. Where is the oil bulb for an oven-control switch located?
10. What precaution must be taken when using the broil element? Why?
11. State two advantages of the self-cleaning oven.
12. Which is the better type of self-cleaning oven? Explain why.
13. How is the high temperature obtained for the cleaning cycle of a self-cleaning oven?
14. Explain why extra insulation is required around a self-cleaning oven.
15. What are the advantages of an electrically timed oven circuit?

Chapter 11

Armored Cable

USES OF ARMORED CABLE (AC)

Armored cable (AC) is a fabricated assembly of insulated conductors in a flexible metallic enclosure. Type AC can be used for both open and concealed wiring systems. Unlike nonmetallic cable (NMC), it may be run on the surface of walls and ceilings in buildings of masonry construction.

Armored cable is widely used as a quick method for distributing power throughout new industrial plants, for installation in existing plants, and for relocating machinery. It can also be used in public and commercial buildings where the possibility of physical damage to the cable makes NMC unsuitable. Armored cable is more flexible than rigid conduit systems. It can be installed in long continuous runs without need for joints and splices. Conduit systems, however, require a box or fitting after each 100 feet (30 m) of conduit and/or after an accumulation of 360° of bend to ease the pulling-in of the conductors.

PROHIBITED USES OF ARMORED CABLE (AC)

Unless permitted by *Exception* in various *Articles* of the NEC, Type AC cable is not permitted for use in such locations as theaters, places of assembly, motion picture studios, commercial garages, and any hazardous areas, or where exposed to corrosive fumes or vapors. (See *Section 333-6*).

ELECTRICAL RATINGS

Unlike NMC, AC has a continuous protective metal covering, so it is approved for use on circuits up to 600 V maximum. It is available in one-, two-, three-, and four-conductor combinations, ranging in size from No. 14 to No. 4/0. Larger conductors may be made to order.

At one time, individual conductors in the cable were covered with a cotton-braid-over-rubber insulation rated for use at a maximum temperature of 60°C (140°F). Changes in the National Electrical Code have resulted in improved insulation of these conductors. Modern AC conductors are insulated with a very durable material called *cross-link* (X-link). X-link has a maximum temperature rating of 90°C (194°F).

TRADE NAMES

Armored cable was once known as *armored bushed cable* (ABC). This name came from the small, antishort bushing inserted into the end of each cable termination. Electricians now often refer to AC simply as BX or BXL. BXL refers to the type of armored cable that has an internal lead sheath over the conductors.

CABLE CONSTRUCTION

Figure 11-1 shows the materials used in the construction of the older style of AC (permitted for use in dry locations only). Figure 11-2 shows modern cable construction.

One type of armored cable, Type ACL, has a lead sheath placed between the conductors and the armor, Figure 11-3. This sheath prevents moisture or chemicals from entering the conductor portion of the cable. The sheath makes it suitable for outdoor use or, when used in raceways, for underground supply lines. (The NEC does not permit the use of ACL or AC for direct burial in the earth.) The aluminum armor on most AC, however, corrodes severely when placed underground. Therefore, steel conduit is used as a replacement on the cable for outdoor and underground use.

To protect the cable from injury, the bends in AC and ACL are to be made so that the radius of the curve of the inner edge of any bend is not less than five times the diameter of the cable.

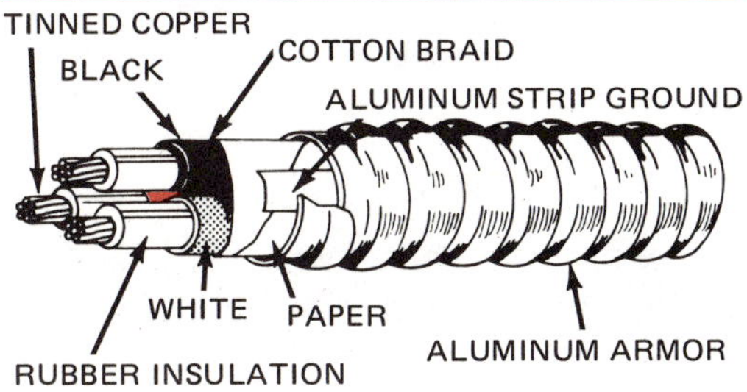

Fig. 11-1 Older style 3-conductor flexible armored cable.

CABLE TERMINATION

The first step in fastening AC to a box or fitting is to remove the armor. This is done with a hacksaw, Figure 11-4. Mark back 8 inches (203 mm), and cut with the hacksaw at an angle of approximately 45°. Take care to remove the correct amount of armor the first time. It is often difficult to hold a short length of cable for a second cut. The insulation on the conductors must not be damaged by the saw during the cutting process. Trim the paper wrapping around the conductors as close to the armor as possible.

After removing the armor and trimming the paper wrapping, fold back the aluminum grounding strip out of the way beside

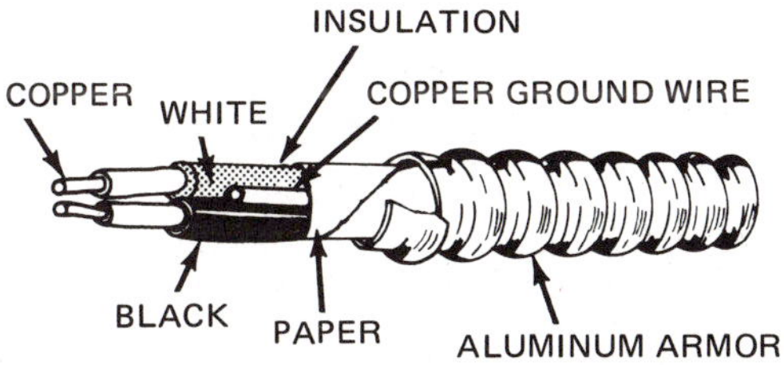

Fig. 11-2 Modern flexible armored cable.

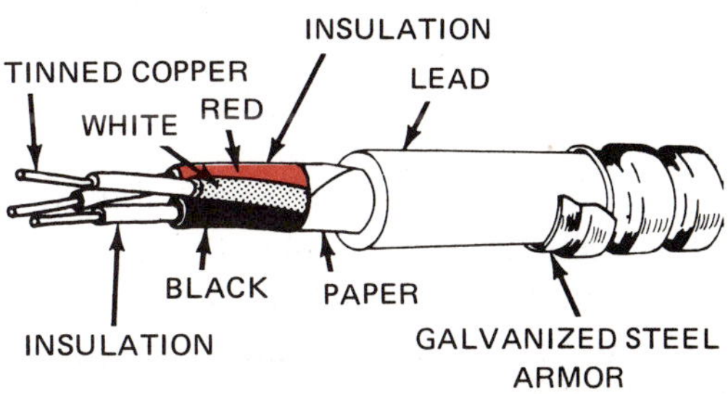

Fig. 11-3 ACL flexible armored cable.

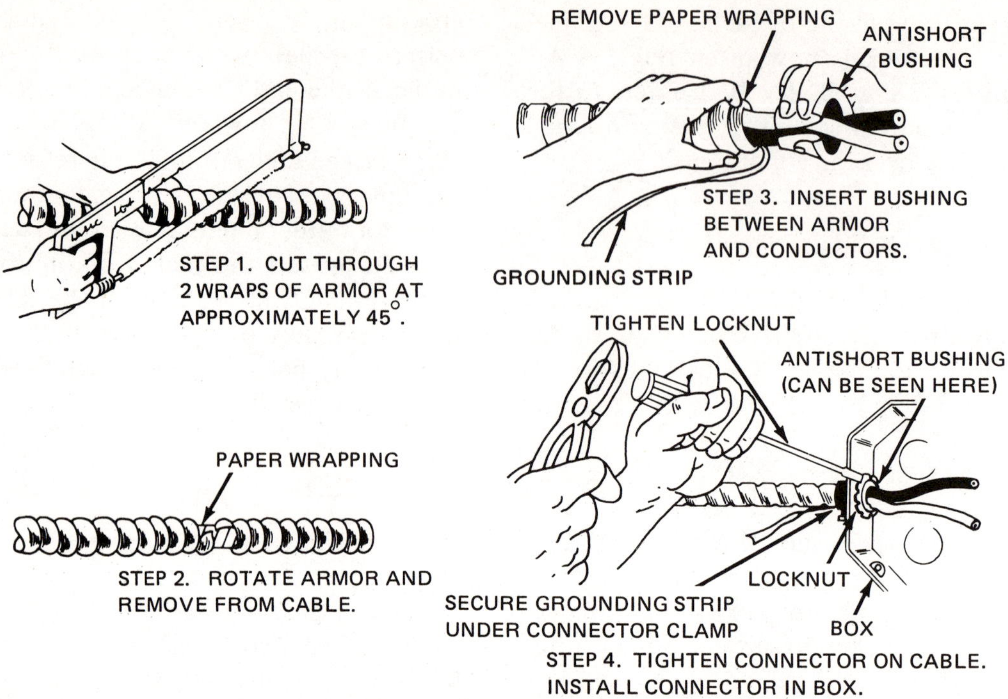

Fig. 11-4 Procedure for terminating armored cable in a box.

the cable. Insert fiber or plastic antishort bushing between the armor and the conductors. This antishort bushing prevents the jagged edges of the armor from cutting into the insulation around the conductors. Once the antishort bushing is in place, insert the cable into an approved box connector. Box connectors hold the antishort bushing in place, secure the cable to the box or fitting, and clamp onto the grounding strip. These connections ensure that the grounding circuit is complete. (See Figures 11-5 and 11-6.)

Cables now being produced have a grounding conductor in place of the aluminum grounding strip. This eliminates the need for folding back and clamping a grounding strip.

CABLE SUPPORTS

Armored cable Types ACL and AC both *must* be supported by an approved device (strap or staple) within 12 inches (300 mm) of *every* box. A maximum of 4 1/2 feet (1.4 m) between the supports on the cable run is allowed. If the cable is run through wooden joists, it *must* be kept back at least 1 1/4 inches (30 mm) from the face. Steel nails driven into the wooden members can easily puncture the aluminum armor. As with NMC, where the AC is closer than 1 1/4 inches (30 mm) to the face of the wooden member, a steel plate is located in front of the cable, (see Chapter 8, Figure 8-10). *Article 333* of the National Electrical Code lists regulations for armored cable installations. Consult local electrical codes

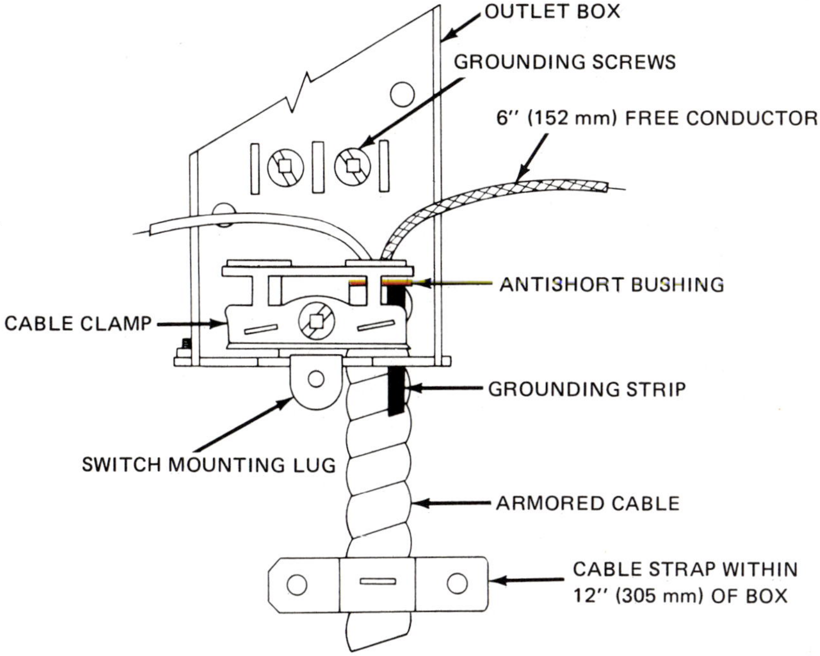

Fig. 11-5 Armored cable secured to box by built-in clamp.

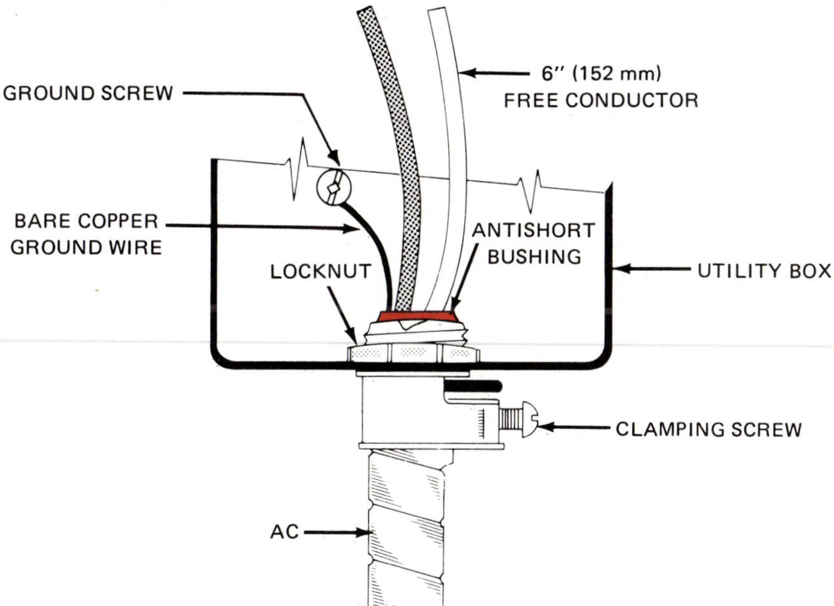

Fig. 11-6 Modern armored cable with ground wire, secured to the box by an AC connector.

for permitted uses of armored cable Types ACL and AC.

FISHING ARMORED CABLE

When installing armored cable in a building that is completed, access to existing outlet boxes is gained by lowering the cable down inside the wall from a hole in the attic or ceiling. This is called *fishing.* Take care to prevent the rough outer armor of the cable from rubbing on and damaging existing cable installations within the wall.

ACL INSTALLATION FOR WET AREAS

When equipped with a lead sheath between the conductors and the armor, ACL armored cable can be used to supply power to outdoor lights, signs, and related equipment. ACL, when installed in raceways, can also be used for underground supply lines to electrical equipment. Pumps, garden lighting, post lights, and supply lines between buildings often require the water-resistant feature of ACL.

Watertight connectors *must* be used to secure this cable to a box or fitting for installation in a wet area, Figure 11-7. Several types of connectors are used; they compress a moistureproof lead sleeve around the cable. (Figure 8-8 shows this method as it is used to supply an outdoor post light.) Figure 11-8 shows a typical outdoor receptacle installation. The National Electrical Code does not allow ACL cable to be used for direct burial in the ground. However, it may be used for underground runs when installed in rigid metallic or rigid nonmetallic conduit as a raceway system. *Note:* ACL does *not* have a grounding strip. It relies on a secure mechanical connection to the box for its ground. The soft lead sheath is allowed to extend past the connector and into the box. This eliminates the need for an antishort bushing.

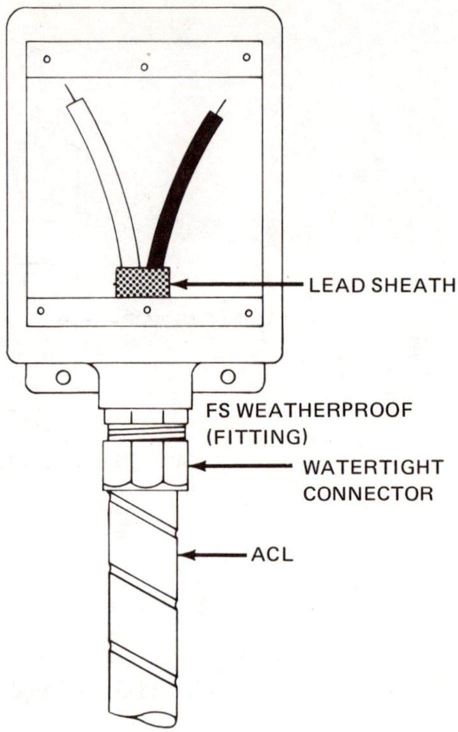

Fig. 11-7 ACL connected to moistureproof fitting.

ARMORED CABLE ACCESSORIES

AC can be bent only so far before the interlocking armor separates. Cable allowed to remain in this condition will not receive the approval of the electrical inspection department. Special connectors providing 45°- and 90°-angles are used where it is necessary to make a sharper angle than the cable will allow, Figure 11-9. The duplex connectors allow a pair of cables to enter one knockout in a box.

Straps and/or staples of the proper size, similar to those used with NMC, are made for supporting armored cable.

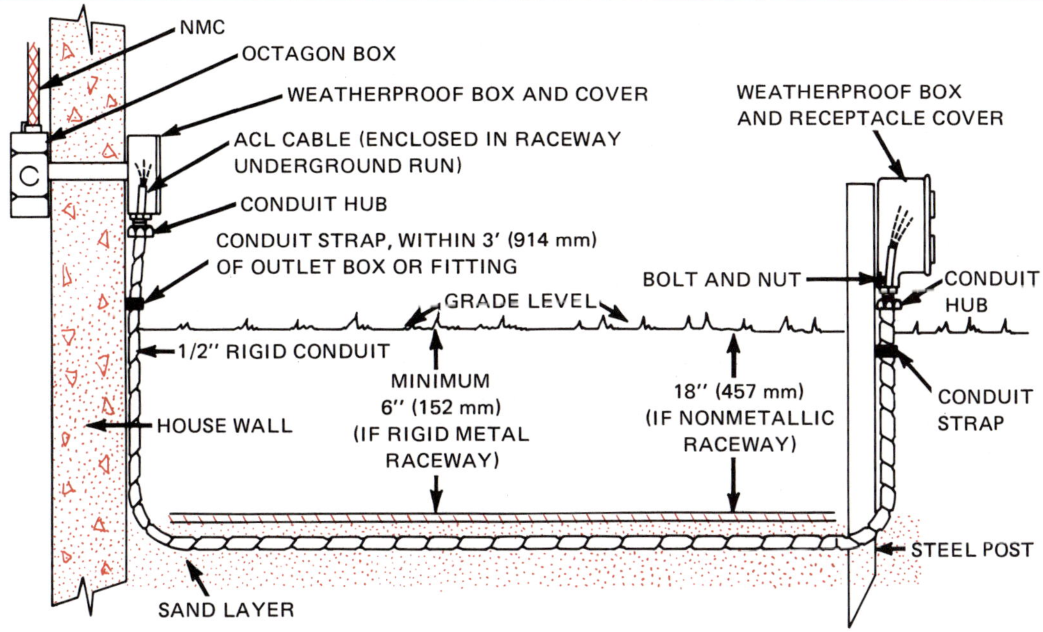

Fig. 11-8 Typical outdoor receptacle installation using ACL enclosed in a raceway underground run. Note: Type ACL cable shall *not* be used for direct burial in the earth (NEC).

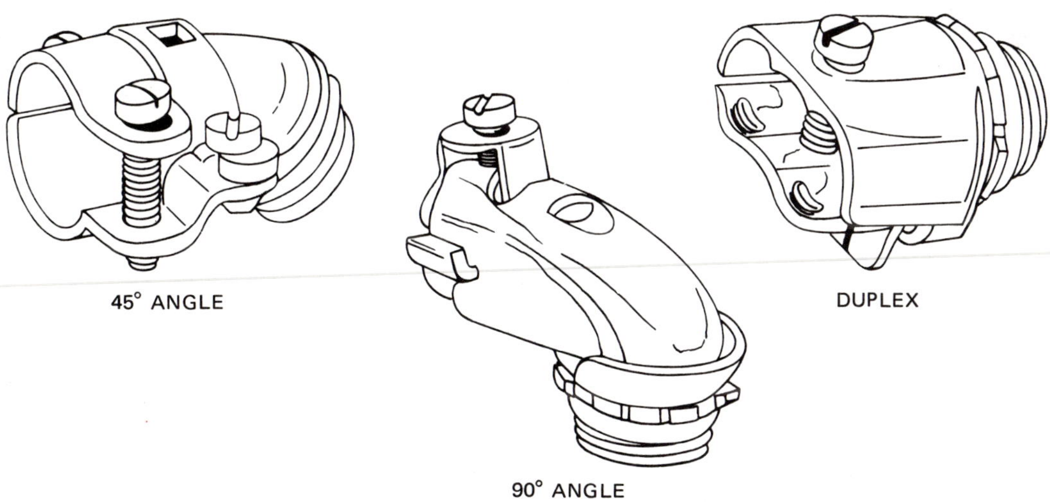

Fig. 11-9 Special AC connectors.

REVIEW

1. Why is AC suitable for both concealed and open wiring systems?
2. Why does AC have a higher voltage rating than NMC?
3. List the AC conductor combinations.
4. Explain in your own words the procedure for terminating AC at a box.
5. What care must be taken when fishing AC?
6. How does ACL for wet locations differ from AC used in dry locations?
7. How is moisture prevented from entering the box or fitting in an outdoor cable installation?
8. How is the ground circuit maintained in AC installations? In ACL installations?
9. Why must care be taken when bending AC?
10. What devices are used for terminating armored cable where a sharp bend must be made?

Chapter 12

Aluminum-sheathed Cable

DESCRIPTION

Aluminum-sheathed cable is a factory-assembled wiring system, consisting of a seamless metal sheath enclosing one or more conductors. The aluminum sheath is both vaportight and liquidtight. The copper conductors within the sheath are usually insulated with a rubber covering rated at a maximum temperature of 90°C (194°F).

Two main types of aluminum-sheathed cable are available: the smooth, snug-fitting sheath type, and the corrugated type.

NEC *Article 334* covers metal-clad cable and its installation.

SMOOTH-SHEATHED CABLE

Smooth-sheathed aluminum cable is basically a cable in a conduit, Figure 12-1. It is assembled and tested at the factory and can be installed simply, which saves considerable time and labor. Smooth-sheathed cable is easy to handle in small conductor sizes, but larger cables are difficult to bend.

CORRUGATED CABLE

The conductors of corrugated aluminum cable are enclosed in an oversized aluminum tube, Figure 12-2. The tube is then passed through a revolving die that compresses two-thirds of the tube onto the insulated conductors. The soft metal arches allow the cable to be bent easily while the work-hardened flat spiral areas support the conductors. Large cables can be bent without the use of special tools, such as those required for a conduit system.

Fig. 12-1 Smooth-sheathed aluminum cable.

Fig. 12-2 Corrugated aluminum cable.

SHEATH COVERING

A polyvinyl chloride (PVC) jacket is extruded over the aluminum sheath. This covering makes the cable resistant to corrosive chemicals and, therefore, suitable for underground installation when approved by local authorities. The jackets are made in various colors to aid identification of circuits and voltages.

ELECTRICAL RATINGS

Aluminum-sheathed cable is made for use on any voltage required. Most of the common sizes of conductor units are readily available with a 600-V rating. The current-carrying capacity of the cables is as allowed by NEC *Tables 310-16* through *310-19.*

The conductors of aluminum-sheathed cable carry a maximum temperature rating of 90°C (194°F).

SINGLE VERSUS MULTICONDUCTOR CABLES

Single-conductor cables have a much higher current rating than multiconductor cables with the same gauge of wire. Therefore, it is more economical to use several single-conductor cables, rather than one large multiconductor cable, for construction of heavy-current circuits.

SHEATH CURRENTS

A conductor carrying electrical current produces a magnetic field around itself. When the current is alternating (ac), the magnetic field rises and falls in strength as the current pulses through the conductor. This produces a moving magnetic field.

Any metal substance within the range of a moving magnetic field will have a current induced into it by the field. When both ends of the cable are grounded on a single-conductor unit, a circuit is formed and the sheath currents circulate. The sheath currents of cables running side by side will circulate between them, if the cables are connected to a common metal box at each end.

As the sheath currents increase in volume, heat is generated in the sheath. This heat can be severe enough to damage the insulation around the conductors within the cable.

When single-conductor cables are run side by side, care must be taken to insulate the cable sheaths at one end, while grounding them at the other end. This prevents the sheath currents from circulating within the cable system. Figure 12-3 shows a typical cable installation (with no sheath currents), and Figure 12-4 shows alternate methods for grounding cables at boxes (with sheath currents).

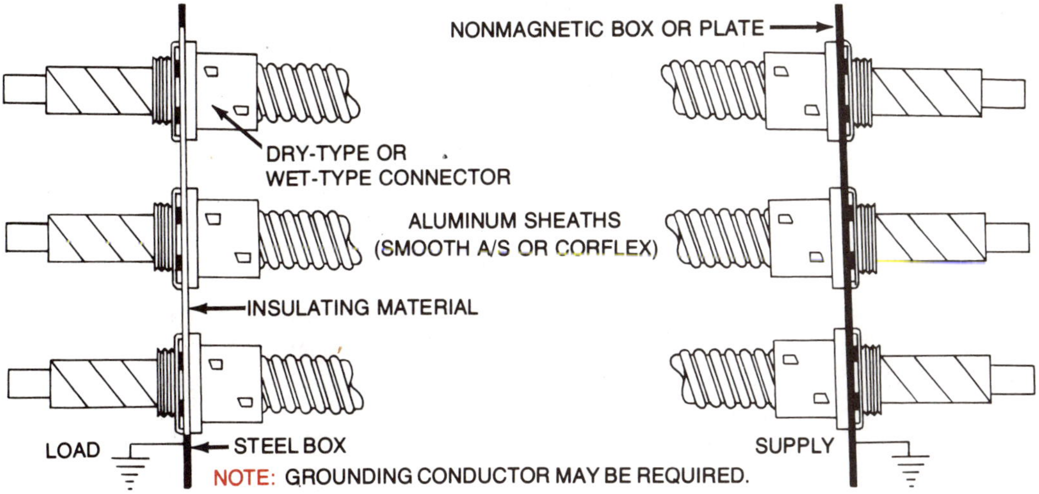

Fig. 12-3 Typical cable installation. (No sheath currents)

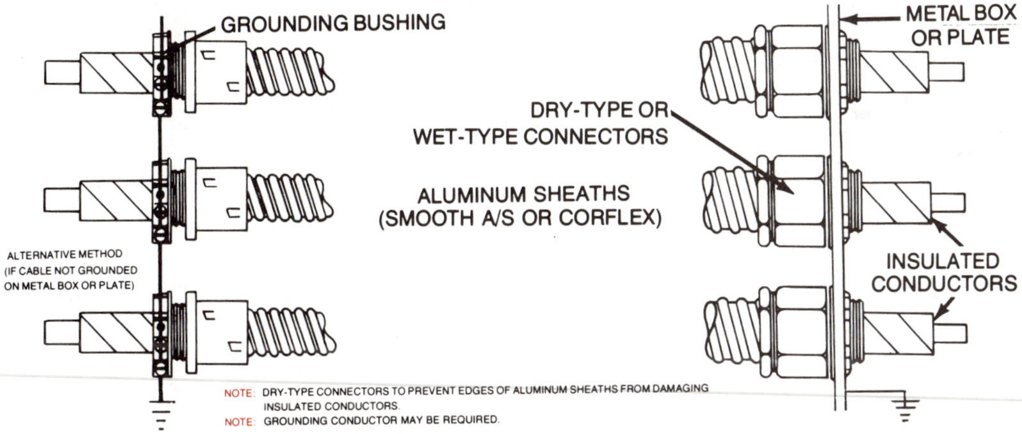

Fig. 12-4 Alternate methods for grounding cables at boxes. (Sheath currents)

Do *not* use magnetic materials, such as steel, for supporting single-conductor cables. These supports can become magnetized and have currents circulating within themselves. These *eddy* currents, as they are called, heat the support and cause damage to the cable insulation.

Connectors used to terminate single-conductor cable must also be made of nonmagnetic material. As the current in a circuit exceeds 200 A, a nonmagnetic plate must be fastened to the box where the cables terminate. Otherwise, serious eddy currents will occur at the box.

Multiconductor cables are more expensive than the single-conductor units required to replace them, but they tend to eliminate the problem of sheath currents. When two or more conductors are enclosed by the same sheath, the magnetic fields around each conductor cancel one another. Multiconductor cables may be encircled or supported by magnetic materials without risk of eddy current damage.

CABLE TERMINATION

Smooth-sheathed cables in the smaller sizes make use of a connector that is similar to an armored cable connector. Typical styles of dry-type and moistureproof connectors are shown in Figures 12-5 and 12-6.

Corrugated cable requires connectors of a design different from that used with the smooth-sheathed cable. Both dry-type and moistureproof connectors are available for corrugated cable, Figures 12-7, 12-8, and 12-9.

PREPARATION OF CABLE FOR TERMINATION

Removing the aluminum sheath from the cable is easy to do. Calculate the quantity of free conductor required in the box, and

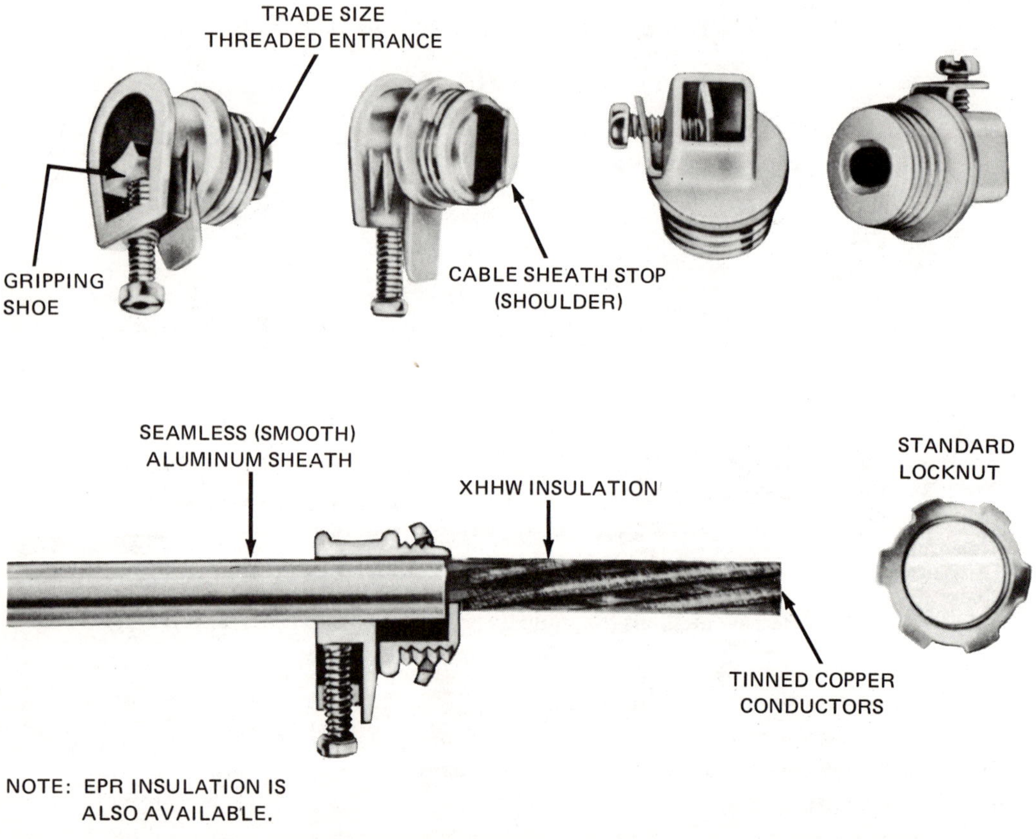

Fig. 12-5 Typical dry-type connectors for smooth-sheathed cable.

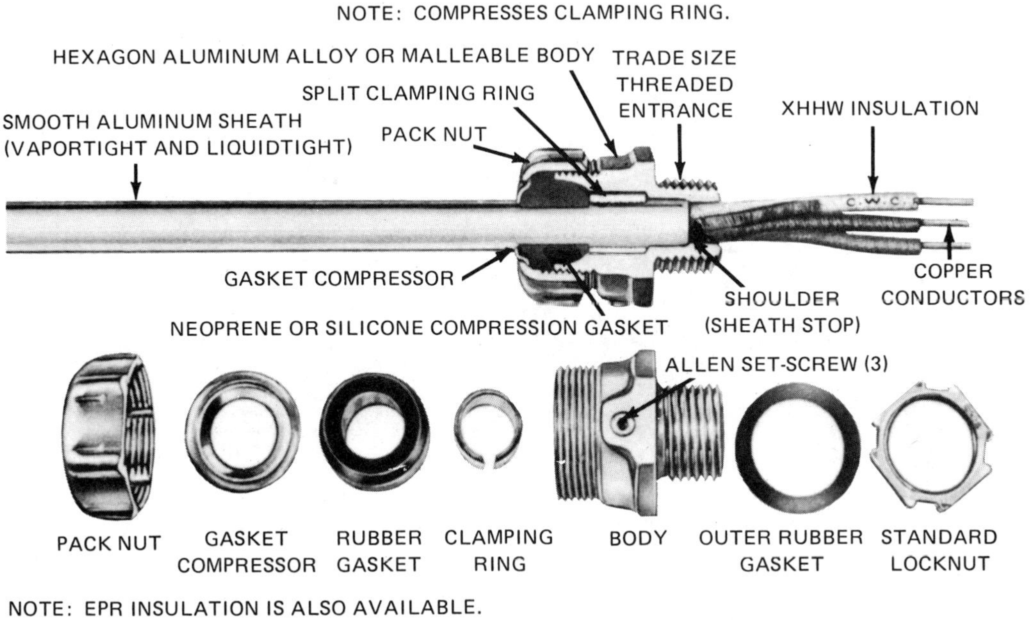

Fig. 12-6 Typical moistureproof connectors for smooth-sheathed cable.

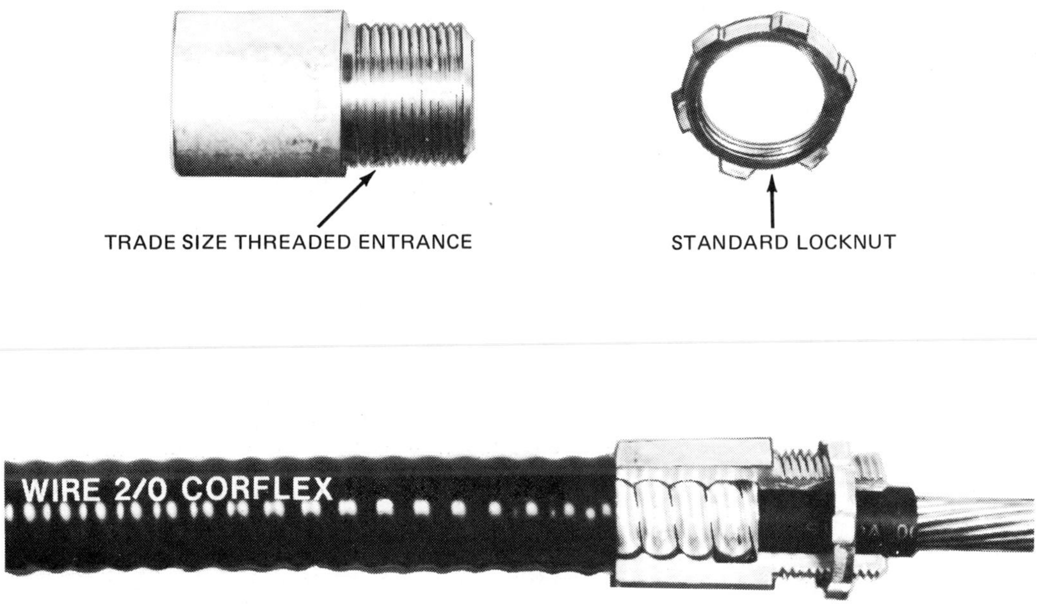

Fig. 12-7 Dry location connector for corrugated cable.

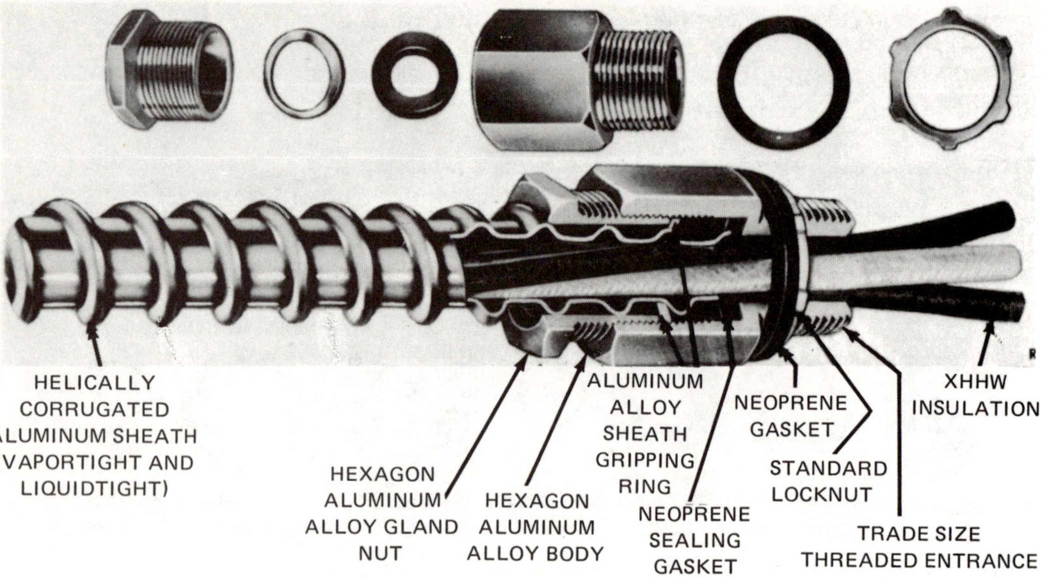

Fig. 12-8 Moistureproof or submersible connector for corrugated cable.

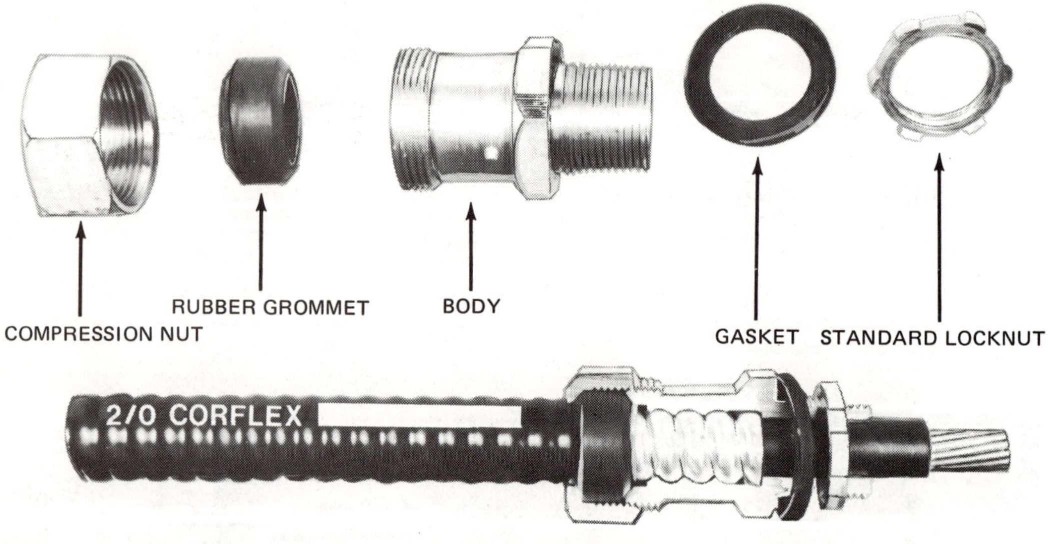

Fig. 12-9 Submersible connector for corrugated cable with PVC jacket.

mark the cable. Score the outer sheath of small cables with a knife. Take care not to cut through the aluminum; just dent the surface. To remove the desired length of sheath, bend the end of it up and down by hand, cracking the sheath at the score line. Pull on the sheath to remove it; it will slip off the conductors easily.

Corrugated cable, with its PVC jacket, is a little more difficult to handle. Figure 12-10 shows how to prepare this type of cable for a connector.

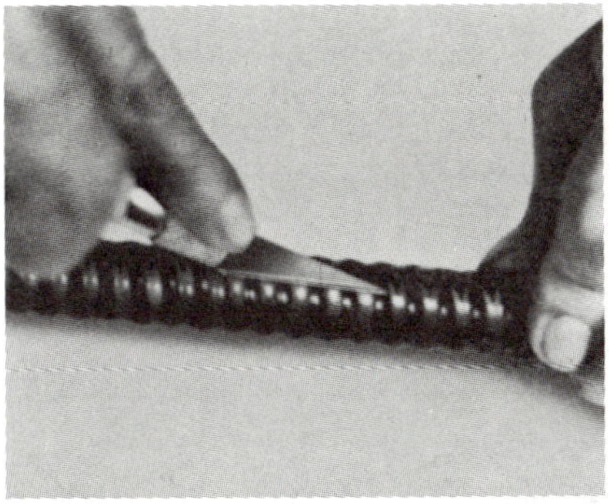

1\.
CUT AND REMOVE SMALL SECTION OF JACKET AT DESIRED LOCATION OF CABLE SHEATH END.

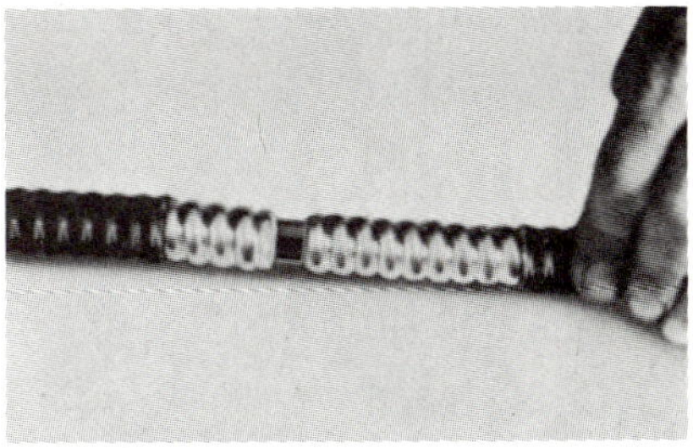

2\.
SCORE AND CUT SHEATH, (AS PER DRY CONNECTOR INSTRUCTIONS). SLIDE UNWANTED SHEATH OFF CONDUCTOR(S).

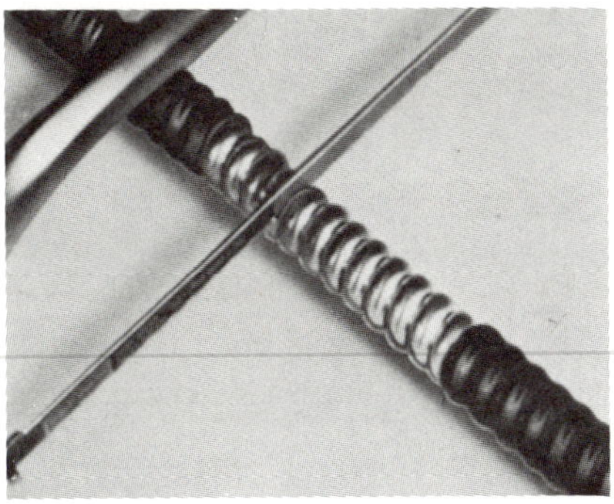

3\.
CAREFULLY CUT THROUGH THE RAISED HELIX USING A FINE-TOOTH HACKSAW (24T). BE CAREFUL NOT TO CUT INTO CABLE INSULATION.

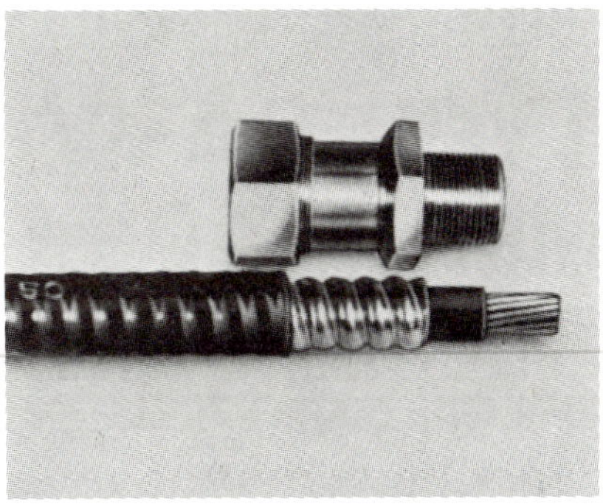

4\.
CUT BACK JACKET FOR LENGTH OF CONNECTOR BODY ONLY.

Fig. 12-10 Procedure for terminating corrugated cable with moistureproof connector.

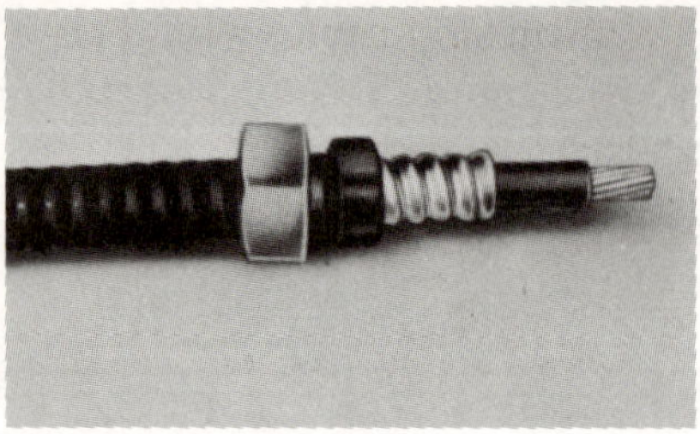

5.
PLACE PACKING NUT AND SEALING GROMMET ONTO CABLE ENSURING THAT GROMMET IS COMPLETELY OVER THE PVC JACKET.

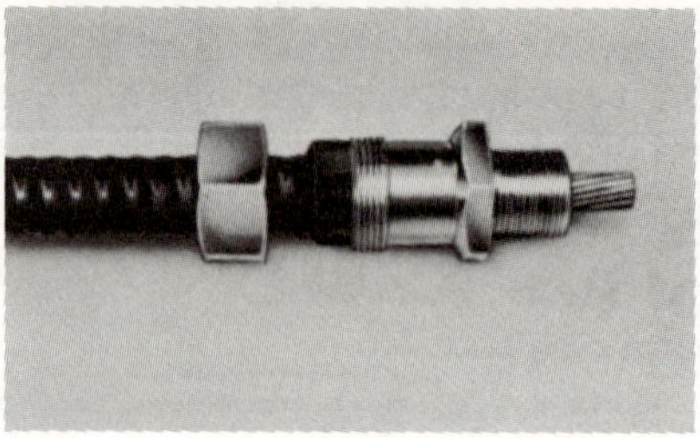

6.
THREAD THE CONNECTOR BODY ONTO THE SHEATH AND TURN BY HAND ONLY UNTIL THE SHEATH BINDS AGAINST THE INTERNAL SHOULDER. TIGHTEN BY HAND ONLY.

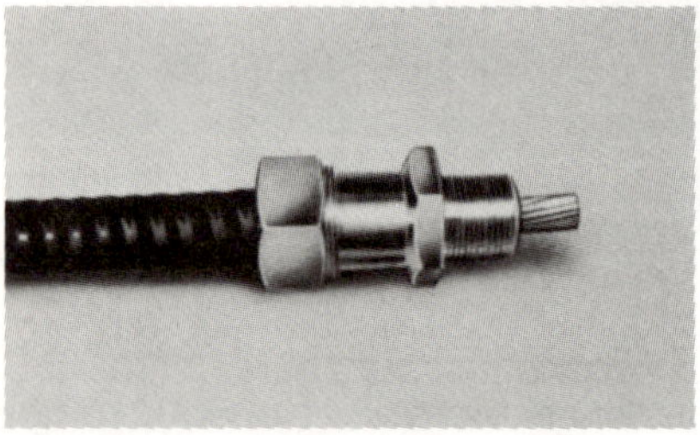

7.
THREAD PACKING NUT ONTO THE BODY AND TIGHTEN SUFFICIENTLY TO BEGIN SQUEEZING GROMMET OUT FROM UNDER THE PACKING NUT.

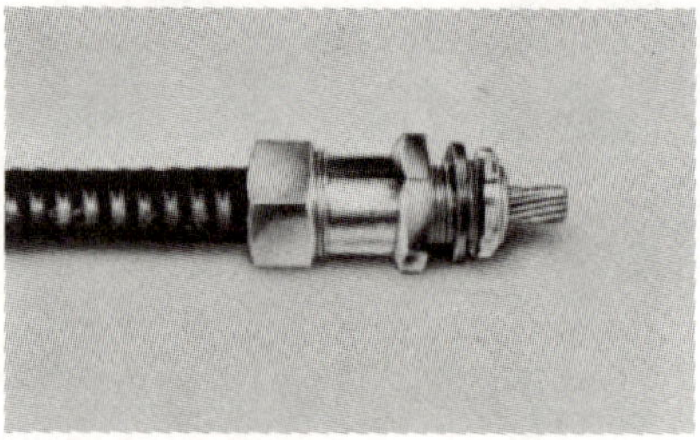

8.
WHEN CONNECTING TO PLATE ENSURE RUBBER GASKET IS PLACED BETWEEN CONNECTOR AND OUTSIDE SURFACE OF PLATE.

Fig. 12-10 Continued

CABLE SUPPORTS

Aluminum-sheathed cable requires nonmagnetic aluminum supports, Figure 12-11. The National Electrical Code requires a strap every 6 feet (2 m), but some sag in the cable may occur in the smaller sizes. One cable manufacturer recommends a strap every 3 feet (1 m) to 6 feet (2 m) on cables up to 1 1/4 inches (32 mm) in diameter. In many industrial installations, multiconductor cables are laid in a trough or rack designed for this purpose. Figure 12-12 shows typical cable applications.

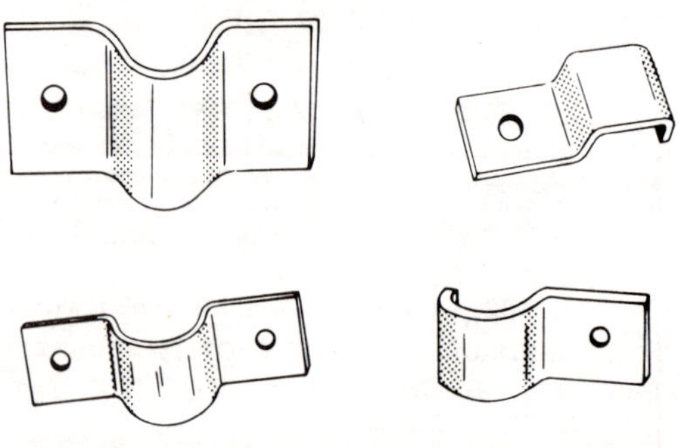

Fig. 12-11 Typical aluminum clips for aluminum-sheathed cable.

MOTOR WIRING

OIL REFINERY

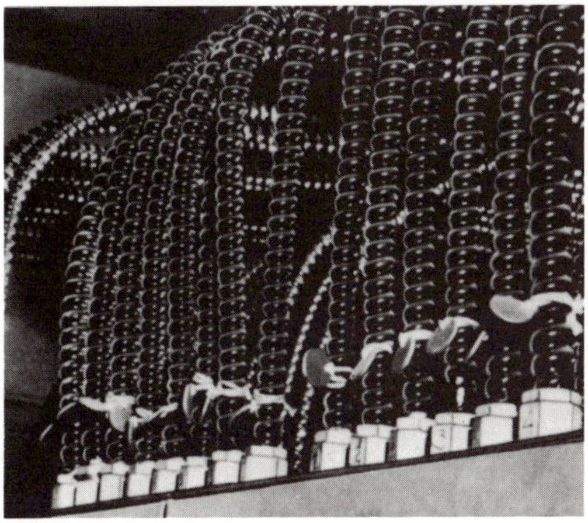

PULP AND PAPER MILL

Fig. 12-12 Cable applications

REVIEW

1. What are the two main types of aluminum-sheathed cable?
2. Which aluminum-sheathed cable is easier to bend in large sizes?
3. What jacket material is used for corrosion-resistant cable?
4. What is the most common voltage rating for aluminum-sheathed cable?
5. What is the maximum temperature rating for the conductors in aluminum-sheathed cable?

6. Why is it more economical to run single-conductor cables, rather than one multiconductor cable?
7. Explain in your own words how sheath currents are produced in aluminum-sheathed cable installations.
8. Why are there no sheath currents in multiconductor cable systems?
9. Magnetic materials must not be used to support or terminate single-conductor cables. Explain.
10. Describe briefly how to remove the aluminum sheath from a cable that is to be terminated in a distribution panel.
11. What spacing is required by the National Electrical Code for straps supporting aluminum-sheathed cable?

Chapter 13

Mineral-insulated (MI) Cable

DEVELOPMENT

Mineral-insulated (MI) cable was developed to provide the electrical industry with a fire-resistant wiring product. This cable cannot cause or contribute to a fire because it contains *no* flammable materials. It was first introduced around 1934. Since then, it has proved to be so versatile that new applications are constantly being devised.

NEC *Article 330* covers the use and installation of mineral-insulated Type MI cable.

FEATURES OF MI CABLE

MI cable resists fire up to the point of the copper melting in and around the cable. It is moistureproof, it resists corrosion, it is not damaged by oil products, and it resists aging. MI cable provides a compact, neat, surface-wiring cable system. It does not sag and it is easy to install. It can be used indoors or outdoors, and it is strong enough for direct burial in the earth (when protected against physical damage or destructive corrosive conditions.) It will withstand an almost unbelievable amount of physical abuse before an electrical breakdown occurs.

HOW MI CABLE IS MADE

Mineral-insulated cable consists of from one to seven high-conductivity copper rods, 30 feet (9-m) in length, which are inserted into a 2-inch (50-mm) tube of seamless copper, Figure 13-1. Magnesium oxide, which is an excellent electrical insulator and conductor of heat, is packed under pressure around the rods. The ends of the tube are sealed, and the 30-foot (9-m) section is drawn through a series of reducing dies. These dies decrease the diameter of the cable and increase its length.

The magnesium oxide insulation has the same density as the copper used in the tube and rods. This means that, as the tube assembly is pulled through the reducing dies, the tube, insulation, and rods are reduced in size simultaneously. The relative spacing between the conductors and the tube is always maintained, because the rods are reduced in direct proportion to the tube itself.

MI cable can be produced in any conductor size by continuing the drawing process until the conductors have been reduced to the desired wire gauge number.

DURABILITY

MI cable can withstand severe physical abuse without electrical failure. This is due to its ability to change, at the same time, the shape of the outer sheath, the mineral insulation, and the internal conductors. Therefore, a crushing blow from a heavy object will merely change the cable's shape, Figure 13-2.

MI cable is as resistant to heat as it is to mechanical injury. Figure 13-3 shows how MI cable compares with other wiring systems under severe heat conditions. The magnesium oxide insulation and the copper sheath tend to conduct heat away from the cable. Fire alarm or pump circuits can use to advantage this cable's ability to operate under high-temperature conditions.

CABLE TERMINATION

Prepackaged termination kits are used to connect MI cable to boxes, cabinets, and fittings. The procedure for a typical MI cable termination is shown in Figure 13-4. Re-

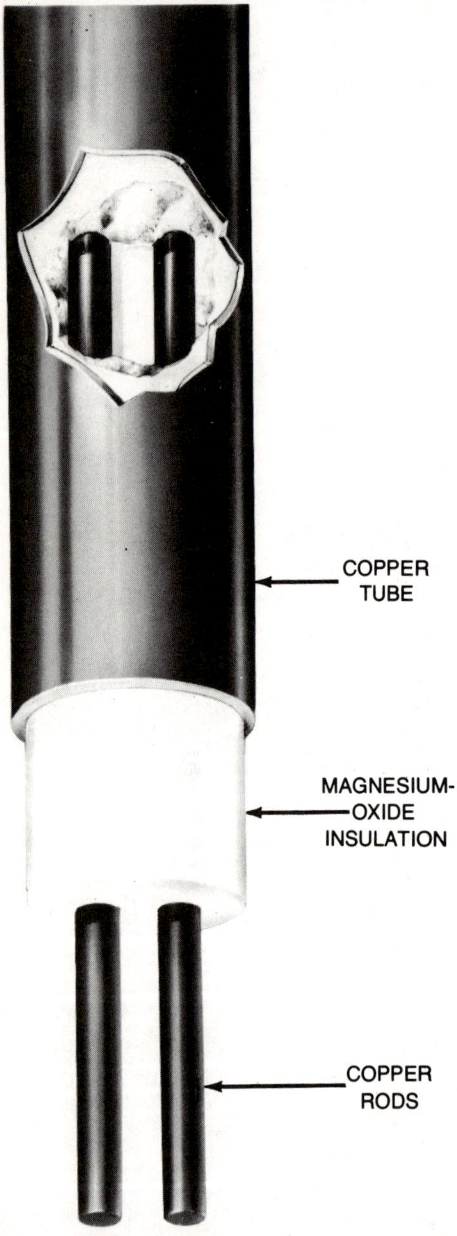

Fig. 13-1 Typical mineral-insulated cable (cutaway view).

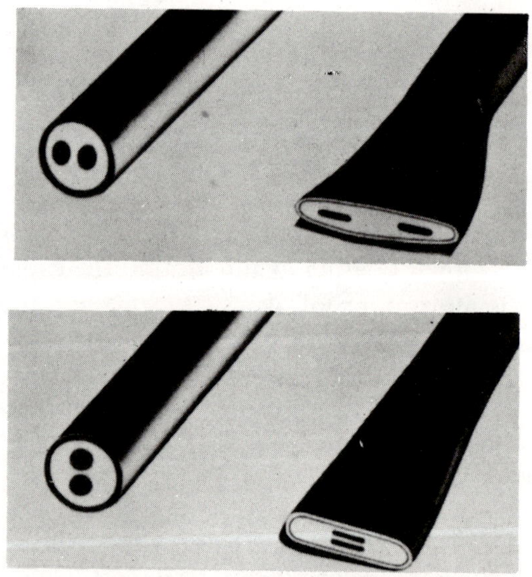

Fig. 13-2 MI cable continues to function when flattened to one-third its original diameter. Only the shape changes.

move the copper sheath with a stripping tool. Place a gland connector on the cable just before threading the self-tapping pot onto the outer sheath. Press plastic sealing compound into the pot. Make sure your hands are clean so that no metal particles or moisture are pressed into the sealing compound. A short circuit will likely occur if metal particles or moisture of any kind are allowed to enter the pot.

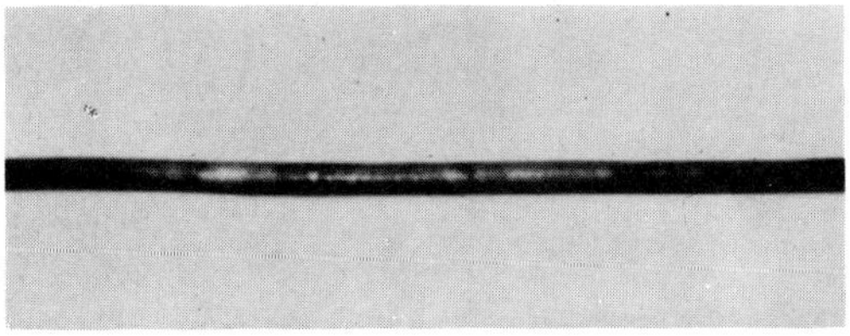

PYROTENAX CABLE (INTACT AFTER 2 HOURS)

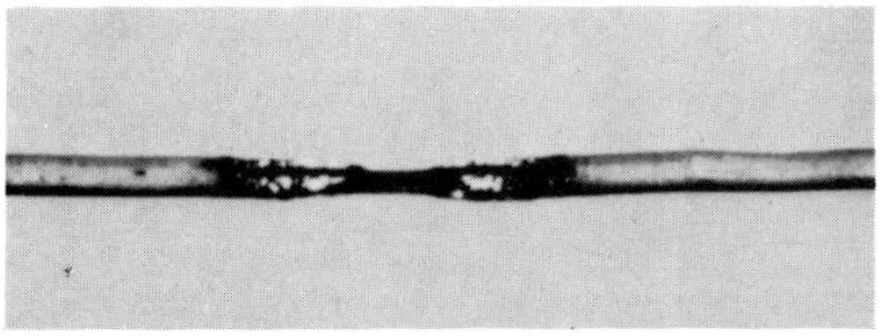

LEAD-SHEATHED RUBBER-INSULATED CABLE (DESTROYED AFTER 25 SECONDS)

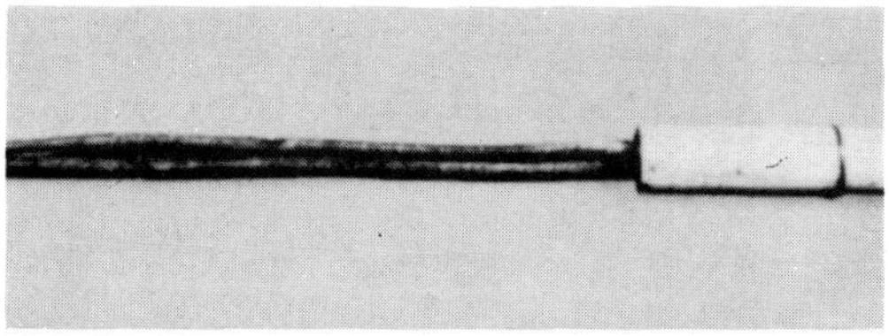

ASBESTOS-INSULATED WIRES IN CONDUIT (GROUNDED AFTER 6 MINUTES)

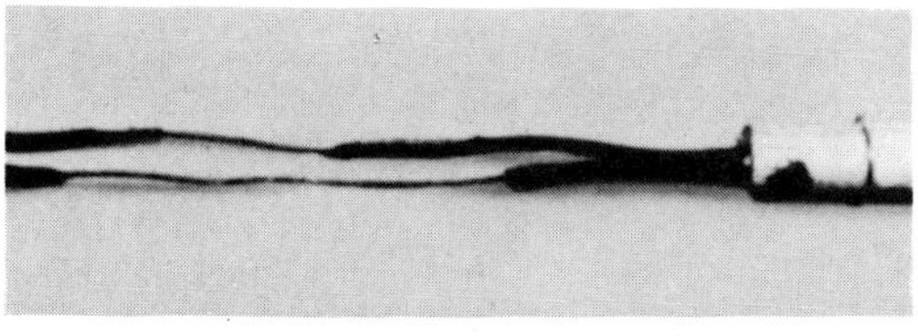

RUBBER-INSULATED WIRES IN CONDUIT (DESTROYED AFTER 2.5 MINUTES)

Fig. 13-3 Fire test of 960°C (1760°F) shows MI cable durability under severe heat conditions.

STEP 1. REMOVE OUTER SHEATH WITH STRIPPING TOOL.

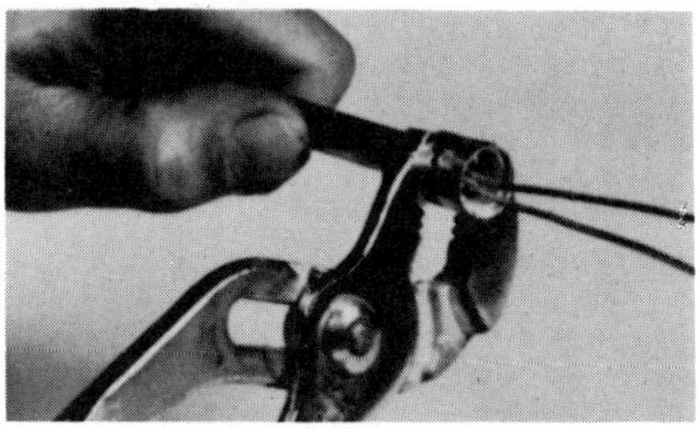

STEP 2. AFTER INSTALLING GLAND CONNECTOR OVER CABLE, THREAD SELF-TAPPING POT ONTO COPPER SHEATH.

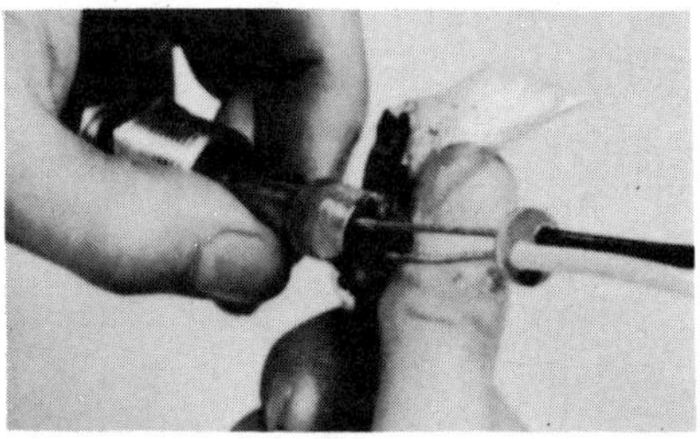

STEP 3. FORCE PLASTIC SEALING COMPOUND INTO POT BY HAND. BE SURE HANDS ARE CLEAN.

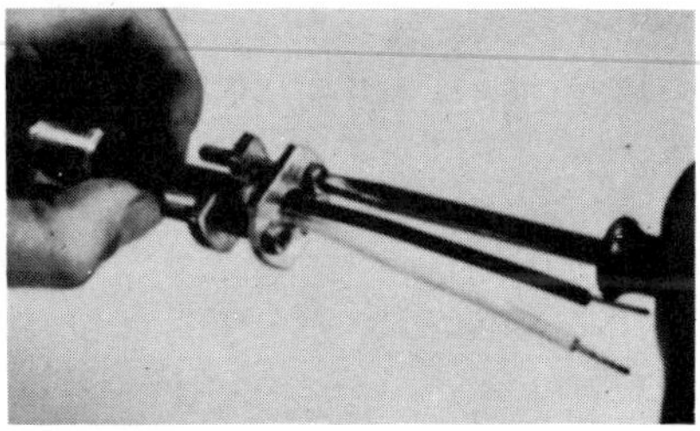

STEP 4. CRIMPING AND COMPRESSION TOOL SECURES SLEEVE SUBASSEMBLY TO POT.

Fig. 13-4 Procedure for typical MI cable termination.

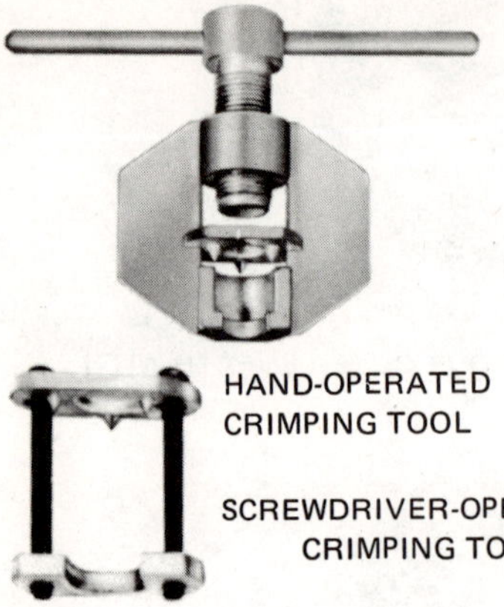

Fig. 13-5 Two types of crimping and compression tools.

Next, install the preassembled insulating sleeves. Use a crimping and compression tool to force the insulating sleeves into place and lock them there.

Figure 13-5 shows a simple screwdriver-operated tool and a more elaborate crimping tool that is also available.

Figure 13-6 shows a cutaway view of a typical MI cable termination, and an assembled unit. These terminal fittings are designed for use at temperatures up to 250°C (482°F). The gland is equipped with a standard electrical conduit thread that fastens to electrical boxes with conduit locknuts or threads into pretapped holes in moisture-proof boxes and fittings.

Corrosion-resistant thermoplastic-jacketed cables and terminations are available, Figure 13-7.

All cable terminations should be checked with an insulation tester. An *insulation tester* is a high-resistance measuring instrument which applies approximately 500 V to the termination and indicates whether or not the termination is safe for use.

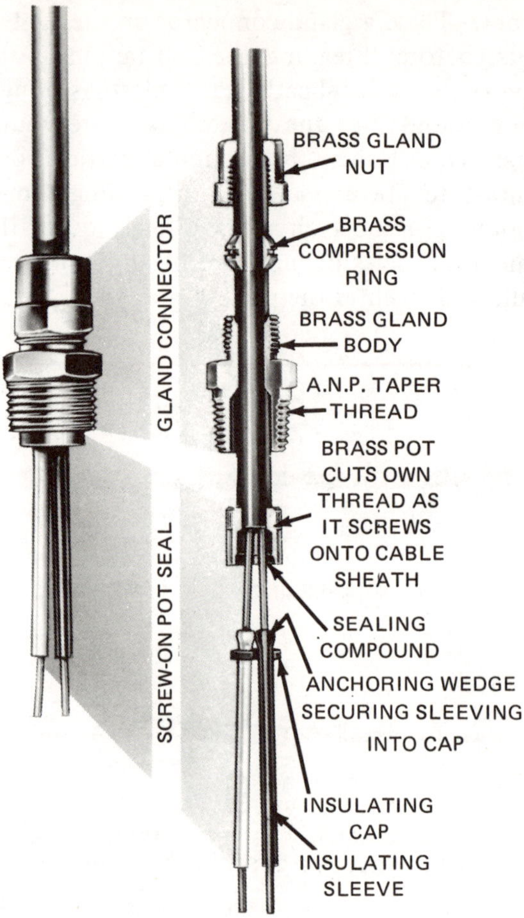

Fig. 13-6 Typical MI cable termination and assembled unit (cutaway view).

The outer sheath, being of a continuous copper construction, provides mechanical protection, a moisture seal, and an adequate path for grounding purposes.

CABLE INSTALLATION

MI cable is supported by copper straps located so as not to exceed every 6 feet (1.83 m) throughout the run. A wooden block and hammer are used to straighten any irregularities in the cable.

Fig. 13-7 Thermoplastic-jacketed cable termination for corrosive areas.

MI cable can be formed into a bend having a radius of not less than five times the diameter of the cable being bent, without placing undue stress on it.

Multiple runs of MI cable require care and precision in planning, Figure 13-8.

Fig. 13-8 MI cable multiple runs.

SHEATH CURRENTS

Single-conductor cables in alternating-current circuits can have induced currents flowing in their outer sheath (see Chapter 12, "Sheath Currents"). These currents are a direct result of electromagnetic fields surrounding nearby cables. Since heat generated by these currents has little effect on MI cables, the cables need not be spaced apart. Instead, MI cables can be grouped together under a common strap, Figure 13-9.

Cables carrying a current in excess of 200 A *must* be fastened to a *nonmagnetic* plate at the box. This plate prevents heat induction from damaging the metal box and its contents.

MI CABLE APPLICATIONS

The versatility of MI cable makes it useful for many purposes and in many situations. MI cable is used in commercial and public buildings, factories, houses and apartments, processing plants, and railway and subway systems. It is also used to heat electrically driveways, pipes and steps; and for communication and transmission systems. It has particular application in all kinds of marine vessels, and in hazardous areas that contain dust, or explosive vapors or liquids. Figure 13-10 shows typical MI cable applications.

Fig. 13-9 Single-conductor MI cables entering distribution panel.

TO SUPPLY MOTOR CONTROL EQUIPMENT

TO SUPPLY ELECTRICITY TO APARTMENT BUILDING

Fig. 13-10 Typical MI cable applications.

REVIEW

1. List three features of MI cable.
2. Neatly sketch a piece of MI cable, and label its parts.
3. Describe how MI cable reacts to crushing blows from blunt objects.
4. What special advantage does MI cable have when used for emergency circuits, such as fire alarms and pump systems?
5. List the steps for terminating an MI cable.
6. What is the high-temperature limitation of MI cable terminal fittings?
7. What type of MI cable is used in corrosive areas?
8. What precautions should be taken to reduce the possibility of damage from sheath currents when fastening MI cable to a box?
9. At what level do sheath currents become a problem for MI cables?
10. List three typical electrical installations where MI cable has an advantage over other types of wiring systems.

Chapter 14

Conduit Wiring

PROTECTIVE FEATURES

A conduit wiring system offers a mechanical protection to electrical circuits that is rare with other wiring methods. Voltage in a conduit system is limited only by the insulation on the conductors within the system. Metallic conduits provide a high degree of fire protection, as well as the ability to safely contain overloaded or short-circuited conductors that could cause or contribute to a fire.

Unlike any other wiring method, conduit wiring allows the conductors within the system to be removed or replaced easily, without dismantling the system. A change in circuit design or equipment often requires conductors of a different size, color, material, or quantity to be installed in the conduit. Consult NEC *Article 300, Wiring Methods,* and the applicable tables in *Chapter 9* of the Code when changing the size or quantity of conductors in a conduit.

CONDUIT APPLICATIONS

Conduit wiring systems are used for surface wiring in apartments, factories, garages, warehouses and public buildings, and for service-entrance equipment for a house.

Conduit can be buried directly in masonry construction. Commercial and industrial buildings constructed of poured concrete often have a conduit wiring system installed before the concrete is poured. Conductors to be installed underground can be adequately protected by a conduit system. *Note:* The NEC prohibits these uses in areas which are subject to severe corrosive influences, unless the conduit is protected and judged suitable for the conditions.

When installed properly, conduit is both watertight and vaportight. Hazardous areas, where explosive liquids, gases, or dusts are present, can be wired safely with conduit and electrical fittings identified for the purpose. Plastic-coated conduits are available for use in areas where corrosive conditions are present. Motors and similar equipment subject to vibration or movement can be connected with flexible conduit. Liquid-tight and dust-tight varieties of flexible conduit are also available.

CONDUIT SIZES

Conduit is usually produced in 10-foot (3.05 m) lengths. This length, regardless of diameter, provides an easy-to-estimate, practical unit for installation and allows for ease of bending and handling. Under normal conditions, these lengths can be quickly assembled into a continuous run. For specific applications, lengths shorter or longer than 10 feet (3.05 m) are available.

The second important dimension of the conduit is internal diameter. This measurement determines the quantity and size of the conductors that can be safely installed in the conduit. Electrical current flowing through the conductor produces heat, and this heat must be given ample free space to dissipate without injury to the conductor's insulation. NEC *Article 300, Wiring Methods,* lists general installation requirements for methods and materials, and allowable conduit capacities. The number of conductors and their type of insulation is shown in *Chapter 9* of the NEC, *Tables 1, 2, 3A, 3B, 3C* and *4*.

Conduit is available in the following trade sizes, as determined by the internal diameter in inches: 1/2, 3/4, 1, 1 1/4, 1 1/2, 2, 2 1/2, 3, 3 1/2, 4, 4 1/2, 5, and 6.

TYPES OF CONDUIT

Many types of conduit are available, Figure 14-1. This chapter discusses rigid (thickwall), EMT (thinwall), IMC (intermediate rigid), rigid aluminum, rigid PVC, flexible metal, and PVC flexible (liquid-tight) conduit.

RIGID CONDUIT (THICKWALL)

Rigid, or thickwall, conduit is produced in aluminum or steel. This thickwall type of conduit provides the greatest amount of mechanical protection to conductors. It is available with a choice of external coatings, such as electroplating, baked-on

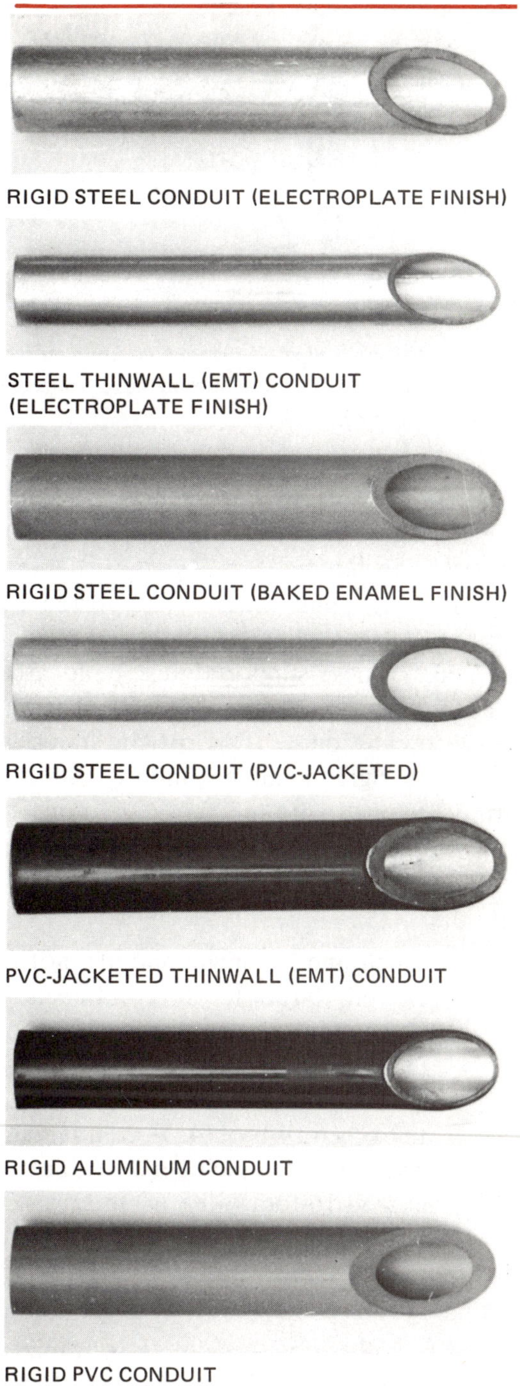

Fig. 14-1 Electrical conduit raceways. (Ends cut at an angle to relate thickness of conduit.)

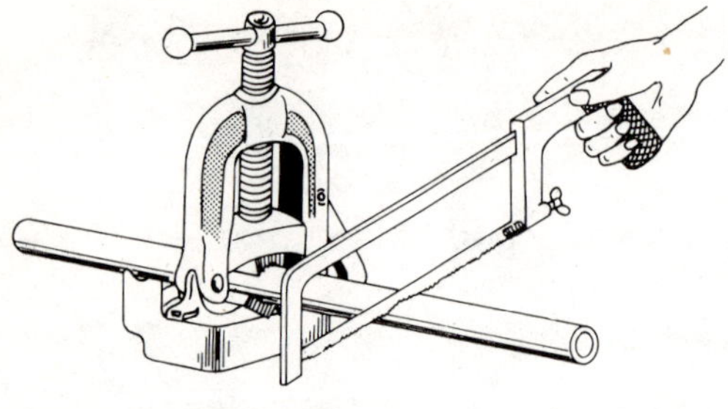

Fig. 14-2 Using a pipe vise to cut conduit with a hacksaw. Note: Hacksaw should have 24 teeth per inch (25 mm).

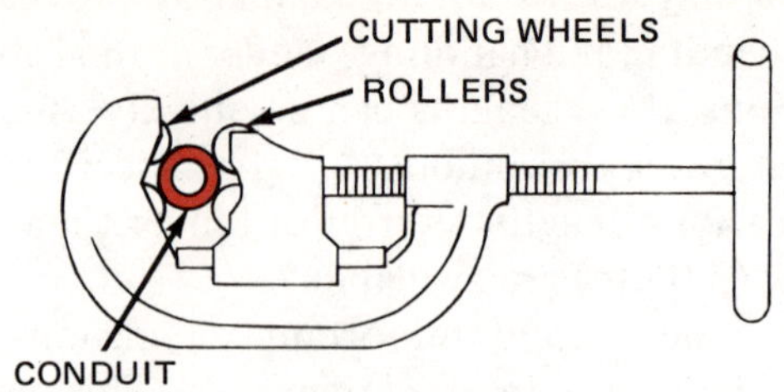

Fig. 14-3 Typical pipe cutter.

enamel, or polyvinyl chloride, to reduce the damaging effect of corrosive chemicals in certain installations.

Rigid conduit must be supported by approved straps, clips, or hangers at regular intervals. These supports must be located in accordance with *Article 346* of the National Electrical Code. This article states in part that rigid metallic conduit must be firmly fastened within 3 feet (914 mm) of each outlet, junction box, cabinet or fitting, and:

- 1/2-inch and 3/4-inch conduit: not to exceed 10-foot (3.05 m) intervals
- 1-inch conduit: not to exceed 12-foot (3.66 m) intervals.
- 1 1/4-inch and 1 1/2-inch conduit: not to exceed 14-foot (4.27 m) intervals.
- 2-inch and 2 1/2-inch conduit: not to exceed 16-foot (4.88 m) intervals.
- 3-inch and larger conduit: not to exceed 20-foot (6.1 m) intervals.

Generally, steel conduit is galvanized inside and outside. The inside of steel conduit is often coated with an insulating paint or varnish to ease the installation of conductors, to insulate damaged conductors from the metal wall of the conduit, and to prevent internal conduit rusting. Where protected from corrosion solely by enamel, the conduit is permitted for indoor installation where it is not subject to severe corrosive influences. Lengths of rigid conduit must be threaded at each end, and reamed for connection to couplings, boxes, or fittings.

Threading Rigid Conduit. A pipe vise is the best method for holding conduit securely for threading, Figure 14-2. For the best straight cut, hold the hacksaw at an angle of approximately 45° to the horizontal. Figure 14-2 shows the hacksaw being drawn back with one hand. This view is shown so that the pipe vise is visible. In actual practice, the second hand should grip the front portion of the saw. This gives a more secure grip and a straighter cut.

A pipe cutter, Figure 14-3, is also excellent for producing neat cuts for threading. However, using a pipe cutter requires extensive reaming to the inside of the conduit.

Whichever method is used to cut the conduit, a sharp burr, capable of damaging the conductor insulation, will be produced on the inside of the conduit. This burr must be removed with a round file or reamer, Figure 14-4.

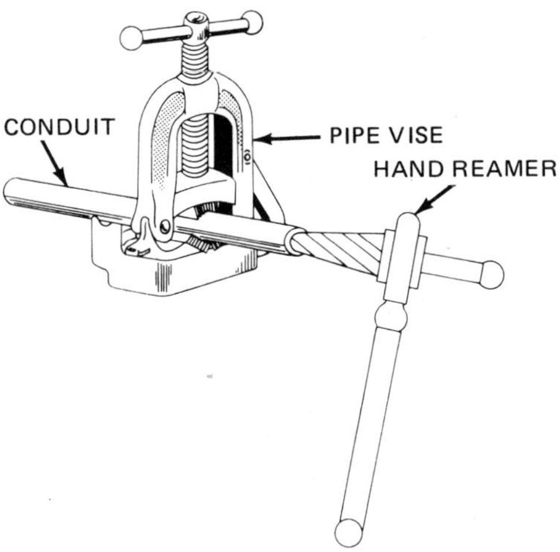

Fig. 14-4 Reaming conduit.

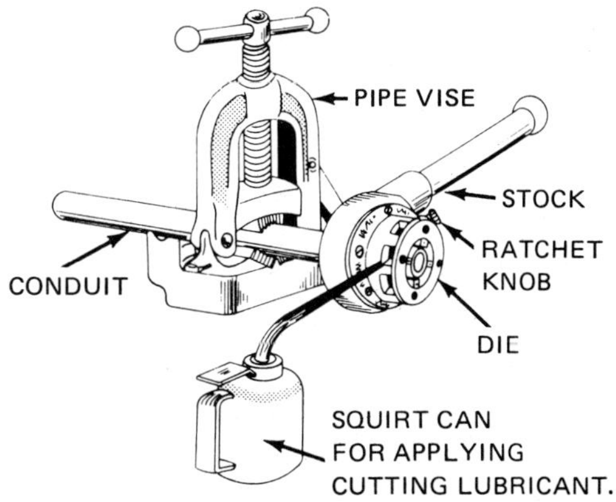

Fig. 14-5 Cutting thread on rigid conduit.

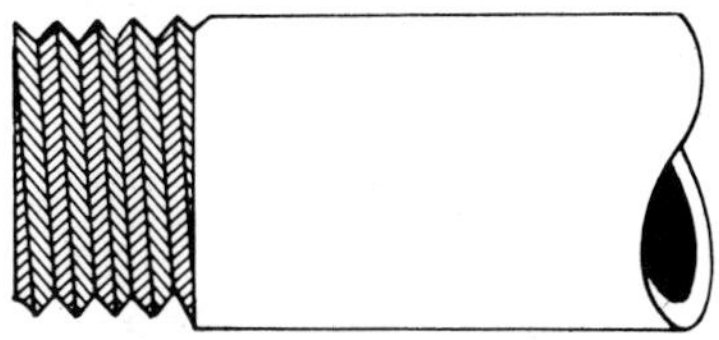

Fig. 14-6 Parallel thread.

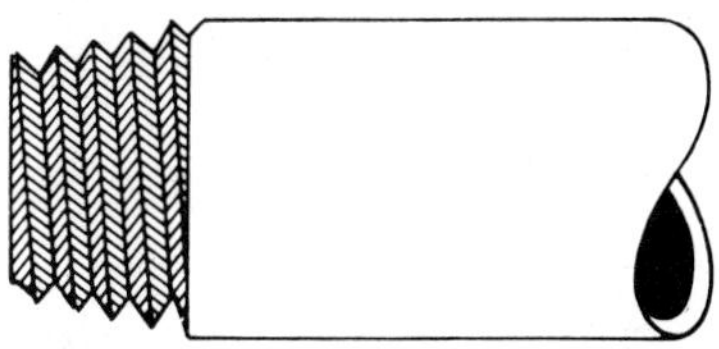

Fig. 14-7 Tapered thread.

It is best to ream the pipe before threading, because some conduit reamers expand the ends of newly threaded conduits. This expanded edge is flared like a bugle and creates a problem when trying to start a threaded conduit fitting onto the new thread.

The actual thread is cut into the prepared end of the conduit by a stock and die set, Figure 14-5. At one time, electrical stocks and dies cut parallel threads. Modern threading tools produce tapered threads providing a 3/4-inch (19 mm) taper per foot (305 mm), similar to those used with water-pressure systems (see Figures 14-6 and 14-7). This allows conduit fittings to be installed securely on the new thread. Couplings are threaded internally with parallel threads. Matching the tapered threads of the conduit to the straight threads of the coupling tightly joins two sections of conduit.

Cut sufficient thread into the conduit to engage completely the threads available on the conduit fitting. This provides the greatest mechanical security and ground continuity. A standard-length cut should be made, avoiding excessive threads. Threading removes the protective coating from the conduit, exposing the threads to rust.

Application of a cutting liquid-lubricant eases the physical effort required to cut a thread, and also extends the life of the die's cutting edges. Several excellent threading

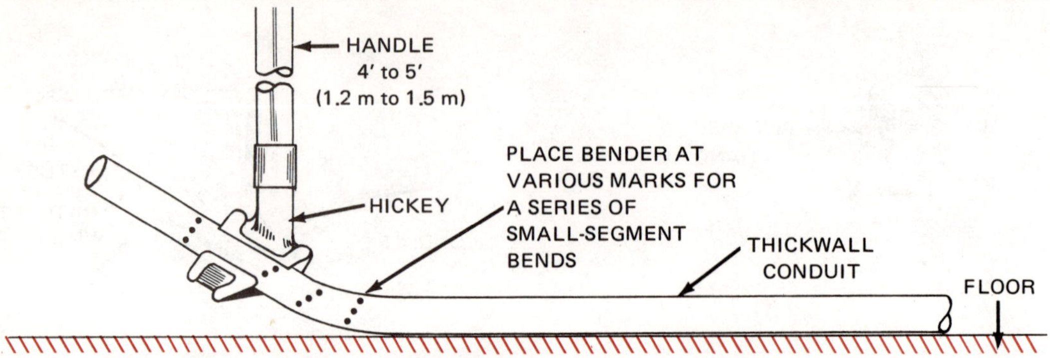

Fig. 14-8 Application of a manual conduit bender (hickey).

lubricants are made for this purpose, but oil or liquid soap can be used when the proper lubricant is not at hand. The simplest method of applying the cutting lubricant is with an oil squirt can.

Bending Rigid Conduit. This section outlines the basic methods of bending rigid conduits. Many hours of practice are needed to master the art of bending and forming conduit systems.

Conduits up to 1 inch (25 mm) in internal diameter are usually formed manually with the help of a bending tool called a *hickey,* Figure 14-8. Hickeys are available in standard conduit sizes. For best results, a hickey of the proper size for the conduit should be used. Usually, conduit of a larger trade size than the hickey does not fit into the bending tool. Conduit smaller than the tool is designed to handle often slips in the tool, producing kinks, inaccurate bends, and/or bruises on the operator.

Bends must be made without reducing the internal diameter of the conduit. A kink in the bend will make installation of conductors much more difficult and might even damage the insulation, Figure 14-9. Figure 14-10 shows two safe methods for using the manually operated hickey.

Conduit can be bent to go around corners, pass over obstacles, or enter a box at the proper angle. Figure 14-11 shows five typical bends and their uses.

The *offset* bends are used to enter a box or cabinet, and can also be used to bring the conduit from one level to another on any surface, such as a wall or ceiling.

The *square saddle* bend is used to bypass protrusions in a wall or to pass over several conduits located side by side. The *round saddle* bend is used, for example, to pass over a pipe. Saddle bends take the most time to form. Care and practice, however, result in smooth, well-aligned bends.

The *90°* bend is used for the desired stub-ups and to go around inside corners. Commercially produced bends are available from the manufacturer of the conduit. An *outside corner* bend requires a conduit fitting of one of the types shown in Figure 14-24.

The *45° offset* bends are often used to overcome obstructions in the path of the conduit.

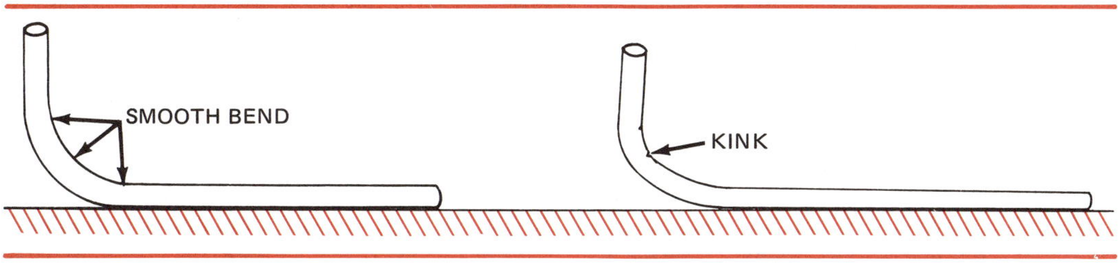

Fig. 14-9 Smooth conduit bends are important. Conduit bends with kinks are *not* suitable for conductor installation.

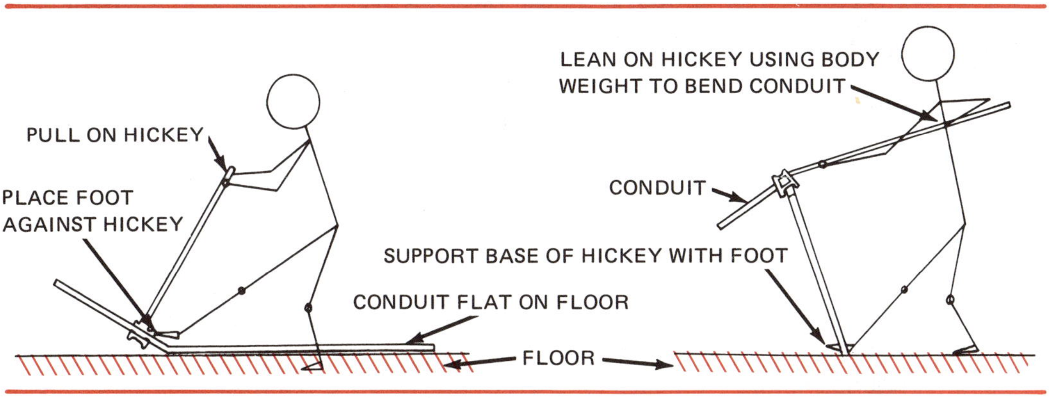

Fig. 14-10 Methods for bending conduit manually.

The operator who uses body weight rather than arm strength can greatly reduce the strain of bending conduit. A 1-inch (25 mm) diameter conduit can be very resistive at times – a challenge to the operator. Good technique and practice are very important. The emphasis is on matching the bender to the conduit. Electricians, however, often have to use a building support beam, a hole in a wall, or some other means when a bend must be made and no hickey is available. Take great care not to kink the conduit in such conditions.

When bending thickwall and Type IMC conduit, the operator should *not* try to make a full bend (90°) with a single grip of the bender. The head of the hickey should be moved away from or towards the operator approximately 1 inch (25 mm) per grip. This process makes the completed bend a series of small, continuous curves, Figure 14-12. It also results in a more symmetrical bend with less chance of kinking the conduit. The accuracy of the bend can be easily checked by placing the vertical part of the conduit against a wall, and also by measuring the desired length of the stub-up.

Conduits larger than 1 inch in diameter are usually formed with a hydraulic conduit bender. These units are available with hand- or motor-driven pump attachments, Figures 14-13 and 14-14. Take care to set the conduit in the bender carefully, because enormous pressure is exerted by the hydrau-

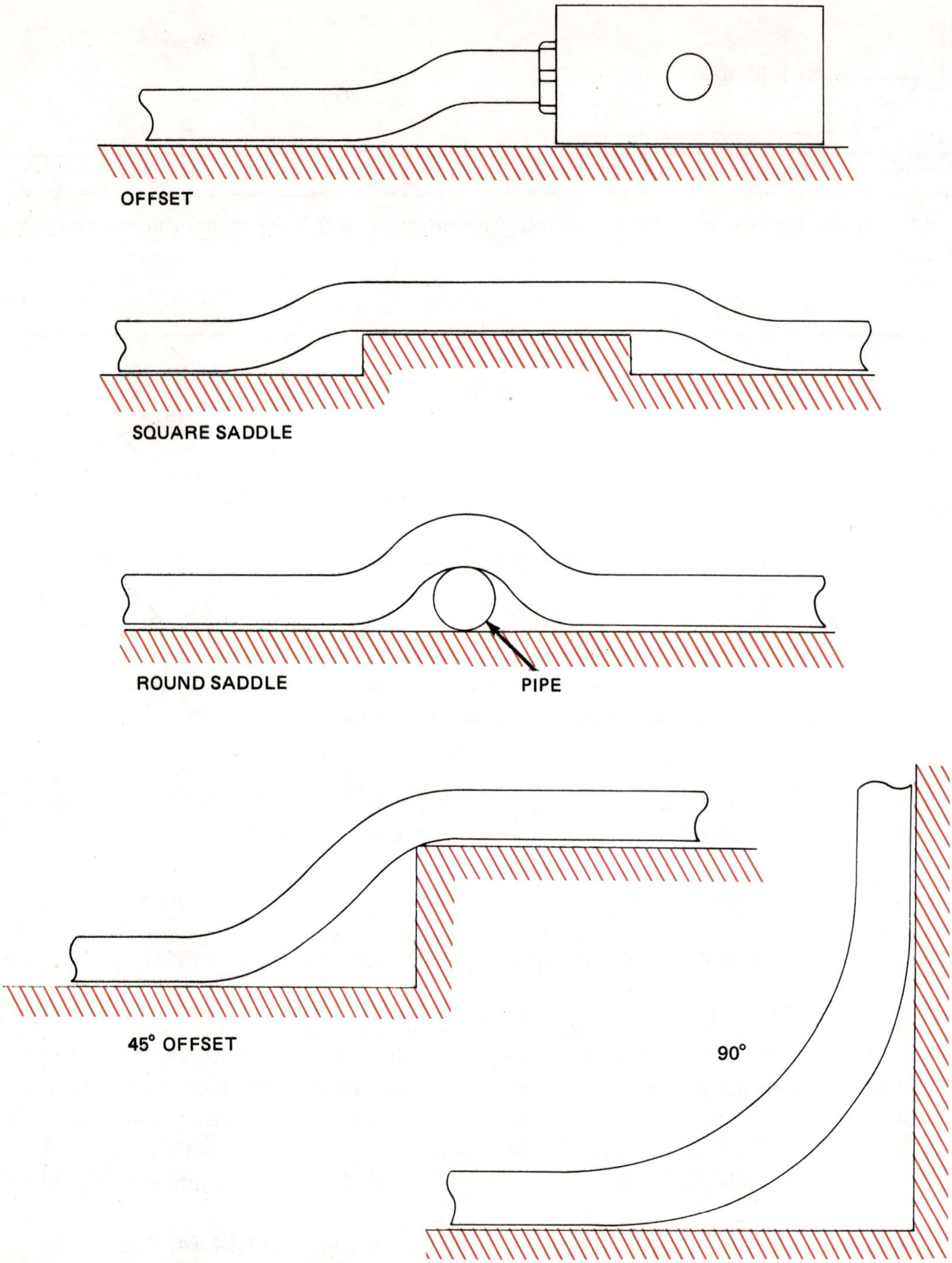

Fig. 14-11 Examples of conduit bends.

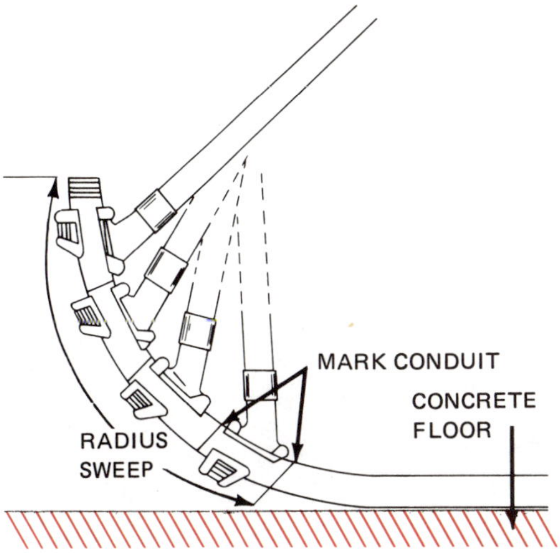

Fig. 14-12 Making a 90° bend. (*Note:* Many small curves required for smooth bend.)

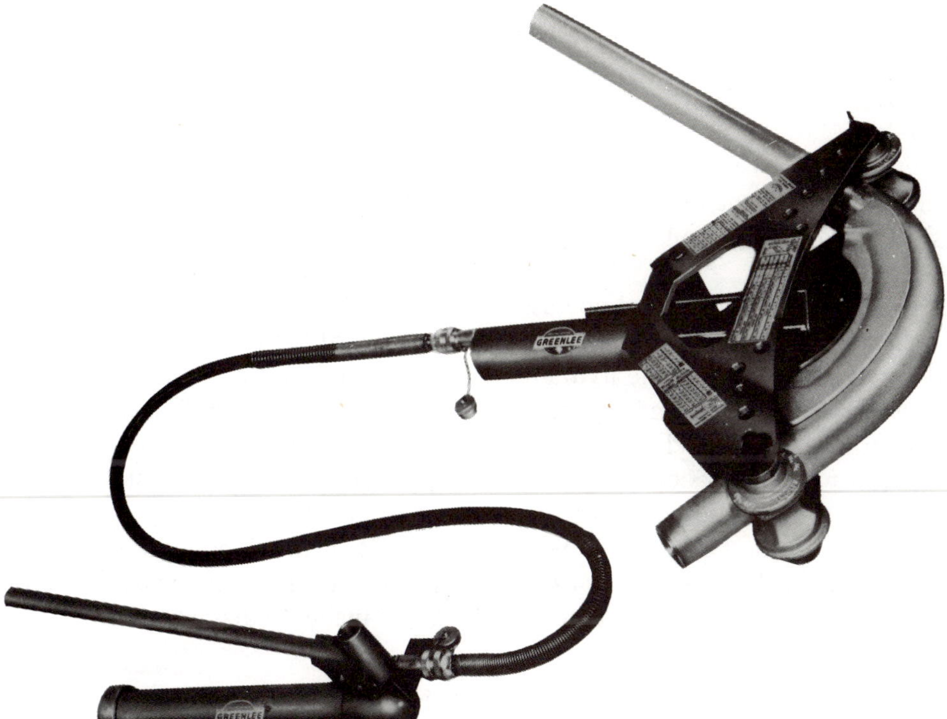

Fig. 14-13 A hand-operated hydraulic bender simplifies forming large-diameter conduits.

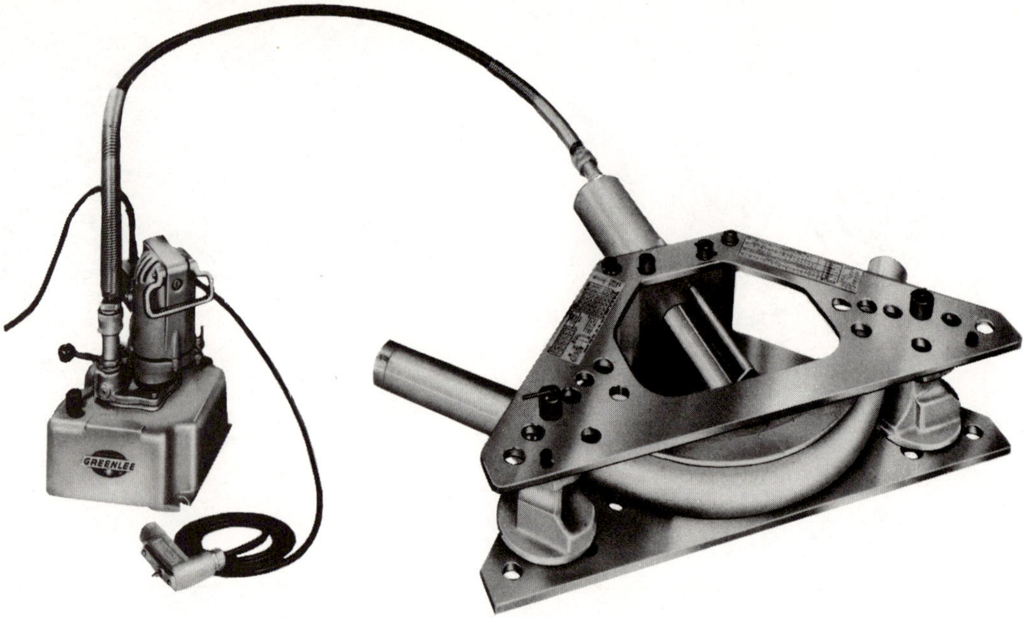

Fig. 14-14 A motor-driven hydraulic bender speeds forming large-diameter conduits.

lic system. The use of the hydraulic bender requires detailed instructions and much practice.

As in the case of the hand-operated hickey, bends should be made up of a series of small curves. When using the hydraulic bender, release the pressure and reposition the conduit for each curve. Place a series of equally spaced pencil marks on the conduit to assist in locating the pressure points needed to produce a smooth, symmetrical bend. Since it is very difficult to remove an excess of bend from large conduits, take care not to overbend. If the conduit is underbent, it is easier to place the conduit back in the bender and add a few degrees of bend than it is to take out the bend. If, however, a small amount of bend must be taken out, drop the conduit onto the base of the bend from about 2 feet (600 mm) off the floor. The weight of the conduit and the force of hitting the floor will straighten the bend slightly, Figure 14-15. Take care not to flatten the bottom of the bend or damage the floor.

Conduit systems are often installed in large commercial buildings before the concrete is poured. Figure 14-16 shows a typical rigid-steel conduit system for a high-rise office building.

Mechanical conduit benders are used to help form medium-size conduits, up to 1 1/2 inches in diameter, Figure 14-17.

INTERMEDIATE METAL CONDUIT (IMC)

Intermediate metal conduit (IMC) is a metal raceway of circular cross section consisting of couplings, connectors, and fittings identified for the installation of electrical conductors.

Construction. The construction of IMC is very similar to that of rigid (thickwall) conduit and EMT (thinwall) conduit. The wall thickness of IMC is in-between that of rigid conduit and EMT, as shown by comparison of conduits in the 1/2-inch trade size:

Type	Wall Thickness
Rigid Thickwall	0.109 inch
IMC	0.070 inch
EMT	0.042 inch

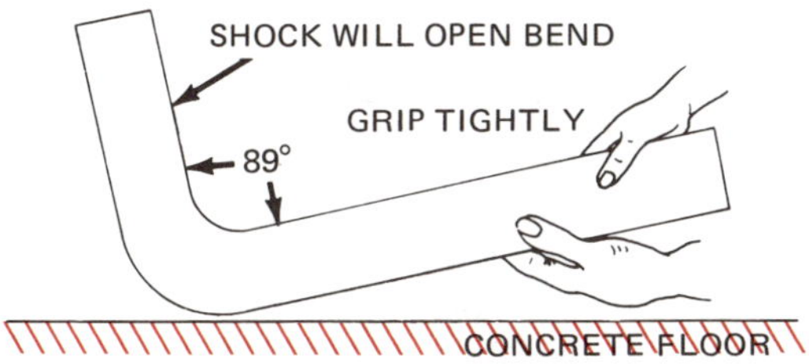

Fig. 14-15 Opening a bend.

Fig. 14-16 Installing conduit before pouring concrete.

Uses. IMC can be used wherever rigid thickwall conduit is permitted. As with rigid conduit, it can be threaded with standard pipe dies, and it can be used with all conduit fittings used with rigid thickwall. Various local codes may differ on the use of IMC, particularly for corrosive locations and direct burial. It is recommended that the local inspection authority be consulted for a decision before installing IMC.

Supports. IMC conduit must be firmly fastened within 3 feet (914 mm) of each outlet box, junction box, cabinet, or fitting. It must be supported at least every 10 feet (3.05 m).

Size. IMC is available in trade sizes 1/2 inch to 4 inches.

Fig. 14-17 Typical mechanical conduit bender.

EMT (THINWALL) CONDUIT

The proper name for thinwall conduit is electrical metallic tubing (EMT). Although EMT does not provide the same degree of mechanical protection as rigid conduit or IMC, it does have several important advantages.

EMT is made of lightweight steel tubing, so it does not require as much physical strength on the part of the installer as does rigid conduit when assembling the conduit system, Figure 14-18. Because of its lightweight thinwall construction, it is not practical to thread EMT. This feature alone saves a great deal of time and work when installing EMT. Because EMT is not threaded into a fitting or box, the job does not depend upon the electrician's ability to rotate a length of conduit, which may be full of bends and in an awkward location, until the threaded end is secure in the box or fitting. Mechanical connectors and couplings eliminate this chore.

Article 348 of the National Electrical Code lists the conditions under which EMT can be substituted for rigid conduit.

Fittings for EMT. Mechanical fittings are used to couple lengths of EMT and/or to connect the conduit to boxes and cabinets. These fittings are made of steel or zinc alloy (die cast). Take care not to subject the die-cast zinc fittings to great pressure or stress, as they break easily. Zinc alloy is used because it is easily die cast and is moderately priced, but a person stepping on a run of conduit coupled with die-cast fittings is likely to crack a coupling. For this reason, die-cast zinc fittings should not be used in poured concrete buildings where they can be damaged (and go unnoticed) before the pouring of concrete. Raintight and concrete-tight compression-type fittings must be used.

Two main types of EMT fittings are: the set-screw type and the compression type, Figures 14-19 and 14-20. The set-screw type is used in dry locations; only a screwdriver is needed to secure the connector to the conduit. Glandular (compression) connectors are fastened to the conduit by using adjustable pliers or a wrench. These raintight connectors are used outdoors, in poured concrete installations, or in other damp areas.

PVC-jacketed EMT. Some EMT is jacketed with polyvinyl chloride (PVC), which resists corrosive chemicals and vapors, Figure 14-21. Take care not to cut or damage this outer jacket when forming bends. Liquid PVC is applied to the connectors and couplings of such installations to give complete protection in corrosive areas.

Fig. 14-18 Steel thinwall conduit (EMT) with electroplated finish.

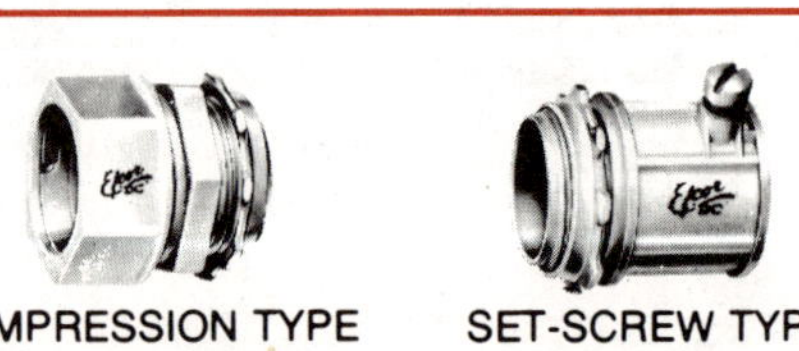

Fig. 14-19 EMT connectors.

Bending EMT. A special conduit bender (hickey) is used to form offsets and bends in EMT, Figure 14-22. The hickey forms the conduit with a radius that will not tend to kink the conduit where it has been bent. The hickey also supports the sides of the conduit as it bends, further reducing the possibility of kinking. The operator places a foot on the hickey to make sure that the conduit remains within the hickey during bending. This reduces the tendency of the EMT to kink.

Because the bender fits the conduit very closely, a separate hickey is needed for each size of conduit from 1/2 inch to 1 1/4 inch. EMT 1 1/4 inch in diameter and larger is usually formed in a hydraulic bending unit, Figure 14-23. The hickey completely encloses the EMT at the point of pressure, virtually eliminating kinks in the conduit.

Both manual and hydraulic benders can form a 90° bend in one continuous movement of the bender. Unlike rigid conduit, EMT does not have to be bent in a series of small curves. These benders can be very accurate: a skilled operator, following the measurements provided on the unit, can produce a 90° bend that is within 0.08 inch (2 mm) of the required dimension. Such accuracy is necessary because EMT is very difficult to straighten and rebend without kinking. Even though EMT does not require as much physical strength to bend as does rigid conduit, considerable practice is needed before professional results will be obtained.

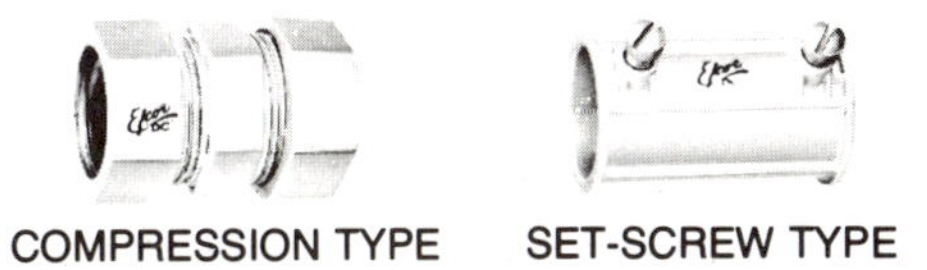

Fig. 14-20 EMT couplings.

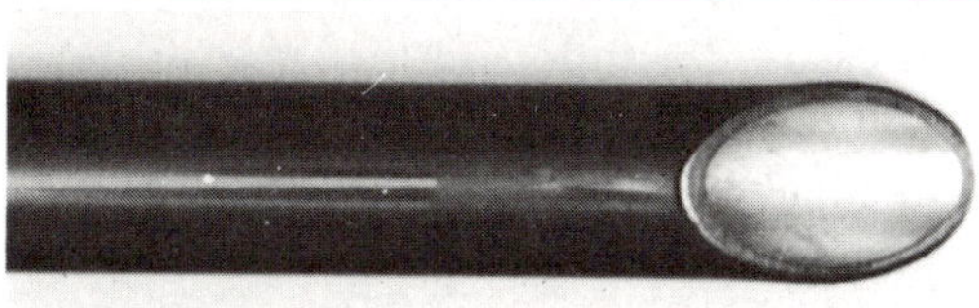

Fig. 14-21 PVC-jacketed thinwall conduit (EMT).

CONDUIT FITTINGS

During a conduit installation, it is often necessary to make a sharper bend than the

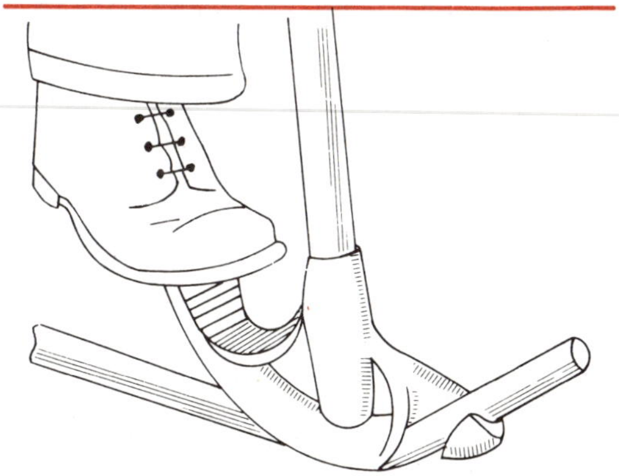

Fig. 14-22 Typical EMT bender (hickey).

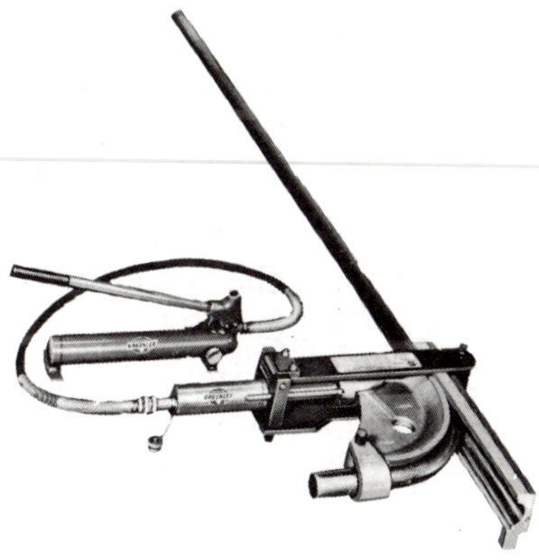

Fig. 14-23 Typical hydraulic EMT bender.

conduit will allow. A conduit fitting is then used. These fittings, which are often called *condulets,* allow the conduit installer to go around corners, and provide access to the conduit for the installation and removal of conductors, Figure 14-24.

Threadless, set-screw fittings are available for direct application to EMT. The standard fitting with its internal conduit thread is suitable for use with rigid conduit or connector-equipped EMT.

Some fittings are listed with the prefix L in combination with another letter; for example LB, LL, or LR. The first letter of the prefix indicates the shape of the fitting. For example, the L of LB indicates that the fitting is shaped like the capital letter L.

The second letter of the prefix indicates the location of the conductor access opening in relation to the shorter conduit opening; that is, the direction in which the short arm of the fitting is pointing when the fitting is held with the conductor access opening facing the viewer, and the longer arm pointing upwards. For example, when an LR fitting is held with the conductor access opening facing the viewer, and the longer arm is pointing upwards, the short arm of the fitting will be pointing to the right. An LL fitting held the same way will have its short arm pointing to the left. An LB fitting will have its short arm pointing backwards.

Some fittings are designed to provide a branch path from the main conduit run to some device, Figure 14-25.

The National Electrical Code requires that access to the conduit be provided every 100 feet (30 m) of conduit, or every 360° of accumulated bends, whichever occurs first. Special fittings are made for this purpose, Figure 14-26.

Condulets can be capped with covers made from aluminum, steel, porcelain, or a fiber composition. When excessive moisture is present, a cork or rubber gasket must be used between the cover and the fitting, Figure 14-27.

Electrical boxes for conduit systems are manufactured in many forms. Manufacturers produce such a variety that complete catalogs are necessary to list them. Figure 14-28 shows a number of typical electrical boxes for conduit fittings. Examples illustrated include FSC with switch, FSX with switch, and FS with explosionproof outlet and cord connectors.

TERMINATING CONDUIT

When rigid conduit is brought into boxes through knockout holes, locknuts and bushings are used to secure it to the box, Figures 14-29 and 14-30. These fasteners must be installed tightly enough to provide both mechanical security (strength) for the installation, and grounding circuit continuity.

While it is always wise to use a tool for its intended purpose, common practice among electricians is to tighten the locknut and bushing by placing the blade of a screwdriver against the unit, and then striking the handle of the screwdriver with the flat side of a pair of pliers. (See Step 4, Figure 11-4.) A hammer and cold chisel are often needed with very large conduits.

Locknuts are usually made of steel and have sharp teeth on one side to bite into the box and establish the grounding circuit. Some conduit installations call for a double locknut; that is, a locknut on each side of the box.

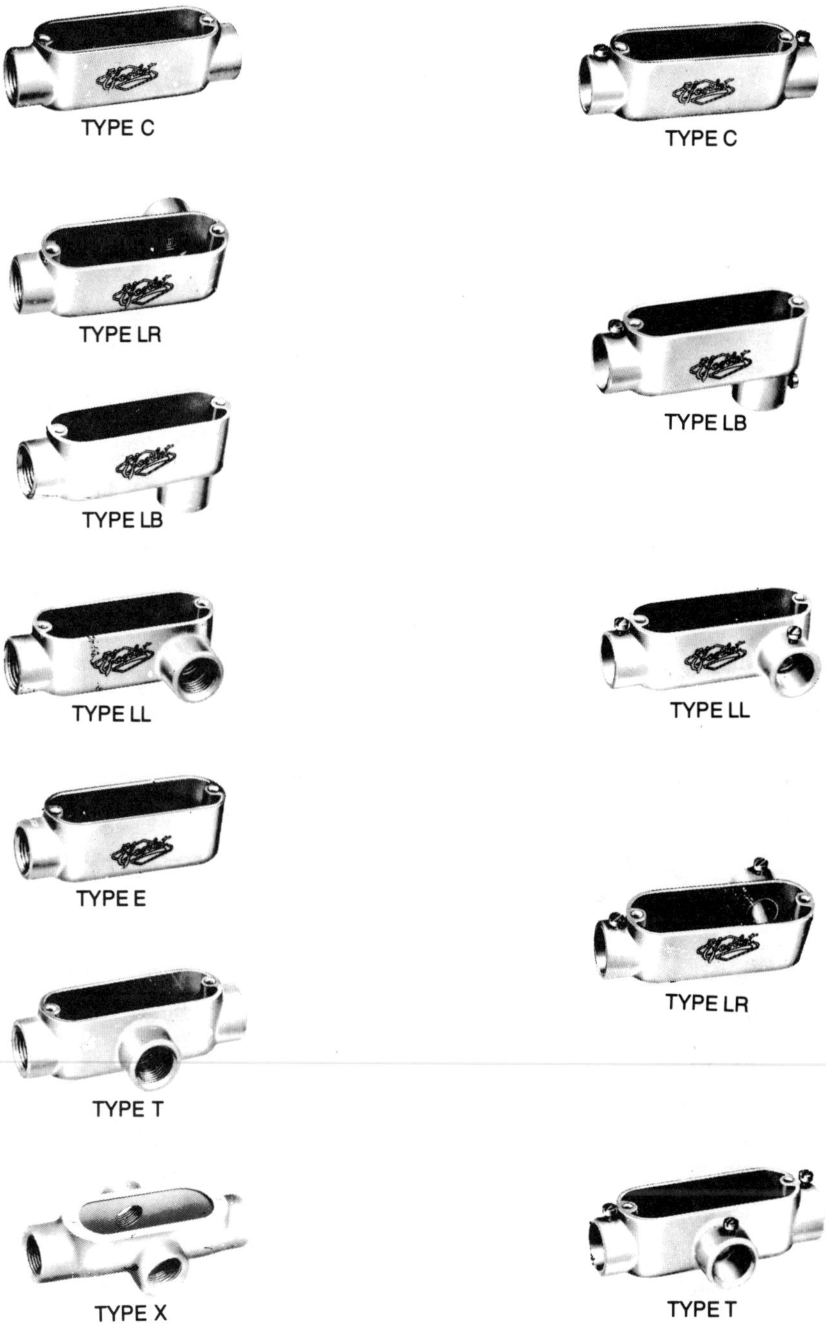

Fig. 14-24 Types of conduit fittings (condulets).

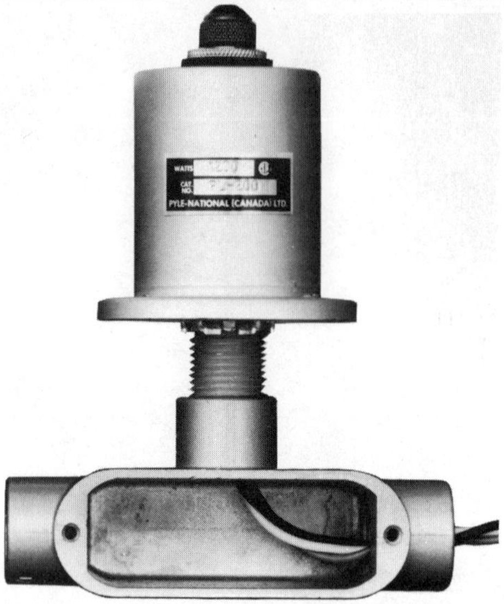

Fig. 14-25 Type "T" conduit fitting provides branch path from main run.

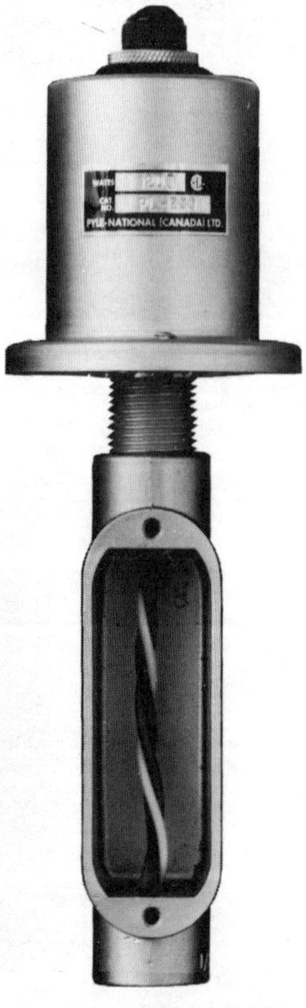

Fig. 14-26 Type "C" conduit fitting provides access to conduit for installing or removing conductors.

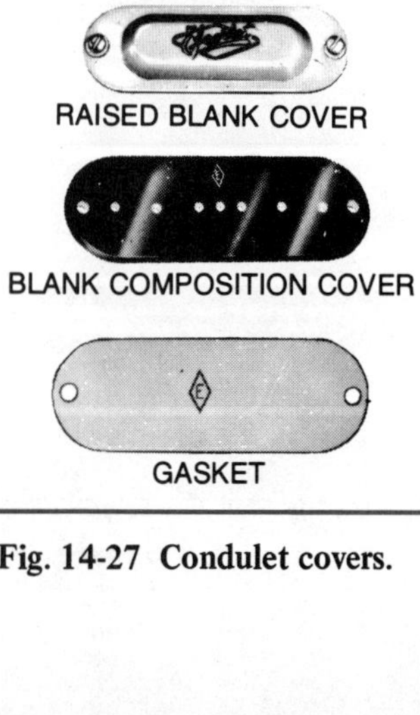

Fig. 14-27 Condulet covers.

EXPLOSION-PROOF MOTOR STARTING UNITS

FS FITTING WITH EXPLOSION-PROOF OUTLET AND CORD CONNECTOR

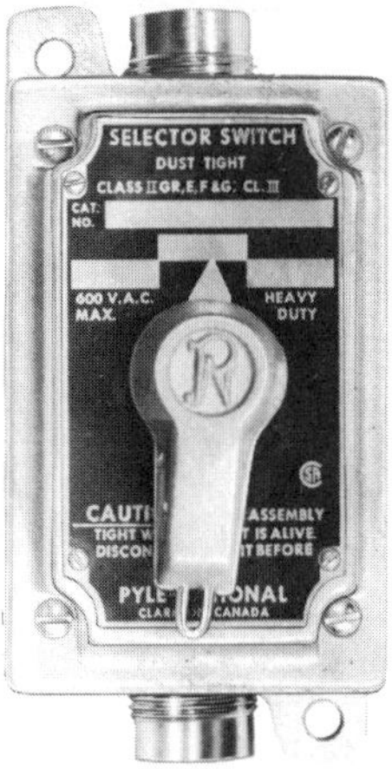

FSC FITTING WITH SWITCH (2 CONDUIT OPENINGS)

FS FITTING (SINGLE CONDUIT OPENING)

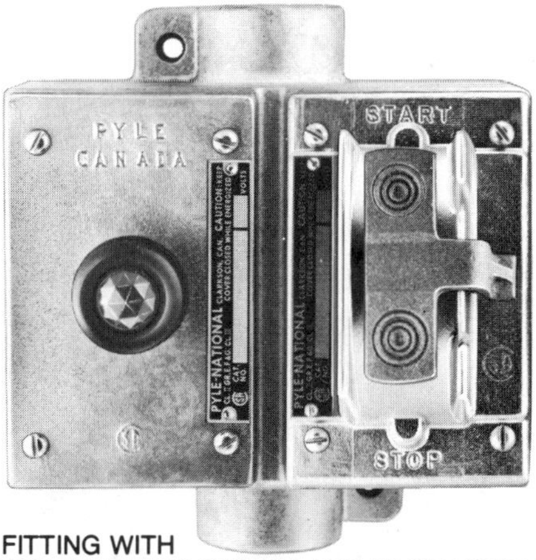

FS FITTING WITH MOTOR STARTING SWITCH AND PILOT LIGHT (2 GANG)

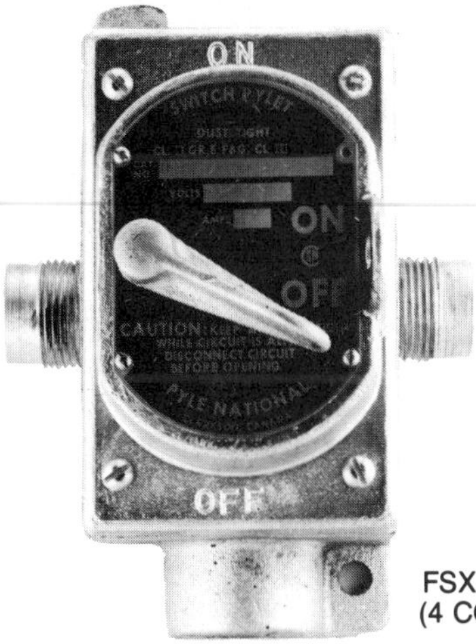

FSX FITTING WITH SWITCH (4 CONDUIT OPENINGS)

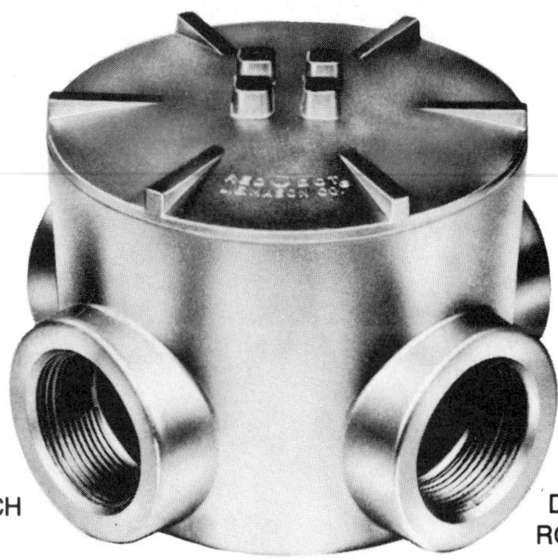

100 mm DIAMETER ROUND BOX

Fig. 14-28 Typical electrical boxes for conduit fittings.

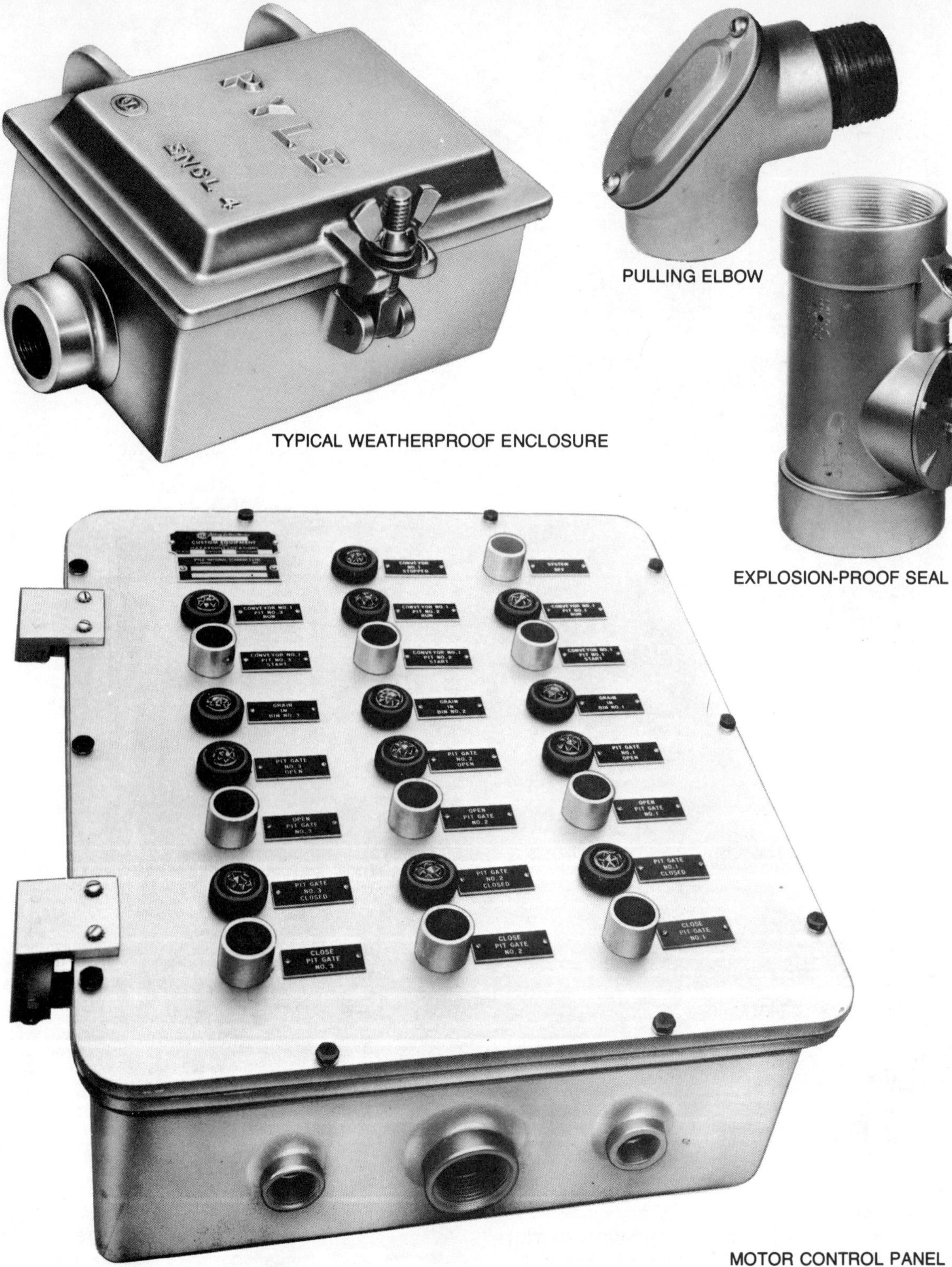

TYPICAL WEATHERPROOF ENCLOSURE

PULLING ELBOW

EXPLOSION-PROOF SEAL

MOTOR CONTROL PANEL

Fig. 14-28 Continued

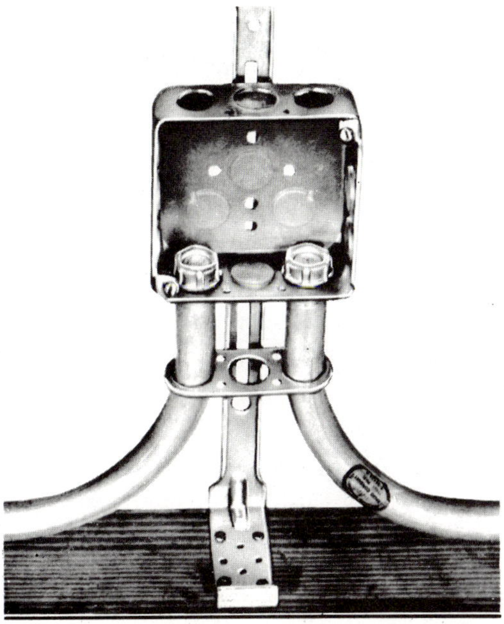

Fig. 14-29 Conduit connected to outlet box.

Fig. 14-30 Locknut and bushing terminating devices.

Bushings, which provide a smooth opening through which the conductors can enter the box, are made of die-cast aluminum, steel, or nylon. Steel bushings with a nylon insert are also available.

Figure 14-31 shows how conduits are secured to a box with a single locknut and bushing. When securing conduits for circuits of over 250 V to ground, or where No. 4 AWG or larger conductors enter a raceway, a system of two locknuts and a plastic bushing, or an insulated grounding bushing, is used. This is done because insulation on large and heavy cables can be punctured or pierced by the weight of the cable pressing on a solid metal bushing. A metal bushing with a plastic insert is also available. This double-locknut method is also used on conduit systems that have an applied voltage in excess of 250 V, Figure 14-32.

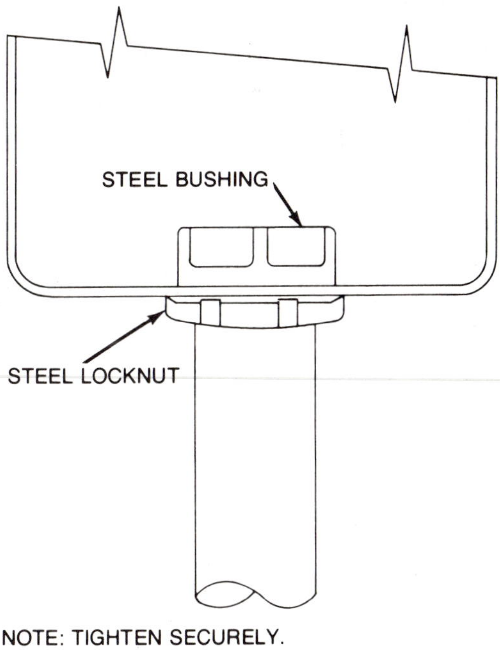

Fig. 14-31 Single locknut and bushing installation.

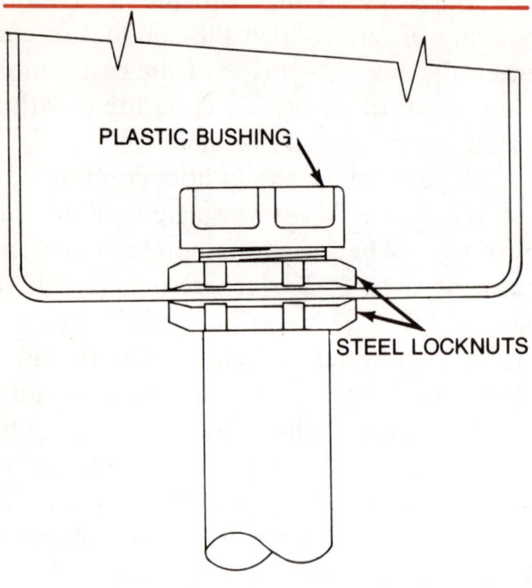

Fig. 14-32 Double locknut and bushing installation.

RIGID ALUMINUM CONDUIT

Lightweight, easily handled rigid aluminum conduit is often used to enclose residential service-entrance conductors, Figure 14-33. Aluminum is rustproof and will not stain or streak the surface on which it is mounted. The high-electrical conductivity of aluminum provides a safe grounding circuit for the system. As a nonsparking metal, aluminum is safe for use near explosive gases or vapors. It is also ideal for alternating-current systems that require a nonmagnetic substance to enclose the conductors.

Bending Rigid Aluminum Conduit. Bending aluminum conduit does not require much physical effort. A hickey, or some other forming device that will not kink the bend, must be used. Standard rigid conduit hickeys can be used if extreme care is taken not to flatten the bend with too small a radius.

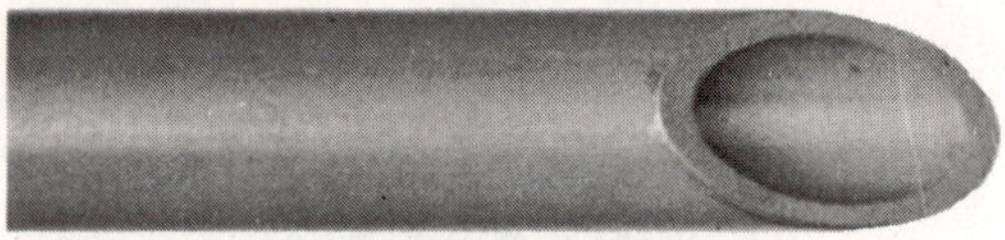

Fig. 14-33 Rigid aluminum conduit.

EMT benders one size larger than the conduit are often used successfully, and hydraulic benders usually produce satisfactory bends.

Threading Rigid Aluminum Conduit. Standard dies used for threading aluminum conduit should be sharp and well-lubricated while cutting the thread. Chips of metal should be cleared from the die occasionally; otherwise, they will lodge between the die and the conduit and tear the newly cut threads. When reaming rigid aluminum conduit, take care not to flare and expand the end because flaring makes it difficult to attach conduit fittings.

Installing Rigid Aluminum Conduit. It is important to avoid cross-threading aluminum conduit when attaching couplings, locknuts, bushings, condulets, or similar devices. The soft aluminum will lose its threads very quickly if an attempt is made to remove a cross-threaded fitting. Putting lubricating oil on the threads before assembly helps to eliminate this problem.

Rigid aluminum conduit should be supported by straps placed at regular intervals, as is done with rigid steel conduit. The light weight of the aluminum is very noticeable when this conduit is being installed from a ladder or some other unsteady position.

If aluminum conduit is to be embedded in concrete belowground, and there is a chance that wetting will occur every so often, a bituminous-base paint or pitch should be applied to the conduit. This will protect it from any moisture in the concrete that may attack and corrode the metal.

RIGID PVC CONDUIT

Rigid PVC (polyvinyl chloride) conduit protects conductors in the most severely corrosive locations, Figure 14-34. Moisture and/or condensation has no effect on this durable product. Rigid PVC conduit does not conduct electricity and, because it is nonmagnetic, it is not affected by sheath currents experienced in some other wiring systems. A grounding conductor must be installed where required by NEC *Article 250.* Rigid PVC conduit has no voltage limitations, and it resists aging, exposure to ozone, sunlight, and underground environments. It is immune to electrolytic action, and heavy blows to it do not cause permanent damage.

Light Weight Advantage of Rigid PVC Conduit. Rigid PVC conduit is approximately five times lighter than steel conduit, and twice as light as aluminum conduit of the same size. As a result, the installer finds it much easier to handle, especially when working from ladders or scaffolds.

Applications of Rigid PVC Conduit. Rigid PVC conduit is used in areas where corrosion is a problem; for example, in paper mills, meat-packing plants, chemical and electroplating plants, barns and animal shelters, hospitals, and food preparation plants. It is also being used more and more for residential service-entrance conduit, because it is easily handled, and is nonrusting and nonstaining. It is advisable to check UL listings for the intended application of rigid nonmetallic conduit.

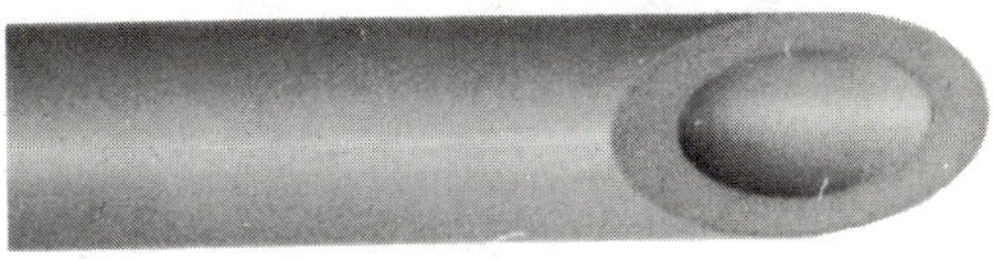

Fig. 14-34 Rigid PVC conduit.

Joining and Terminating Rigid PVC Conduit. A metal-cutting hacksaw or carpenter's wood saw can be used to cut the conduit to length. A pocketknife will quickly remove any burr on the inside of the cut end. Lengths of PVC are assembled to their fittings by using a form of solvent welding. No threads are required. Dirt and grease from handling are removed with a PVC cleaner. As shown in Figure 14-35, solvent cement is brushed onto the prepared end of the conduit and the inside of the fitting to be joined. The conduit is then pushed into the fitting, given a quarter turn, and set aside. After a few minutes, the connection will withstand the strain of use.

Bending Rigid PVC Conduit. Rigid PVC conduit is made of a thermoplastic material that softens at temperatures between 115°C (239°F) and 130°C (266°F). If proper care is taken, the conduit will not flatten and/or kink while being bent. All bends should have a radius of at least 10 times the diameter of the conduit. Commercially produced bends are available from the manufacturer of the conduit.

An approved type of PVC heater is used to direct a heated airstream at the conduit. This airstream is capable of raising the temperature of the conduit to the desired level

Fig. 14-35 Apply cement to pipe and fitting. Insert pipe into fitting and give it a quarter turn.

for bending. Open flame will damage the conduit and should not be used. Cold-water quenching or natural cooling will maintain the shape of the bend. The conduit should be overbent slightly to allow for spring-back as the conduit is cooling. (See Figures 14-36, 14-37, and 14-38.)

Rigid PVC Conduit Expansion. Whenever the temperature variation exceeds a range of 15°C (59°F), PVC conduit will expand and contract enough to warrant the use of expansion joints. An *expansion joint* is simply one PVC tube telescoped within another. A 100-foot (30-m) length of con-

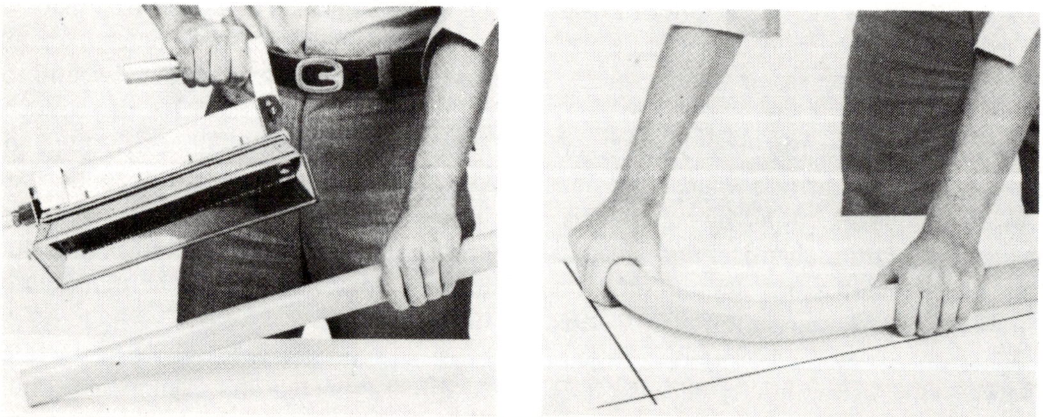

Fig. 14-36 Use a PVC heater to soften pipe for bending. Use guidelines to establish proper bend angle.

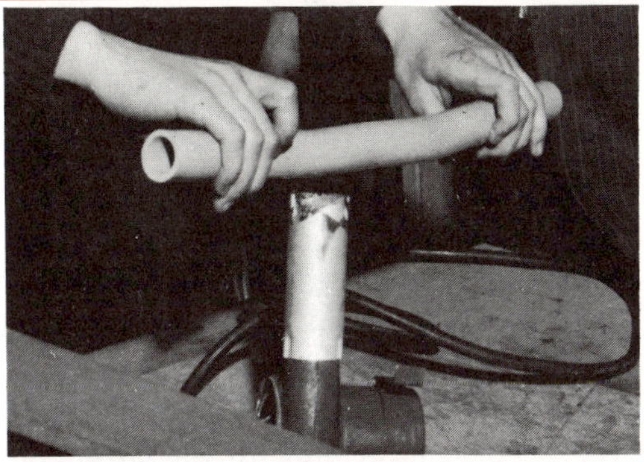

Fig. 14-37 Using a PVC heater before bending rigid PVC conduit.

duit will expand approximately 3 1/2 inches (90 mm). Expansion joints are available from the manufacturers of PVC conduit.

Rigid PVC Conductor Installation. Less physical effort is needed to pull conductors into rigid PVC conduit than is needed with other types. This is due primarily to the smooth, friction-free interior of the conduit.

Rigid PVC Conduit Supports. Supports for rigid PVC conduit need not be as rugged as those for other forms of conduit. They should be spaced at regular intervals and the possibility of snow, ice, and wind loads should be taken into account. According to the NEC, the required support is as follows:

Conduit Size (Inches)	Maximum Spacing Between Supports (Feet)
1/2 – 1	3 (914 mm)
1 1/4 – 2	5 (1.52 m)
2 1/2 – 3	6 (1.83 m)
3 1/2 – 5	7 (2.13 m)
6	8 (2.44 m)

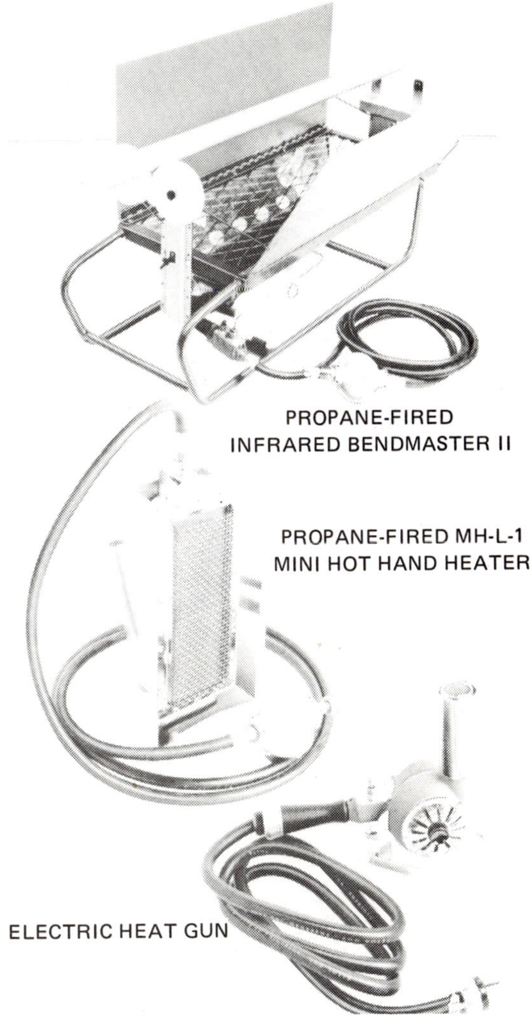

Fig. 14-38 Heat sources for bending PVC conduit.

FLEXIBLE METAL CONDUIT

Flexible metal conduit (flex) combines mechanical protection with maximum flexibility for nonhazardous locations. This versatile raceway is made of an interlocking steel or aluminum strip, Figure 14-39. Flex is available in the following lengths and internal diameters:

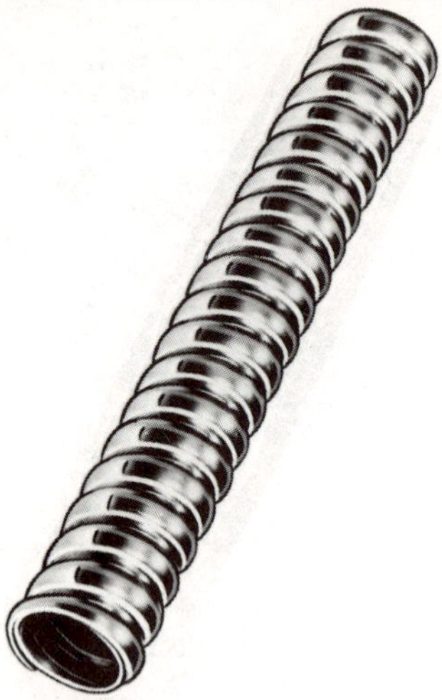

Fig. 14-39 Flexible metal conduit.

Internal Diameter (Inches)		
1/2	3/4	1
1 1/4	1 1/2	2
2 1/2	3	4 (aluminum)

Standard Length of Coil (Feet)			
25 (7.6 m)	50 (15 m)	100 (31 m)	250 (76 m)

Flex Applications. Conduit protection is sometimes needed in a location where rigid conduit cannot be formed to the contours required because of close working conditions. An example is a wiring system that must pass over and around steel girders, existing machinery and equipment, or be fished through masonry walls and ceilings. Using flex greatly simplifies the installation work. Another example is motors and/or machines with vibrating moving parts. These parts must be provided with a power supply system that will not allow metal fatigue to set in and fracture the raceway, because ground continuity can also be lost if the metallic raceway is broken or separated.

Fig. 14-40 Flexible conduit strap (2 hole).

For practical purposes, the minimum size flexible metal conduit to use is 1/2-inch electrical trade size. Sometimes, there is a tendency to use 3/8-inch flexible conduit in the wrong places and for the wrong purpose. However, several NEC exceptions do permit the use of 3/8-inch electrical trade size. Some of these exceptions are:

- as permitted for underplaster extensions.
- as permitted for motors.
- permitted in lengths of not more than 6 feet (1.83 m) as part of an approved assembly, for tap connections to lighting fixtures, or for lighting fixtures.

Flex Supports. Approved flex supports are similar to those used with armored cable, Figure 14-40. They must be installed within 12 inches (305 mm) of each box or cabinet and at regular intervals of not more than 4 1/2 feet (1.4 m) throughout the run. Where flex is fished or in lengths of up to 3 feet (1 m) requiring flexibility, supports are not needed.

Flex Ground Continuity. Flexible metal conduit is permitted to be used as a grounding means when both the conduit and fittings are approved for grounding, when it does not exceed 6 feet (1.83 m) in length, and where the conductors are protected for 20 amperes or less. Where an equipment bonding jumper is required, it must be installed in accordance with NEC *Section 250-79.*

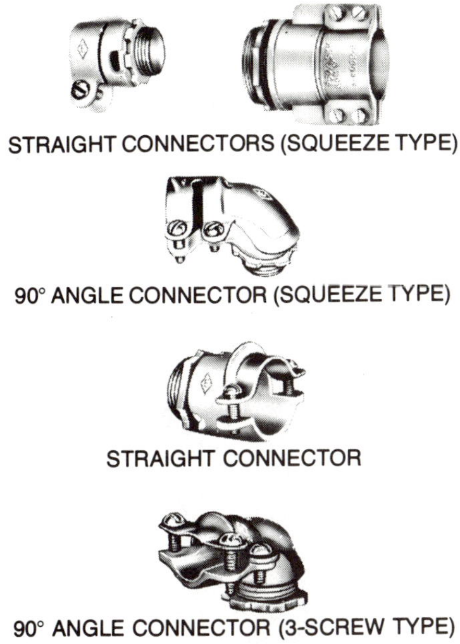

Fig. 14-41 Flexible conduit connectors.

Fig. 14-42 Liquidtight flexible conduit.

Cutting the Flex. Cut flex to length with a hacksaw in the same manner as for armored cable.

Terminating the Flex. Flexible conduit connectors are used to secure the flex to boxes, cabinets, condulets, and/or to couple with other forms of conduit, Figure 14-41. An insulating sleeve, similar to the antishort bushing used with armored cable, can be inserted between the conductors and the armor to prevent chafing of the conductors. Some modern flex connectors

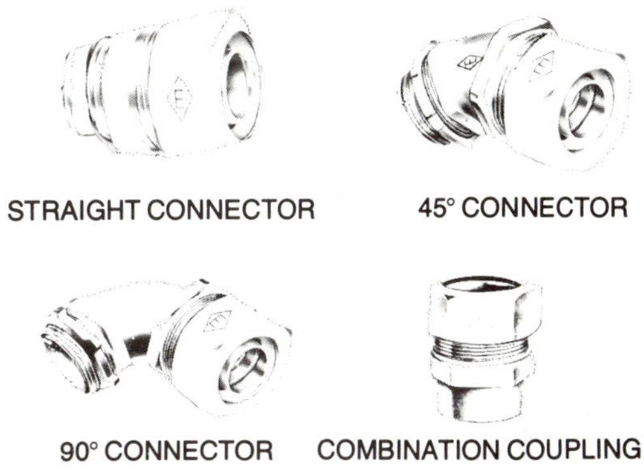

Fig. 14-43 Liquidtight flexible conduit connectors and coupling.

have a plastic insert built into the threaded end of the unit.

Flex connectors should be fastened securely to the box or cabinet with a conduit locknut to provide mechanical security and to assist in ground continuity.

LIQUIDTIGHT FLEXIBLE CONDUIT

The PVC-jacketed liquidtight flexible conduit is excellent for use in damp locations, corrosive areas, or around machines where coolants and/or cutting and lubricating liquids are likely to splash onto the flex, Figure 14-42. Moistureproof connectors are used to terminate it, Figure 14-43. These connectors make use of a nylon compression ring to grip the outer jacket, rather than the metallic clamping device used on regular flex connectors.

Take care not to puncture the PVC jacket either during installation or later when it is in use.

The National Electrical Code requires that liquidtight flexible conduit not be used where temperatures are in excess of 60°C (140°F). These temperatures could damage the PVC jacket. On the other hand, take care not to use it in temperatures that are low enough to cause injury to the jacket when it flexes.

Table 1 of NEC *Chapter 9* lists guidelines for the type and size of conductors allowed in liquidtight flexible conduit, in accordance with the percentage fill.

Grounding. Liquidtight flexible metal conduit is permitted as a grounding conduit in sizes 1 1/4 inches or less, where:

1. both conduit and fittings are identified for grounding, and
2. It does not exceed 6 feet (1.83 mm) in length.

In applications where an equipment bonding jumber is required around liquidtight flexible metal conduit, it must be installed according to NEC *Section 250-79*.

KNOCKOUT CUTTERS

It is often necessary to increase the size of a knockout already provided in a box or cabinet. Using a reamer or file is often clumsy, time consuming, and harmful to the worker's hands. A knockout cutter should be used instead.

Knockout cutters are made in all standard conduit sizes. For example, using a 3/4-inch (20 mm) knockout cutter produces an opening that accepts a 3/4-inch (20 mm) conduit or connector. The actual diameter of the hole will be approximately 1 1/16 inches (27 mm).

The hand-operated knockout cutter has a hardened-steel cutter which is drawn through the metal box by a wrench-tightened bolt, Figure 14-44. Hydraulic-powered cutters use the same type of hardened-steel cut-

Fig. 14-44 Hand-operated knockout cutters.

Fig. 14-45 Hydraulic-powered knockout cutter.

ter, but require much less physical effort on the part of the operator, Figure 14-45.

INSTALLING THE CONDUCTORS

The National Electrical Code requires that conductors be drawn into the conduit system *after* it has been completely assembled. This method prevents damage to the conductors while bending, forming, or fastening the conduit to boxes. It also ensures that the conductors may be replaced should the need arise.

A flattened, tempered-steel wire, which is produced in lengths of 25 feet (7.5 m), 50 feet (15 m), 100 feet (30 m), and 200 feet (60 m), is pushed into the conduit between boxes and fittings. This steel wire, called a *fish tape,* is available in 1/8-inch (3 mm), 3/16-inch (5 mm), and 1/4-inch (6 mm) widths to suit light- and heavy-conductor installations.

The tape is usually inserted into the conduit by means of a series of short, frequent pushes. It is important to keep the tape in motion, because the constant pulse and vibration is needed to ease the tape around bends and offsets in the conduit.

A loop (or end) that is almost closed helps guide the tape around the bends, couplings, and fittings, Figure 14-46. Often, a second fish tape is pushed into the conduit from the opposite end, hooked onto the first tape by rotating the first tape several times to engage the end loops, and then used to help draw the first tape through the conduit, Figure 14-47. This method is especially useful on conduit runs that have a number of 90° bends.

When pulling large conductors or many conductors into a large conduit, the fish

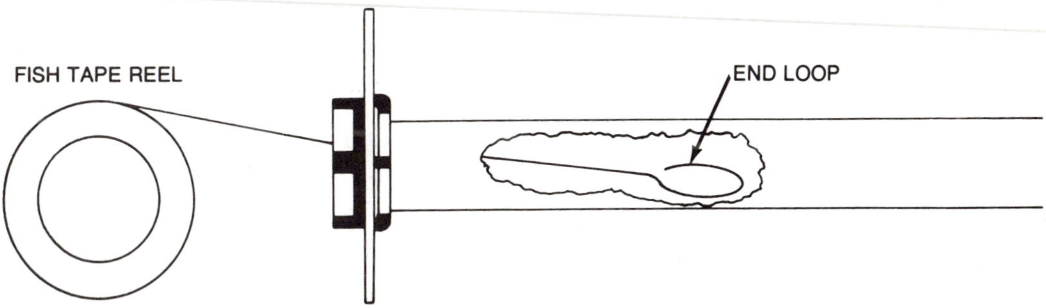

Fig. 14-46 Conduit fishing (one loop).

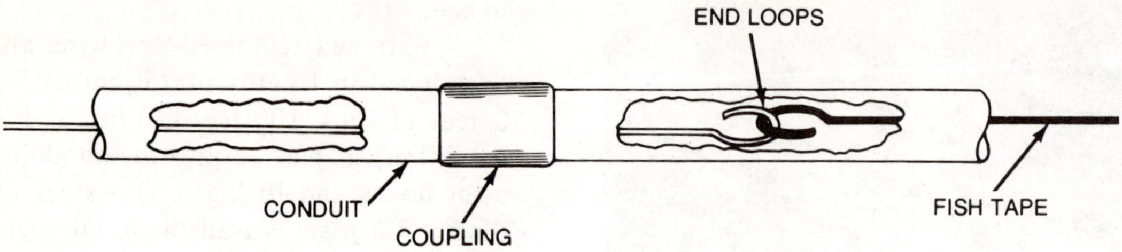

Fig. 14-47 Conduit fishing from both ends of run (two loops).

tape is often used to draw a rope or cable into the conduit. The rope is then used to pull the conductors through.

Wire-pulling lubricants are available. When a generous amount of lubricant is applied to the conductors, the physical strain of installation is greatly reduced. In fact, many other ways are used to reduce the physical strain. Every experienced electrician has devised some unusual apparatus for pulling conductors into a conduit system. A pulley-leverage system or an electric motor-driven winch-type puller may be used.

Securing Conductors to Fish Tape. When two or more conductors are to be pulled through a conduit at the same time, take care to fasten them securely to the fish tape. Much time and effort will be wasted if any of the conductors break away from the tape. Figure 14-48 shows two methods for securing the conductors to a tape.

COMPRESSED-AIR FISHING

Compressed air can be used to force a lightweight ball through a conduit, Figure 14-49. The ball can be almost as large as the inside diameter of the conduit. A length of strong string is attached to the ball. Once the string reaches the opposite end of the run, it is then used to draw in a heavier fish tape or rope. The compressed air can be obtained from a tank or compressor. A hose and nozzle is used to control the rate of airflow. This method is very simple and saves a great deal of time and effort.

CONDUIT FILL

As discussed in Chapter 7, conductors require air space for cooling. For the same reason, the *number* of conductors in any given size of conduit or tubing must be limited. The *Tables* in NEC *Chapter 9* specify the number of conductors allowed. This number depends upon the conductor internal size and type of insulation. (See NEC *Table 3A*, as shown in Table 14-1).

To determine the number of conductors allowed, Tables 14-1 and 14-2 are used together with Table 5-1, Chapter 5. For example, a No. 14 gauge conductor with THW insulation has a diameter of 0.131 inch. (See Table 5-1). To determine how many of these conductors are allowed in 1/2-inch tubing, refer to NEC *Chapter 9, Table 3A*. then, looking under Type THW, and moving to the right under 1/2 inch, it is found that 6 conductors can be pulled in.

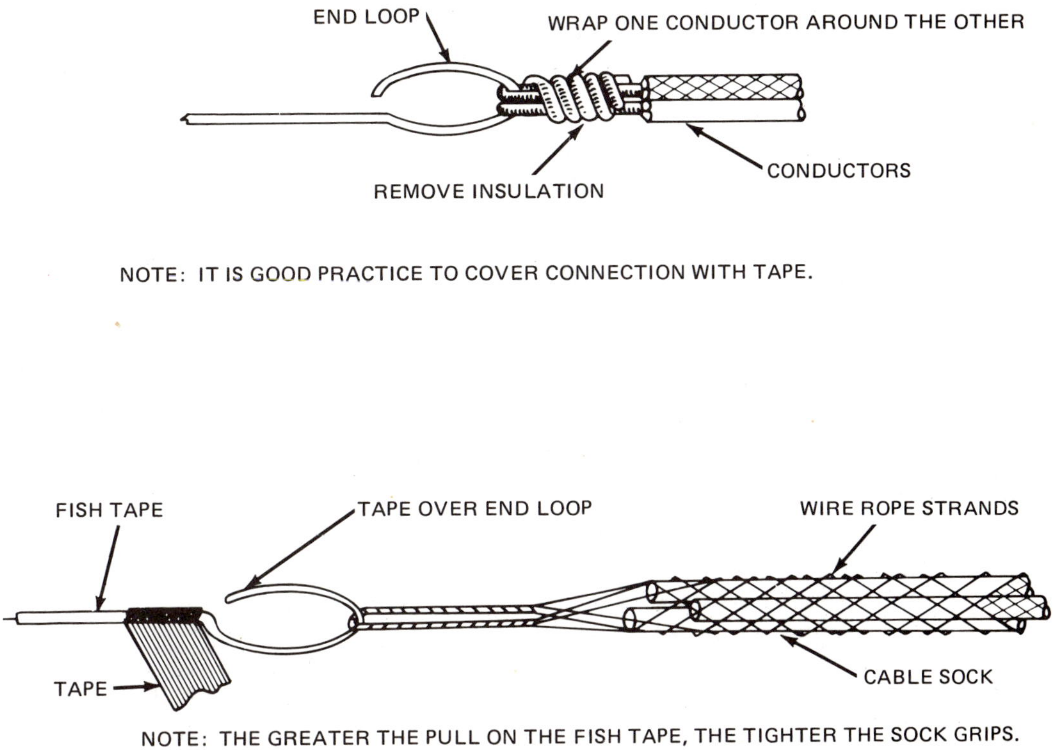

Fig. 14-48 Methods for securing conductors to fish tape.

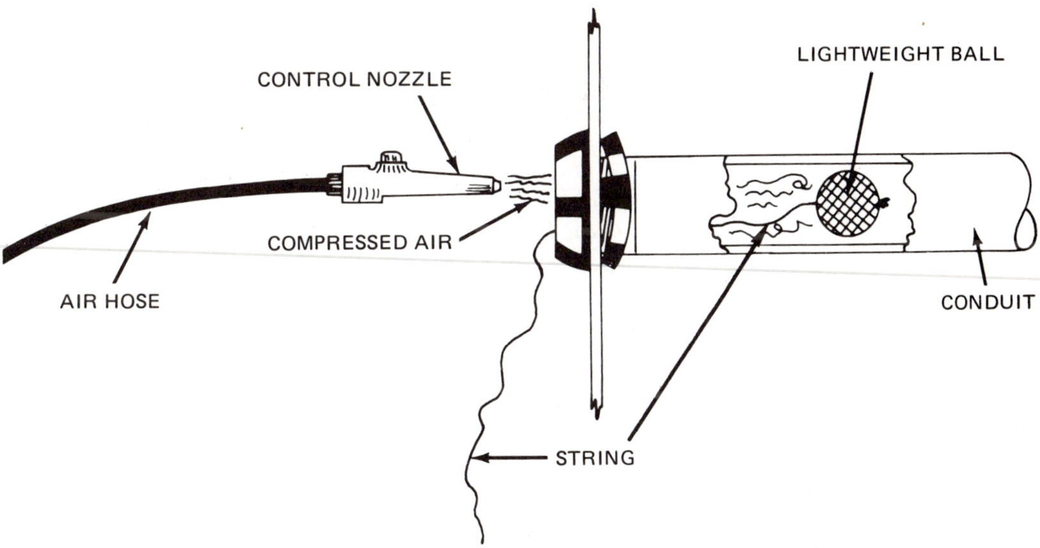

Fig. 14-49 Conduit fishing using compressed air.

TABLE 14-1
Maximum Number of Conductors in Trade Sizes of Conduit or Tubing

Conduit Trade Size (Inches)		½	¾	1	1¼	1½	2	2½	3	3½	4	4½	5	6
Type Letters	Conductor Size AWG, MCM													
TW, T, RUH,	14	9	15	25	44	60	99	142						
RUW,	12	7	12	19	35	47	78	111	171					
XHHW (14 thru 8)	10	5	9	15	26	36	60	85	131	176				
	8	2	4	7	12	17	28	40	62	84	108			
RHW and RHH	14	6	10	16	29	40	65	93	143	192				
(without outer	12	4	8	13	24	32	53	76	117	157				
covering),	10	4	6	11	19	26	43	61	95	127	163			
THW	8	1	3	5	10	13	22	32	49	66	85	106	133	
TW,	6	1	2	4	7	10	16	23	36	48	62	78	97	141
T,	4	1	1	3	5	7	12	17	27	36	47	58	73	106
THW,	3	1	1	2	4	6	10	15	23	31	40	50	63	91
RUH (6 thru 2),	2	1	1	2	4	5	9	13	20	27	34	43	54	78
RUW (6 thru 2),	1		1	1	3	4	6	9	14	19	25	31	39	57
FEPB (6 thru 2),	0		1	1	2	3	5	8	12	16	21	27	33	49
RHW and	00		1	1	1	3	5	7	10	14	18	23	29	41
RHH (with-	000		1	1	1	2	4	6	9	12	15	19	24	35
out outer	0000			1	1	1	3	5	7	10	13	16	20	29
covering)	250			1	1	1	2	4	6	8	10	13	16	23
	300			1	1	1	2	3	5	7	9	11	14	20
	350				1	1	1	3	4	6	8	10	12	18
	400				1	1	1	2	4	5	7	9	11	16
	500				1	1	1	1	3	4	6	7	9	14
	600					1	1	1	3	4	5	6	7	11
	700					1	1	1	2	3	4	5	7	10
	750					1	1	1	2	3	4	5	6	9

Note: Additional conductor insulation types may be found by referring to NEC *Table 3B* and *Table 3C* in *Chapter 9.*

TABLE 14-2
Table 4. Dimensions and Percent Area of Conduit and of Tubing

Areas of Conduit or Tubing for the Combinations of Wires Permitted in Table 1, Chapter 9.

		Area – Square Inches								
		Not Lead Covered				Lead Covered				
Trade Size	Internal Diameter Inches	Total 100%	2 Cond. 31%	Over 2 Cond. 40%	1 Cond. 53%	1 Cond. 55%	2 Cond. 30%	3 Cond. 40%	4 Cond. 38%	Over 4 Cond. 35%
1/2	.622	.30	.09	.12	.16	.17	.09	.12	.11	.11
3/4	.824	.53	.16	.21	.28	.29	.16	.21	.20	.19
1	1.049	.86	.27	.34	.46	.47	.26	.34	.33	.30
1 1/4	1.380	1.50	.47	.60	.80	.83	.45	.60	.57	.53
1 1/2	1.610	2.04	.63	.82	1.08	1.12	.61	.82	.78	.71
2	2.067	3.36	1.04	1.34	1.78	1.85	1.01	1.34	1.28	1.18
2 1/2	2.469	4.79	1.48	1.92	2.54	2.63	1.44	1.92	1.82	1.68
3	3.068	7.38	2.29	2.95	3.91	4.06	2.21	2.95	2.80	2.58
3 1/2	3.548	9.90	3.07	3.96	5.25	5.44	2.97	3.96	3.76	3.47
4	4.026	12.72	3.94	5.09	6.74	7.00	3.82	5.09	4.83	4.45
4 1/2	4.506	15.94	4.94	6.38	8.45	8.77	4.78	6.38	6.06	5.56
5	5.047	20.00	6.20	8.00	10.60	11.00	6.00	8.00	7.60	7.00
6	6.065	28.89	8.96	11.56	15.31	15.89	8.67	11.56	10.98	10.11

REVIEW

1. List three main advantages that a conduit wiring system has over other wiring systems.
2. List three areas where a conduit wiring system is used to protect the power supply.
3. In what length is conduit usually produced? In what diameters?
4. How is conduit measured to determine its trade size?
5. List five main types of conduit.
6. Why is the inside of rigid conduit generally galvanized and coated with insulating paint or varnish?
7. Explain in your own words the steps for cutting and threading rigid conduit.
8. What is the main disadvantage of using a pipe cutter to prepare an end of rigid conduit for threading?
9. What can be used as a substitute for thread-cutting lubricant?
10. What are the two types of conduit bender used for rigid conduit? Which is used for the smaller sizes of conduit? Which is used for larger sizes?
11. Why must care be taken when forming a bend in rigid conduit?
12. For what three purposes are bends in rigid conduit made?
13. Explain in your own words how to make a 90° bend in a length of 3/4-inch rigid conduit.
14. Why are EMT mechanical fittings made of zinc-alloy? What is the main disadvantage of using zinc-alloy?
15. What are the two main types of EMT fittings? Where is each used?
16. What protection does a PVC jacket give to EMT conduit?
17. What is the main difference between a rigid conduit bender and one used for thinwall conduit?
18. List three uses for condulets.
19. List five different shapes of condulet fittings.
20. Of what materials are condulet covers made?
21. Describe two methods for securing rigid conduit to boxes. Why are plastic bushings sometimes used?
22. List four advantages of rigid aluminum conduit.
23. What is the main advantage of flexible conduit?

24. List three examples where flex is useful.
25. How is ground continuity maintained in a flex system?
26. How is flex terminated? How is PVC-jacketed flex terminated?
27. Why must care be taken when installing PVC-jacketed flex?
28. What is a *knockout cutter*? How and why is it used?
29. Why is it important to keep the fish tape in motion when inserting it into a run of conduit?
30. What method is used to help draw a fish tape into a run of conduit with many bends?
31. What are the advantages of compressed-air fishing?
32. Why is it necessary to limit the number of conductors in conduit or tubing?

Chapter 15

Residential Service Wiring

THREE-WIRE DISTRIBUTION SYSTEM

A three-wire distribution system is used to supply residences with 120 V and 240 V. The two main methods for bringing the three-wire system into a house are by overhead wiring from a pole transformer and by underground wiring from a distribution transformer mounted below or above grade level. Chapter 1 discusses both methods in greater detail.

ELECTRICAL UTILITY

The conductors carrying current from the transformer – whether it is pole-mounted, at grade level, or below grade level in a transformer vault – belong to the local electrical utility. It is the utility's responsibility to install, maintain, and service the conductors, which are known as *service drop conductors* (electric supply lines).

The cost of installing service drop conductors is assumed by the local electrical authority, providing the residence is within 100 feet (30 m) of a distribution pole or transformer. Home owners beyond the 100-foot (30-m) distance often have to pay the cost of placing a pole or running conductors from their houses to the 100-foot (30-m) margin.

CONSUMER'S SERVICE

The consumer's service includes all service boxes and related equipment, up to and including the point at which the electrical utility makes its connections. However, the meter usually is furnished and owned by the electrical utility.

Figure 15-1 shows a typical residential service with overhead supply lines. *Article 230* of the National Electrical Code provides up-to-date regulations for the installation of a consumer's service.

SERVICE SIZE AND CAPACITY

At one time, a residence was considered to be well equipped if it had a 60-A main switch, with an 8-circuit-plus-range (stove fuses) distribution panel. The meter was usually located inside the house between the service boxes. Figure 15-2 shows this older, 60-A service installation method.

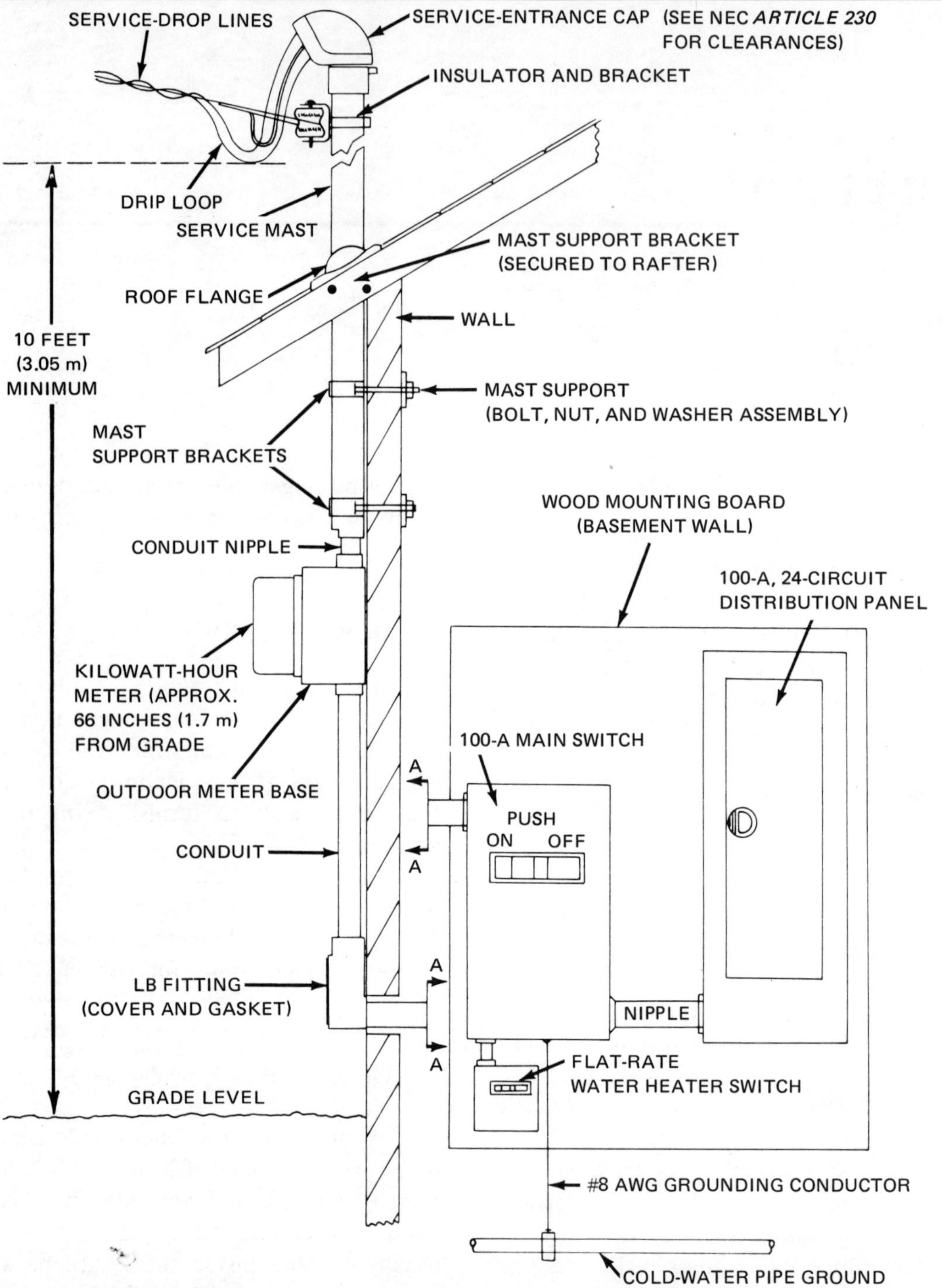

Fig. 15-1 Typical 100-A service (conduit) installation with overhead supply lines.

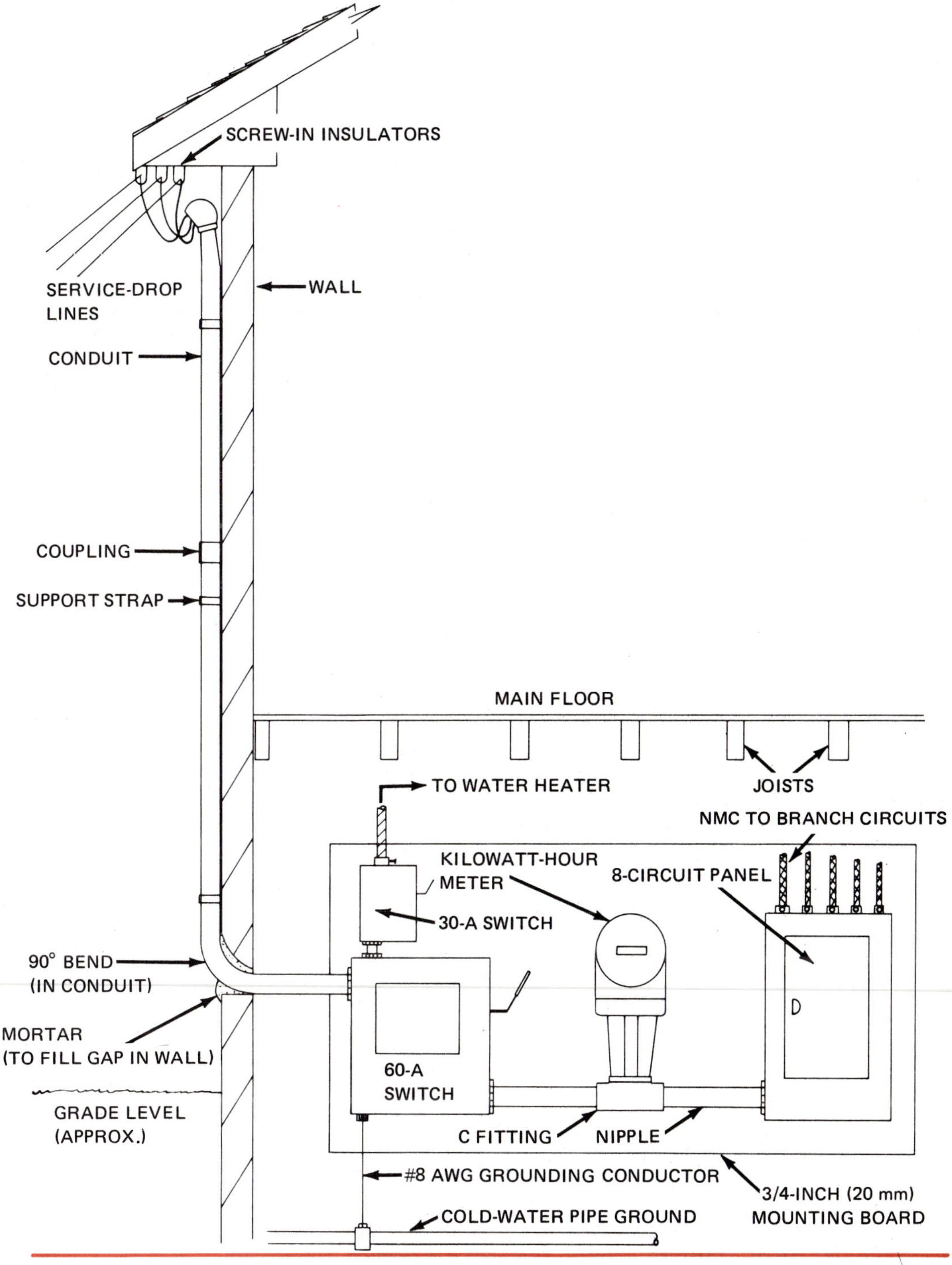

Fig. 15-2 Typical older method 60-A service installation with overhead supply lines (indoor meter method).

As electrical appliances became more popular, the 60-A service installation became inadequate. In addition, local utility employees, who had to enter homes to read the meter so that the customer could be billed, often found it difficult to gain access.

The need for more circuits in the home and, therefore, more fuses in the distribution panel made the 100-A, 20-circuit panel a logical choice. The extra circuits (fuses) available allowed appliance receptacles in the kitchen and laundry areas to be fused separately. This feature was necessary because of the high current demands of modern appliances in these two areas. At the same time, the meters were moved outdoors, eliminating the access problem for meter readers.

The NEC specifies that, in addition to the general-purpose receptacle and lighting circuits, two or more 20-A branch circuits must be provided for the outlets in the kitchen, pantry, family room, and dining room. Furthermore, such circuits must not be used to supply any other outlets. No. 12 AWG wire is required for these circuits, because they carry a higher amperage than standard lighting and receptacle circuits. These circuits are computed at 1500 W each.

The 100-A service can be installed with a separate main switch and distribution panel (as shown in Figure 15-1) or in the newer combination unit shown in Figure 15-3. The switch and panel combination unit reduces installation time and costs by eliminating the nipple, locknuts, and bushings required when the units are installed separately. Since most combination units are factory-prewired (between the main switch and the distribution panel) even

Fig. 15-3 Combination service-entrance panel with circuit-breaker main disconnect and circuit-breaker branch circuits.

more time is saved. Clothes dryers and electric water heaters need separate circuits, however, so manufacturers were encouraged to produce 24-circuit panels.

For a time, the 100-A service was considered the ultimate in residential equipment. But, as more spacious houses were designed, and often equipped with electric heating and air-cooling units, the 100-A service went the way of the 60-A installation. The 200-A service panel, with up to a maximum of 42 circuits took its place, Figures 15-4. Figure 15-5 shows the circuit schedule for this type of load.

Normally, a building or other structure is served by only one service. By exception, when more than one service is permitted, the requirements of NEC *Article 230-2* should be consulted.

As larger houses are designed, the need for more circuits and the heating and air-

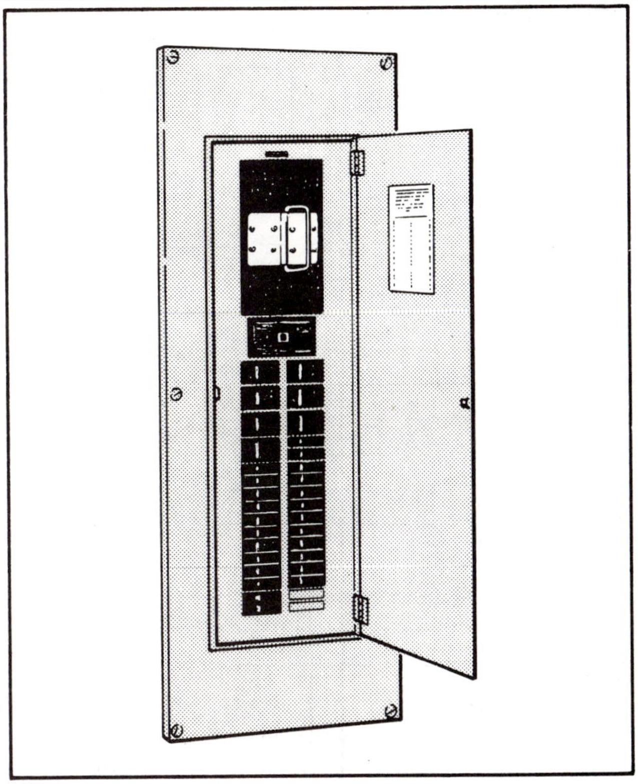

Fig. 15-4 Typical 200-A combination service-entrance panel.

conditioning demands increase. Today, it is common to find a single-phase and, sometimes, a three-phase, 400-A, 600-A, or even an 800-A service in a private residence. While these services may seem incredibly large to the average home owner, they are increasing in popularity as more and more electrical conveniences are installed in modern houses.

INSTALLATION TECHNIQUES

Service-entrance equipment must be installed in a location suitable to both the local electrical utility and the inspection department. Utility authorities are usually quite willing to visit homes, upon request, and offer advice on the location of service equipment. This can be of great help to the installer, who otherwise might mount equipment in a manner unacceptable to the authorities, resulting in expensive and time-consuming equipment changeover. Mast height, meter height and location, point of entry into the building, and box location can be selected quickly by an inspector on a service location visit. Refer to NEC *Article 230, Services.*

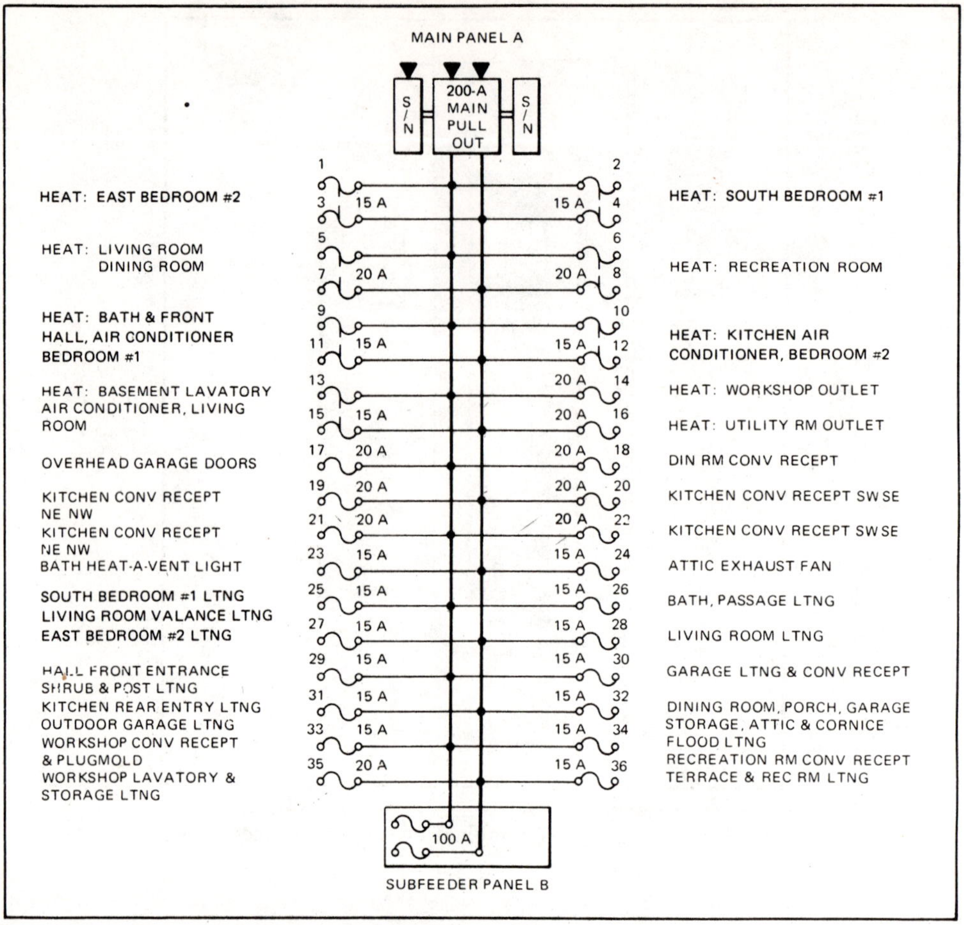

Fig.15-5 Circuit schedule for the 200-A service panel.

SERVICE MAST

The National Electrical Code requires that the lowest service-drop conductors be located a minimum of 10 feet (3.05 m) above grade. Other clearances for service-drop conductors not in excess of 600 V are, for example:

- 15 feet (4.57 m) over residential driveways. 12 feet (3.66 m) if voltage is 300 V or less to ground.
- 18 feet (5.49 m) over public streets or commercial driveways.
- 8 feet (2.44 m) above all points of a roof accessible for walking.
- 3 feet (914 mm) above a pitch roof with a slope of not less than 4 inches (102 mm) in 12 inches (305 mm).
- 18 inches (457 mm) over the roof, provided the conductors do not pass over more than 4 feet (1.22 m) of the roof overhang, and are attached to a through-the-roof raceway or approved support.
- 3 feet (914 mm) from windows, doors, porches, fire escapes and

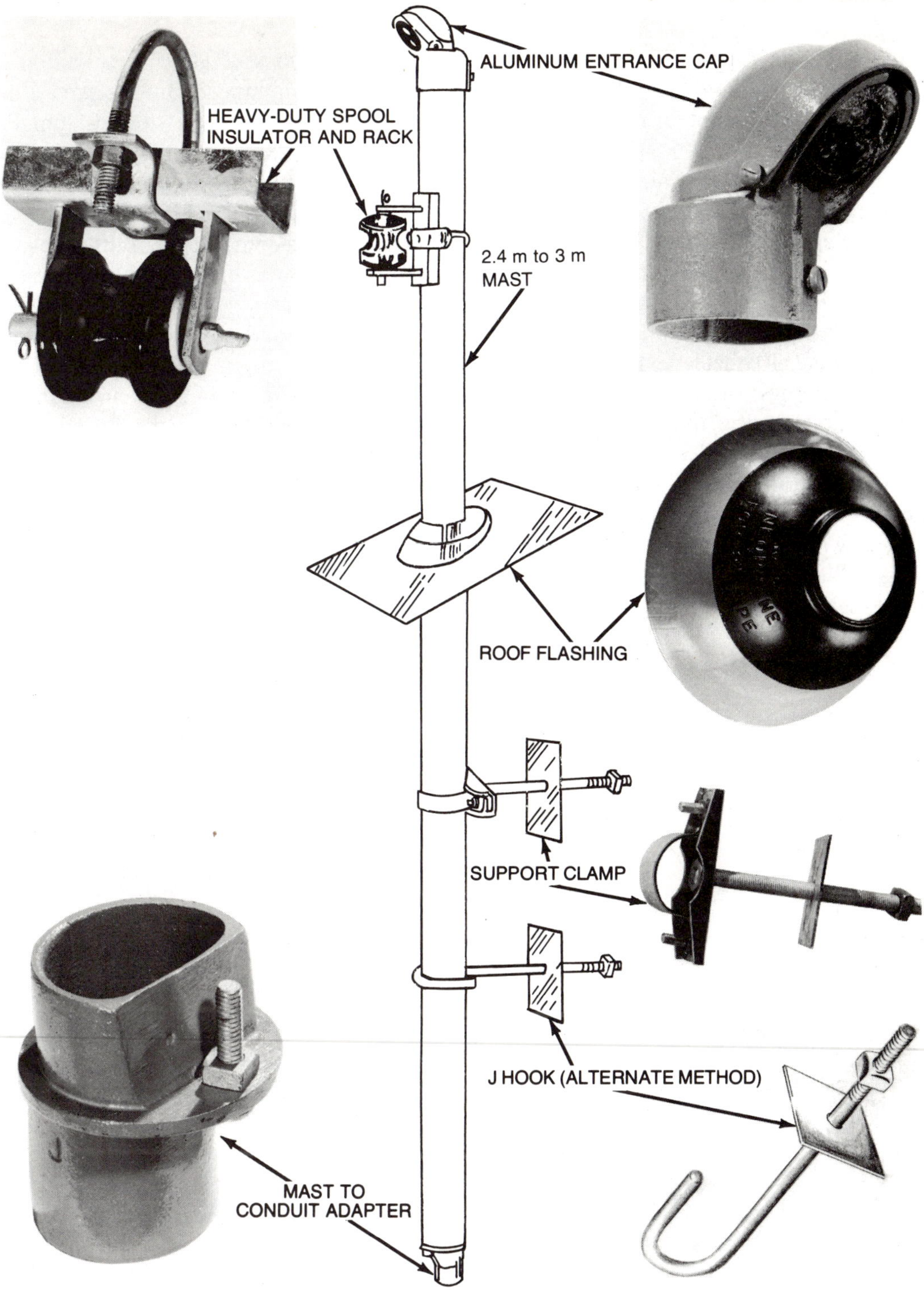

Fig. 15-6 Typical service mast and components.

similar locations. Conductors above the top level of a window are considered to be out of reach from the window.

Clearance allowances are calculated to prevent damage to electrical supply lines by large vehicles, and to reduce the possibility of a person coming into contact with the wires.

The entrance cap should be located not more than 30 feet (9 m) above grade on high buildings; otherwise, it would be difficult for electrical utility personnel to reach the cap. Figure 15-6 shows a typical mast unit and its individual components (identified for use on masts), which are available with 100-A and 200-A service fittings. Where a service mast is used to support service-drop conductors, it must have sufficient strength or bracing to safely bear the strain imposed by the service drop.

The service mast, which is sometimes called the *standpipe,* must be securely fastened to the building to prevent the service-drop lines from pulling it loose. Wind, rain, snow, ice, and contraction of the wire in cold weather often exert tremendous pressures on the mast.

Several methods are used for installing the mast. (See Figures 15-6, 15-7, and 15-8.) One common method for securing the mast is to use long bolts that pass completely through the wall and fasten over a wooden member on the inside of the house. Large square washers and nuts are used to complete the system (as shown in Figure 15-1).

The mast is usually located about 3 feet (1 m) from the corner of the building

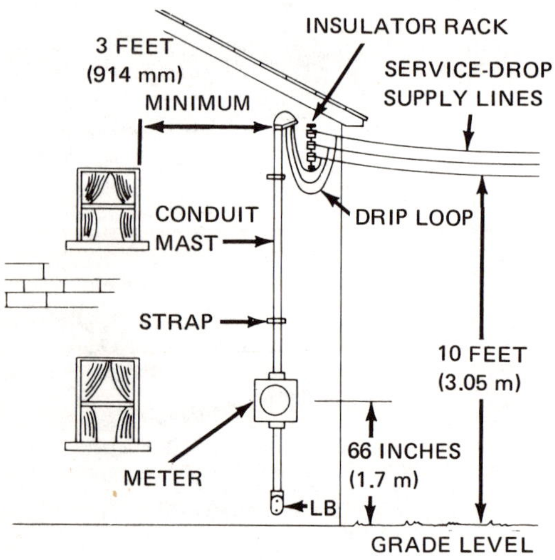

Fig. 15-7 Mast installation on two-story house.

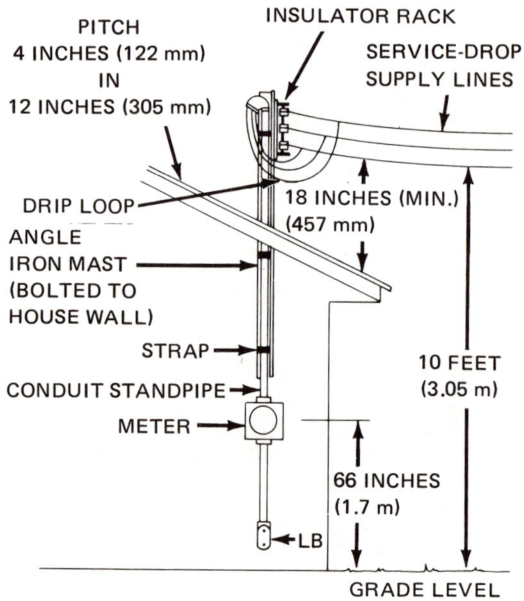

Fig. 15-8 Mast assembly on one-story house.

and in such a way that no supply lines will pass within 3 feet (1 m) of a window or any other point that might provide access to the wires. This is to prevent a person from touching the wires accidentally.

A 30-inch (800-mm) length of free conductor must be left at the entrance cap. This functions as a drip loop, so that rain will not run back into the mast, and also gives the utility authority enough conductor for the connection.

METER SOCKET

A metal enclosure, complete with threaded hubs for conduit, is installed at approximately 5 1/2 feet (1.7 m) above grade. This enclosure is designed so that the service conductors to be metered are attached to solderless lugs (mechanical termination) within the enclosure. The meter is then plugged into the enclosure in much the same way as a cord cap is plugged into a receptacle. In fact, this is where the name meter socket originated.

Meter sockets, which are often called *meter bases,* are available in several styles. (See Figures 15-9, 15-10, 15-11, and 15-12.) Threaded hubs are sized to accept the rigid conduit required for the service installation; for example, 1 1/4 inches (32 mm) for a 100-A service, and 2 inches (51 mm) for the 200-A unit. Some meter sockets have hubs that bolt onto the actual metal box of the socket.

One type of meter socket is equipped with a pair of special solderless lugs on the line side of the socket. These lugs allow a pair of No. 10 gauge, red-insulated conductors to be brought into the main switch to supply the flat-rate water heater. The smallest conductor allowed in a service conduit is a No. 10 gauge wire. Since these conduc-

Fig. 15-9 Typical 100-A service meter socket.

Fig. 15-10 Typical 200-A service meter socket.

Fig. 15-11 Typical meter socket assembly for use on a two-family dwelling.

tors are attached to the line side of the meter, any current passing through them will not register on the meter.

FLAT-RATE WATER HEATER SYSTEM

Some local utilities operate on a flat-rate water heater supply system. This means that the consumer pays a set amount each month and may use as much power for heating water as is required. Large families, which need a lot of hot water, find this an economical system. Small families, however, are usually better off if the water heater, together with the lights and other appliances, operate through the meter.

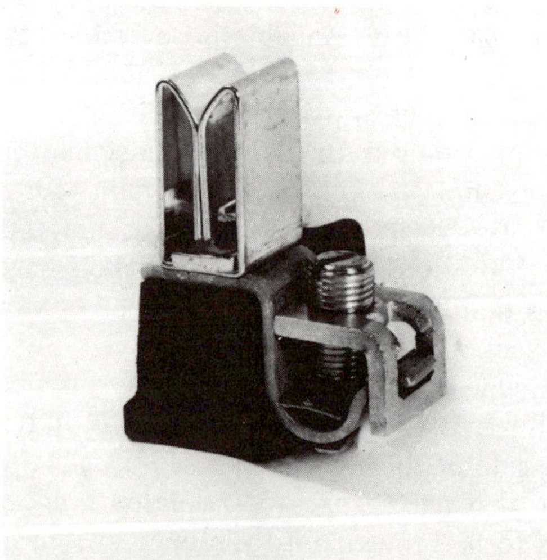

Fig. 15-12 Standard terminal block for 100-A and 200-A meter sockets.

With this system, they pay only for the amount of power used.

The flat-rate water heater system is good for both the consumer and the local utility. The demand for power is greatest during the hours of 7 AM to 9 AM and 3 PM to 6 PM. During these times, the utility can turn off all the flat-rate water heaters in the area. Power from the water heater units can then be directed to industrial customers. This is done by sending a high-frequency signal through the supply lines, which is picked up by the relay in the basement of the house. The relay then turns off the power to the heater. The same process is used to restore the power to the water heater. This same energy conservation system is used by some utility authorities to control air conditioners, water heaters, and heating devices during peak-demand periods.

Some communities require the two red-insulated conductors to be enclosed in the service conduit and mast all the way up to the weatherhead fitting. A 30-inch to 36-inch (762-mm to 914-mm) length of drip loop is left for connection to a fourth conductor, which is brought to the home from the closest utility pole (on overhead service installations). This fourth conductor carries the high-frequency signal to one red wire, while the second red conductor is connected to one of the incoming electrical supply lines for the house. Underground service (laterals) installations normally use the meter socket with the special lugs.

SERVICE-ENTRANCE ELBOWS

A 90° corner (or Type LB condulet) is used to bring the standpipe (service mast) from the meter socket through the wall and into the service boxes. A 90° bend can be made in the service conduit, but a groove must be chiseled in the wall for approximately 12 inches (300 mm) above the hole to allow the conduit to fit tightly against the wall, Figure 15-13. While this takes time, it is the best method to use when the service conduit is in a driveway or similar area, where the conduit may be damaged by vehicles.

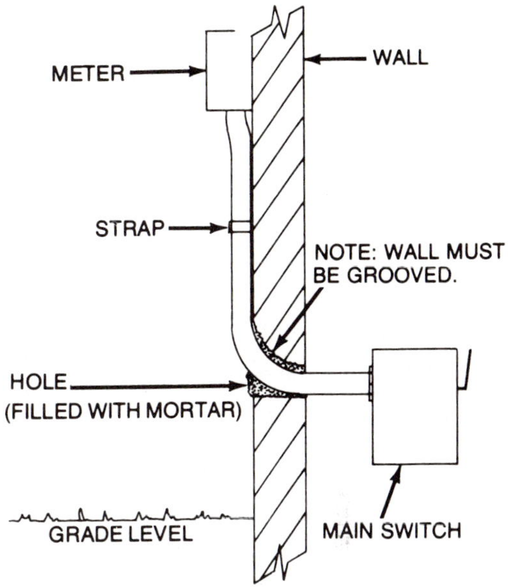

Fig. 15-13 Method for driveway installation of standpipe (90° bend).

An alternate method of handling the 90° bend is shown in Figure 15-14. This method does away with the need for grooving the wall. The disadvantage of this method is that the standpipe protrudes from the wall a few extra inches (several centimetres). This *gooseneck-bend* method of entering the building was once used widely. With the increase in conduit size and the development of condulet fittings, however, this difficult bend has become less popular.

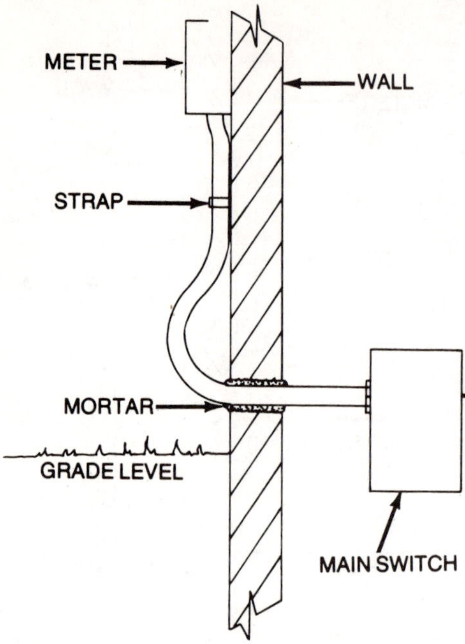

Fig. 15-14 Alternate method for driveway installation of standpipe (gooseneck bend).

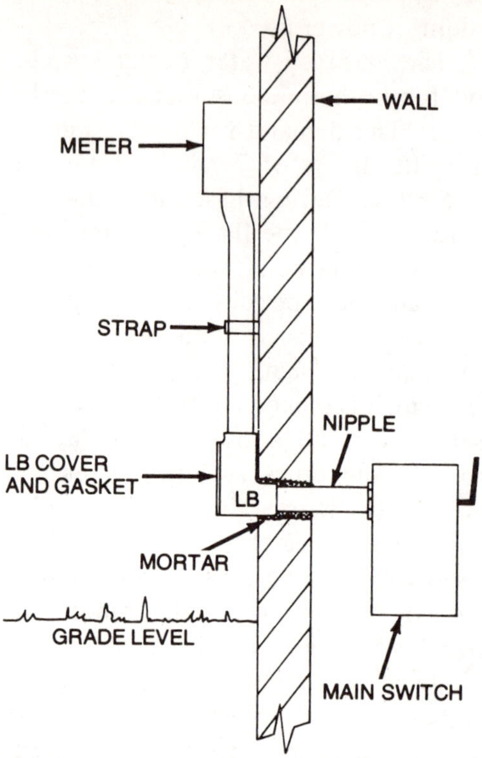

Fig. 15-15 LB condulet method for entering building.

A condulet fitting, such as the Type LB condulet, is approved for use where there is little chance of mechanical injury cracking the casting, Figure 15-15.

The hole in the wall must be made slightly larger than the fitting to allow the short arm of the condulet to enter the wall. A weatherproof gasket and cover fastened to the LB condulet seal out any moisture. The LB condulet fitting provides access to the conduit and makes it easier to pull in the large conductors required for modern houses.

Fig. 15-16 Service ell fitting with cover and gasket.

A special fitting, called the *service ell*, can also be used for entering a building. (See Figures 15-16 and 15-17.) The main advantage of a service ell is that the short arm does not extend into the wall. Because the hole in the wall does not need to be made any larger than the conduit nipple, this type of installation is more weatherproof than other types.

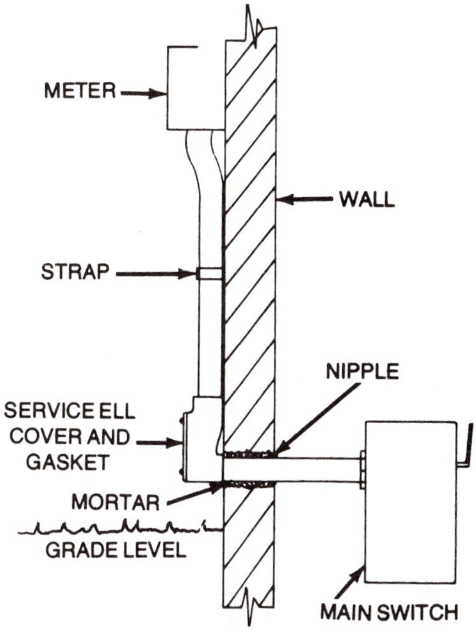

Fig. 15-17 Service ell installation.

No matter which method is used to enter the building, one National Electrical Code requirement *must* be met. That is, to fill all openings around the area where the conduit enters the wall in order to prevent water damage to the house and/or the service equipment. Mortar, or a similar material, trowelled into place will stop moisture or cold air from entering.

UNDERGROUND SERVICE (SERVICE LATERALS)

Most new subdivisions have underground service (service laterals) installations. Getting rid of overhead wiring helps give a neat, uncluttered appearance to the streets. The distribution transformer can be located underground in a transformer vault or on a concrete pad. The vault or pad can be out near the street or behind the house, with the electrical supply lines coming in under the back yard, Figure 15-18.

LOCATION OF SERVICE BOXES

Section 230-72 of the NEC lists the places where service boxes shall be located. In general, service boxes should be mounted on a wall about 5 feet (1.5 m) above the floor and as close as possible to the point where the service conduit enters the building. If there is any doubt about where to install a box, the local inspection authority should be consulted.

SERVICE MOUNTING BOARD

Service boxes are usually supported on a wooden mounting board approximately 3/4 inch (20 mm) thick. Plywood does an excellent job, but regular 3/4-inch (20-mm) lumber can be used. A 100-A service requires a square panel, 3 feet (1 m) on each side. A 200-A service will likely need a square panel, 4 feet (1.2 m) on each side. Larger services require a full 4-foot x 8-foot (1200-mm x 2400-mm) sheet of plywood. The board should be fastened securely to the wall with such devices as concrete nails, toggle bolts, or masonry plugs and screws.

The wooden panel allows branch-circuit conductors to be supported close to the distribution panel. Future equipment can easily be secured to the wooden panel.

SERVICE BOXES

Service boxes were once arranged as two separate units, (as shown in Figures 15-1 and 15-2). Modern service equipment, in

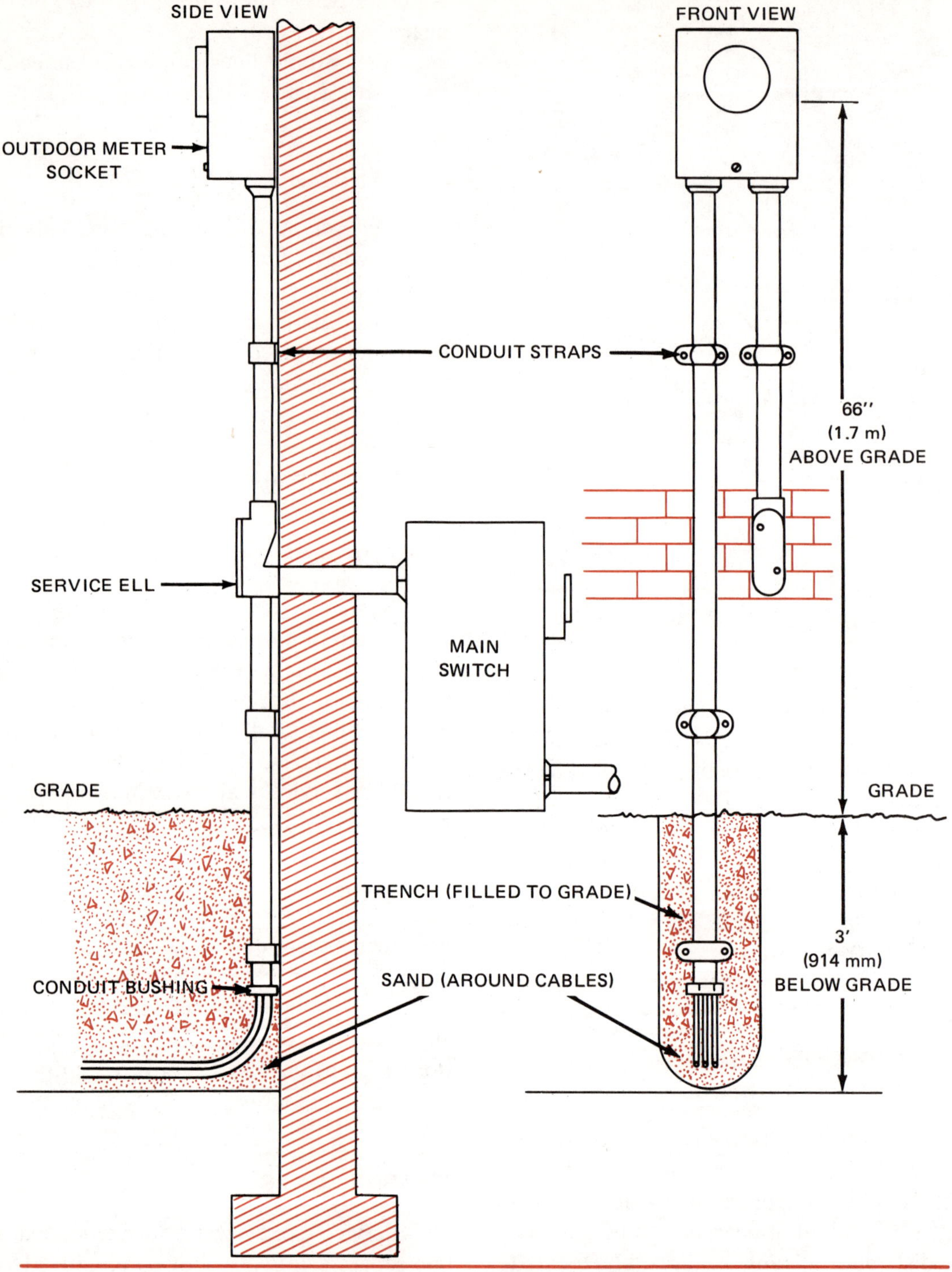

Fig. 15-18 An underground service installation (front and side views).

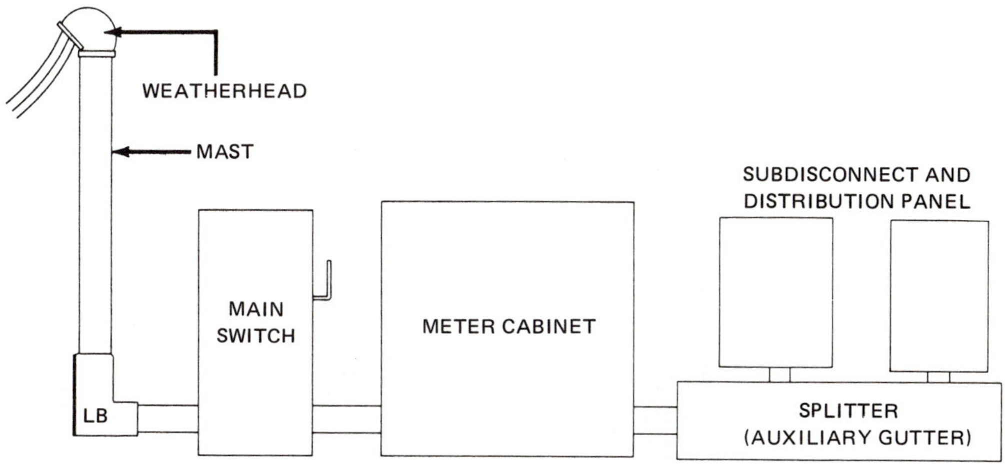

Fig. 15-19 Common arrangement of large service boxes.

100-A and 200-A ranges, is usually arranged in the form of combination units (as shown in Figures 15-3, 15-4, and 15-5). Larger service boxes are usually arranged as shown in Figure 15-19.

SERVICE WIRING

Table 5-2 lists the size of copper and aluminum conductors to be used, with their various ampacities. Once conductor size has been determined, Tables 5-1 and 14-1 are used to decide what size of conduit is required.

When an outdoor meter is used, the conductors must first pass through the meter terminals (lugs). The neutral wire may be clamped or bolted in the meter socket. It then passes directly to the neutral block in the main switch. A neutral wire that is not making a good contact will cause uneven voltage distribution in an unbalanced system, which can be *dangerous.*

The two *live* wires in the service conduit, after passing through the meter, are connected directly to the *line* terminals in the main switch. The line terminals are nearly always located at the *top* of the switch and are designed so that when the switch is *off,* only these two terminals are *live.* Some manufacturers design their equipment so that the line terminals are covered by an *insulating barrier* to prevent a person from accidentally touching them, Figure 15-20. The *load* side of the switch (bottom terminals) is then connected to the distribution panel.

Figure 15-21 shows the wiring of a typical 200-A service. The 100-A service uses smaller conductors, but is wired in the same way.

Figure 15-22 shows the box arrangement and wiring for a two-family house. Figure 15-23 shows a similar box arrangement and wiring for a two-family house with separate metering.

Fig. 15-20 A main switch with line terminals at top and load terminals at bottom.

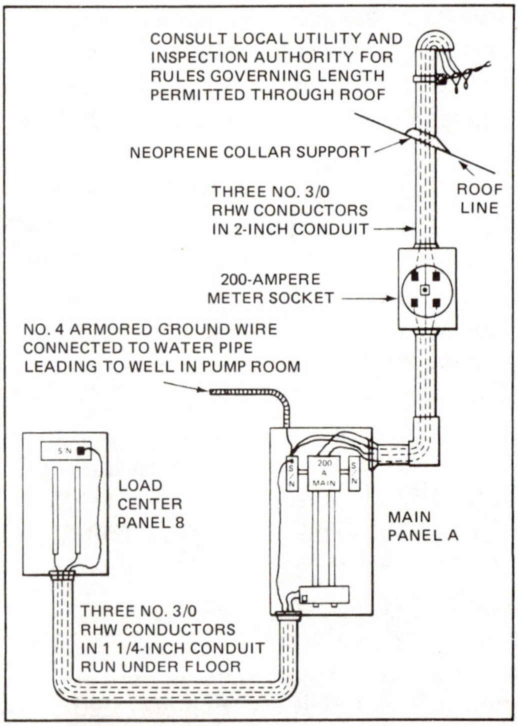

Fig. 15-21 Typical 200-A combination main circuit-breaker panel.

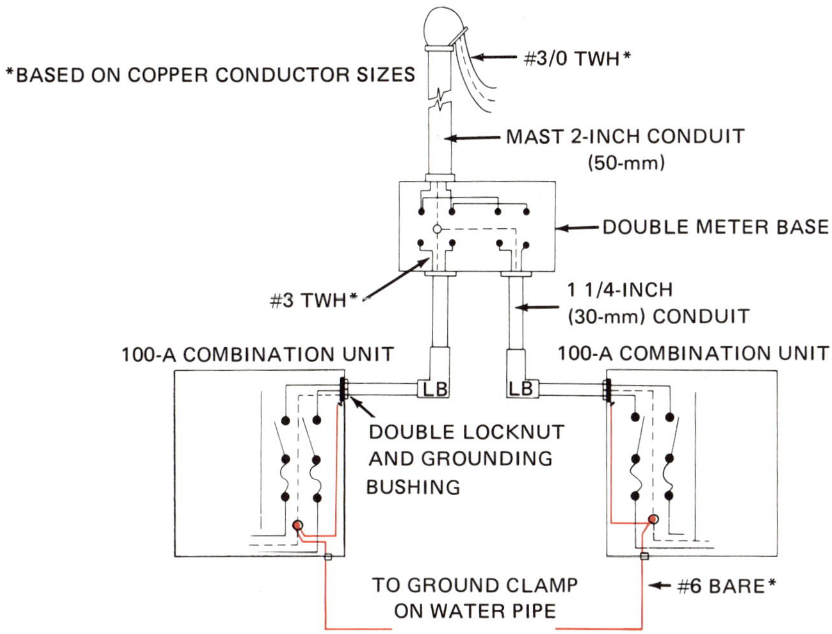

Fig. 15-22 Typical two-family dwelling service installation.

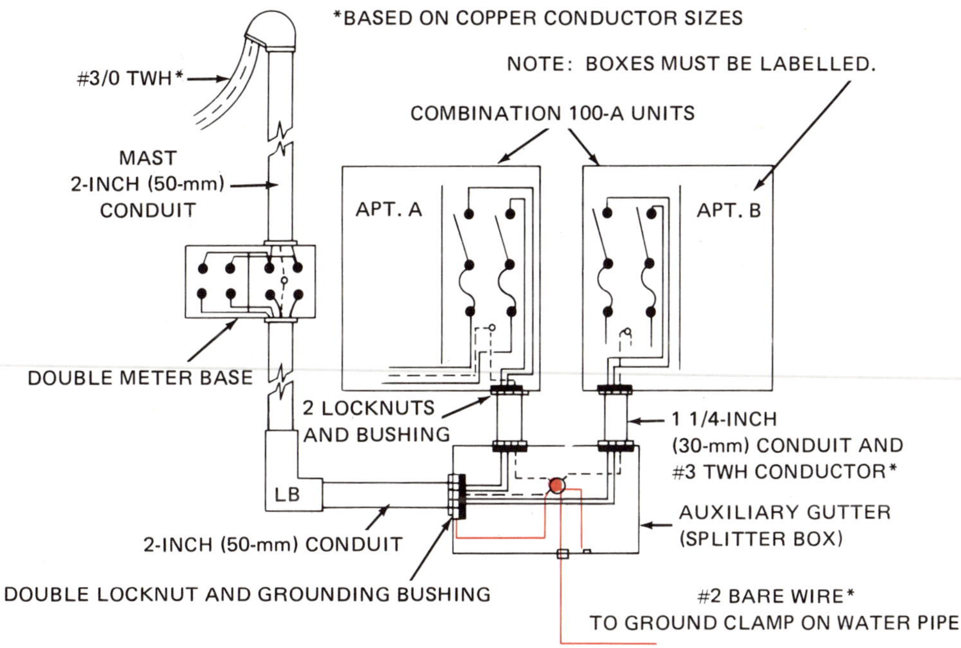

Fig. 15-23 Typical two-family dwelling service installation (separate metering).

Apartment buildings, where the apartments are metered separately, must be handled differently. Figure 15-24 shows a typical service installation for a five-unit apartment building. Large services of this type may have indoor metering facilities.

When a single-meter system is needed for a service with a capacity larger than 225 A, the current flow must be metered without actually passing full line current through the meter. Since most meters have a maximum current capacity of 225 A, a system of current-reducing transformers is used. These current transformers carry the full line current in a main (primary) winding, which usually consists of a solid bar of copper. The line current is reduced to a safe level by the secondary winding, and then passed on to a specially designed meter. Figure 15-25 shows an older style indoor metering service-entrance system.

The current transformers and meter may be housed in a metal meter cabinet. Several cabinet sizes are available; the smallest measuring about 3 feet (1 m) square and about 12 inches (300 mm) deep. Figure 15-26 shows a typical indoor meter for use in a meter cabinet. Figures 15-27 and 15-28 show typical current transformers. Transformers are available in a variety of sizes and ampere capacities for indoor locations and outdoor pole-mounted units.

SERVICE GROUNDING

Electrical services must be grounded for two reasons. The first is that the steel mast, which rises 15 feet to 30 feet (4.6 m to 9 m), is an attractive target for lightning. Grounding the mast reduces the chance of lightning striking the house. If lightning

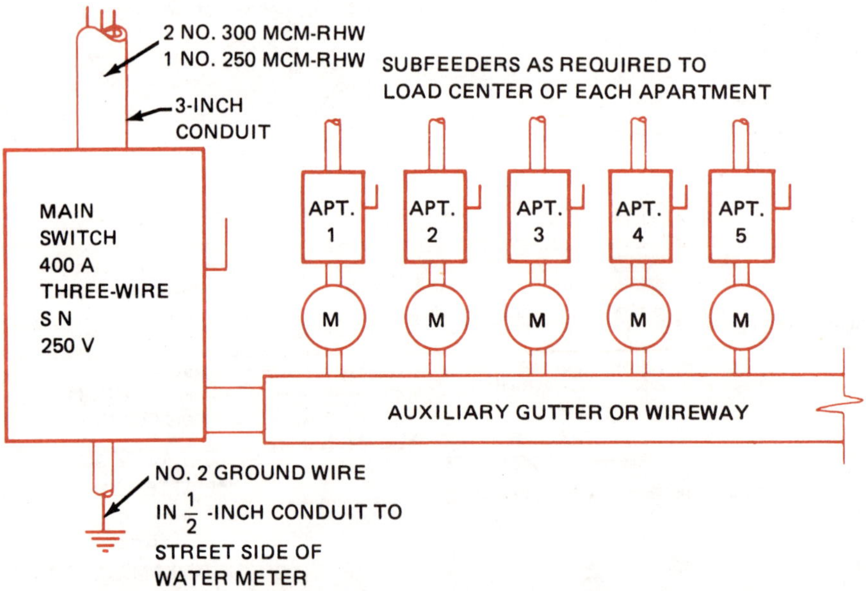

Fig. 15-24 Typical apartment service installation (separate metering).

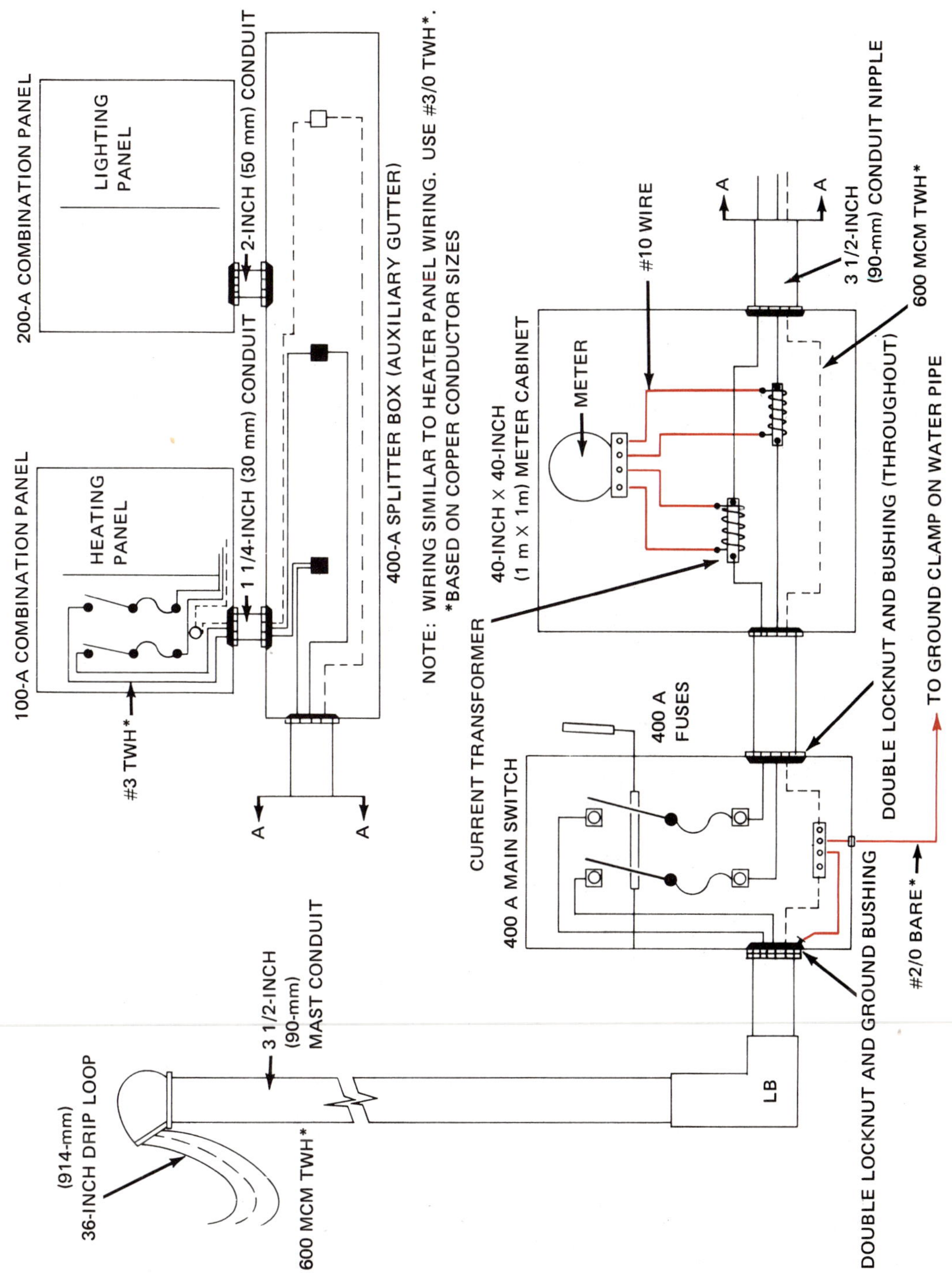

Fig. 15-25 Typical older style single-metering, high-ampere service installation.

does strike, grounding provides a direct path to the earth.

The second reason for grounding electrical services is that grounding of the boxes and cabinets gives another important form of protection. If one of the live wires in the system comes into contact with any of the metal enclosures, a short circuit will occur. The excessive short-circuit current will flow along the ground conductor and pass harmlessly into the earth. The usual result is a blown fuse. Once the fault has been located and repaired, the fuse can be replaced easily. If the boxes were not grounded, all of the conduit, boxes, and cabinets would become alive and *dangerous.* A person standing on moist earth or a damp floor and touching part of the metallic system would receive a 120-V shock. In a large building, with no blown fuse on an ungrounded system, the fault may not show up for some time and then be almost impossible to locate.

The most common method for providing a ground for a residential service is to connect a grounding conductor from the neutral block in the main switch to the cold-water supply pipe as it comes out from the basement floor. The neutral block is connected to the switch box by a brass bolt. This ensures that both the box and neutral conductors are grounded. The ground clamp must be fastened onto the water pipe ahead of the meter (or bonded around the meter), because a leakproof compound (a noncon-

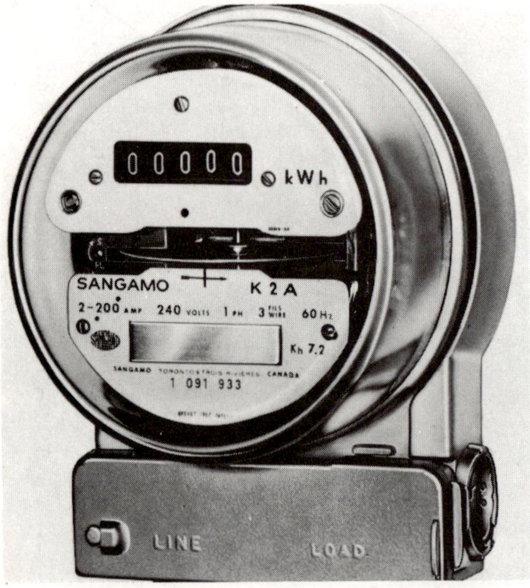

Fig. 15-26 Indoor meter for use in metering cabinet.

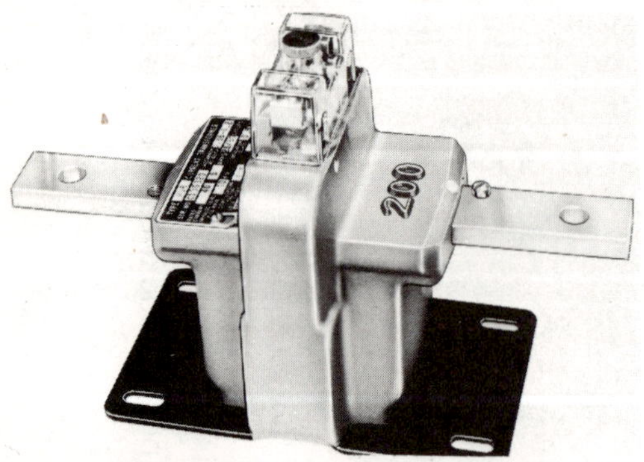

Fig. 15-27 A 200-A capacity current transformer.

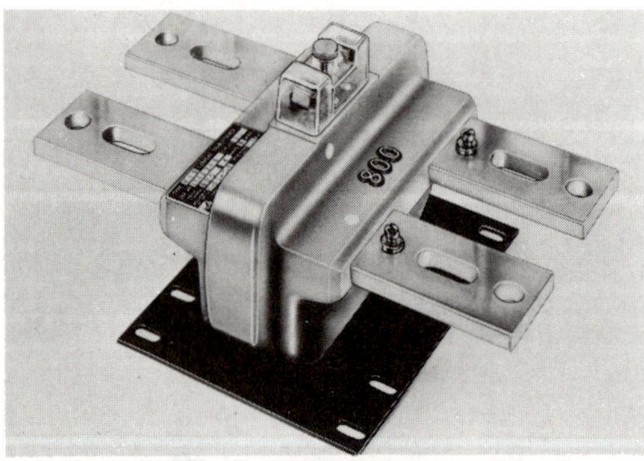

Fig. 15-28 An 800-A capacity current transformer designed to operate with both live wires of system.

ductor of current) is applied to the plumbing threads before assembly, Figure 15-29. The grounding wire from the main switch box is usually bare copper or covered with a white insulation. (Table 16-1 lists the conductor sizes to be used.)

When a house has copper plumbing which is properly soldered at the fittings, the grounding conductor can be run from the main switch to the closest cold-water pipe. The hot-water tank interferes with the electrical continuity of the hot-water pipe and should not be used. Since the water meter with its waterproof connections is still a problem, a jumper of the same gauge conductor as the grounding conductor must be installed to bypass the meter, Figure 15-30. Then, if the water meter is removed at any time, ground continuity is maintained.

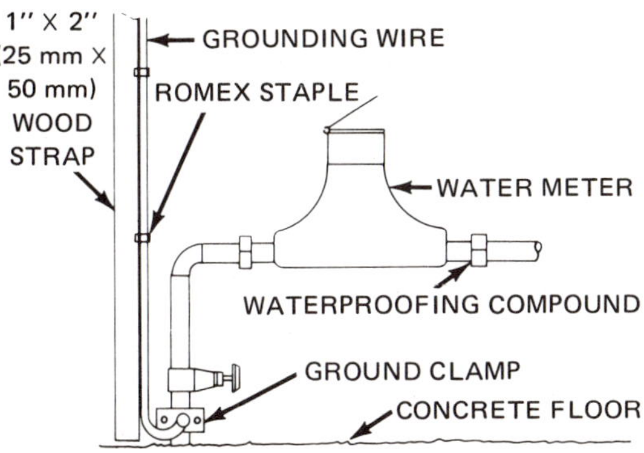

Fig. 15-29 Ground clamp connected to water pipe ahead of water meter.

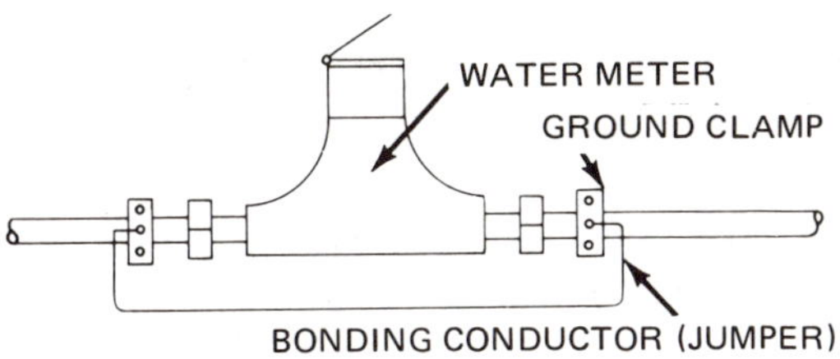

Fig. 15-30 Typical water meter bypass.

Fig. 15-31 Typical ground clamp for use in residential service.

In some areas, houses do not use a water meter. In these cases, the grounding conductor should be connected by a ground clamp to the cold-water pipe where it first enters the basement, Figure 15-31. NEC *Article 250* requires that a metal underground water pipe have an additional supplementary grounding electrode, and permits this electrode and all piping (hot- or cold-water lines) and gas lines to be bonded together.

Two precautions must be taken to make sure that the mast is grounded effectively. First, two securely tightened locknuts, one inside and one outside the box, are used. Second, a special grounding bushing, with provision for a grounding conductor, is threaded onto the conduit. This conductor, which should be about No. 6 gauge, joins the grounding bushing to the neutral block. Figure 15-32 shows a typical grounding bushing.

Rural communities usually do not have a cold-water supply pipe, such as that used for grounding services in urban communities, because plastic or PVC water lines may be used. Several different methods for grounding are available for these areas. The most common method is to drive a 3/4-inch (20-mm) diameter, galvanized steel rod (grounding electrode)

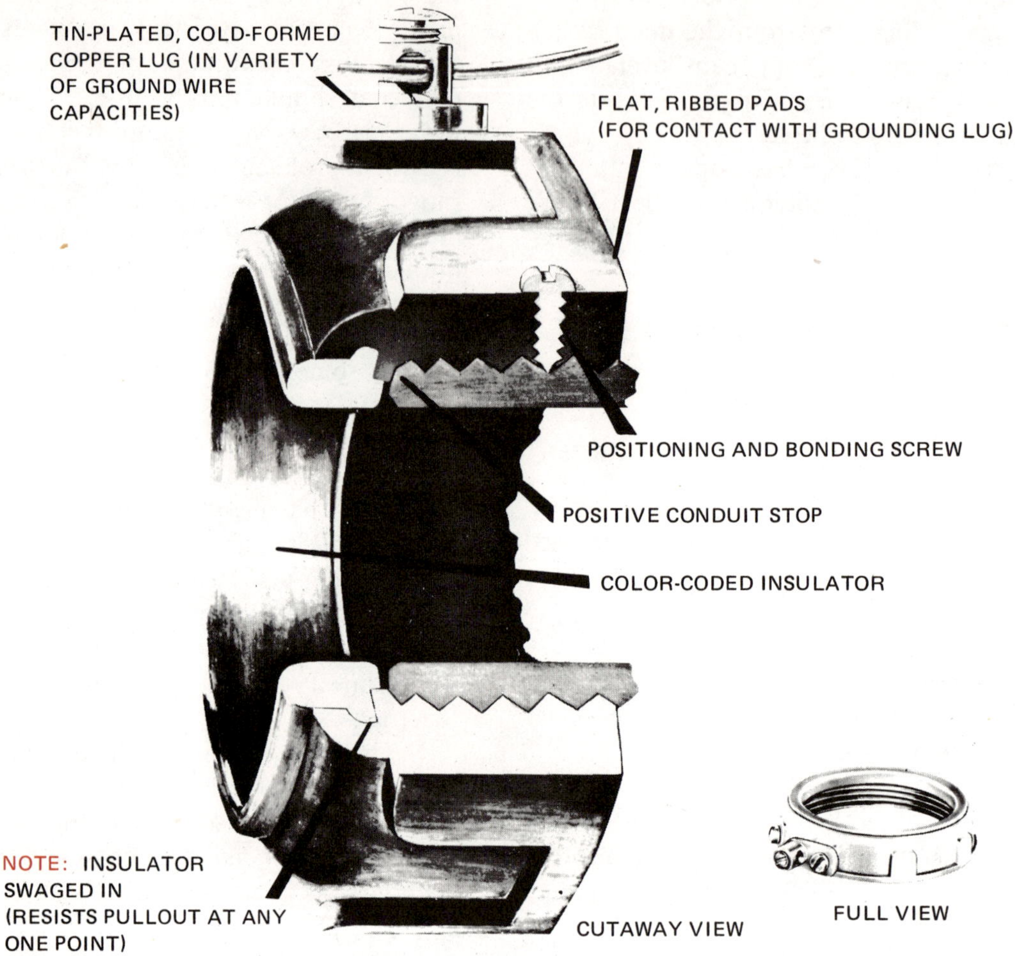

Fig. 15-32 Grounding bushing (full and cutaway views).

into the earth, Figure 15-33. If additional rods are needed, another rod can be threaded onto the first rod and driven into the ground, or three rods can be placed in a triangular pattern. The rods must be spaced not less than 6 feet (1.83 m) apart and driven in a minimum of 8 feet (2.4 m). The 8-foot (2.4-m) depth is needed to place the rods in contact with permanently moist earth. This ensures a high-quality ground contact.

Commercial ground rod drivers are available, but most often a sledge hammer is used to drive the rods into the earth. Layers of clay and stones or rocky terrain are problems to the installer. Sometimes the rods, striking a layer of rock, gradually turn and move back to the surface, several feet (metres) from the starting point. Inspection authorities allow the rods to be driven in at an angle of not more than 45° from the vertical if the terrain is hard and difficult.

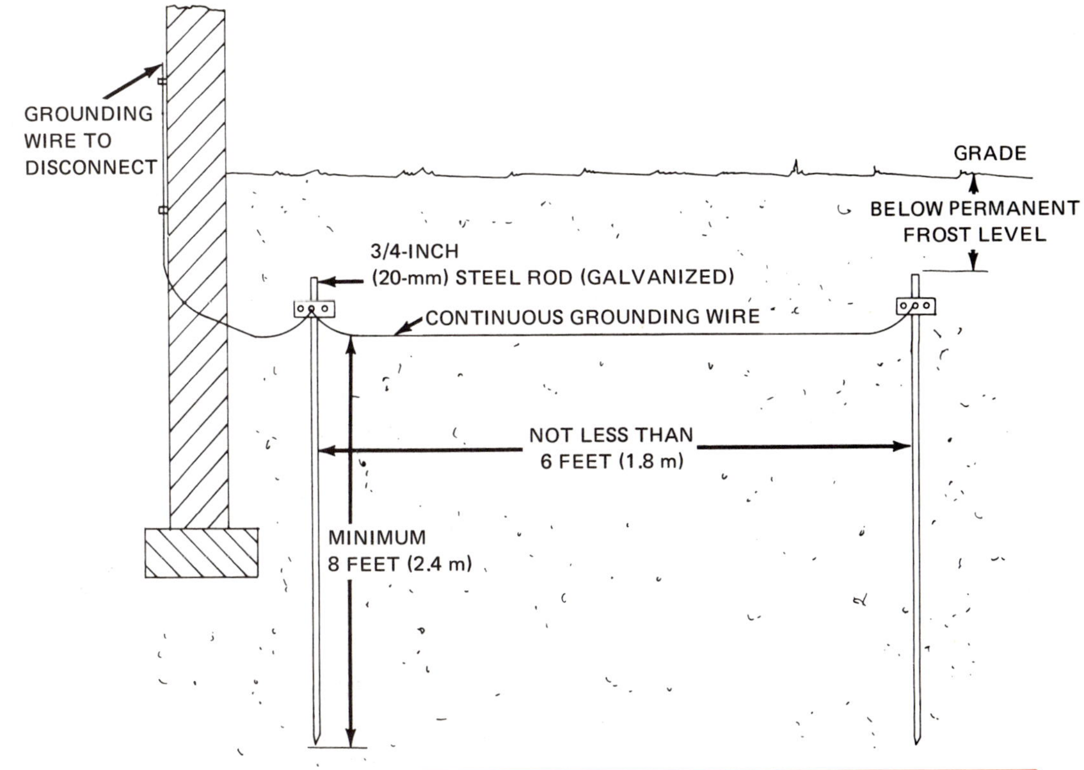

Fig. 15-33 Typical ground rod installation.

A grounding conductor *must* run in a continuous length from the neutral block in the main switch to the first rod. Here it passes unbroken through a clamp on the rod, and then travels to the second (and third) grounding electrode some 6 feet (1.83 m) or more away. Take care to check with the inspection authority for up-to-date information about the type of clamp allowed on the ground rod. The clamping device must make a secure electrical connection to both rod and conductor. Once the rods and grounding conductor are covered with earth, it may be many years before they are checked again. Some local electrical inspectors require the grounding clamps to be above grade level for inspection, and then driven below the surface for protection.

In very rocky soil, it may be impossible to drive in ground rods. In these cases, a steel plate, exposing not less then 2 square feet (0.186 m^2) of surface to the soil and a minimum of 1/4 inch (6.35 mm) thick, can be buried in the earth, Figure 15-34. The plate should be installed as deep as possible to ensure a good ground contact.

Often, when grounding rods (artificial grounding) are used, inspection authorities require that a grounding conductor the same size as the main grounding conductor be run from the neutral block of the main switch to a clamp on the copper plumbing.

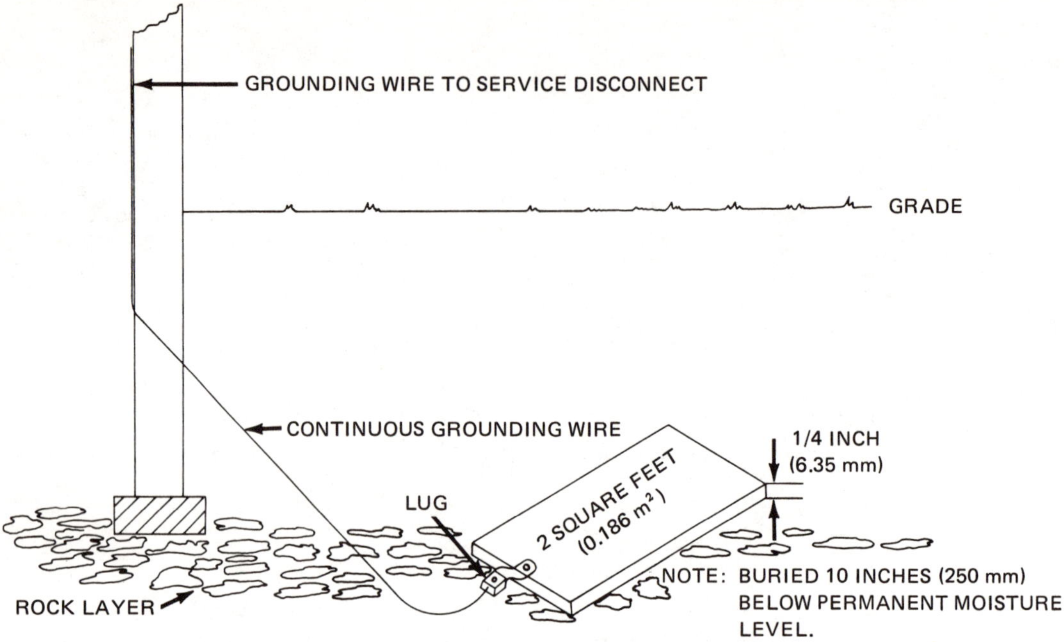

Fig. 15-34 Typical plate electrode grounding method.

This installation is typical in rural communities and is used to prevent the chance of a voltage difference existing between the earth and the plumbing system in the house. If a plastic (PVC) water system is used throughout the house, this extra bonding conductor is not required.

METERS

The meter records the number of kilowatt-hours (1 kW is equal to 1000 W) of power consumed. One kilowatt-hour (kW·h) is a unit of energy equal to that expended by one kilowatt in one hour. Some meters give a direct reading the way odometers do on cars (see Figure 15-26). Many other types of meters are in use, however. These other types of meters have five small dials to indicate the consumption, Figure 15-35. Moving from right to

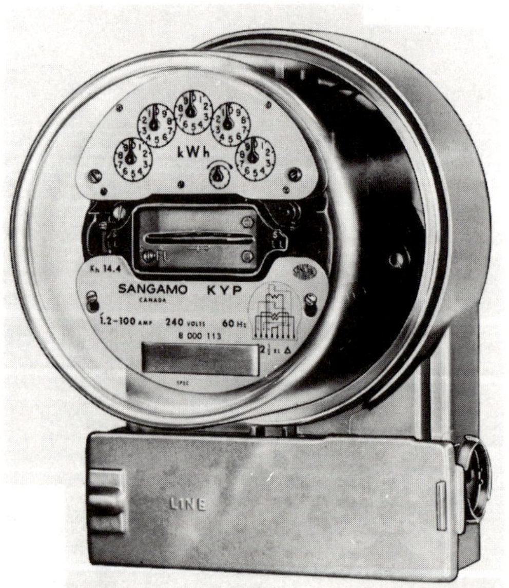

Fig. 15-35 A five-dial meter.

Fig. 15-36 Typical five-dial meter reading (26 422 kW·h).

left across the dials, the first dial shows single units, the second tens, the third hundreds, the fourth thousands, and the fifth tens of thousands. Figure 15-36 shows a typical five-dial meter reading.

The consumer pays the local utility for energy at a rate of so many cents per kilowatt-hour. The number of kilowatt-hours of energy used during a given period is read from the meter. The number is multiplied by the rate per kilowatt-hour, and the customer is billed. Some utilities also add a fuel-adjustment charge to the bill.

CIRCUIT BREAKERS

This chapter has concentrated on fuse-protected switches and panels. However, more and more houses are being equipped with circuit breakers. The subject of circuit-breaker protection is covered in Chapter 16 (Industrial Services), because circuit breakers were first used in industry.

TEMPORARY SERVICE

Power is needed on construction sites for portable drills, saws, concrete mixers, and other equipment. A temporary service is usually erected on a pole, Figure 15-37. This unit is designed with a main switch, a small distribution panel, several duplex receptacles, and a weather-resistant cabinet to protect it. Mast height can be as low as 15 feet (4.6 m). A single ground rod is usually enough for this type of installation. Often, the local utility charges a flat rate or fee for service, and a meter is not installed. A temporary service must be removed after the construction work is completed.

INSPECTION PERMIT

An application for inspection must be filed with the local inspection authority for all new installations of electrical equipment or major repairs. A residential service is considered to be an important installation and receives careful attention from the inspection authorities. Power is not usually allowed to be turned on until the inspector is satisfied that every detail is complete and safely installed. This inspection is important, because it guarantees to the owner and the insurance company that every precaution has been taken to ensure a quality installation.

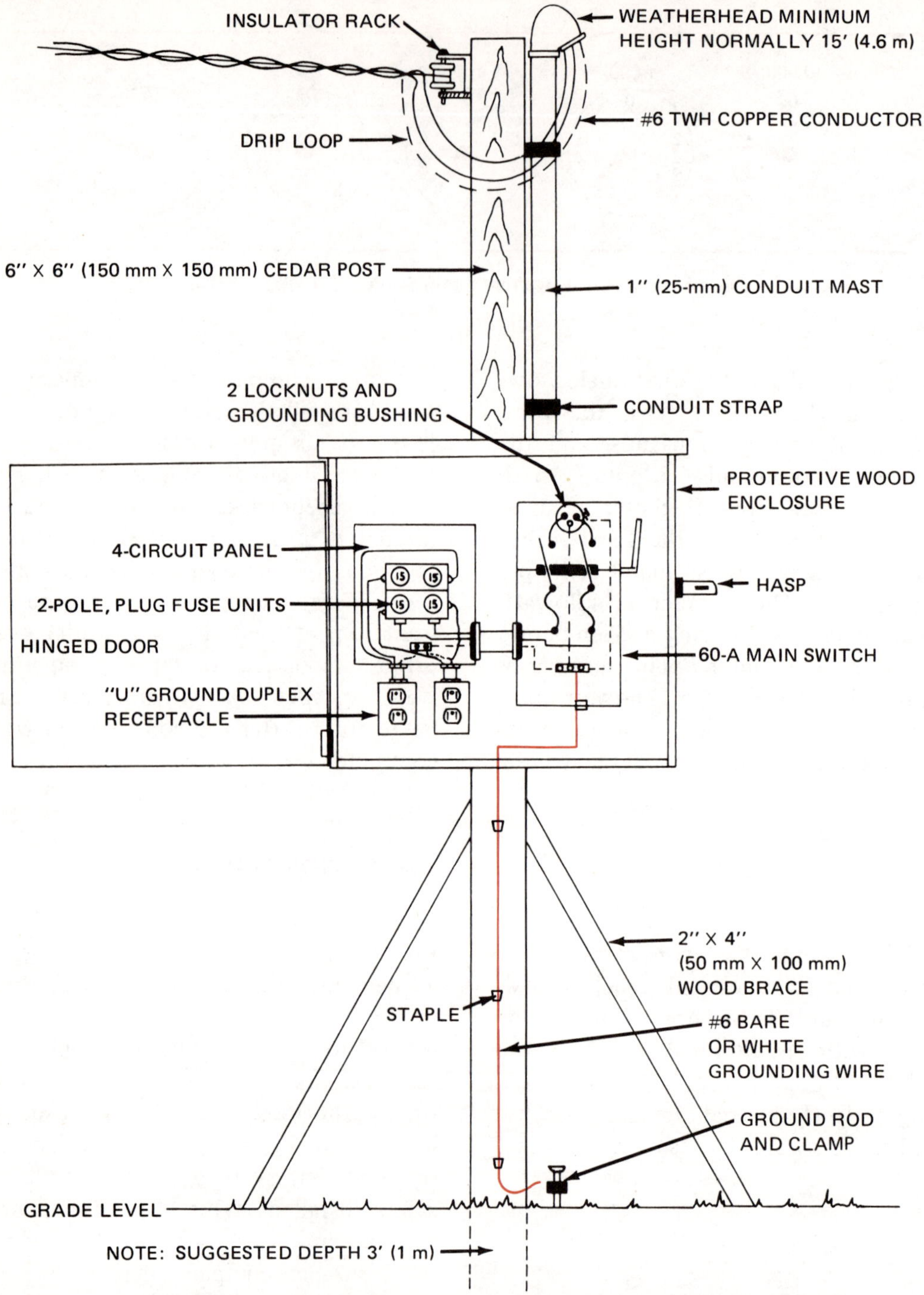

Fig. 15-37 A temporary service installation.

REVIEW

1. What are the two main methods for bringing the three-wire distribution system to a house?
2. List the main parts of a consumer's service.
3. Why did the 60-A service become inadequate for modern houses? Give two reasons.
4. Why are combination service panels often used in modern houses?
5. What are the minimum and maximum heights for service-entrance fittings? Why?
6. Describe a drip loop, and explain its purpose.
7. What is the meter base? How high above grade is it mounted?
8. Why is the flat-rate water heater system good for both the home owner and the public utility?
9. Describe the two different methods for supplying current to a flat-rate water-heater system.
10. Where are cast aluminum condulets approved for use? Why?
11. How is moisture stopped from seeping in where a service conduit enters a wall?
12. Where in the National Electrical Code is information given about the size of service conductors and conduit?
13. What danger exists if the neutral wire of the three-wire system is accidentally broken or loosely connected?
14. Explain why current transformers are used with a service larger than 225 A.
15. List two reasons for grounding an electrical service.
16. Why are services with water meters grounded ahead of the water meter? Why are bonding jumpers placed on each side of a water meter to bypass the meter?
17. How is a service grounded when a house has copper plumbing throughout?
18. Where in the National Electrical Code is information given about the size of grounding wire to be used on a residential service?
19. What two precautions must be taken when grounding a service mast?
20. Describe in your own words two different methods for grounding a service in a rural community where there is no municipal water supply system.
21. Why are plastic bushings or metal bushings with plastic liners used on service conduits?
22. Why is electrical inspection important to the home owner?

Chapter 16

Industrial Services

SERVICE VOLTAGES

Small factories and commercial buildings are sometimes equipped with the 120/240-V service used in residential installations. The larger capacity 120/240-V services (400 A, 600 A, and 800 A) have more than enough circuits for the lights, receptacles, and motors these small, limited-production facilities need.

Industrial buildings usually have motor-driven equipment. These motors often need far more current than do lighting circuits. When being started, a motor draws three to five times its normal operating current, which places an even greater load on the service equipment. A motor designed to operate on a higher voltage will produce the same amount of power but with a decrease in line current. For example, a 1-hp (750-W) motor operating at 115 V (motor operating voltage) requires approximately 16 A. The same motor, internally connected to operate on 230 V, requires about 8 A.

Motors and equipment designed to operate on even higher voltages need still less current. When the current is reduced, the wire size in the motor windings can also be reduced along with the branch-circuit conductor and size of conduit to the motor. This allows, for example, a 550-V motor to be much smaller in size than a 1-hp (750-W), 115-V motor.

POLYPHASE (THREE-PHASE) SYSTEM

Special voltage systems are available for commercial and industrial use. These are known as *polyphase* (three-phase) systems. Three-phase motors and equipment are more efficient, lower in cost, and smaller in size, and usually need less current than equivalent units designed for use with a single-phase supply system.

Three-phase voltages are produced by the alternators (ac generators) at the power station, where they can be stepped up to much higher voltage levels for long-distance distribution. These voltages can then be transformed by the consumer to any one of several common voltage levels. The most useful three-phase voltage levels are 550 V, 440 V, 240 V, and 208 V.

The 550-V and 440-V power systems use 3 live wires without a neutral conductor, Figure 16-1. The 240-V and 208-V systems

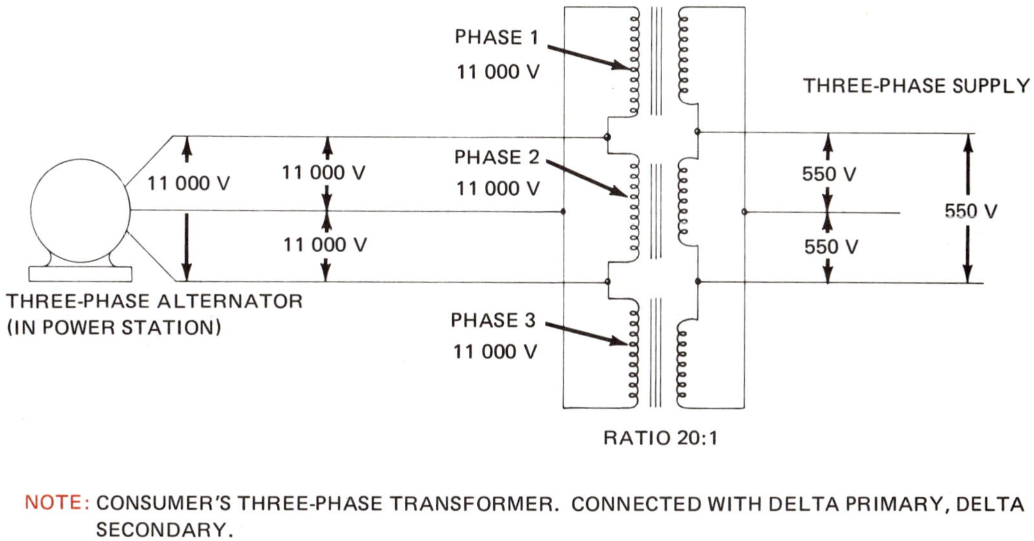

Fig. 16-1 Typical three-phase 550-V or 440-V supply system.

are available in 3-wire (3 live conductors) and 4-wire (3 live and 1 neutral conductor) combinations. A typical three-phase 208Y/120-V supply system is shown in Figure 16-2. The advantage of this system is that there is a 120-V supply available between the neutral conductor and any one of the three live conductors for lighting and receptacles.

Another advantage of the three-phase system is that motors can be reversed easily. A polyphase motor will rotate in the opposite direction if any two of the three live wires are interchanged. To reverse a single-phase motor, the motor often has to be dismantled and the internal connection of the windings changed.

When working on a three-phase service-entrance, take care *not* to interchange any of the main feeder live wires. If the live wires are interchanged, *every* motor in the building will have its direction of rotation reversed.

550-V AND 440-V SYSTEMS

The 550-V and 440-V three-wire services are used primarily for motors and their controls. The meter is usually located outside although, on older services, it may be located within the building in a metal meter cabinet. These three-phase services are available in sizes ranging from 100 A and upward. The main service switch boxes and distribution panels have terminals for the three live wires, but do not provide for a neutral conductor, since there is no neutral conductor.

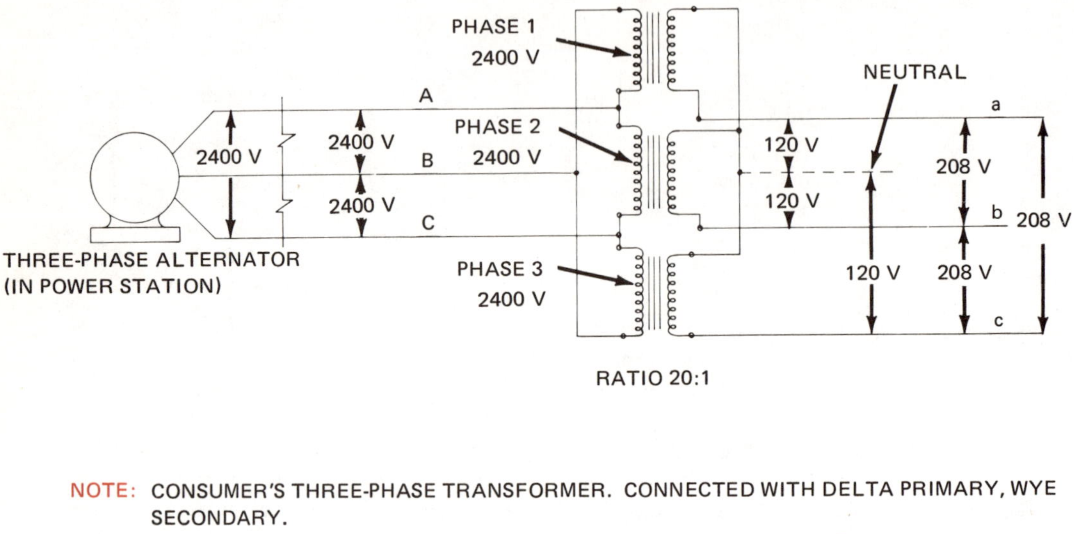

Fig. 16-2 Typical three-phase 208Y/120-V supply system.

Conductor and conduit sizes are chosen in the same way as they are for the residential service. Mast height, meter height, and box height and location are also the same as they are for a residential service. Figure 16-3 shows a typical 550-V or 440-V service installation. The three live conductors are often all one color, usually black, red, or blue. Because of magnetic inductance, the three live conductors must be run together in the same conduit.

Grounding is limited to the boxes, cabinets, panels, and conduits of the system. This is done by placing a grounding conductor between the grounding lug in the main switch box and the cold-water supply pipe. NEC *Table 250-94*, shown in Table 16-1, lists main grounding conductor sizes. Bonding jumpers are sized according to NEC *Table 150-95*, shown in Table 16-2.

METERING

The three-phase system needs a special meter, containing two voltage coils and two current coils, for recording power. This unit is like two meters in one. The total amount of energy used is equal to the sum of both meter sections. This total is shown as a single reading on the instrument's dials, Figure 16-4. Current transformers and potential (voltage) transformers are used with the three-phase meter, Figure 16-5.

Method of Distribution. As Figure 16-3 shows, the service conductors enter a splitter box (junction box/auxiliary gutter) after leaving the meter cabinet. Subfeeder disconnect switches are then joined to the splitter. Conductors of the same ampacity rating as the disconnect switch then dis-

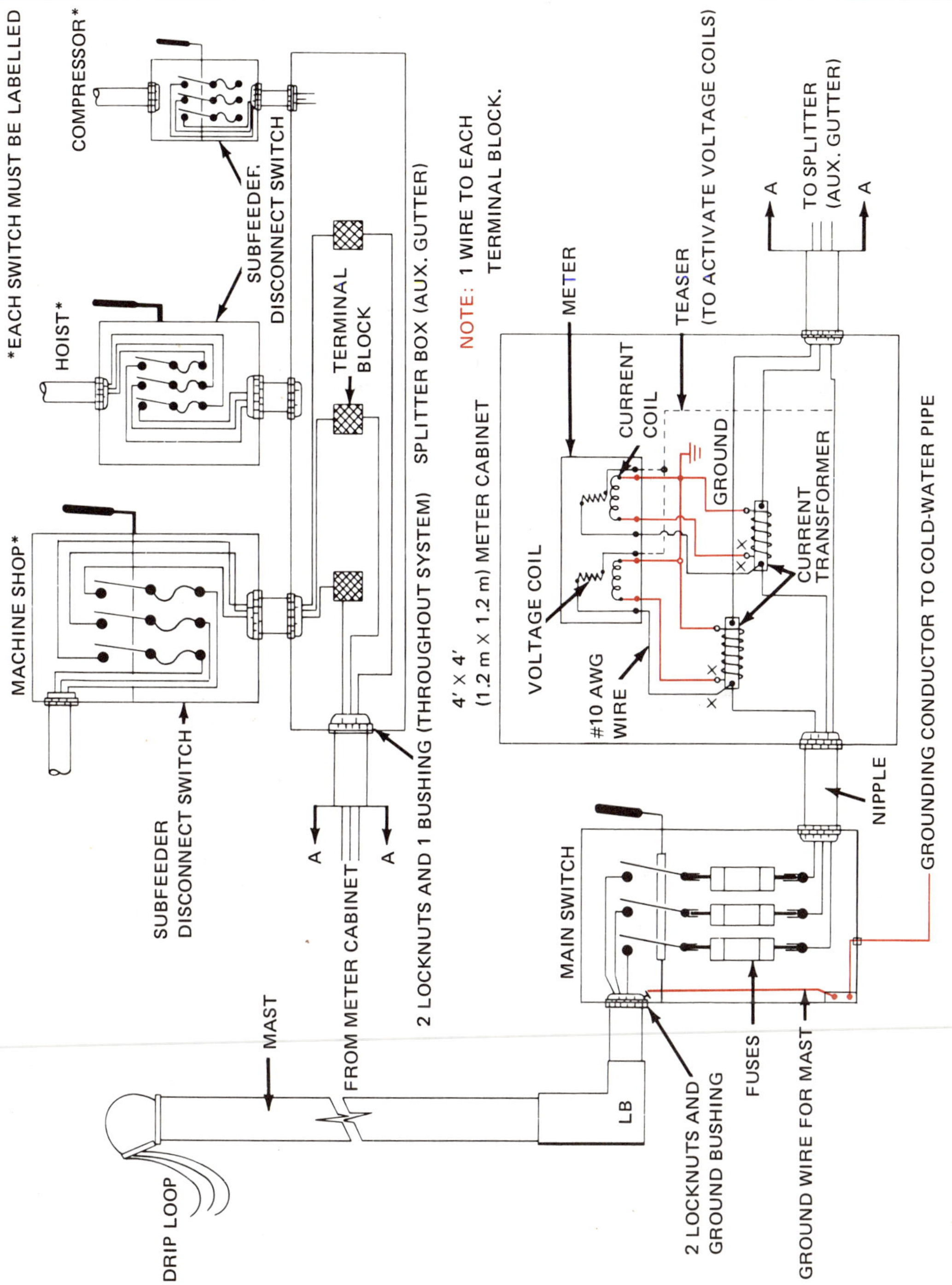

Fig. 16-3 Typical three-phase 550-V or 440-V service installation. (Older style indoor metering.)

Table 250-94. Grounding Electrode Conductor for AC Systems

Size of Largest Service-Entrance Conductor or Equivalent Area for Parallel Conductors		Size of Grounding Electrode Conductor	
Copper	Aluminum or Copper-Clad Aluminum	Copper	*Aluminum or Copper-Clad Aluminum
2 or smaller	0 or smaller	8	6
1 or 0	2/0 or 3/0	6	4
2/0 or 3/0	4/0 or 250 MCM	4	2
Over 3/0 thru 350 MCM	Over 250 MCM thru 500 MCM	2	0
Over 350 MCM thru 600 MCM	Over 500 MCM thru 900 MCM	0	3/0
Over 600 MCM thru 1100 MCM	Over 900 MCM thru 1750 MCM	2/0	4/0
Over 1100 MCM	Over 1750 MCM	3/0	250 MCM

Where there are no service-entrance conductors, the grounding electrode conductor size shall be determined by the equivalent size of the largest service-entrance conductor required for the load to be served.

* See installation restrictions in Section 250-92(a).

See Section 250-23(b).

TABLE 16-1

Table 250-95. Minimum Size Equipment Grounding Conductors for Grounding Raceway and Equipment

Rating or Setting of Automatic Overcurrent Device in Circuit Ahead of Equipment, Conduit, etc., Not Exceeding (Amperes)	Size	
	Copper Wire No.	Aluminum or Copper-Clad Aluminum Wire No.*
15	14	12
20	12	10
30	10	8
40	10	8
60	10	8
100	8	6
200	6	4
300	4	2
400	3	1
500	2	1/0
600	1	2/0
800	0	3/0
1000	2/0	4/0
1200	3/0	250 MCM
1600	4/0	350 "
2000	250 MCM	400 "
2500	350 "	600 "
3000	400 "	600 "
4000	500 "	800 "
5000	700 "	1200 "
6000	800 "	1200 "

* See installation restrictions in Section 250-92(a).

TABLE 16-2

Fig. 16-4 Three-phase meter.

Fig. 16-5 Current transformer.

tribute the three-phase supply to various parts of the building. Each subfeeder disconnect must be labelled correctly to show exactly which area or what equipment it controls.

A second subfeeder disconnect switch is usually mounted on or near the equipment. This is a safety measure for persons servicing the equipment. These switches are designed so that padlocks can be placed on the switch, to lock it in the "open" position. A warning tag, *"Danger – Do not operate,"* is usually placed at the switch. The padlocks keep the switch out of service until the repairs are completed and all service personnel are clear of the machine. Each padlock is removed as each person completes service. When the last padlock has been removed, and the warning tag removed, then the service may be restored safely.

Grounding. 550 V will both severely *shock* a person and cause *serious burns.* As a rule, none of the three-phase 550-V or 440-V conductors is grounded. This isolated system prevents a person in contact with grounded equipment from being injured if and when a live conductor is also touched. Many services are equipped with three indicating lights (ground detectors) to warn that a live conductor has come in contact with a metal box or associated piece of equipment. All metal boxes, panels, conduits, and fittings, however, are grounded.

Take great care when working with a 550-V or 440-V system, because all three conductors are *alive* and *dangerous.* Sometimes current leaking from faulty equipment establishes a partial ground on one conductor. In such cases, any protection given by isolating the system is lost.

208Y/120-V SYSTEM

While 550-V and 440-V systems are excellent sources of power for motors and related equipment, they are not ideal for general lighting and power requirements. Often the 550-V system is reduced through a transformer to a 240-V/120-V delta connection and then distributed to lighting and general power circuits. Only one of the three transformers of the delta bank is center-taped for the neutral. Only the two phase leads from this transformer are used for lighting loads, giving 120 V to the neutral. The third phase conductor gives a much higher voltage to the neutral, and *cannot* be used for lighting loads. This can be expensive, because of the cost of the equipment required. The three-phase, 4-wire, 208Y/120-V supply system combines the best qualities of both three-phase and single-phase systems, and is ideal for lighting and general power.

The three-phase, 208-V system is available for motor-driven equipment. The fourth (neutral) conductor in the system provides 120 V between itself and any of the three live phase conductors for lighting and general power requirements. (Figure 16-2 shows how the two voltages are obtained.)

Service-entrance equipment for the 240-V/120-V delta system or the 208Y/120-V system looks very much like the equipment for 550/440-V units. The difference is that in the 240-V/120-V delta system or the 208Y/120-V system, allowance must be made for the neutral conductor. It is treated in much the same way as the neutral conductor in a single-phase residential system.

Since 120 V are available between the neutral and live conductors, a group of three current transformers must be used to record the power accurately. Figure 16-6A shows a typical 208Y/120-V service installation. (indoor meter). Figure 16-6B shows a high-voltage metering system (outdoor meter). These services are usually available in sizes ranging from 225 A and up.

Often, a 550-V, three-phase, 3-wire supply system is brought into a building, and then changed to a 208Y/120-V, 4-wire system inside, Figure 16-7. In other installations, dry transformers for reducing the voltage are mounted on the wall above the auxiliary gutters and subfeeder disconnect switches, Figure 16-8.

Figure 16-9 shows a three-phase dry transformer with the three sets of windings for reducing the voltage in each phase of the system. Large industrial service installations often come in the form of a cabinet or switchboard system and are ordered from the manufacturer for each individual job or installation. Figure 16-10 shows a small unit. Figure 16-11 shows a larger unit with room for step-down transformers in the screened cabinets (on the right-hand side of the photograph).

Conductors are often run to these switchboard systems from a vault or tunnel built under the unit. Instead of using one very large, hard-to-handle conductor, several smaller wires or cables with a combined ampacity to match the larger cable are installed in parallel runs (see NEC *Article 310).* Figure 16-12 shows this type of installation. Note how the crimp-on solderless lugs are used to terminate the cables.

Grounding. The neutral wire of the 240-V/120-V delta system or the 208Y/120-V system is grounded to the water supply pipe in much the same way as it is in the residential service. This connection gives 120 V between any one of the live conductors and ground of the 208Y/120-V system. (Check Table 16-1 for sizes of the service grounding conductors.)

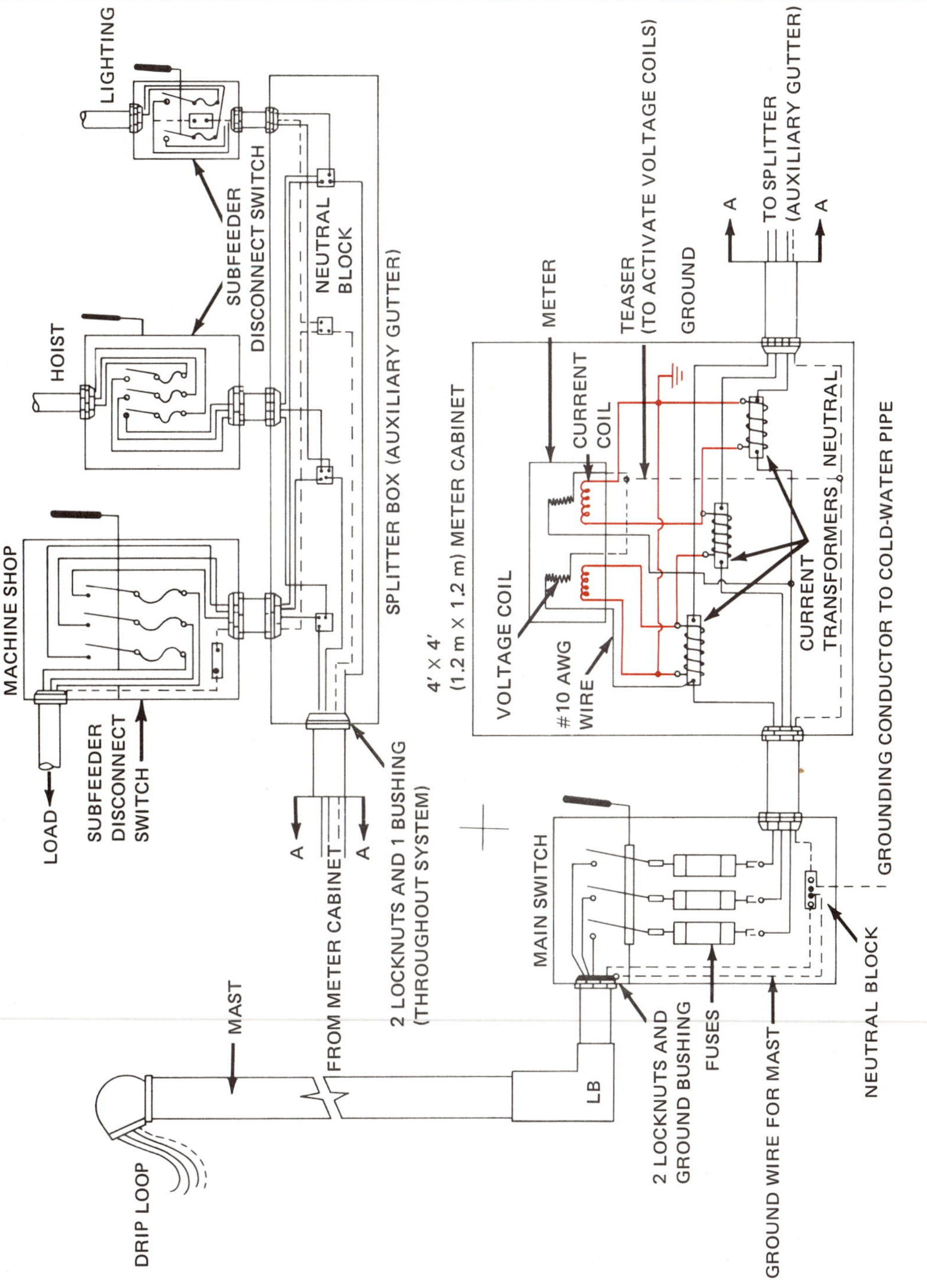

Fig. 16-6A Typical three-phase, 208Y/120-V service installation. (Older style indoor metering.)

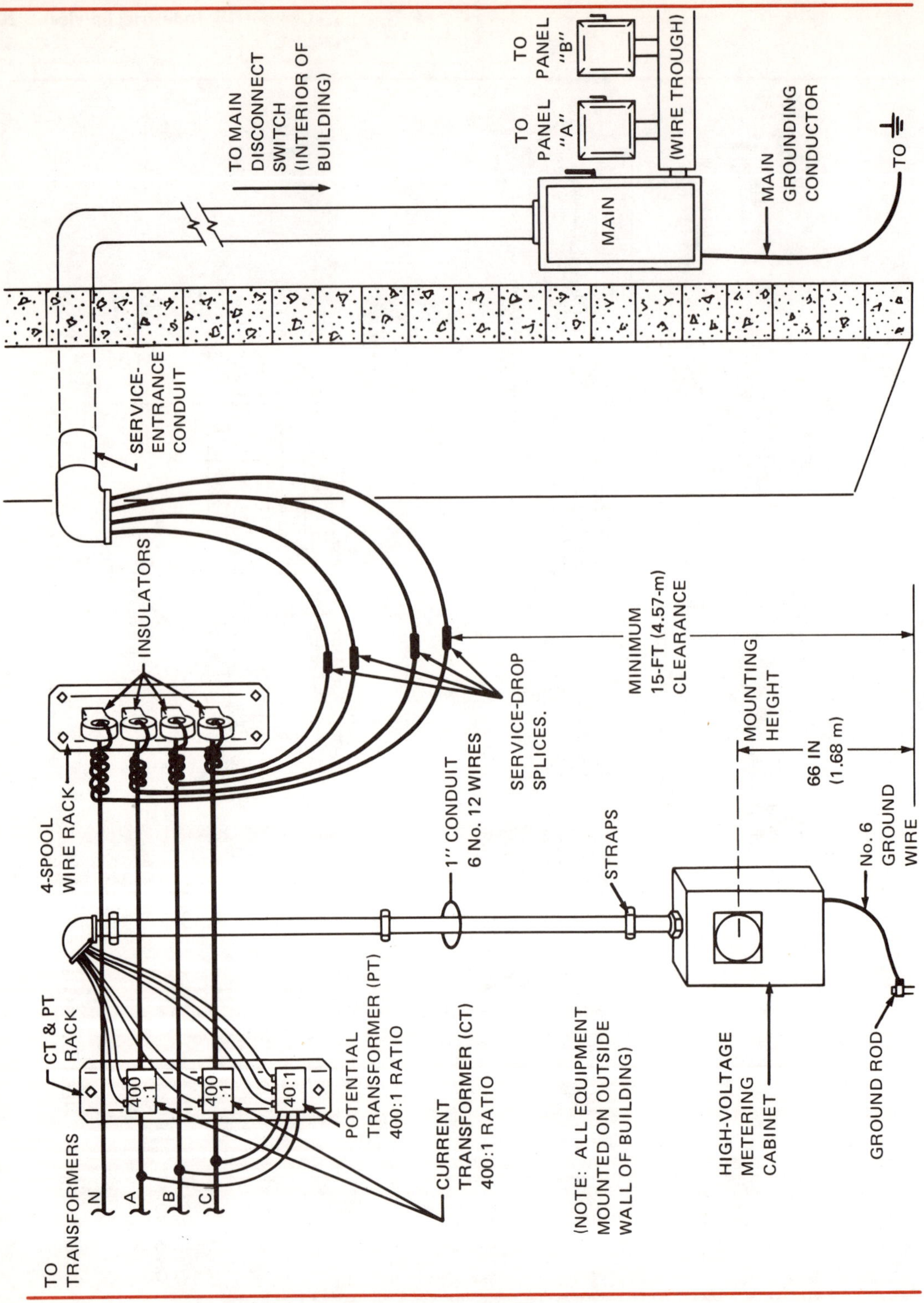

Fig. 16-6B High-voltage metering system (outdoor meter).

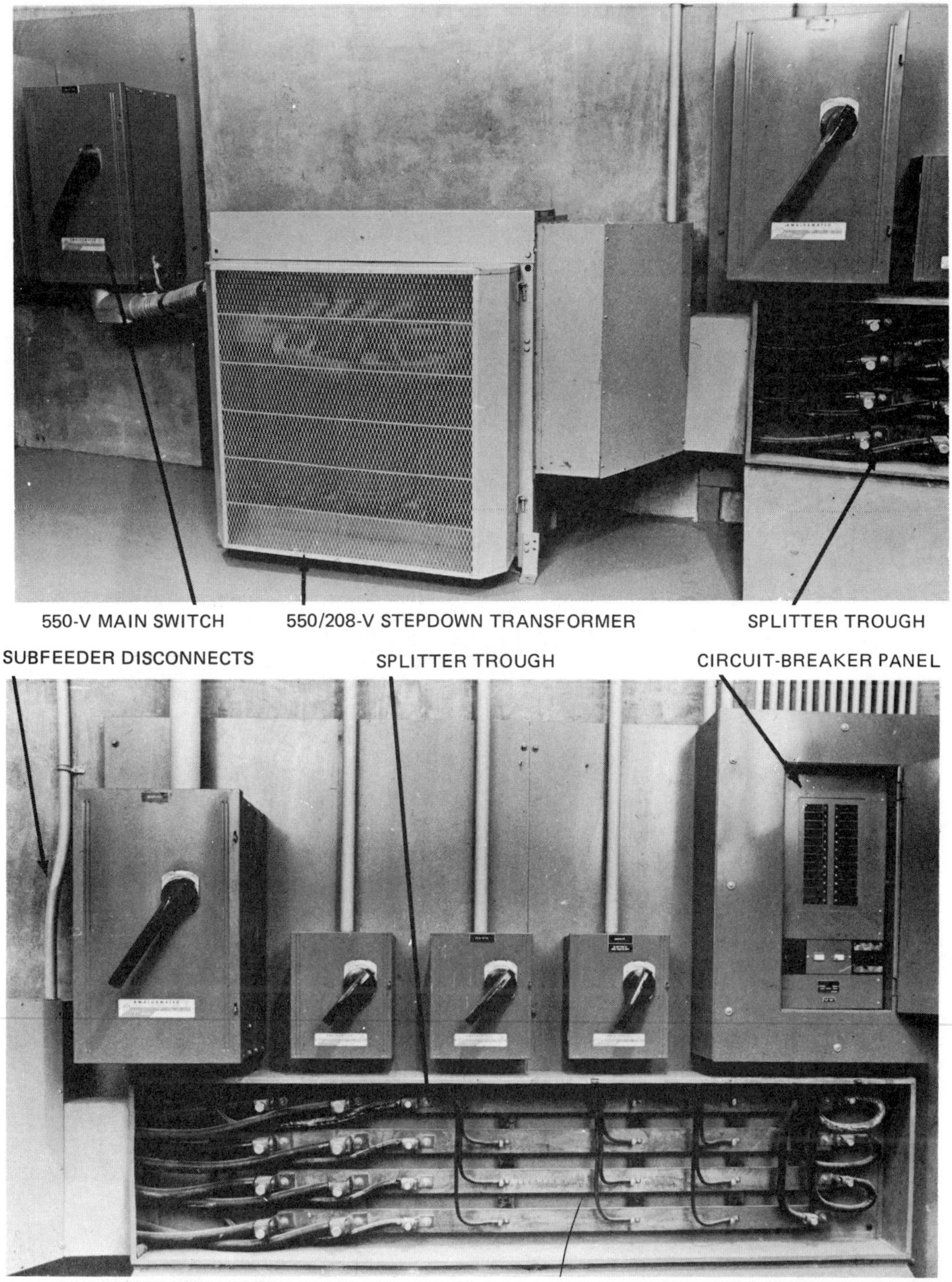

Fig. 16-7 Typical 208Y/120-V three-phase, 4-wire service (older installation).

Fig. 16-8 A main switchboard room installation.

Fig. 16-9 A fry-type, core and coil assembly transformer used in single-ended unit substation.

Fig. 16-10 Typical fusible-type small distribution switchboard.

CONDUCTOR AND CONDUIT SIZE

NEC *Table 310-16* (shown in Table 5-2) and Tables through *310-19* with *Notes* list the current-carrying capacity of the various conductors, when no more than three are enclosed in one conduit. The 208Y/120-V system requires four conductors. *Note 8* of NEC *Table 310-16* states that when four through six conductors are used in a conduit, the *ampacity* of the conductors listed in *Tables 310-16* through *310-19* must be reduced to 80 percent of the current value listed in the tables. Other numbers of conductors are also shown, with their respective derating percentages. Conduit size for this or any service is determined by using the tables in NEC *Chapter 9*.

DEMAND METERS

Industrial users of electricity are assessed in two ways for energy used. The first is by calculating the total energy con-

Fig. 16-11 A large main switchboard, with step-down transformers.

sumption. This is calculated in much the same way as it is for residential customers.

The second way is a charge for the maximum amount of energy drawn from the utility at any given time. This charge is based upon what is called the *demand factor*. The demand factor is recorded by a special indicator on the meter. The demand indicator stays at the highest reading reached during a given period of time, until the meter reader takes the reading and turns it back to zero. The demand factor charge is used so that an industrial user will not place unreasonable loads on the utility's electrical supply system for short periods of time. Such loads force the utility to install large, expensive

Fig. 16-12 Installation of conductors to main switchboard.

pieces of equipment just for that industrial user.

CIRCUIT BREAKERS

Circuit breakers and fuses both have advantages. This section covers only the advantages of circuit-breaker protection.

One of the greatest advantages of the circuit breaker is that it can be a manually operated switch. If an overcurrent condition causes the breaker to open the circuit, it can be reset by hand and put into operation again (after the problem has been cleared) without replacing any parts. The built-in switch mechanism allows the breaker to be used as a control switch for a circuit. Often these breakers are used to control lighting circuits or motor circuits, as well as to provide overcurrent protection. When used to control 120-V fluorescent lighting, the circuit breaker must be approved for switching and marked "SWD." Circuit breakers are more expensive than control switches, but they save on the material and labor that are needed if separate control switches are installed, or if a blown fuse must be replaced. Figure 16-13 shows a typical circuit breaker.

A thermomagnetic circuit breaker has an advantage when short-circuit conditions occur. Since a short-circuit condition reduces the electrical resistance in a circuit to a very low value, the current flow in the circuit rises to a very high level in a very short time (in microseconds; that is, millionths of a second). As the alternating current rises to a maximum value in its cycle (1/240 second), the circuit breaker senses this drastic increase and opens the circuit safely.

The circuit is opened in two ways. A heat-sensitive thermal element (bimetal strip) rises in temperature and triggers the circuit breaker when excessive current flows. A prolonged overload condition also will heat up and operate the thermal element. In the sudden overload (short-circuit or ground-fault) situation, however, a latch-on, magnetic trip assembly quickly operates the circuit breaker. Under short-circuit conditions, the sudden excess in current flow activates this magnetic trip assembly and opens the circuit in microseconds. Fuses do not have this magnetic tripping device.

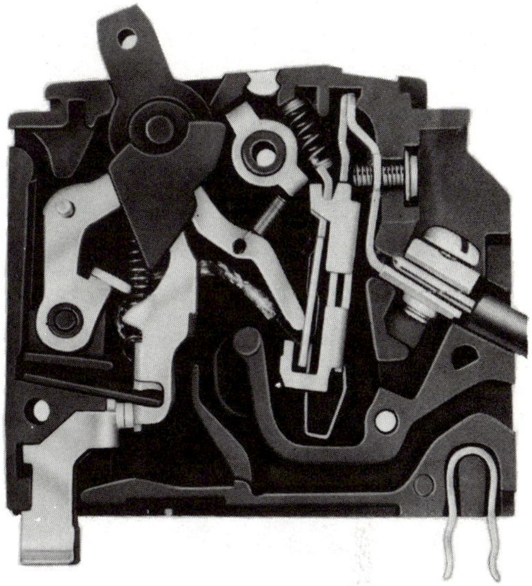

Fig. 16-13 Circuit breaker (open view).

Another advantage of circuit breakers is that they are much smaller in size than fused disconnect switches. Figure 16-14 shows a 30-A, 2-pole fused disconnect switch. Figure 16-15 shows the much smaller 30-A, 2-pole circuit breaker, which will be mounted in the distribution panel shown in the inset. The distribution panel is then mounted in a wall and fitted with a

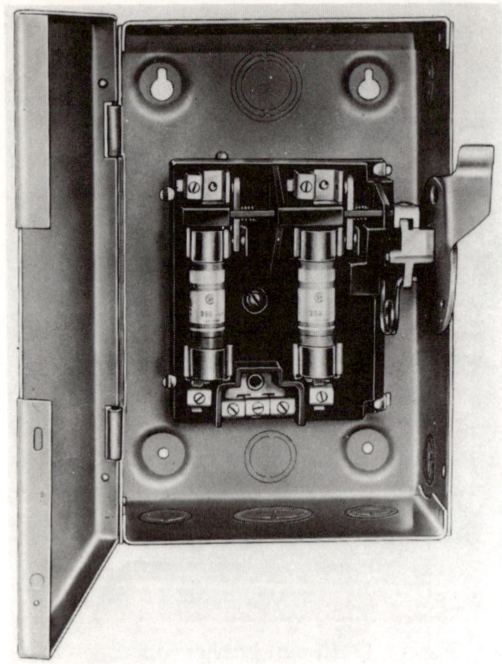

Fig. 16-14 Typical 30-A fused disconnect switch.

Fig. 16-15 Compact circuit breaker for installation in distribution panel.

finished cover once the wall has been covered in, or it may be surface mounted. (See Figures 16-16 and 16-17.)

Large industrial load centers are built with circuit breakers throughout, and the breaker units are assembled into switchboards, Figure 16-18.

Industrial circuit breakers are made in single-pole, double-pole, and three-pole units, with voltage and current ratings to match the circuit conductors. (*Caution:* Do not install a circuit breaker with a lower voltage rating in a higher-voltage circuit.) Some of the larger units are tripped by compressed air to speed up tripping time and help reduce arc damage to the breaker's contacts. A shunt-coil trip unit for circuit breakers is sometimes used for remote-location, emergency-circuit *open* control.

Fig. 16-16 Distribution panel mounted flush in wall.

GROUND-FAULT CIRCUIT INTERRUPTERS (GFCI)

The ground-fault circuit interrupter (GFCI) is a relatively new device adapted

Fig. 16-17 Distribution panel with cover.

from the basic circuit breaker, Figure 16-19. It provides what one manufacturer calls *people protection*, and is designed both to prevent dangerous electrical shocks and to provide overcurrent protection.

Sometimes a piece of electrical equipment develops a *ground fault,* which is a slight leakage of current from any live conductor to the metal frame or housing of a tool or appliance. A few of the common causes of a ground fault are a wire pinched between two parts, insulation breakdown, moisture in the equipment, or metal shavings in the unit. The higher the voltage of the system, the greater the possibility of a ground fault. The leakage current, however, is often so low that it will not trip the normal circuit breaker (or blow a fuse). But it is high enough to *electrocute, cause serious harm,* or *transmit a painful shock* to anyone who comes into contact with the faulty equipment. Portable tools often have

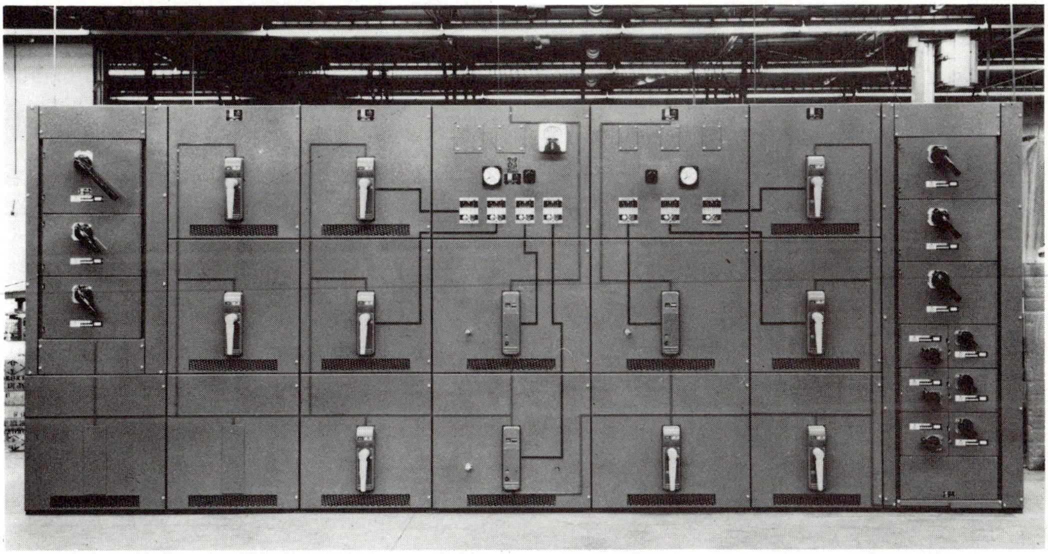

Fig. 16-18 Industrial switchboard circuit-breaker assembly installation.

ground faults and cause many electrical shocks.

The GFCI detects leakage currents as low as 2 mA (0.002 A). It then opens the circuit to protect the operator of the equipment.

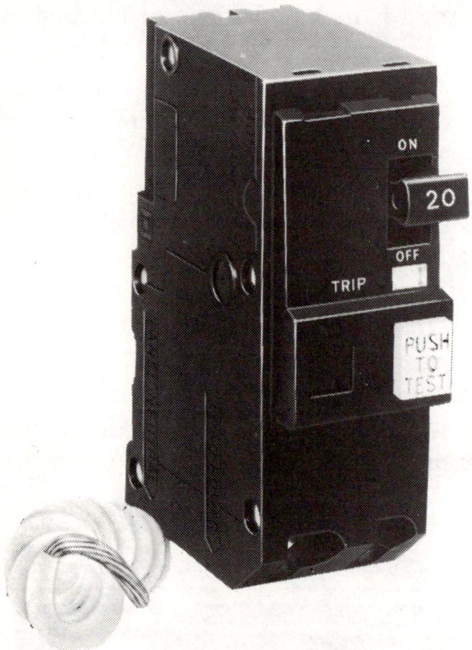

Fig. 16-19 Ground-fault circuit-interrupter.

The average person who receives a shock of 20 mA suffers great pain and loss of muscular control. A *loss of life* can occur at approximately 300 mA. Obviously, the current required to trip a normal 15-A circuit breaker can seriously injure a person. The average power tool equipped with a 3-prong plug and operating on a 15-A circuit is a potentially *dangerous* item if the ground prong is removed for any reason.

Operation of the GFCI. The GFCI is a self-contained unit that may fit directly into a distribution panel. Other types may require special enclosures. Figure 16-20 shows the operation of a GFCI.

The GFCI operates on the principle that the current *leaving* a circuit is equal to the current *entering* that circuit *(Kirchhoff's Current Law).*

Both supply conductors of the circuit pass through a highly developed transformer. When there is no leakage current, the magnetic fields around the supply conductors cancel one another. No voltage is produced in the transformer.

If a leakage current develops, more current is entering the circuit on one supply conductor than is leaving on the other (part

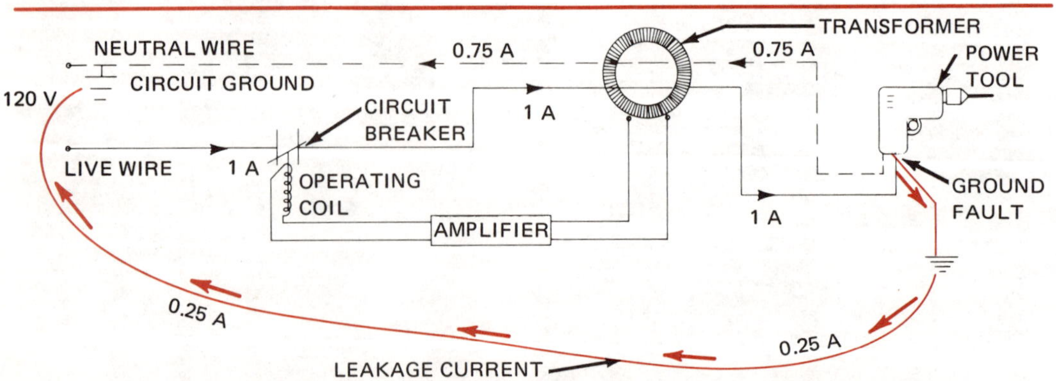

Fig. 16-20 Schematic diagram for typical ground-fault circuit-interrupter.

of the current is short-circuited to ground). This magnetic imbalance causes a voltage to be induced into the transformer coils. An amplifier increases the strength of this voltage and uses it to trip the circuit breaker. Leakage currents as low as 2 mA (0.002 A) will trip the GFCI breaker.

GFCI units are made in a variety of current ratings to suit most circuits. They are recommended for both home and industrial protection. A special (and more sensitive) unit is made for hospitals. *Article 680* of the NEC requires that swimming pools, decorative pools, or other pools with lighting units be protected with a GFCI.

A receptacle-type GFCI is also available. This unit does not open the branch circuit, but offers operating protection to any electrical device plugged into the receptacle.

REVIEW

1. What are three advantages of the polyphase, or three-phase, voltage system in industrial or manufacturing buildings?
2. How is a three-phase service metered?
3. What is the size of a grounding conductor for a 200-A service? A 400-A service? A 600-A service?
4. List the boxes or enclosures used in a three-phase service.
5. How is a three-phase disconnect switch different from a single-phase disconnect?
6. How is the grounding of a three-phase, 550-V service different from that of a single-phase service?
7. Explain why current transformers are used.
8. What advantage does the 4-wire, 208Y/120-V service have over the 3-wire, 550-V or 440-V installation?
9. When installing a three-phase, 4-wire service, how is the ampacity of the conductors determined?
10. What is a *demand meter?* Why and how is it used?
11. List three main advantages a circuit breaker has over a fuse.
12. List two types of circuit protection provided by circuit breakers.
13. In what two ways does a circuit breaker trip a circuit when under short-circuit conditions?
14. Why is it important to trip a circuit breaker in microseconds when a short circuit occurs?
15. What method is used to trip large industrial circuit breakers? Why?
16. What is a *ground fault*?

17. List three ways in which ground faults often occur.
18. Why is a standard circuit breaker (or fuse) sometimes useless when a ground fault occurs in a circuit?
19. What is the lowest current level at which a GFCI in the circuit can operate?
20. Explain in your own words how a GFCI operates.

Chapter 17

Fuses

HEAT GENERATION

When an electrical current passes through a conductor, some heat is generated. The more current that passes through a conductor, the more heat that is generated within the conductor.

The amount of heat produced is proportional to the square of the current, expressed as I^2. That is, if the amount of current in a conductor is doubled, the heat will be four times as great. If the amount of current in the same conductor is tripled, the heat will be nine times as great. The length of time that the circuit is energized will affect the heat produced. Therefore, the heat produced is proportional to the square of the current and directly proportional to the time the circuit is energized, expressed as $I^2 t$. It can be seen that a continuous load circuit (one that would be energized for three hours or longer) would produce a greater buildup of heat than a circuit energized for only a few minutes and then allowed to rest.

Several other factors affect the amount of heat produced in any given conductor. The type of metal used in the conductor is a factor. Copper is a better conductor than aluminum, and so it will carry more current. Because it can carry more current, a copper conductor of a given size and current load will generate less heat than an aluminum conductor under the same conditions.

The physical size – the gauge number – of the conductor also helps determine the current-carrying capacity of the conductor and, therefore, the amount of heat that will be generated under certain loads. Small conductors do not carry as much current as large conductors. Therefore, a small conductor trying to carry too large a current load will generate more heat than will a conductor of the right size.

The amount of air surrounding a conductor is another factor, because the air space has a great deal to do with the conductor's ability to cool or give off its heat. The temperature of the air surrounding the conductor is called the *ambient temperature.* If the conductors are crowded into a conduit or box where there is little circulation of air, cooling will be difficult and the ambient temperature will be high. This situation will add to the overall heating of the conductor. Such places as boiler rooms and foundries often have problems with conductors because of high ambient temperatures.

Conductor insulation also has some small effect on the ability of a conductor to give off heat. Some modern plastic insulations tend to retain the heat in the conductor in the same way that insulation in the walls of a building holds in heat. Therefore, the current ratings for these conductors must not be exceeded. If the ratings are exceeded, sufficient heat will be generated and retained to melt or damage the plastic insulation. NEC *Tables 310-16* through *310-19* list the current-carrying capacities of aluminum and copper conductors. Heat rises to a significant level in these conductors when approximately 80 percent of the listed current is passed through the conductor.

DAMAGE FROM OVERHEATING

Overheating of conductors has several effects that cause problems. One effect is that when conductors reach a temperature beyond their normal operating range, there is a softening, called *annealing,* of the conductor. The softening removes any resilience (spring-like action) at the terminal and causes a loosening of the connection. (This effect is similar to the cold flow of metal experienced with aluminum conductors except that, in this case, heat is the cause.)

Another effect is that an increase in temperature brings about a more rapid rate of oxidation on a conductor. Copper is a good conductor, but copper oxide is not. In fact, copper oxide tends to weaken the electrical integrity of a terminal connection. Aluminum oxide causes an even greater problem. It is very close to being an insulator and will cause further heating at the terminals. The oxide establishes an electrical resistance, which results in a loss of voltage at the terminal. Often, enough heat is generated to completely destroy the terminal connection.

Overheating also causes insulation to dry out and become hard and brittle. This means that movement of the conductor will likely cause the insulation to crack and fall off. The exposed conductors may well start a fire if they touch one another or a grounded box. Although it usually takes many months of circuit use to dry out insulation, the problem can easily be overlooked because conductors are seldom checked after installation.

If a very high overload of current occurs, such as with a short circuit, conductors can heat up until they actually glow red. When this overheating happens, the insulation will ignite and start a fire within the conduit, box, or wall of the building. Even if the actual damage from the fire is small, the burning insulation gives off an offensive odor that will linger for a long time.

Obviously, the effects of overheating of conductors can be very serious. The current must be controlled. One effective way to protect a circuit is with a fuse.

WEAKEST THERMAL LINK

A *fuse* is a simple, current-limiting device designed to protect the conductors of a circuit. Protecting the conductors prevents serious damage to equipment from overload and fire.

Fuses contain a strip of heat-sensitive, alloy metal that has a slightly higher electrical resistance and lower melting point than the copper or aluminum circuit conductors. The metal used in a fuse can be a zinc alloy, copper alloy, or silver, depending upon the fuse type and use. Copper and silver links have a small amount of tin added to their composition to reduce their melting temperature. Because of its higher resistance, the link heats up sooner than the conductors in the circuit. The fuse link then automatically opens a circuit when excessive current is passed through the conductors. Under short-circuit conditions (that is, two live wires shorted together or one live wire shorted to ground), the fuse link opens (melts) in just a fraction of a second.

The fuse is designed to be a "weak link in the chain." Its job is to break (open) the circuit before any damage is done to the circuit conductors. For this reason, the fuse is often called the weakest thermal link in the circuit.

Since a fuse is heat-sensitive, ambient temperature plus the heat generated in the fuse determines when the fuse will melt. Some fuses are designed to be more heat-sensitive than others. Remember that conductors of electricity are often conductors of heat, and so can transfer their heat on to a fuse.

Fuse Ratings. Fuses are rated by the manufacturer in three ways, and expressed in root-mean-square (RMS) values. RMS values (effective values) are those that would be read on a standard voltmeter or ammeter if placed into a circuit of the alternating-current type.

The *continuous current rating* is the amount of circuit current the fuse will carry without blowing and interrupting the circuit. This rating should be matched to the current rating of the circuit conductors.

Maximum rated voltage indicates to the installer the type of circuit and voltage conditions under which the fuse can safely operate. The higher the voltage, the larger the size (length) of the fuse or distance between the contact points of the fuse. *Do not* use a lower voltage rated fuse on a circuit operating at a higher voltage source. See NEC *Section 110-9*.

Interrupting capacity is the amount of current that the fuse can safely interrupt in the circuit under short-circuit conditions. Current flow can be very high when a short circuit occurs, and so the type of fuse used must match the availability of current flow from the circuit's source. The available short-circuit current of the source depends upon: 1) the supply transformer's rating in kilovolt-amperes (kVA) and its percent impedance, 2) the size of the service conductors, and 3) the length of the conductors. See NEC *Section 110-9*.

EDISON-BASE OR PLUG FUSES

The most common type of fuse used in residential buildings is the *Edison-base* fuse, which is usually called the *plug* fuse, Figure 17-1. Plug fuses are used for all lighting and receptacle circuits operating at a maximum voltage of 125 V between conductors. Standard plug fuses are made in current ratings of 3 A, 6 A, 10 A, 15 A, 20 A, 25 A, and 30 A. (See NEC *Section 240-6* for standard ampere ratings.) Time-delay plug fuses are available in many different fuse ratings. For the operation of the time-delay fuse, see the section in this chapter on dual-element fuses.

All plug fuses are made in about the same physical size, but each must have its ampere rating marked so that it will be plainly visible after the fuse has been installed. Some manufacturers stamp the ampere rating on the center contact. The ordinary plug fuse (symbol W) is satisfactory when only incandescent lighting or heating appliances are connected. Time-delay plug fuses (symbol T) are excellent for motor-driven appliance circuits. Plug fuses

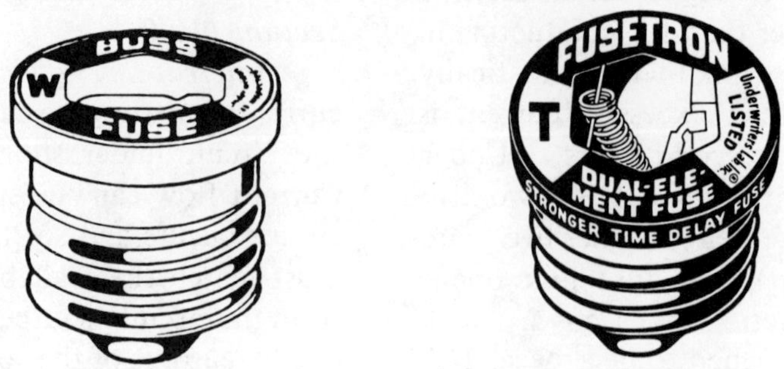

(A) ORDINARY PLUG FUSE, NONTIME-DELAY, EDISON-BASE TYPE

(B) FUSETRON DUAL-ELEMENT, TIME-DELAY PLUG FUSE, EDISON-BASE TYPE

Fig. 17-1 Typical Edison-base plug fuses.

of the older type, with a standard Edison screw base, are now permitted for replacement use only (NEC *Section 240-51*.) New plug fuses must be Type S with screw bases not interchangeable with Edison-base fuses, and with ranges of 0 to 15 amperes, 16 to 20 amperes, and 21 to 30 amperes. The Type S base will not permit a fuse of a higher-ampere range to be interchanged with a fuse of a lower-ampere range.

All plug-type fuses rated at 15 A or less have some sort of prominent hexagonal shape.

The electrician will calculate the various circuits for their ampere load when installing fusible equipment. Based on this rating, an adapter of the proper size is inserted into the Edison-base fuseholder. The proper Type S fuse is then inserted into the adapter. Once installed the adapter cannot be removed, thereby making the fuseholder nontamperable and noninterchangeable. Adapters are ranged for 0 to 15 amperes, 16 to 20 amperes, and 21 to 30 amperes group rating. A Type S fuse from another group rating will not be interchangeable.

Often a branch circuit has a motor load. With a motor on the circuit, the ordinary plug fuse is not recommended because it has very little time lag, and the starting current of any motor may open the fuse (particularly if the circuit is already partially loaded). The Type S fuse may have a time-delay feature. If it does, it is known as a dual-element fuse. Momentary overloads do not cause a dual-element fuse to blow. However, a dual-element fuse opens rapidly if there is a heavy overload or a short circuit. Motor branch circuits require a time-delay fuse or a dual-element fuse for protection.

Circuit Fault Indications. The see-through window of the plug fuse is a great help for finding what caused a fuse to blow. If an overloaded circuit or a motor starting up has caused the fuse to open, the glass or mica face will still be clear and only a small part of the link will have melted away, Figure 17-2. If a short circuit or violent overload has caused the fuse to open, the inside of the glass or mica face of the fuse will usually blacken. It will often be impossible to see the remains of the fuse link, because most of the link will have melted from the sudden heat. When this happens, the cause of the problem should

Fig. 17-2 Circuit fault indications.

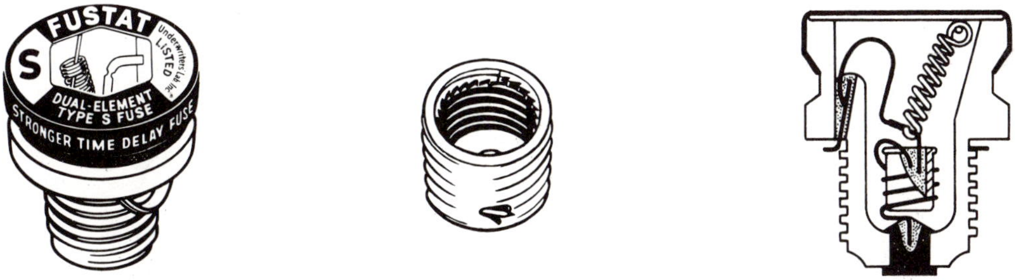

Fig. 17-3 Type S fuses and adapter.

be located and corrected before replacing the fuse.

When a fuse operates close to its rated current value for any length of time, it starts to warm up. The circuits in a distribution panel can be checked simply by running a finger over the face of the fuses. Any fuse that feels warm is carrying current close to its rated value. Standard plug fuses will not open until the rated current is reached. If a load of current close to the rating of the circuit is carried for a long time, it can cause serious heat damage to the fuseholder and distribution panel. Always be sure that plug fuses are screwed in tightly.

TYPE S FUSES

Type S fuses, Figure 17-3, are of the plug type, and are classified at not over 125 volts and 0 to 15 amperes, 16 to 20 amperes, and 21 to 30 amperes. As classed by amperes, these fuses may not be interchanged with others in a lower ampere classification. A Type S fuse is designed so that it can be used only in a Type S fuseholder, or a fuseholder with the correct Type S adapter for the purpose. Type S adapters are made so that they cannot be removed once they have been installed. Type S fuses, fuseholders, and adapters are made to be interchangeable, regardless of the manufacturer.

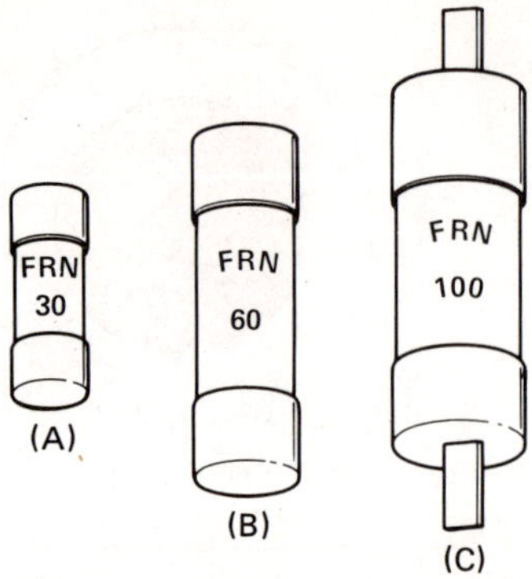

Fig. 17-4 Comparing the physical sizes of cartridge-type fuses. (A) and (B) are ferrule-contact fuses. (C) is a knife-blade contact fuse.

Fig. 17-5 Example of knife-blade cartridge fuse.

FERRULE-CONTACT CARTRIDGE FUSES

Residential equipment such as furnaces, electric ranges, clothes dryers, water heaters, and electric heaters are rated to operate at 230 V and often need a different type of fuse. For these applications, the 250-V rated cartridge fuses are used. Units rated at 300 V and 600 V are also available for industrial and commercial use. Figure 17-4 shows cartridge-type fuses and a relative comparison of their sizes.

The 600-V ferrule-contact cartridge fuse is generally larger in physical size than the 250-V unit. It is larger in diameter and longer in length than a 250-V fuse of the same current rating. The diameters and lengths of the regular 250-V and 600-V fuses were established many years ago when the state of the art of fuse design was limited to the replaceable-link type. The replaceable-link fuse, often referred to as the *renewable* fuse, did not contain an arc-quenching filler material as do today's modern fuses that have both high-quality tube material and arc-quenching filler material. This new high-technology design permits fuses to be much smaller in diameter and much shorter in length than the older style fuses, but still fully able to handle the circuit's full-voltage and current-rating demands. This is discussed later in this chapter.

KNIFE-BLADE CARTRIDGE FUSES

When currents in excess of 60 A are flowing in a circuit, the larger, more rugged knife-blade cartridge fuse is used to protect the conductors, Figure 17-5. The 250-V, 300-V, and 600-V fuses, in all ampere ratings, are used in residential, commercial, and industrial applications.

As with the ferrule-contact cartridge fuses, the 600-V knife-blade cartridge fuse is generally larger in physical size than the 250-V knife-blade fuse. The ferrule-type cartridge fuses are made in two physical sizes: 0–30 A, and 31–60 A. The knife-blade type cartridge fuses are made in the following physical sizes: 61–100 A, 101–200 A, 201–400 A, and 401–600 A. The length and diameter of the fuses in each group increase as the current rating increases. For example, the fuses in the 61–100 A group are smaller in length and diameter than those in the 101–200 A group.

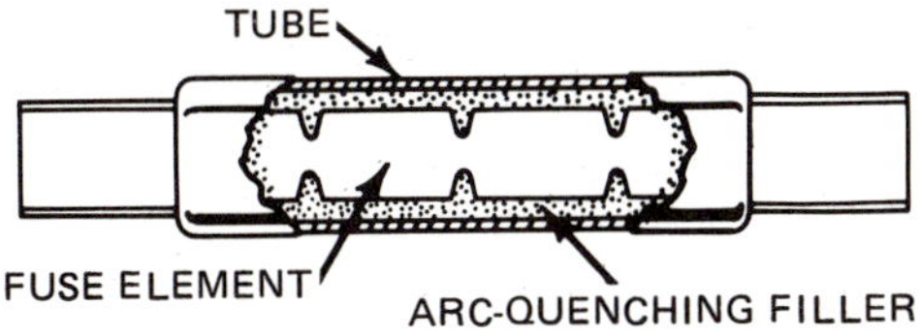

Fig. 17-6 Knife-blade cartridge fuse construction.

ARC-QUENCHING MATERIAL

Most of the large cartridge fuses are filled with an arc-quenching material. This fire-extinguishing material, made of gypsum powder or silica sand, is designed to prevent an arc from continuing or bursting through the jacket of the fuse. In large fuses, a short circuit often results in heat and flame from the fuse link, and so arc-quenching material is important. Figure 17-6 shows a knife-blade cartridge fuse with arc-quenching filler, fuse element, and tube.

ONE-TIME CARTRIDGE FUSES

The common one-time fuse used by many home owners has, for many years, been made with a zinc alloy link having an interrupting capacity of 10 000 amperes. This type of fuse, originally required by the Code and referred to as the Code fuse or NEC fuse, is now classified by UL as the Class H fuse. This is to distinguish these fuses from newer, higher-rated fuses of similar dimensions. If a short circuit or overload condition causes this fuse to blow, the entire fuse unit has to be replaced.

The standards of the 250-volt and 600-volt dimensions of ferrule-type fuses to 60-amperes, and blade-type fuses from 61 amperes to 600 amperes, were set in 1904 and still apply today. These fuses have continuous current ratings up to 600 amperes and a short-circuit rating of 10 000 amperes. UL does not require the label to state either Class H or the interrupting rating if it is not more than 10 000 amperes. Recent developments have produced a one-time fuse with a copper element or link having an interrupting capacity of 50 000 amperes. The copper element is nondeteriorating, unlike the old style zinc alloy, and provides much added protection for a small increase in cost. The label of these fuses is marked K5, and shows the interrupting rating as well.

All continuous current ratings to 600 amperes are available for either 250-volt or 600-volt systems, with different dimensions for each voltage rating. They can be either one-time type, with or without time delay, or renewable type, without time delay. The one-time fuse is discarded entirely when blown. The renewable fuse has a reusable outer cartridge.

SHORT CIRCUITS

A short-circuit current is distinguished from an overload by virtue of the current flowing outside of the normal path. Usually, this current is much larger than overload values, ranging from tens to hundreds of times that which is normal. Great damage can be expected quickly under short-circuit conditions, in the form of thermal, magnetic, and arcing damages. Time is critical in the removal of short circuits from the system.

CURRENT LIMITATION

A current-limiting fuse is a fuse which, when interrupting current in its current-limiting range, reduces the short-circuit current flowing in the faulted circuit to a much lesser current than that which would flow if the current-limiting fuse were not in the circuit.

A current-limiting fuse protecting a component is required to limit the energy a short circuit can deliver to a let-through current value not to exceed the component's short-circuit withstand rating.

Current-limiting fuses can restrict the short-circuit current before the first damaging peak current is reached, and can clear the circuit in one-half cycle or less.

RENEWABLE-LINK CARTRIDGE FUSES

In industrial installation, fuses often need to be replaced. Renewable-link fuses are shown in Figures 17-7 and 17-8. Renewable fuses are listed by UL as Class H fuses, having an interrupting rating of 10 000 A maximum.

When the fuse is blown, the cartridge is opened and the burned-out fusible element, or link, is replaced with a new link. It is very easy (accidentally or intentionally) to replace the link with one of a different, often higher, rating, or, to insert two links in place of one. This is a violation of the Code and should never be allowed. Uninformed maintenance personnel sometimes use double-linking to prevent fuse blowing, which can result in a destructive fire or equipment damage.

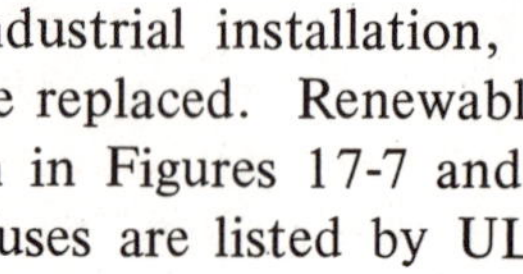

Fig. 17-7 Typical renewable-link cartridge fuse.

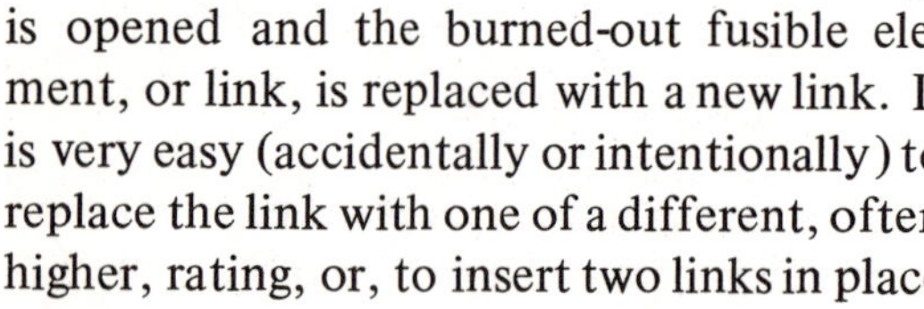

Fig. 17-8 Typical renewable-link knife-blade cartridge fuse.

Renewable fuses have been in existence for many years. It was once thought that using renewable-link fuses reduced maintenance costs. This is no longer true. The cost of labor, and the possibility of replacing the link with one of the improper size, not replacing it mechanically tight, or doubling up on the links, has brought about numerous problems in the field as evidenced by the actual charring of the fuse tube.

For these reasons, most large commercial and industrial users install Class RK fuses that have a 200 000-A interrupting rating, and superior overcurrent protection (as covered later in this chapter).

When reassembling the fuse parts, care must be taken to see that all connections are securely tightened. A loose connection will cause excessive heat.

CODE REQUIREMENTS

The National Electrical Code does not provide detailed specifications for material and equipment. However, it is required that all equipment used in electrical installations be adequate and suitable for the intended use. The Code requires cartridge fuses to be plainly marked, either by printing on the fuse barrel or by a label attached to the barrel, showing the ampere rating, the voltage rating, and the interrupting rating where it is other than 10 000 amperes. The fuse must be marked *current limiting* where applicable, and it must bear the manufacturer's name or trademark.

The Code further requires compliance with the manufacturer's instructions and those that are provided with any accompanying *listed* or *labeled* equipment when installing such equipment. The use of such

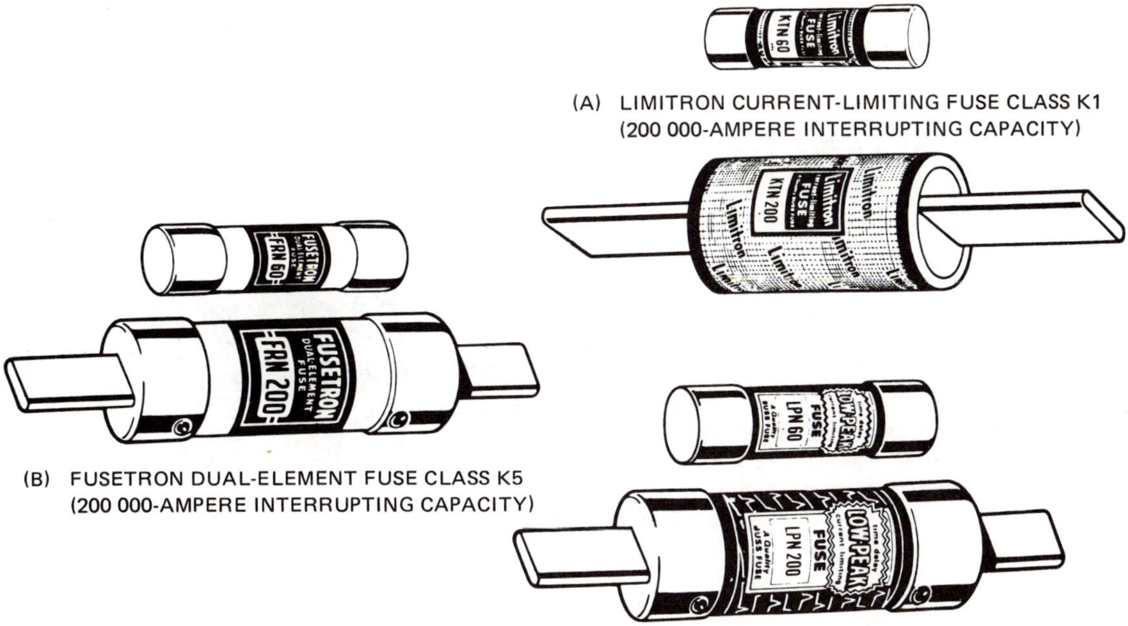

(A) LIMITRON CURRENT-LIMITING FUSE CLASS K1 (200 000-AMPERE INTERRUPTING CAPACITY)

(B) FUSETRON DUAL-ELEMENT FUSE CLASS K5 (200 000-AMPERE INTERRUPTING CAPACITY)

(C) LOW-PEAK DUAL-ELEMENT CURRENT-LIMITING FUSE, CLASS K1 (200 000-AMPERE INTERRUPTING CAPACITY)

Fig. 17-9 Examples of high-interrupting capacity fuses.

equipment must also comply with the instructions.

HIGH-INTERRUPTING CAPACITY FUSES (HIC)

High-interrupting capacity fuses must be used when the available short-circuit current exceeds 10 000 A rms symmetrical. They are available in 250 V, 300 V, and 600 V, and in ampere ratings of 0–6 000 A. These are the UL-listed Class G, J, K, L, RK1, RK2, and T fuses. See Figure 17-9.

The term *high-interrupting capacity* fuse indicates a fuse that has an interrupting rating of 50 000, 100 000, or 200 000 rms symmetrical amperes, according to the most recent UL standard. The terms interrupting capacity and interrupting rating are used interchangeably in the electrical industry.

HIC fuses are very practical in motor circuits. They provide rapid short-circuit protection and time delay on moderate (temporary) overloads, which allow for the extra current needed during the motor's starting period.

The HIC fuse does not deteriorate because it has a moistureproof melamine or ceramic tube. It is filled with high-quality silica sand for arc quenching. These fuses, in most cases, contain silver links, although some may contain copper links.

The Canadian equivalent of the U.S. HIC fuse is referred to as the *high-rupture capacity fuse* (HRC). This fuse is available in two basic types: the Form I fuse and

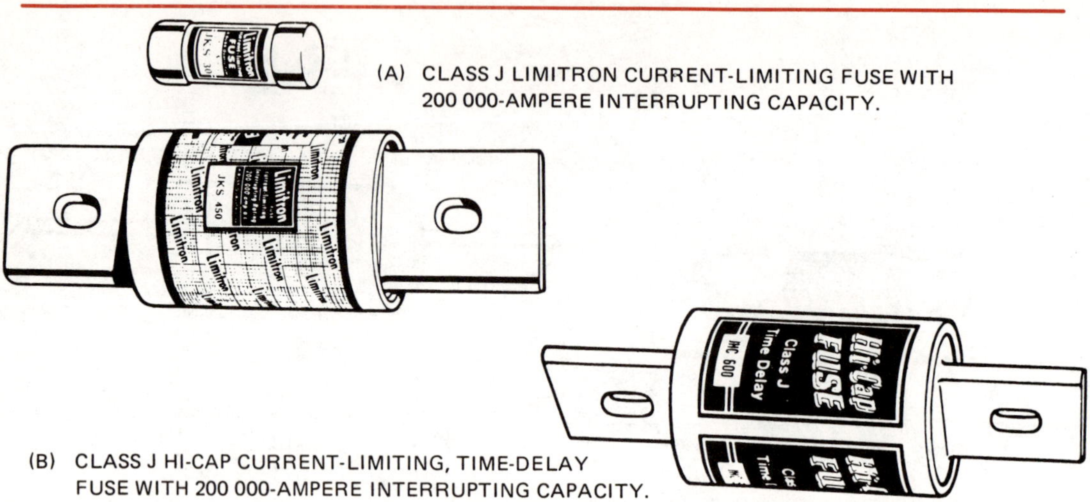

(A) CLASS J LIMITRON CURRENT-LIMITING FUSE WITH 200 000-AMPERE INTERRUPTING CAPACITY.

(B) CLASS J HI-CAP CURRENT-LIMITING, TIME-DELAY FUSE WITH 200 000-AMPERE INTERRUPTING CAPACITY.

Fig. 17-10 Typical Class J fuses.

Form II fuse (known as HRC-I and HRC-II).

Form I (HRC-I). These fuses provide both overload and short-circuit protection. They have a high-interrupting capacity (100 000 amperes and 200 000 amperes) and are made in a variety of types and classes for use in protecting cables and equipment.

Form II (HRC-II). This version of the HRC fuse provides short-circuit protection *only* and must be used in conjunction with some other overload device such as a circuit breaker or overload relay. A Form II fuse must never be used to replace a Form I fuse, but a Form I fuse may be used in place of a Form II fuse since it provides both types of protection.

Class K. Class K fuses are grouped into three categories: K1, K5 and K9. These fuses may be UL listed with interrupting ratings in symmetrical rms amperes in values of 50 000, 100 000 or 200 000 amperes. For each K rating, UL has assigned a maximum level of peak let-through current (I_p) and energy as given by I^2t. Class K fuses have varying degress of current-limiting ability, depending upon the K rating. Class K1 fuses have the greatest current-limiting ability and Class K9 fuses the least ability. Class K fuses may be listed as time-delay fuses as well. In this case, UL requires that the fuses have a minimum time delay of 10 seconds at 500% of the rated current. Class K fuses are available in ratings ranging from 0–600 amperes at 250–600 volts ac. Class K fuses have the same dimensions as Class H fuses.

Class J. Class J fuses are current limiting and are so marked, Figure 17-10. They are listed by UL with an interrupting rating of 200 000 symmetrical rms amperes. Class J fuses are physically smaller than Class H fuses. Therefore, when a fuseholder is installed to accept a Class J fuse, it will be impossible to install a Class H fuse in the fuseholder.

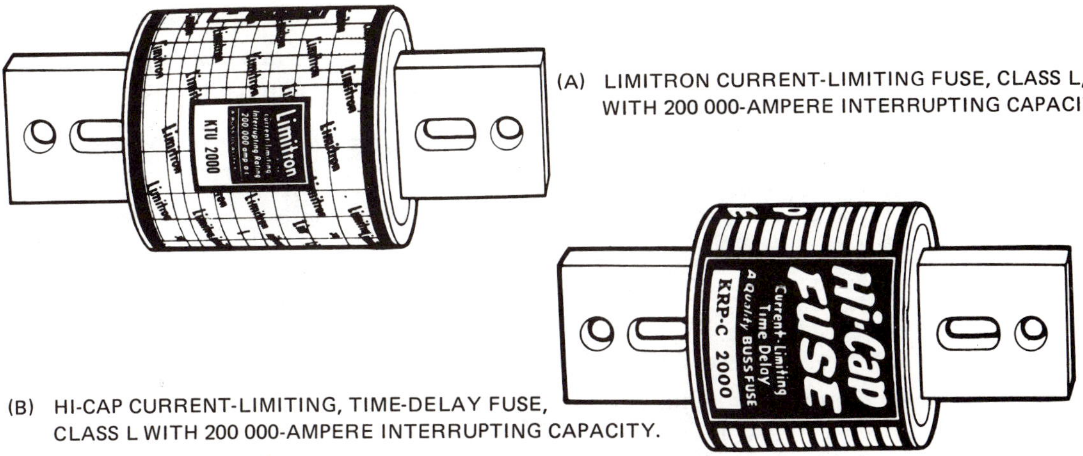

(A) LIMITRON CURRENT-LIMITING FUSE, CLASS L, WITH 200 000-AMPERE INTERRUPTING CAPACITY.

(B) HI-CAP CURRENT-LIMITING, TIME-DELAY FUSE, CLASS L WITH 200 000-AMPERE INTERRUPTING CAPACITY.

Fig. 17-11 Typical Class L cartridge fuses

The UL has assigned maximum values of I^2t and I_p which are slightly less than those for Class K1 fuses. Both fast-acting, current-limiting Class J fuses and time-delay, current-limiting Class J fuses are available in ratings ranging from 0–600 amperes at 600 volts ac. UL standards for time-delay Class J fuses have not been established. However, the time-delay Class J fuse is listed by the Canadian Standards Association (CSA).

Class L. Class L fuses, Figure 17-11, are listed by UL in sizes ranging from 601–6 000 amperes at 600 volts ac. These fuses have specified maximum values of I^2t and I_p. They are current-limiting fuses and have an interrupting rating of 200 000 symmetrical rms amperes. These bolt-type fuses are used in bolted pressure contact switches. Class L fuses are available in both a fast-acting, current-limiting type and a time-delay, current-limiting type. Both types of Class L fuses meet UL requirements.

Class T. The Class T fuse is a recent addition to the UL listing for fuses. Class T fuses are current-limiting fuses and are so marked, Figure 17-12. These fuses are UL listed with an interrupting capacity of 200 000 symmetrical rms amperes. Class T fuses are physically smaller than Class H or Class J fuses. The configuration of this type of fuse limits its use to fuseholders and switches that will reject all other types of fuses.

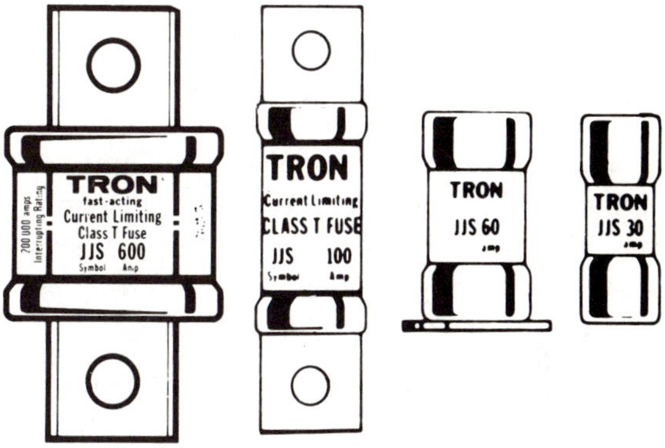

Fig. 17-12 Class T cartridge fuses

Class T fuses have electrical characteristics similar to those of Class J fuses and are tested in a similar manner by UL. These

Fig. 17-13 Class G fuses.

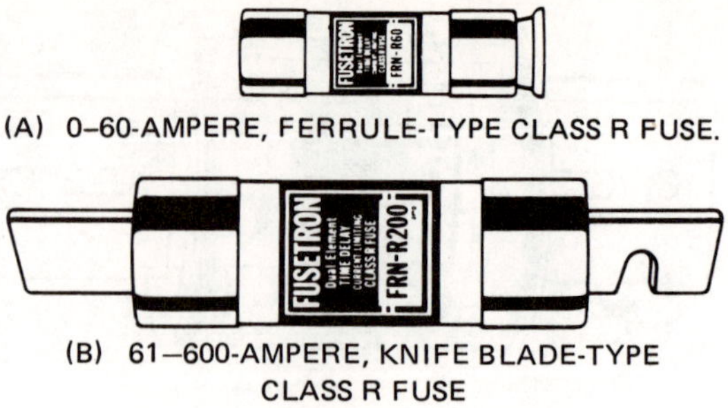

Fig. 17-14 Typical Class R fuses.

fuses are available in ratings ranging from 0–600 amperes at both 250 volts and 600 volts ac. The 300-volt classification is pending UL listing.

Class G. Class G fuses are cartridge fuses with small physical dimensions, Figure 17-13. They are used on circuits of 300 volts or less to ground. Class C fuses are available in sizes ranging from 0–60 amperes and are UL listed at an interrupting capacity of 100 000 symmetrical rms amperes. Commonly called Type SC fuses, they are essentially cartridge Type S fuses. To prevent overfusing, Class G fuses are size limiting within the four categories assigned to their ampere rating. Therefore, a fuseholder designed to accept a 15-ampere Type SC fuse will not accept a 20-ampere Type SC fuse; and a fuseholder designed to accept a 20-ampere Type SC fuse will not accept a 30-ampere Type SC fuse; and so on for the four categories.

Class R. The Class R fuse, Figure 17-14, is another recent development in the UL standards listing for fuses. This fuse is a nonrenewable cartridge type and has an interrupting rating of 200 000 rms symmetrical amperes. The peak let-through current I_p and the total clearing energy (I^2t) values are specified for the individual case sizes. The values for I^2t and I_p are specified by UL based on short-circuit tests at 50 000, 100 000, and 200 000 amperes.

Class R fuses are divided into two subclasses: Class RK1 and Class RK5. The RK1 fuse has characteristics similar to the Class K1 fuse. The Class RK5 fuse has characteristics similar to the Class K5 fuse. These fuses *must* be marked either Class RK1 or RK5. In addition, they are marked to be current limiting.

The ferrule-type Class R fuse has a rating range of 0–60 amperes. The knife-blade type Class R fuse has a rating range of 61–600 amperes. When a fuseholder is designed to accept a Class R fuse, it will be impossible to install a standard Class H or Class K fuse (these fuses do not have the annular ring or slot of the Class R fuse). Class R fuses can be installed in older style fuse clips on existing installations.

Electrical equipment manufacturers will provide the necessary rejection-type fuseholders in their equipment which is then tested with a Class R fuse at short-circuit current values of 50 000 A, 100 000 A, or 200 000 A. Each piece of equipment will be marked accordingly.

It is most important that large, high-ampere fuses be held very securely in the

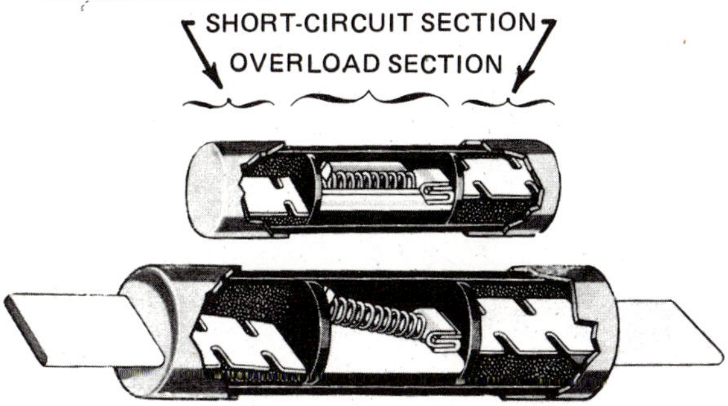

Fig. 17-15 Cutaway view of a Fusetron dual-element fuse. On overloads, the spring-loaded trigger assembly in the center chamber opens the fuse. The copper elements in both end chambers open when a short circuit occurs.

fuse panel. Any loose contact with these fuses will result in tremendous heat being developed at the point of contact, which could severely damage the fuse and its panel. For this reason, certain of fuses rated at 70 A to 600 A, and all Class L fuses rated at more than 600 A, are bolt-in connection types to assure a positive, firm contact with the fuse panel.

DUAL-ELEMENT FUSES

The versatile dual-element fuse is made in current ranges up to 600 A at 250 V and 600 V, and is ideal for motor, heater, appliance, and lighting circuits, Figure 17-15.

As the word dual suggests, this fuse gives two kinds of protection. The first kind is due to the time delay built into the fuse. A motor that needs extra current for starting can take advantage of it. For example, a circuit with a normal 30-A fuse will probably blow if a motor with a 25-A running current is started. This is because motors tend to draw three to five times their normal running current when starting. The time delay of a 30-A, dual-element fuse will allow the motor to start safely. If a minor overload is placed on the motor for a short period of time, the dual-element fuse will allow the motor to continue operating. This eliminates the nuisance of unnecessary equiment shut-downs. An extended overload or short circuit causes the fuse to open, however, and this prevents damage to the circuit conductors.

The second kind of protection is due to the dual-element fuse being heat-sensitive. Electric heaters and heating appliances often operate at a current level near the capacity of the circuit conductors. These types of loads known as *cyclic* heating loads, are required by the Code to be protected by a Type D dual-element fuse or a high-interrupting capacity (HIC) fuse. Most dual-element fuses in use today are of the Class RK type, having an interrupting rating of 200 000 A rms symmetrical. Some lighting circuits are heavily loaded and also near capacity. The heat generated in the conductors often travels along the conductor to the fuse. Since the dual-element fuse is heat-sensitive, it opens the circuit when too much heat accumulates in the conductors. As a result, inspection authorities often insist upon dual-element fuse protection in circuits containing heat-producing electrical equipment.

Some fuse manufacturers offer time-delay fuses that really are not true dual-element fuses. Time delay in these fuses is accomplished by using a single-element fuse link which has a heavy slug of solder on it at one point. This slug acts as a sort of *heat sink,* which absorbs heat, thus adding time delay to the fuse. This type of loaded-link design is not thermally sensitive to external heat, and depends entirely upon the heat developed within the fuse to cause the link to melt.

Metal fatigue of the fuse link in some types of fuses can cause a fuse to open for no apparent reason. This usually happens when the fuse is very heavily loaded, or is subjected to repetitive current surges. Fuses manufactured to high-quality standards generally do not experience link fatigue.

Dual-element fuses are made in plug, ferrule-contact cartridge, and knife-blade cartridge styles. Although slightly more expensive than other types of fuses, the dual-element fuse should be considered when choosing fuses for a circuit.

REVIEW

1. What is the effect on a conductor as an electrical current passes through?
2. Why is air circulation around a conductor important?
3. List and explain the types of damage that overheating conductors can cause.
4. What is the purpose of a fuse in an electrical circuit?
5. Which material is best for fuse link construction? Why?
6. Explain the meaning of the term "weakest thermal link."
7. List three different types of fuses, and give their current and voltage limitations.
8. What is a Type S fuse adapter? What is its purpose, and how is it used?
9. List the circuit faults that can cause a fuse link to open. How does each show on the fuse?
10. Describe a simple method for deciding whether a fuse is operating close to its rated current value.
11. Why are some cartridge fuses filled with a powder?
12. What is the main advantage of the renewable-link fuse?
13. What are the disadvantages of the renewable-link fuse?
14. Describe the HIC fuse. What main advantage does it have over standard cartridge fuses?
15. List and explain the two types of protection offered by a dual-element fuse.
16. How many types or classes of HIC fuses are there? Name them.
17. What is the purpose of rejection tabs or notches in the ends of HIC fuses?
18. Dual-element fuses are generally sized at 125% of a motor's full-load current rating. What size dual-element fuse would you install on a 40-A motor?

19. Class RK fuses may be used on electrical circuits capable of delivering not over ________ A rms symmetrical, whereas the Class H (ordinary, one-time and renewable fuses) are limited to use on circuits capable of delivering not over ________ A rms symmetrical.

20. Why are large-ampere fuses bolted into the fuse panel rather than held in place by spring clips?

Chapter 18

Residential Electric Heating and Cooling

RESIDENTIAL COMFORT SYSTEMS

Using electricity for residential comfort systems has several advantages compared with using gas or oil.

ADVANTAGES OF ELECTRICAL HEATING

One major advantage of an electrical heating system is that the temperature of each room can be controlled independently. This is done by placing a thermostat in a room to control the heating units for that room only. In the event of equipment failure, heat from adjoining rooms will flow into the room with the defective heater unit. With heating systems that use furnaces as a heat source, on the other hand, the entire building cools off when equipment fails. The increased demand for central air-conditioning systems, however, makes this type of heating unit most acceptable.

Although one type of electric furnace does use a metal duct system similar to that used for gas or oil heat, most electrical heatting systems do not need either a duct system or a furnace. This is a second major advantage of an electrical heating system. Furnace and/or oil storage units are not needed, so the floor space can be used for other purposes. The absence of ducts makes it easier to finish a basement with ceiling tile and wall panels. Because there are no fumes or hot gases to be removed, a chimney is not required.

Electrical heating systems have other advantages: they are noiseless and odor-free, they do not give off combustible fumes, and they are clean to operate. In addition, any house designed for use with an electrical heating system will likely be well built, because building standards are high for this type of home and are rigidly controlled.

COSTS OF ELECTRIC HEATING

Heat is often calculated in terms of British thermal units (Btu). Approximately 3.41 Btu equal 1 watt of electrical energy (3.41 Btu = 1 W). When heat is calculated in terms of joules, 1 watt is equal to 1 joule per second (1 W = 1 J/s).

Local utilities that supply electrical energy usually agree that the electrically produced heating source is one of the more expensive heating units. When comparing the cost of heating systems, however, other factors must be taken into account. For example, a house with an electrical heating system must be above average quality in construction. Windows, doors, and other places where air can enter or leave must be properly fitted, sealed with caulking compound, and equipped with storm or double glass window units. Some communities have heating inspectors who check all such installations. Such an inspection is in addition to the inspection made by the electrical authority.

Another factor to be considered is the absence of furnace and ductwork (and their maintenance costs). This frees capital that can be spent on the individual heating units required for each area of the home. (Some heating manufacturers make electric furnaces that take advantage of existing ductwork, which allows the customer to convert to electrical heating with a minimum of inconvenience and expense.)

The average consumer who is using electrical energy to heat a house uses much more electricity than would otherwise be used. As a result, the *end rate* (lower costs to consumer) is reached much sooner, and both heating and lighting energy are obtained at the cheaper end rate. The end rate is also applied to cooking appliances and water-heating units. Local utility rate schedules must be examined to compare energy costs.

Electrical energy is in good supply, which is not always the case with other forms of energy. This should be considered.

When a cost analysis is prepared and all factors are taken into account, in terms of price, a properly installed electrical heating system is usually found to be competitive with other forms of heating.

INSULATION

The main function of insulation is to trap *dead air* (air that is stationary) between the fibers or cells. Dead air prevents heat from escaping. Winter heat travels from hot to cold (that is, from inside the house to outside). Storm doors and windows simply trap dead air in the space between the layers of glass. It is this dead-air layer that is responsible for the heat-saving feature of the window units.

Insulation for the electrically heated house must be of a high quality and properly installed to a minimum thermal resistance of R-12 (R-2.1), approximately 3.5 inches (90 mm) in the walls and R-28 (R-4.9) with reference thickness of 8.5 inches (215 mm) in the ceiling (based on fiberglass batts). See figure 18-1. *Note:* To change English R values to metric RSI values, multiply by 0.175. For example, English R-28 equals metric R-4.9. To change metric R values to English R values, multiply by 5.7.

The R value is assigned by the manufacturers and refers to resistance to heat flow. It is an indication of the heat-retaining ability of the insulation. The higher the R value, the better the heat-retention quality of the insulation. Table 18-1(A) lists the thicknesses required for specific R values. Heat-loss factors are shown in Table 18-1(B). Each building-inspection department has its own requirements.

Insulation is available in several forms: batts (blanket insulation), rolls, pouring and/or blowing, and rigid sheets (slabs).

Batts. Batt (blanket) insulation is produced in 4-foot (1.2-m) lengths. It is made in various widths to form a pressure fit between studs and joists. It is available in R-8 (R-1.4), R-10 (R-1.8), R-12 (R-2.1), R-20 (R-

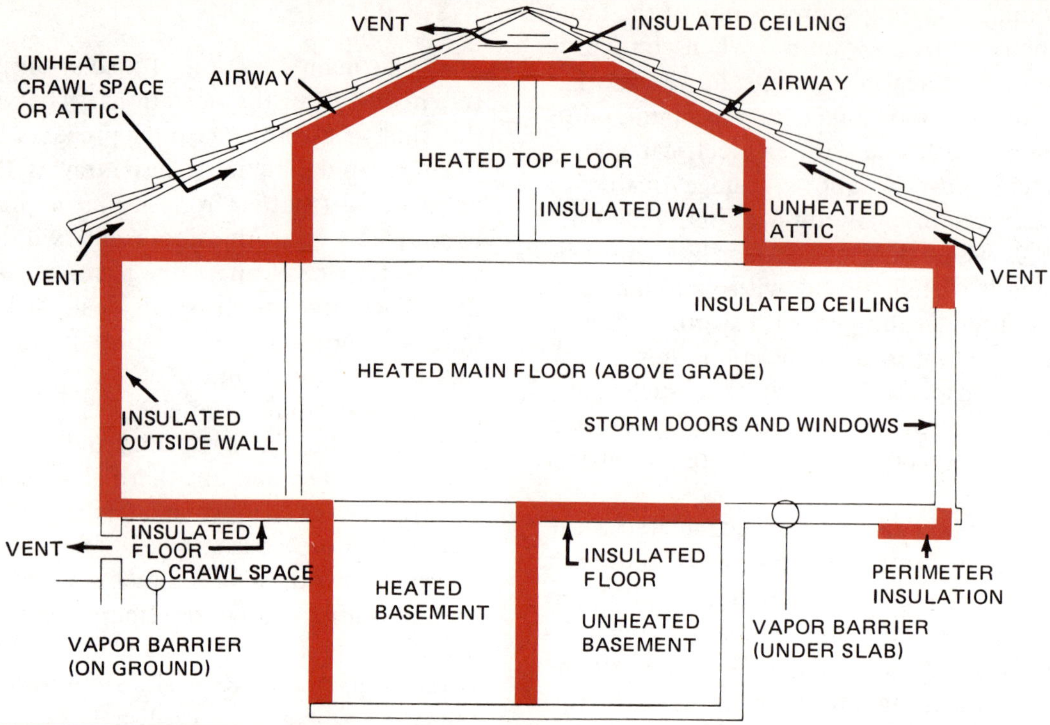

Fig. 18-1 Installing residential thermal building insulation. (*Note:* **All walls, ceilings, roofs, and floors separating heated from unheated spaces should be insulated.)**

TABLE 18-1 Blanket Insulation R Value Data

(A) REFERENCE THICKNESS BY MANUFACTURER

**"R" DESIGNATION	FIBERGLAS CANADA LTD.		JOHNS-MANVILLE J.M. FIBER GLASS	HOLMES INSULATIONS LTD.	CANADIAN GYPSUM CO.
	*Stripes	inches	inches	inches	inches
R-8	—	—	—	2	2
R-8	2	3	3	2.2	2.2
R-10	3	3.5	3.5	3	3
R-12	4	3.5	3.5	3.4	3.4
R-12	—	4	4	—	—
R-20	5	6	6	5.5	6
R-20	—	—	7	—	—
R-28	—	8.5	9	—	—

*Number of brown stripes branded on Friction batts by Fiberglas Ltd. Other manufacturers show R number.
**All R values quoted are English. Unidentifiable batts use R-2.

The following table lists average total resistance (Rt) and average heat-loss factors in watts per square foot per degrees Celsius for the structural materials most commonly used.

*The total average resistance (Rt) includes a correction for the effect of wood studs and framing where used. For rigid insulation, the Rt has no studding correction as it is assumed that the material will be applied with mechanical fasteners to the wall.

TABLE 18-1 Continued
(B) HEAT-LOSS FACTORS
Using Batt, Blanket or Loose Fill Insulation per square foot
per degree Celsius temperature difference

	Metric Values		English Values	
	Rt	W/m²·°C	Rt	W/ft²·°C
Through ceilings				
Lath & Plaster No Insulation	0.352	2.90	2.00	31.21
R-7	1.46	0.68	8.32	7.31
R-10	1.87	0.53	10.66	5.70
R-12	2.15	0.466	12.26	5.01
R-20	3.33	0.301	18.98	3.23
R-28	4.51	0.221	23.66	2.37
R-32	5.93	0.170	33.80	1.82
***Brick Veneer Walls**				
Inside Lath & Plaster No Insulation	0.704	1.42	4.01	15.28
R-8	1.69	0.591	9.63	6.36
R-10	2.01	0.50	11.46	5.38
R-12	2.20	0.45	12.54	4.84
***Frame Walls – Stucco Exterior**				
Inside Lath & Plaster No Insulation	0.67	1.48	3.82	15.92
R-8	1.65	0.602	9.41	6.47
R-10	1.94	0.517	11.06	5.56
R-12	2.13	0.471	12.14	5.06
***Frame Walls – Wood Siding**				
Inside Lath & Plaster No Insulation	0.79	1.25	4.50	13.45
R-8	1.80	0.556	10.26	5.98
R-10	2.04	0.488	11.63	5.25
R-12	2.25	0.443	12.83	4.76
***With R-8 Exterior**				
Insulation R-8 Interior	2.90	0.341	16.53	3.67
Replacing R-10	3.15	0.318	17.96	3.42
Sheathing R-12	3.33	0.295	18.98	3.17
With 2" x 6" Framing				
R-20 Insulation	3.33	0.295	18.98	3.17

Note: To convert English R values to metric R values, multiply the English R value by 0.175.

To convert square feet (ft²) to square metres (m²), multiply ft² by 0.093.

To convert inches to millimetres, multiply inches by 25.4.

3.5), and R-28 (R-4.9) for use with both wood and steel studding. As the thermal resistance value of the batt changes, so does the reference thickness. The batt form of insulation is most often made of fiberglass, and is available with a factory-applied vapor barrier. It is generally recommended, however, that a continuous 0.008-inch (0.2-mm) polyethylene air-vapor barrier be used on site as well, Figure 18-2.

Rolls. Blanket insulation in roll form is produced in long rolls of various widths to fit stud and joist patterns. It can be cut to the desired length with a sharp knife, and is available in R-8, R-10, and R-12 values. (See Figures 18-3, and 18-4.)

Pouring Insulation. Pouring insulation is loose insulation made of mineral (rock) wool, fiberglass, vermiculite, or cellulose.

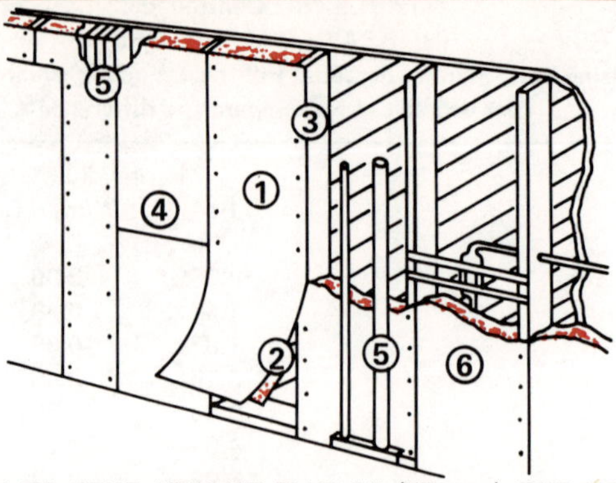

STEP 1. CUT INSULATION TO STUD HEIGHT PLUS 3" (75 mm) FOR TOP AND BOTTOM NAILING FLANGES. USE MARKINGS ALONG EDGE OF BLANKET TO EASE MEASURING. (OR APPLY INSULATION FROM ROLL. CUT OFF AT BOTTOM TO FIT.)

STEP 2. FLUFF BLANKET TO FULL THICKNESS, OPEN NAILING FLANGES, AND PRESS BACK INSULATION TO LEAVE END NAILING FLANGE EXPOSED.

STEP 3. IF ROLLS ARE USED, START AT STOP OF STUD SPACE AND WORK DOWN. BATTS CAN BE INSTALLED FROM FLOOR UP.

STEP 4. STAPLE OR NAIL FLANGES TO STUD AT 5" to 6" (130 mm TO 150 mm) SPACINGS. TAKE CARE AIR-VAPOR BARRIER FACES BUILDING INTERIOR ON WARM (IN WINTER) SIDE.

STEP 5. TAKE CARE TO INSULATE COLD SIDE OF PIPES AND DRAINS. FIT BLANKET INSULATION BEHIND PIPES OR USE EXTRA PIECES OF BLANKET OR POURING INSULATION.

STEP 6. FILL ALL NARROW SPACES WITH BLANKET OR POURING INSULATION, OR SCRAPS FROM BATTS AND/OR ROLLS. CUT AND APPLY AIR-VAPOR BARRIER TO PROVIDE VAPOR PROTECTION.

Fig. 18-2 Procedure for installing batt and roll (blanket) insulation.

It is most suited for installation in horizontal spaces, such as an attic area, and can be hand poured from bags to a suitable depth. Some companies make a mineral glass or cellulose fiber insulation that can be blown into place with air pressure through a long hose.

Pouring of insulation is often used by home owners to boost the R value of their existing insulation. Care should be taken to choose a material that will maintain its R value over the years. Cellulose insulation is a form of paper product (similar to newsprint) that is chemically treated (usually with dry chemicals) to provide a degree of moistureproofing and fireproofing. This type of insulation, when dry, is a good insulating material. Like many other paper products, however, the cellulose insulation has a tendency to absorb moisture after it has been installed, thus drastically reducing the R value. Due to its comparatively low initial cost, many home-insulating companies select this material for residential use. If the moistureproofing chemicals are not properly applied, and allowance is not made for adequate ventilation (soffit and peak areas), the installation of this material is virtually a waste of time and money. Figure 18-5 illustrates pouring insulation.

Rigid Sheets. Polystyrene and polyurethane insulation is produced in slabs or rigid sheets. The slabs can vary in size. One very

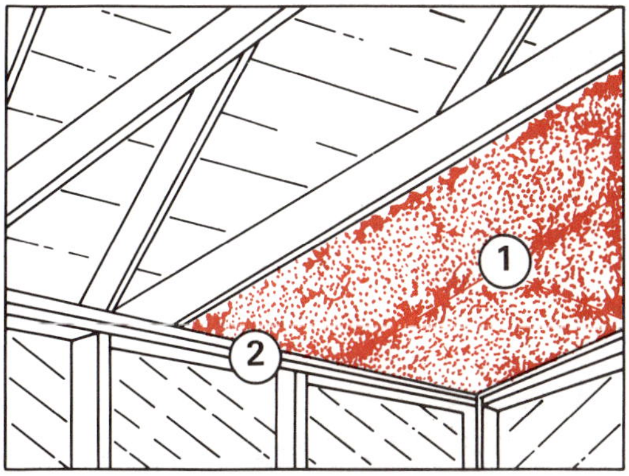

STEP 1. APPLY BATT OR ROLL INSULATION OVERHEAD BEFORE PLASTER BASE IS ATTACHED. THEN APPLY CONTINUOUS POLYETHYLENE AIR-VAPOR BARRIER.

STEP 2. CUT ROLL INSULATION TO LENGTH BEFORE APPLICATION OR APPLY FROM ROLL AND CUT WITH SHARP KNIFE AT END OF JOIST.

Fig. 18-3 Procedure for installing roll or blanket insulation.

common unit is the 4-foot by 8-foot (1200-mm by 2400-mm) sheet in thicknesses of 1/2 inch (13 mm), 1 inch (25 mm), 2 inches (50 mm), 3 inches (75 mm) and 4 inches (100 mm). The rigid form of insulation is often used for commercial buildings of masonry construction and basement walls of houses. An adhesive is used to attach the slabs to the wall. An air-vapor barrier is not required because polystyrene resists all moisture migration. (See Figures 18-6 and 18-7.)

Polystyrene insulation supports combustion, and gives off a dangerous gas when ignited. For this reason, a protective surface of 1/2-inch (13-mm) gypsum board or similar fire-resistant product must be installed directly over the polystyrene insulation.

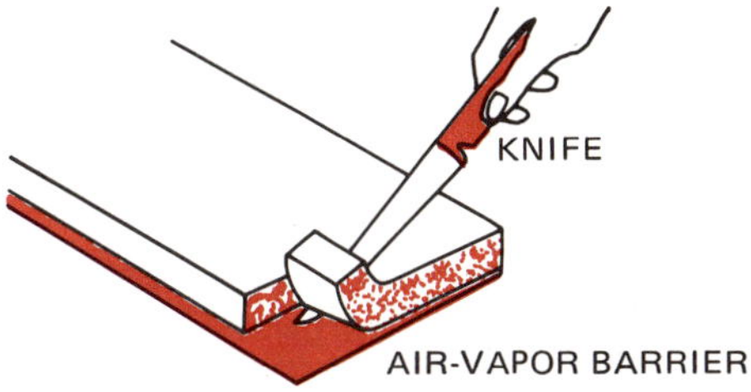

NOTE: LAY INSULATION ON FLAT SURFACE AND CUT TO MAKE END FLANGE.

Fig. 18-4 Procedure for cutting roll or blanket insulation.

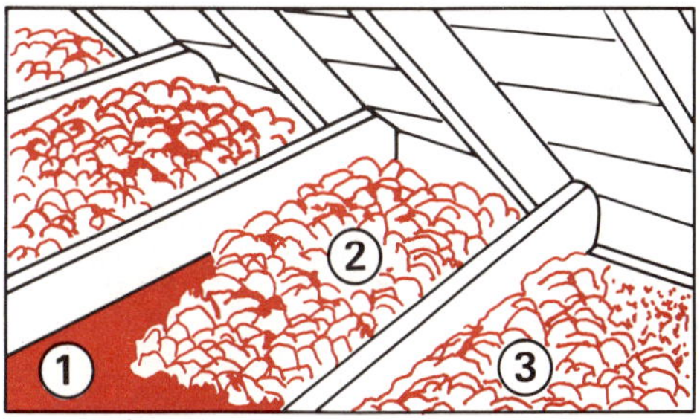

STEP 1. PLACE AIR-VAPOR BARRIER AT BOTTOM OF SPACE BETWEEN JOISTS. LAP TO ENSURE GOOD BARRIER TO FLOW OF AIR AND VAPOR.

STEP 2. FOR REQUIRED THERMAL RESISTANCE VALUE, POUR OR BLOW INSULATION BETWEEN JOISTS TO DEPTH RECOMMENDED BY MANUFACTURER. TAKE CARE INSULATION DOES NOT COVER SOFFIT OR UNDER-EAVE VENTILATORS. THE NEC REQUIRES THAT THERMAL INSULATION BE INSTALLED NOT CLOSER THAN 3 INCHES (76 mm) FROM A RECESSED FIXTURE ENCLOSURE. WIRING COMPARTMENT, OR BALLAST; AND THAT IT NOT BE INSTALLED ABOVE THE FIXTURE SO AS TO ENTRAP HEAT.

STEP 3. RAKE OR SMOOTH INSULATION TO ENSURE EQUAL THICKNESS OVER ENTIRE AREA. TO OBTAIN PROPER SETTLED DENSITY AND THICKNESS, FOLLOW THE MANUFACTURER'S INSTRUCTIONS.

Fig. 18-5 Procedure for installing loose insulation.

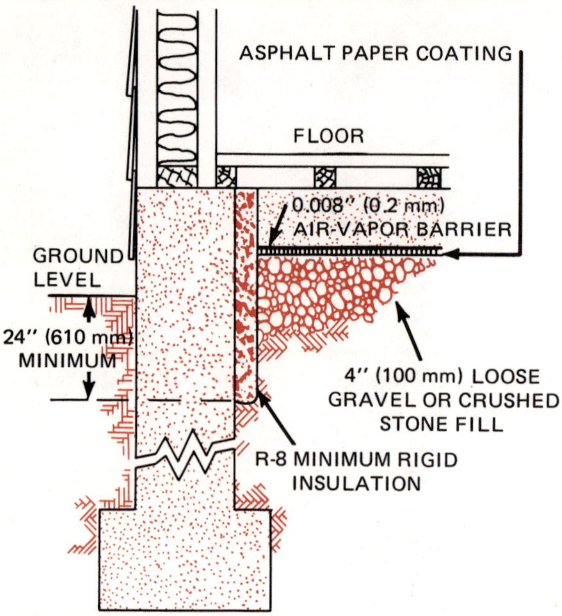

Fig. 18-6 Interior vertical installation. (Concrete floor on grade.)

AIR-VAPOR BARRIER

Insulation entraps dead air only when it is dry and free of moisture. If the insulation is damp, it will conduct heat out of the building very quickly. The greatest threat is from excess moisture inside the building trying to escape to the outside through the walls. The moisture given off by cooking, bathing (showers), and people must be removed safely. Otherwise, it will condense on the walls, form drops, and run down, which often damages wall surfaces.

To keep insulation reasonably dry, an air-vapor barrier must be installed between the plaster and the insulation, Figure 18-8. Since electrically heated buildings rely heavily on quality insulation to keep the cost of heating low, the insulation must be well protected. A continuous, 0.008-inch (0.2-mm) polyethylene air-vapor barrier is recommended for use on the warm (in winter)

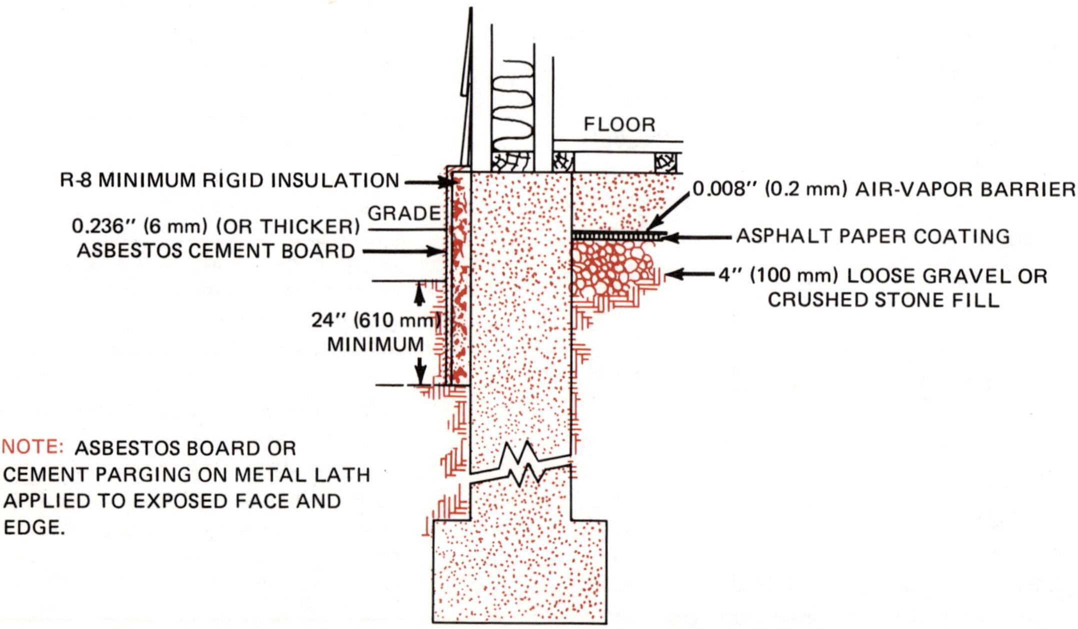

Fig. 18-7 Exterior vertical installation. (Concrete floor on grade.)

side to protect the insulation from internal moisture sources. This moisture must then be removed artificially.

VENTILATING DEVICES

Ventilating fans are placed over stoves and in bathrooms to remove excess moisture. They also remove some of the heat from the house, however, so they should not be used continuously. These fans are an important part of the heating installation and are looked for by electrical heating inspectors.

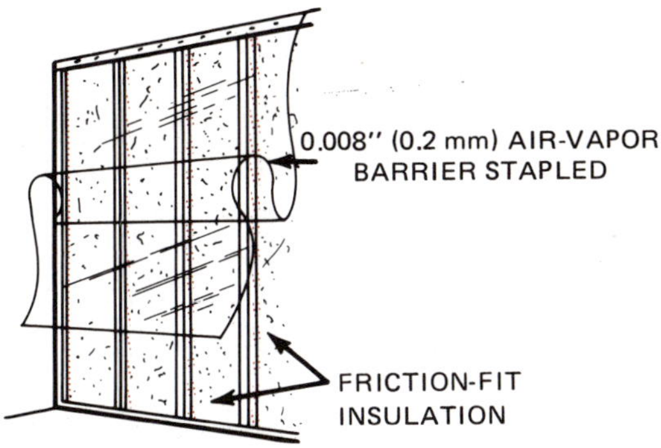

Fig. 18-8 Air-vapor barrier must be installed over insulation on the warm side.

Attic insulation must also be kept moisture-free. This is done by a system of vents placed in various parts of the attic. A venting area of 10.8 ft² (1 m²) of free unobstructed ventilating area for each 1613 ft² (150 m²) insulated attic area and/or 3226 ft² (300 m²) of insulated ceiling is required. (See Figure 18-9.) The circulation of air tends to keep the insulation dry and moisture-free. Remember that a cold attic during the winter indicates that the insulation is indeed working and keeping heat in the house.

BASEBOARD HEATER

The baseboard heater is one of the most common types of electrical heating

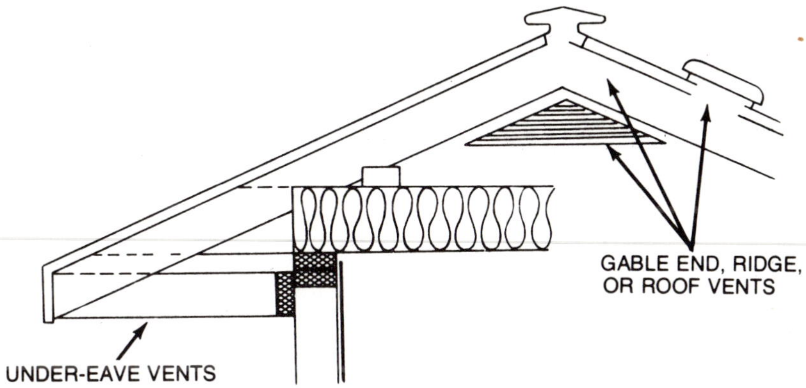

Fig. 18-9 Attic ventilation. Provide minimum 10.8 ft² (1 m²) free (unobstructed) ventilating area for each 3226 ft² (300 m²) insulated ceiling area. Cathedral-type and low-pitched roofs require 10.8 ft² (1 m²) ventilation for each 1613 ft² (150 m²) insulated attic area. Distribution of vents must provide cross-ventilation in attic from end to end and from top to bottom.

units. It consists of an enclosed heating element which is fitted with heat-radiating metal fins. This assembly is supported in a metal frame. It operates on the principle that air heated by the element will rise and flow out of the top of the frame, while cooler air will be drawn from the floor area into the bottom of the heater. This process is called *convection of air,* Figure 18-10.

Baseboard heaters are rated in both wattage and voltage. The most common unit for residential use is the 240-V heater, simply because a 240-V heater of a given wattage requires half as much current as a 120-V heater of the same wattage. This reduces the need for extra-large conductors in the heater circuits.

Baseboard heaters are made in low- and standard-wattage densities. This means that a 1000-W heater in a low-watt density is approximately 6 feet (1.8 m) long, and low enough in surface temperature so that it may be installed safely under drapes and other heat-sensitive materials. This makes it suitable for use under some synthetic fabrics which are heat-sensitive, and shrink, lose their shape, or discolor when exposed to heat for a prolonged period of time.

A 1000-W heater in a standard-watt density is approximately 4 feet (1.2 m) long and reaches a higher surface temperature than the low-watt unit. The 1000 W of heat is more concentrated in this type of heater, so it should *not* be used under or near delicate fabrics. This type of heater is excellent, however, for use in areas where space is limited and higher surface temperatures are not a problem.

Draperies should be hung with the nearest fold at least 2 inches (50 mm) away from the heater and 1 1/2 inches (40 mm) off the floor. These clearances allow the proper air circulation through the heater.

Baseboard heaters are produced in a variety of wattage ranges, starting as low as 250 W and going as high as 3000 W. Manufacturers provide specifications for their own products. Two types of baseboard heaters are shown in Figures 18-11 and 18-12. The type shown in Figure 18-12 is a baseboard heater with a receptacle and switch for use with a window-type air-conditioner unit.

Installation of Baseboard Heaters. Baseboard heaters should be mounted at the finished floor level to allow the cool air to enter the heater easily. They also must be mounted on a flat surface, without bending or distorting the heater frame. If the frame is distorted in any way, the heater will produce some noise as it expands and contracts during its heating cycles.

The heater is secured to the wall with several wood screws. It needs nearly no maintenance, except for an occasional vacuuming to remove lint collected on the heater fins. Lint slows the convection of air through the heater and lowers its efficiency.

RADIANT-HEATING CEILING CABLE

Another type of electrical heating unit is the radiant-heating ceiling cable, Figure 18-13. (See NEC *Article 424.*) This slender, insulated wire must be installed carefully on the ceiling before the plaster is applied. It is purchased in a calibrated length, with a minimum of 7 feet (2.1 m) of nonheating leads. The cable must be identified within 3 inches (76 mm) of the terminal end. Operating voltages are color coded:

Color	Voltage
Yellow	120 V
Blue	208 V
Red	240 V
Brown	277 V

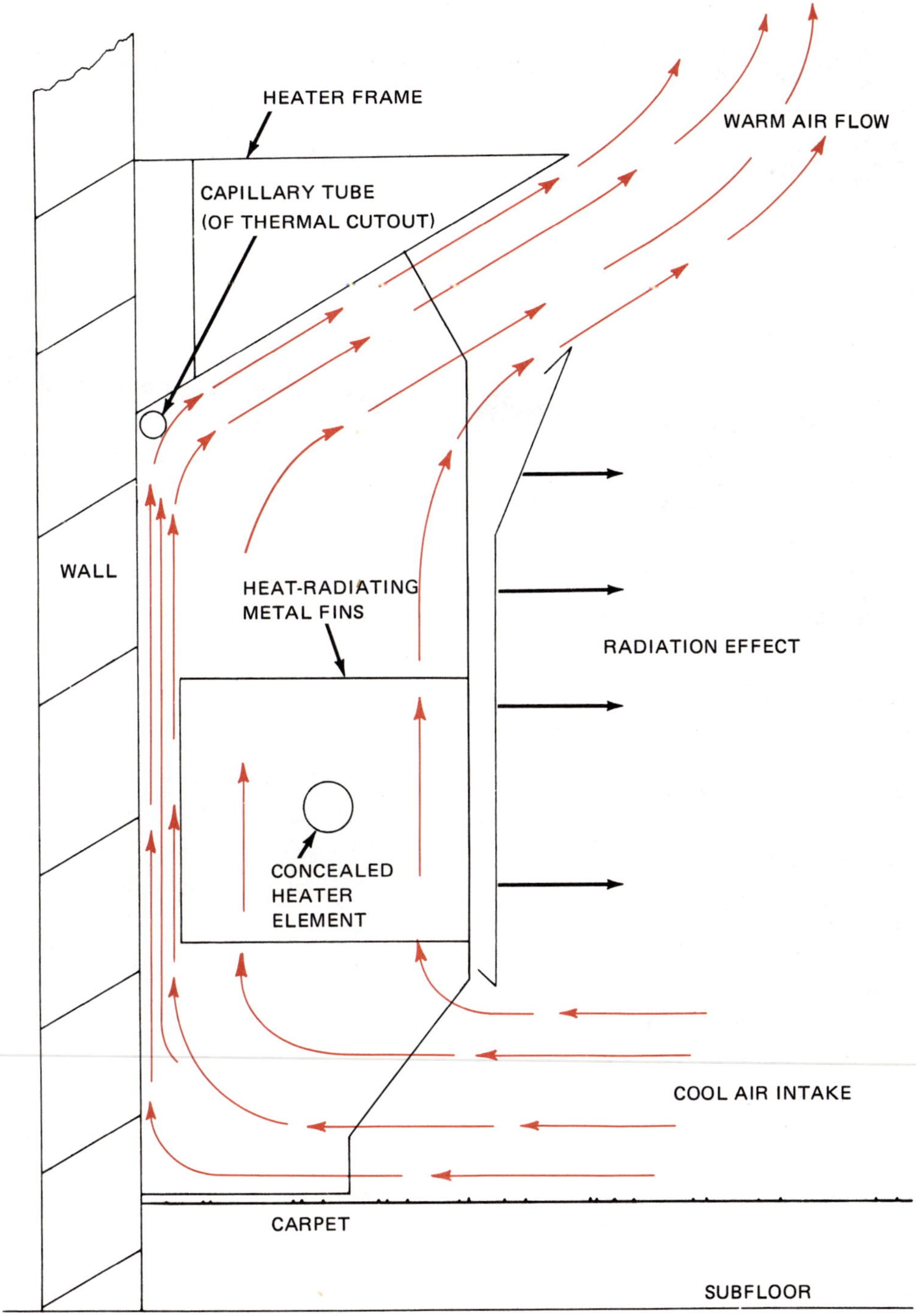

Fig. 18-10 Air convection in baseboard heater.

Fig. 18-11 Typical baseboard heater with built-in thermostat.

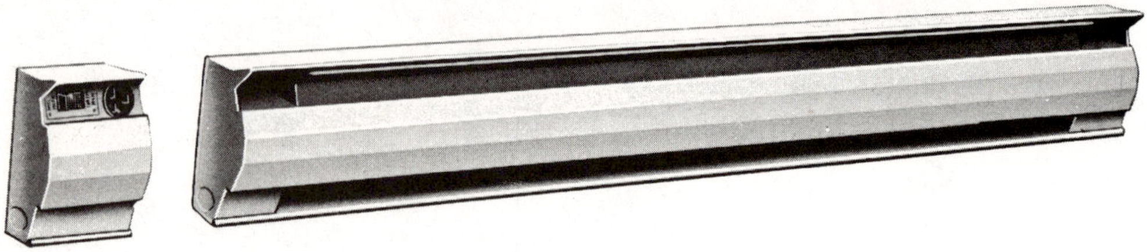

Fig. 18-12 Typical baseboard heater with air conditioner switch and receptacle attachment.

It must not be cut or shortened in any way, but used at the voltage source indicated on the cable as it comes from the supplier. Any change in length will change the output and the current flow in the cable, and also might establish a failure point at the new splice. However, nonheating leads may be cut. Ceiling cable radiates its heat energy into the room, warming any heat-absorbing object in the room. The entire ceiling, therefore, becomes a heat source.

Fig. 18-13 Reel of radiant-heating ceiling cable.

Installation of Ceiling Cable. If a layer of dry-wall plaster has been installed first, the ceiling cable can be stapled directly to the dry wall with special staples. The staples must be fastened not more than 16 inches (406 mm) apart. Take care not to staple through the cable or damage it in any way. The cable should be kept a minimum of 8 inches (200 mm) away from ceiling light fixtures and about 6 inches (150 mm) from walls. Parallel runs must be installed not less than 1 1/2 inches (38 mm) on centers. These are safety measures to prevent a person from driving a hook or fastener into the cable when hanging drapes or installing new lighting fixtures.

If a building has poured concrete walls and ceilings or a similar masonry construction, the ceiling cable will need hanger strips. *Hanger strips* are plastic strips which are run across the ceiling at each end of the room, secured to the ceiling with adhesive, and then fitted with the heating cable. The cable is looped from end to end. The plastic supporting strips provide the proper spacing between the cable runs. Cables should be spaced at least 2 inches (50 mm) on the heated ceiling, depending upon cable spacing calculations. Manufacturers provide the proper spacing requirements for their cables. Often, adhesive tape is needed at intervals in a long room to prevent sagging of the cables before plastering.

Each cable set comes with a minimum length of 7 feet (2.1 m) of heavily insulated conductor (nonheating leads) at each end of the cable. This conductor is called the *cold lead.* The cold leads are run from the ceiling down to an electrical box on the wall where the thermostat and/or power supply is located. The length of these leads should not be changed.

The sand-base gypsum plaster (a minimum thickness of 1/2 inch [13 mm]) must be installed carefully, so that the cable is not nicked or damaged with a trowel. The plaster must be allowed to dry fully before the cable is energized. Take care to check the cable for continuity both before and after plastering. Using the cable to dry the plaster will only result in shrinking and/or cracking of the plaster around the heating cable.

Once the plaster is dry (allow at least a week), the cable can be energized and allowed to heat the room. Increase the temperature setting of the thermostat slowly, and do not pass the midpoint setting of the thermostat for a least two weeks. This is a safety precaution in case the plaster is not thoroughly dry in all areas of the ceiling.

HEAT-LOSS AREAS

Window areas are one of the major heat-loss areas in a house. Proper caulking around window frames will stop infiltration of cold air and drafts and save a great deal of heat over a period of time.

Electrically heated houses must have storm windows. The aluminum-framed type of storm window is neat, easily removed, and needs little or no maintenance. The aluminum frame, however, conducts some heat away from the home. Wood-framed storm windows need more maintenance than the aluminum type, and often are difficult to open, close, or remove. However, they retain heat better than those made of aluminum.

Double-pane glass, which is hermetically sealed to maintain a dry-air space between the two panes, is of some benefit. These units are often used for picture windows. The dry-air space is approximately 1/4 inch (6 mm) wide and does not control heat loss as well as storm windows.

Fireplaces are another cause of heat loss in a house. They give off a lot of heat when operating. When not operating, however, a great deal of heat is drawn out of the room, up the flue, and out the chimney. A brisk wind across the chimney will speed heat removal from the room. To prevent this, the fireplace damper should always be well fitted and kept closed when the fireplace is not in use. (Approximately 1000 W of additional heat should be installed in a room with a fireplace to compensate for the heat being drawn up the chimney.)

If a house is built on a concrete slab or pad, a rigid exterior type of insulation must be placed around the edge and about 3 feet (914 mm) under the slab. This is to prevent heat from radiating out of the slab and into the cool earth..

BASEMENT HEATING

Residential basements are given special treatment when calculating heat loss. For the purposes of heat-loss calculation, the basement is divided into two parts, Figure 18-14.

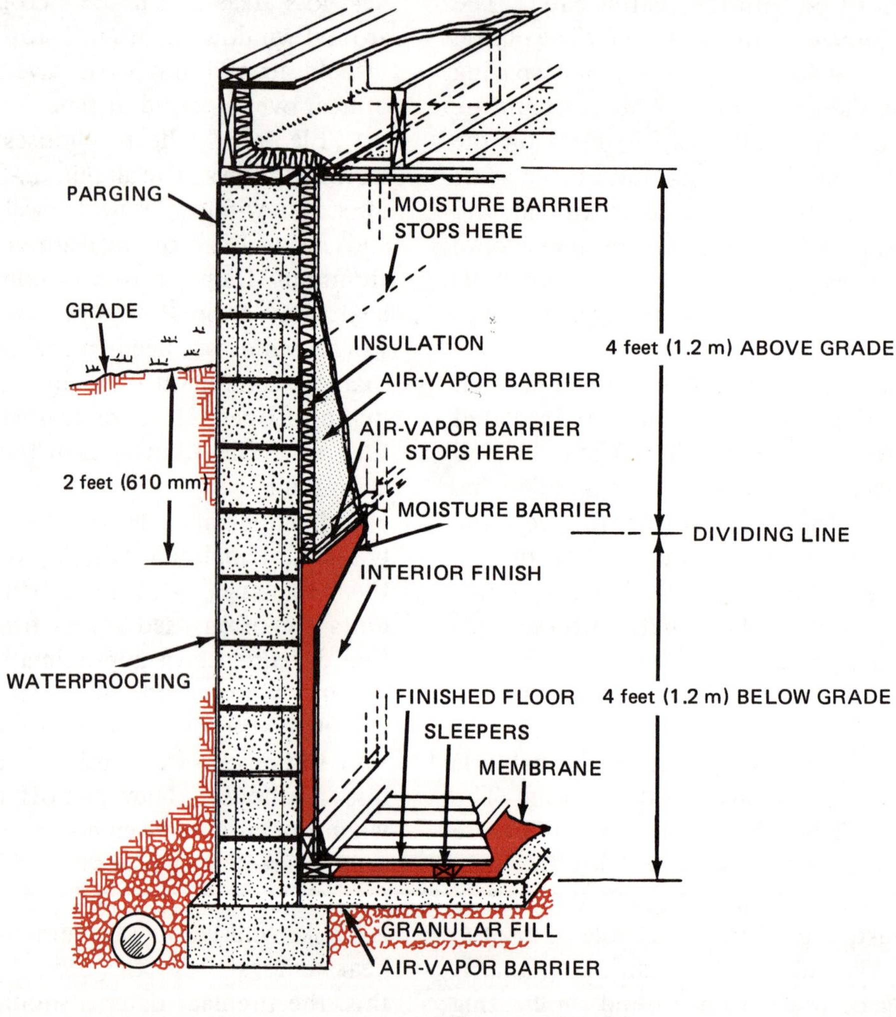

NOTE: BASEMENT WALL ABOVE GRADE MUST MEET SAME REQUIREMENTS AS ANY OTHER EXTERIOR WALL EXPOSED TO WEATHER. WALL BELOW GRADE SHOULD BE MOISTUREPROOFED ON INSIDE. THERMAL INSULATION SHOULD BE APPLIED ON EXPOSED WALL TO DISTANCE OF 2 FEET (610 mm) BELOW GRADE (WITH AN AIR-VAPOR BARRIER). VAPOR BARRIERS AND DAMP-PROOFING TO BE IN ACCORDANCE WITH REQUIREMENTS DETAILED IN THE LOCAL RESIDENTIAL STANDARDS.

Fig. 18-14 Typical basement wall treatment.

In a typical basement with an 8 foot (2.4-m) ceiling, the area from the ceiling down to 2 feet (610 mm) below ground level (grade) is called the above-grade part. This is because it is thought that earth to a depth of 2 feet (610 mm) is approximately the same temperature as the open air surrounding the part of the basement extending aboveground. The rest of the basement is called the below-grade part.

Local building codes often require basement walls that are 50 percent above grade to have a minimum of R-12 (R-2.1) value insulation installed over the entire wall. Basement walls that expose less than 50 percent of their surface to above-grade temperatures must be insulated with R-12 (R-2.1) value insulation to a depth of 2 feet (610 mm) below grade level.

Energy-conscious home owners, however, may wish to insulate the complete basement wall, thus reducing heat loss and lowering energy costs for this area of the home. If there is any chance of water leaking into the basement area, the insulation should be kept approximately 1 foot (300 mm) from the floor to prevent the insulation from soaking up the water.

A check of local building codes is advisable to determine the exact amount of below-grade insulation required in a given area.

Often, rigid insulation is fastened with an adhesive to the above-grade walls. Wooden strapping and R-8 (R-1.4) value batts can also be installed for heat retention. Sometimes, small amounts of heating cable are placed in the floor to help the main heaters warm up the area.

HEAT-LOSS CALCULATIONS

The First Step: Heat-Retaining Walls and Ceilings. To determine the amount of heat needed for a house, the heat-retaining ability of the walls and ceilings must be calculated first. The ability of building products and techniques to retain heat varies. Figure 18-15 shows a typical cross-section of a wall and the R numbers (resistance to heat) for each of the materials used. Figure 18-16 shows a typical ceiling installation and its R values. The total R in each figure represents the heat-retaining ability of the materials used without the aid of insulation.

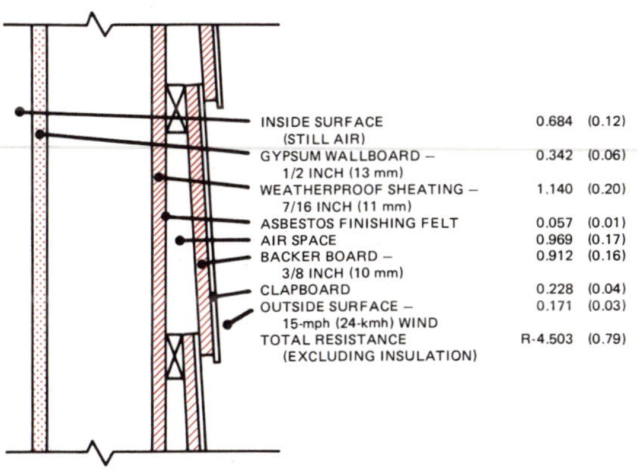

Fig. 18-15 Typical wall section.

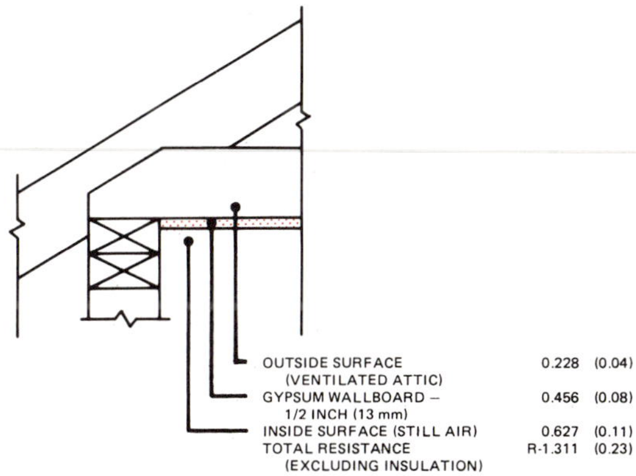

Fig. 18-16 Typical ceiling section.

Table 18-1B lists the average total resistance (Rt) required for walls and ceilings in Ontario, Canada, regardless of the type of heating system used in the building. An average Rt, corrected for wood framing, is used since there will be a different rate of heat loss through the studs compared with heat loss through the insulation.

A table such as that shown in Table 18-1B is used to find the R value of the insulation required. In the same way, the average Rt of various types of construction and location are found in such tables. (Insulation with an R-12, or R-2.1, designation is now commonly used for residential walls.) Additional information on quality and R value of the insulation can be obtained from manufacturers upon request.

Table 18-2 shows the R value for various building products. More complete tables can usually be obtained from the local electric utility or building supplier.

The Second Step: Heat Loss through Walls and Ceilings. The second step in determining the amount of heat needed for a house is to calculate the heat lost through the walls and ceilings. To do this, the R total of the wall or ceiling must be mathematically converted to a U factor. The *U factor* (overerall coefficient of heat transfer) is the amount of heat transmitted through a heat barrier in one hour per square metre of surface for each 1°C of temperature difference between the inside and outside of the barrier.

For example, using metric values, a frame wall with wood siding, having inside lath and plaster with R-2.1 insulation, has a total average Rt of 2.25 (see Table 18-1). Therefore, the U factor equals 1/2.25, which equals 0.444 W/m^2 for each 1°C of design temperature difference.

The outside design temperature is found by locating the approximate region of the house, Tables 18-3A, B, and C. The Design Temperature column of Table 18-3A lists the outdoor temperature for selected regions in Canada, based on the temperatures experienced for 2.5 percent of the January outdoor temperatures that can occur.

For example, if the house is in Omaha, Nebraska, the outdoor design temperature for winter would be −10°F (−23°C). Consider that the occupants want an indoor temperature of 73°F (23°C).

The design temperature difference in Fahrenheit is 73°F − (−10°F) which equals 83 degrees. The U factor for each degree difference on the Fahrenheit scale is 0.084 Btu per hour per square foot. The heat loss for 83 degrees on the Fahrenheit scale is 83 × 0.084 which equals 6.972 Btu per hour per square foot. Rounded off, the U factor is 7 Btu per hour per square foot.

The design temperature difference in Celsius is 23°C − (−23°C) which equals 46 degrees. The U factor for each degree difference on the Celsius scale is 0.444 W/m^2. The heat loss for 46 degrees on the Celsius scale is 46 × 0.444 W/m^2 which equals 20.424 W/m^2. Rounded off, the U factor is 20 W/m^2.

The heat lost through the ceilings is calculated in much the same way as it is for heat lost through the walls.

To find the total heat loss through walls of a room, the total area of wall exposed to cold air must be calculated. This is done by multiplying the height of the room by the length of the exposed wall. (Remember that a corner room has two exposed walls.) Figure 18-17 shows a wall with a gross area 8 feet (2.4 m) high times 15 feet (4.6 m)

TABLE 18-2 Resistance Values for Various Building Materials

Building Material	Thickness (inches)	Resistance Factor	Heat Loss Factor (W/ft² ·°C)
Insulating Fiberboard Sheathing	1	2.38	25.61
	5/8	1.49	41.11
	1/2	1.19	51.44
Gypsum Board	1/2	0.32	192.11
Plywood	1	1.25	48.86
	1/2	0.63	97.83
Hardwoods (Maple, Oak, etc.)	1	0.91	67.26
Softwoods (Pine, Fir, Spruce, etc.)	1	1.25	48.86
Western Cedar (18% moisture content)	1	1.56	39.17
Loose Fill Insulation			
Macerated Paper *(Cellulose Fibre)	1	3.57 to 4.01	10.76 to 24.30
Mineral Wools (2.0–5.0 lb/ft³ density)	1	3.34	18.40
Vermiculite (Expanded Mica) (7.05 lb/ft³ density)	1	2.08	29.38
Air Spaces (in walls)	3/4 to 4	0.97	62.96
Foamed Styrene (Density 1.62 lb/ft³) (See manufacturer's specifications)	1	3.46 to 4.01	10.76 to 25.06
Polyurethane	1	5.92 to 6.04	10.33 to 10.11
Flooring (Hardwood)	3/4	0.68	89.65
	3/8	0.34	179.31
Brick (clay or shale)	4	0.42	145.40
Concrete Block (3 oval core)			
Sand and Gravel	4	0.71	86.10
	8	1.11	55.21
	12	1.28	47.78
Cinder Block	4	1.11	55.21
	8	1.72	35.51
	12	1.89	32.28

Values shown above are taken from the Acceptable Building Material Systems & Equipment — Central Mortage and Housing Corporation.

*Illinois Institute of Technology.

Note: To convert English R values to metric R values, multiply the English R value by 0.175.

To convert square feet (ft²) to square metres (m²), multiply ft² by 0.093.

To convert inches to millimetres, multiply inches by 25.4.

To convert lb/ft³ to kg/m³, multiply lb/ft³ by 16.018.

TABLE 18-3(A) Design Data for Selected Locations in Canada

Province and Location	Design Temperature 2.5% °C	Degree Days Below 18°C
Ontario		
Ailsa Craig	−17	3980
Ajax	−20	4080
Alexandria	−24	4580
Alliston	−23	4520
Almonte	−26	4740
Ansonville	−33	6220
Armstrong	−39	6892
Amprior	−27	4800
Atikokan	−34	6040
Aurora	−21	4300
Bancroft	−27	4960
Barrie	−24	4470
Barriefield	−22	4240
Beaverton	−24	4580
Belleville	−22	4190
Belmont	−17	3980
Bowmanville	−20	4130
Bracebridge	−26	4800
Bradford	−23	4410
Brampton	−19	4200
Brantford	−17	3920
Brighton	−21	4240
Brockville	−23	4300
Brooklin	−20	4240
Burks Falls	−26	5070
Burlington	−17	3700
Caledonia	−17	3920
Cambridge	−18	4130
Campbellford	−23	4410
Camp Borden	−23	4470
Cannington	−24	4580
Carleton Place	−25	4690
Cavan	−22	4470
Centralia	−17	3940
Chapleau	−35	5950
Chatham	−16	3530
Chelmsford	−28	5290
Chesley	−19	4240
Clinton	−17	4130
Coboconk	−25	4740
Cobourg	−21	4190
Cochrane	−34	6230
Colborne	−21	4190
Collingwood	−22	4580
Cornwall	−23	4470
Corunna	−16	3810
Deep River	−29	5180
Deseronto	−22	4080
Dorchester	−18	4030
Dorion	−33	5890
Dresden	−16	3700
Dryden	−34	6080
Dunbarton	−19	4030
Dunnville	−15	3810
Durham	−20	4620
Dutton	−16	3750
Earlton	−33	5866
Edison	−34	6000
Elmvale	−24	4580
Embro	−18	4130
Englehart	−33	5950
Espanola	−25	5070
Exeter	−17	4080
Fenelon Falls	−25	4690
Fergus	−20	4610
Fonthill	−15	3700
Forest	−16	3830
Fort Erie	−15	3590
Fort Frances	−33	5830
Gananoque	−22	4240
Georgetown	−19	4250
Geraldton	−35	6550
Glencoe	−16	3810
Goodrich	−16	4190
Gore Bay	−23	4910
Graham	−37	6470
Gravenhurst	−26	4740
Grimsby	−16	3580
Guelph	−19	4220
Guthrie	−24	4520
Hagersville	−16	3920
Haileybury	−32	5830
Haliburton	−27	4920
Hamilton	−17	3710
Hanover	−19	4350
Hastings	−23	4470
Hawkesbury	−25	4800
Hearst	−34	6500
Honey Harbour	−24	4580
Hornepayne	−37	6580
Huntsville	−26	4760
Ingersoll	−18	4030
Jarvis	−16	3860
Jellicoe	−36	6450
Kapuskasing	−33	6366
Kempville	−25	4540
Kenora	−33	5932
Killaloe	−28	4940
Kincardine	−17	4240
Kingston	−22	4266
Kinmount	−26	4800
Kirkland Lake	−33	6150
Kitchener	−19	4110
Lakefield	−24	4630
Lansdowne House	−39	7110
Leamington	−15	3560
Lindsay	−24	4580
Lions Head	−19	4350
Listowel	−19	4630
London	−18	4068
Lucan	−17	4030
Maitland	−23	4300
Markdale	−20	4690
Martin	−36	6330
Matheson	−33	6220
Mattawa	−29	5340
Midland	−23	4580
Milton	−18	4080
Milverton	−19	4520
Minden	−26	4850
Mississauga	−18	3810
Mitchell	−18	4400
Moosonee	−36	6931
Morrisburg	−23	4410
Mount Forest	−21	4755
Muskoka Airport	−26	4837
Nakina	−35	6540
Napanee	−22	4130
Newcastle	−20	4130
New Liskeard	−32	5830
Newmarket	−22	4350
Niagara Falls	−16	3740
North Bay	−28	5318
Norwood	−24	4520
Oakville	−18	3640
Orangeville	−21	4650
Orillia	−25	4610
Oshawa	−19	4130
Ottawa	−25	4673
Owen Sound	−19	4220
Pagwa River	−34	6330
Paris	−17	4030
Parkhill	−16	3980
Parry Sound	−24	4620
Pembroke	−28	4960
Penetanguishene	−23	4580
Perth	−25	4520
Petawawa	−29	5010
Peterborough	−23	4520
Petrolia	−16	3750
Picton	−21	4080
Plattsville	−18	4130
Point Alexander	−29	5180
Porcupine	−34	6220
Port Burwell	−15	3810
Port Colborne	−15	3640
Port Credit	−18	3700
Port Dover	−15	3830
Port Elgin	−17	4240
Port Hope	−21	4190
Port Perry	−22	4410
Port Stanley	−15	3810
Prescott	−23	4350
Princeton	−17	4030
Raith	−35	6060
Red Lake	−34	6220
Renfrew	−27	4790
Ridgeway	−15	3590
Rockland	−26	4800
St. Catharines	−16	3550
St. Marys	−18	4130
St. Thomas	−16	3850
Sarnia	−16	3840
Sault Ste. Marie	−25	5180
Schreiber	−35	6070
Seaforth	−17	4240
Simcoe	−17	3962
Sioux Lookout	−34	6180
Smiths Falls	−25	4520
Smithville	−16	3920
Smooth Rock Falls	−34	6280
Southampton	−17	4250
South Porcupine	−34	6220
South River	−27	5180
Stirling	−23	4340
Stratford	−18	4300
Strathroy	−17	3920
Streetsville	−18	4080
Sturgeon Falls	−27	5180
Sudbury	−28	5447
Sundridge	−27	5120
Tavistock	−18	4190
Thamesford	−18	4030
Thedford	−16	3860
Thunder Bay	−31	5746
Tillsonburg	−17	3920
Timagami	−30	5570
Timmins	−34	6189
Toronto	−18	4082
Trenton	−21	4116
Trout Creek	−27	5230
Trout Lake	−38	7680
Uxbridge	−22	4450
Vanier	−25	4690
Vittoria	−15	3860
Walkerton	−18	4160
Wallaceburg	−16	3620
Waterloo	−19	4110
Watford	−16	3810
Wawa	−35	5640
Welland	−15	3640
West Lorne	−16	3750
Whitby	−20	4080

TABLE 18-3(B) Recommended Outside Design Conditions for Localities in the United States

State	City	Winter (heating). dry-bulb, °F	Summer (cooling)	
			Dry-bulb, °F	Wet-bulb, °F
Alabama	Birmingham	10	95	78
	Mobile	15	95	80
Arizona	Flagstaff	−10	90	65
	Phoenix	25	105	76
	Yuma	30	110	78
Arkansas	Little Rock	5	95	78
California. . . .	Bakersfield	25	105	70
	El Centro	25	110	78
	Fresno	25	105	74
	Los Angeles	35	90	70
	San Diego	35	95	68
	San Francisco	35	85	65
Colorado	Denver	−10	95	64
	Pueblo	−20	95	65
Connecticut . .	Bridgeport	0	95	75
Delaware	Wilmington	0	95	78
District of Columbia. .	Washington	0	95	78
Florida	Jacksonville	25	95	78
	Miami	35	91	79
Georgia	Atlanta	10	95	76
	Savannah	20	95	78
Idaho	Boise	−10	95	65
Illinois.	Chicago	−10	95	75
	Springfield	−10	98	77
Indiana	Indianapolis	−10	95	76
	Terre Haute	0	95	78
Iowa	Des Moines	−15	95	78
	Sioux City	−20	95	78
Kansas.	Topeka	−10	100	78
Kentucky. . . .	Louisville	0	95	78
Louisiana. . . .	New Orleans	20	95	80
Maine	Augusta	−15	90	73
Maryland	Baltimore	0	95	78
Massachusetts .	Boston	0	92	75
	Worcester	−5	93	75
Michigan	Detroit	−10	95	75
	Lansing	−10	95	75
Minnesota . . .	Duluth	−25	93	73
	Minneapolis	−20	95	75
Mississippi . . .	Vicksburg	10	95	78
Missouri.	Kansas City	−10	100	76
	St. Louis	0	95	78
Montana	Butte	−20	. . .	. .
	Miles City	−35	. . .	. .
Nebraska	Omaha	−10	95	78
Nevada	Reno	−5	95	65
New Hampshire	Concord	−15	90	73
New Jersey. . .	Newark	0	95	75
New Mexico . .	Albuquerque	0	95	70
New York . . .	Buffalo	−5	93	73
	New York	0	95	75
	Syracuse	−10	93	75
North Carolina	Asheville	0	93	75
	Raleigh	10	95	78
North Dakota .	Bismarck	−30	95	73
Ohio	Akron	−5	95	75
	Dayton	0	95	78
	Toledo	−10	95	75
Oklahoma . . .	Tulsa	0	101	77
Oregon	Portland	10	90	68
Pennsylvania. .	Philadelphia	0	95	78
	Pittsburgh	0	95	75
Rhode Island .	Providence	0	93	75
South Carolina	Charleston	15	95	78
South Dakota .	Sioux Falls	−20	95	75
Tennessee . . .	Nashville	0	95	78
Texas	Austin	20	100	78
	Dallas	0	100	78
	El Paso	10	100	69
	Houston	20	95	78
Utah	Salt Lake City	−10	95	65
Vermont	Burlington	−10	90	73
Virginia	Richmond	15	95	78
Washington . .	Seattle	15	85	65
	Spokane	−15	93	65
West Virginia .	Charleston	0	95	75
	Wheeling	−5	95	75
Wisconsin. . . .	Milwaukee	−15	95	75
Wyoming	Cheyenne	−15	95	65

TABLE 18-3(C) Recommended Outside Design Conditions for Selected Localities Throughout the World

Place	Winter, dry-bulb, °F	Summer Dry-bulb, °F	Summer Wet, bulb, °F	Place	Winter, dry-bulb, °F	Summer Dry-bulb, °F	Summer Wet-bulb, °F
Athens, Greece	30	95	72	Mexico City, Mexico	35	79	61
Berlin, Germany	5	91	70	Montreal, Canada	–10	86	74
Bogota, Columbia	30	70	65	Moscow, Russia.	–40	85	72
Bombay, India		93	82	Osaka, Japan	20	93	81
Buenos Aires, Argentina	30	90	75	Panama, Panama		88	78
Cairo, Egypt.	40	100	80	Paris, France.	15	90	75
Calcutta, India	60	100	82	Rio de Janeiro, Brazil.	55	90	78
Capetown, South Africa	45	88	72	Rome, Italy	30	92	73
Havana, Cuba	60	90	78	Shanghai, China	30	95	82
Honolulu, Hawaii	60	83	73	Stockholm, Sweden	0	83	68
Johannesburg, South Afirca.	40	82	65	Sydney, Australia	50	90	72
London, England.	30	85	70	Tokyo, Japan	20	93	81
Madrid, Spain	25	95	75	Vancouver, Canada	10	80	68
Manila, P.I..		88	78	Vienna, Austria.	5	92	71
Melbourne, Australia	35	95	74	Winnipeg, Canada	–30	90	71

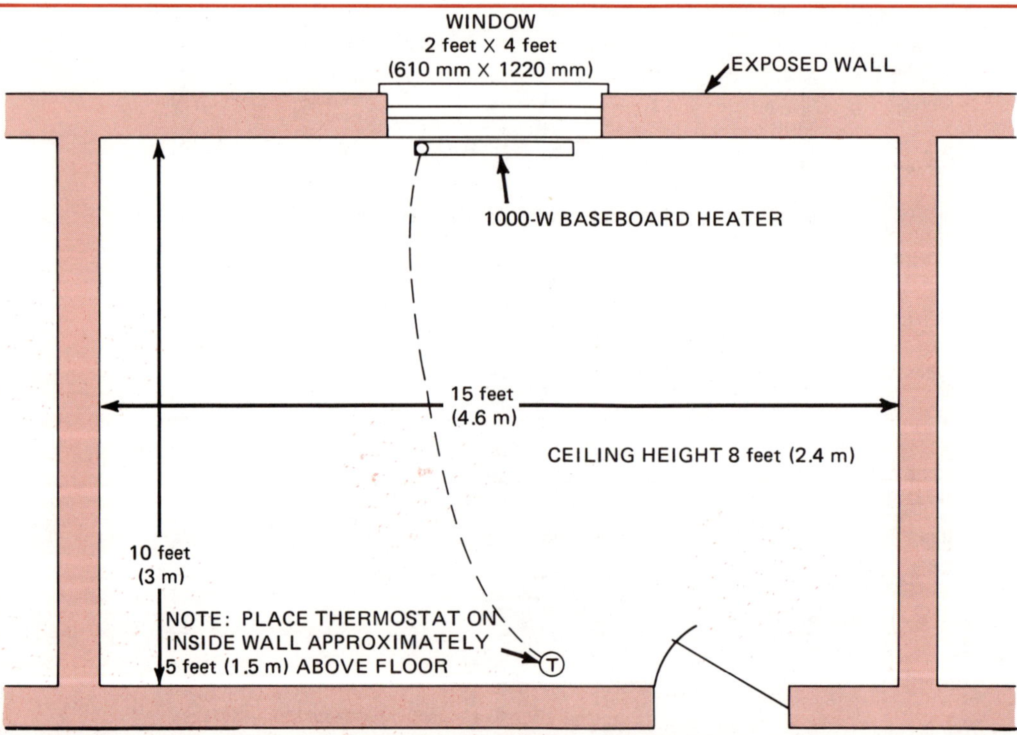

Fig. 18-17 Heating a typical room electrically.

long, which is about 120 ft^2 (11.04 m^2). Since the window occupies 0.72 m^2 of space, the net exposed wall is about 10.3 m^2. The amount of heat loss through this wall in an hour is equivalent to 20 W/m^2 X 10.3 m^2, which is 206 W. *Note:* To convert W/m^2 to W/ft^2, divide W/m^2 by 0.093.

The Third Step: Heat Loss through Doors and Windows. Table 18-4 lists the heat loss per square metre through various types of door and window installations for each degree of design temperature difference. If the room shown in Figure 18-17 has a wood-framed window, that is, single-glazed with a storm, the heat loss will be 2.90 W/m^2 of window for each degree of design temperature difference. The total heat loss through the glass will be as follows. The area of the window equals 0.6 m X 1.2 m, which is 0.72 m^2. Therefore, the heat loss per degree Celsius will be 0.72 m^2 X 2.90 W/m^2, which is 2.09 W/m^2. With a temperature difference of 46°C, the heat loss will be 46 X 2.09 W/m^2, which is 96 W. This heat loss is added to the 206 W lost through the exposed wall.

The Fourth Step: Infiltration Loss. Cold air enters rooms through cracks around doors and windows. Heat loss through such infiltration is calculated by the air change method. Air in a room or a building changes constantly, by the entry and exit of air through doors, windows, cracks, and other areas. The amount of the air change is influenced by the size of room, height of

TABLE 18-4 Heat Loss Factors for Windows and Doors

HEAT LOSS FACTOR PER DEGREE OF TEMPERATURE DIFFERENCE				
	IMPERIAL		METRIC	
Windows and Doors	Rt	$W/ft^2 \cdot °C$	Rt	$W/m^2 \cdot °C$
Single glass — metal frame	0.91	67.26	0.16	6.25
Single glass — metal frame with storm	1.82	34.22	0.32	3.18
Double glass — metal frame (1/4" air space)	1.31	47.68	0.23	4.43
Double glass — metal frame (1/2" air space)	1.42	42.72	0.25	3.97
Triple glass — metal frame (1/2" air space) + storm	2.33	26.26	0.41	2.44
Single glass — wood frame	0.96	62.31	0.17	5.79
Single glass — wood frame with storm	1.99	31.21	0.35	2.90
Double glass — wood frame (1/4" air space)	1.59	37.88	0.28	3.52
Double glass — wood frame (1/2" air space)	1.71	25.41	0.30	3.29
Triple glass — wood frame (1/2" air space) + storm	2.96	20.77	0.52	1.93
Skylight — metal frame — single glass	0.79	74.58	0.14	6.93
Skylight — metal frame — double glass (1/4" air space)	1.19	51.33	0.21	4.77
Skylight — wood frame — single glass	0.91	67.26	0.16	6.25
Skylight — wood frame — double glass (1/4" air space)	1.53	40.36	0.27	3.75
Doors				
Patio Doors — single glass — metal frame	0.91	69.09	0.16	6.42
Patio Doors — double glass — wood frame	1.14	53.16	0.20	4.94
Solid wood — 1 1/2"	2.05	29.92	0.36	2.78
Solid wood 1 1/2" (wood storm)	3.70	16.46	0.65	1.53
Solid wood 1 1/2" (metal storm)	3.02	20.12	0.53	1.87
Metal (Flat)	0.91	69.09	0.16	6.42
Metal (Corrugated)	0.74	85.56	0.13	7.95

Note: To convert English R values to metric R values, multiply the English R value by 0.175.

To convert square feet (ft^2) to square metres (m^2), multiply ft^2 by 0.093.

To convert inches to millimetres, multiply inches by 25.4.

ceiling, and ventilating devices, such as fans. Table 18-5 lists various room conditions and the expected air change over a period of one hour.

If, for example, the room shown in Figure 18-17 has a 2.4 m ceiling height and no doorway that is exposed to outside temperatures, then 0.75 air changes per hour can be expected. Table 18-6 lists the heat loss factors (watts per square metre) for various air changes and temperature differences.

Under the 0.75 Air Change heading, with a 2.4 m ceiling height and a design temperature difference of 46°C, the loss is 28.5 W/m² of floor space. The floor area of the room is 4.6 m X 3 m, which is 13.8 m². The infiltration loss is 13.8 m² X 28.5 W/m², which is 393 W.

This heat loss is added to the wall and window losses.

The Fifth Step: Heat Loss through the Ceiling. If, as according to Table 18-1, the total average Rt, corrected for wood framing, is 4.51, the R designation of the required insulation is easy to find. An insulation with an R-4.8 designation is used to provide maximum heat retention for the room's ceiling.

The U factor for the ceiling is 1/4.51, or approximately 0.221 W/m² per degree Celsius of temperature difference. If the temperature difference is 46°C, the heat

TABLE 18-5 Air Changes per Hour for Various Rooms and Conditions

INFILTRATION FACTORS	AIR CHANGE METHOD
NEW HOUSES (Weatherstripped, storm doors and storm windows)	
Rooms without exposed doors	0.75 air changes/hour
Rooms with exposed door/s — Entrance Area	1.0 air changes/hour
Basements	
Walls less than 50% above grade	0.5 air changes/hour
Walls more than 50% above grade or fully exposed	0.75 air changes/hour
EXISTING HOUSES (not weatherstripped, storm doors and storm windows)	
Rooms without exposed doors	1.0 air changes/hour
Rooms with exposed door/s — Entrance Area	1.5 air changes/hour
Basements	
Walls less than 50% above grade	0.75 air changes/hour
Walls more than 50% above grade or fully exposed	1.0 air changes/hour

COEFFICIENTS FOR INFILTRATION PER SQUARE FOOT PER DEGREE CELSIUS TEMPERATURE DIFFERENCE, 8-FOOT (2.4 m) CEILING HEIGHT

	No Exposed Doors		With Exposed Doors Entrance Area	
New Houses	W/m²·°C	W/ft²·°C	W/m²·°C	W/ft²·°C
Areas other than basements	0.61	6.56	0.82	8.82
Basements	0.41	4.41	0.61	6.56
Existing/Conversion Houses				
Areas other than basements	0.79	8.50	1.23	13.23
Basements	0.62	6.67	0.79	8.50

Note: To convert square feet (ft²) to square metres (m²), multiply ft² by 0.093.

loss is 46°C X 0.221 W/m²°C, which is 10.2 W/m² for each hour of operation.

The ceiling area of the room shown in Figure 18-17 is 13.8 m². This is the area through which heat can pass. The total heat loss through the ceiling in one hour is 13.8 m² X 10.2 W/m², which is 141 W.

This heat loss is added to the loss through walls, windows, and infiltration.

The Sixth Step: Total Heat Loss for the Room. To calculate the total heat loss for the room over a period of one hour, the loss through each of the four areas must be added. For example:

Heat loss through the wall:	206 W
Heat loss through the window:	96 W
Infiltration loss:	393 W
Heat loss through the ceiling:	141 W
Total heat loss for the room:	836 W

DETERMINING HEATER SIZE

Some winters are colder than others and often have more days below the outdoor design temperature than listed for the area. (See Tables 18-3A, B, and C.) To compensate for this, choose the next heater size above the calculated heat load for the area.

The sample room shown in Figure 18-17 has a calculated heat loss of 836 W.

TABLE 18-6 Heat Loss Factors per Square Metre Floor area and Air Change Values (Infiltration)

Ceiling Height (m)	Values for Various Temperature Differences — Heat Loss Factors Per Square Metre (W/m²)							
	1°C	38°C	41°C	44°C	47°C	50°C	53°C	56°C
0.5 Air Change								
2.4	0.41	15	17	18	19	20	22	23
2.7	0.465	18	19	20	22	23	25	26
3.0	0.50	19	21	22	24	25	27	28
0.75 Air Change								
2.4	0.62	24	25	27	29	31	33	35
2.7	0.70	27	29	31	33	35	37	39
3.0	0.775	29	32	34	36	39	41	43
1.0 Air Change								
2.4	0.81	31	33	36	38	41	43	46
2.7	0.93	35	38	41	44	47	49	52
3.0	1.0	39	42	45	48	51	54	58
1.5 Air Change								
2.4	1.2	46	50	54	57	61	65	68
2.7	1.4	52	56	61	65	69	73	77
3.0	1.5	58	63	67	72	77	81	86

Note: To convert metres to feet, divide metres by 305.

To convert W/m² to W/f², divide W/m² by 0.093.

A heater of a higher wattage rating than 836 W should be selected for the room. The closest heater unit exceeding the calculated 836 W is the 1000 W unit. It would be reasonable, therefore, to install a 1000 W heater in the room. The extra wattage, above and beyond the calculated heat load, will take care of the cold winters.

Ceiling cable of the same wattage rating as the baseboard heater could be installed as an alternate heat source.

HEATER AND THERMOSTAT LOCATION

Heater units should always be located as close as possible to the major heat loss area of a room. Windows are usually responsible for the most heat loss, and so the baseboard heater is usually located under the window. Warm air rising out of the heater tends to cancel the cooling effect of cold air entering the room around the window. (See Figures 18-18 and 18-19.)

Thermostats are mounted on inside walls. This prevents the cold outer wall from influencing the thermostat. Otherwise, the heater would stay in operation longer than needed to maintain the inside design temperature.

Foyers and entrance halls often receive cold blasts of air when the door is opened. This can be counteracted by installing a forced-air unit in the wall near the entrance. Cold air entering the area will activate the thermostat on the heater. The fan then forces enough warm air through the heater to return the area to inside design temperature.

A forced-air unit consists of a metal enclosure with a detachable fan and heater unit, Figure 18-20. A grille placed over the unit directs the air flow and protects the heater coils.

Sometimes a bathroom with limited wall space at floor level can take advantage of these compace units, Figure 18-21.

HEATING COST

Once the heat loss (heating load) has been calculated for each area, the total heat

Fig. 18-18 Residential baseboard heater installation.

Fig. 18-19 Hallway baseboard heater installation.

load figure can be worked out. The total heat load is the sum of the heat losses calculated for each area. An example for calculating heating costs follows. It is based on estimated heat losses (in round figures) for a 5-room house in Toronto or a similar area.

Heated Area	Heat Loss
Bedroom #1:	1000 W
Bedroom #2:	1000 W
Kitchen:	1000 W
Livingroom and dining room:	2500 W
Bathroom:	500 W
Basement:	3000 W
	9000 W, or 9 kW

Total heat load:

Annual kilowatt-hour consumption =

$$\frac{HL \times DD \times C}{DTD}$$

Where:

HL = heat loss in kilowatts

DD = annual degree days for the area (see Tables 18-3A, B, and C)

C = a constant (15.3 for the Toronto area; consult local utility)

DTD = design temperature difference between indoors and outdoors

Therefore, kilowatt-hour consumption is:

$$\frac{9 \times 4082 \times 15.3}{40} = 14\,052 \text{ kW·h}$$

Typical utility charges are:

First 100 kW·h per month @ 6.9¢/kW·h
Next 400 kW·h per month @ 3.45¢/kW·h
Remaining kilowatt-hours @ 2.4¢/kW·h (end rate)

Customers of electrical heating approach the end rate with lighting and cooking energy. The annual heating cost, therefore, is calculated by multiplying the kilowatt-hour consumption by the end rate.

Total annual cost:

14 052 kW·h × 2.4¢/kW·h = $337.25

OVERHEATING PROTECTION

Baseboard heaters are equipped with a high-temperature safety device which consists of a slender, liquid- or vapor-filled cop-

Fig. 18-20 Typical forced-air heater unit.

Fig. 18-21 Forced air heater unit applications (near cold entry and where wall space is limited).

per tube that travels the length of the heater, Figure 18-22. If for some reason the heater reaches an abnormally high temperature, the expanding liquid or vapor will cause a relay at one end of the tube to open the circuit. This prevents heat damage to the walls of the home.

SNOW-MELTING HEATERS

One type of radiant-heating cable is designed for use in driveways to keep the wheel track areas clear of snow, Figures 18-23 and 18-24. It is installed under the concrete pavement and comes in preassembled lengths from the manufacturer. It is also used to melt snow on ramps and stairways.

On roofs, melted snow that has run into the eaves and become refrozen also causes problems. Eavestrough heaters are used to keep this part of the roof clear, Figure 18-25.

Pipes can be kept from freezing by wrapping them with pipe-heating cable, Figure 18-26. A thermostat that fits along the side of the pipe controls the heating process.

Fig. 18-22 Baseboard heater electrical connection section.

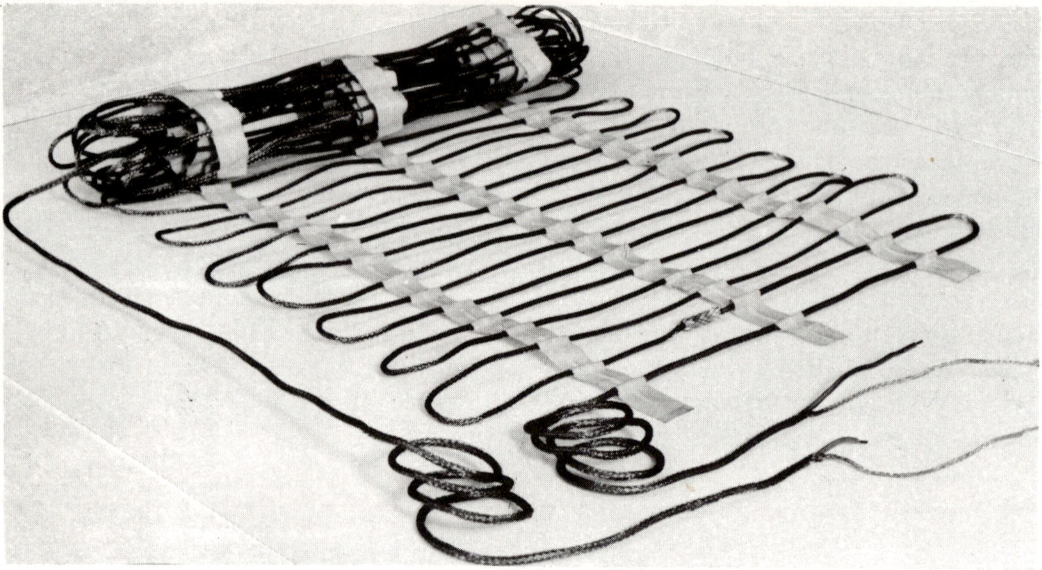

Fig. 18-23 A type of driveway snow-melting cable.

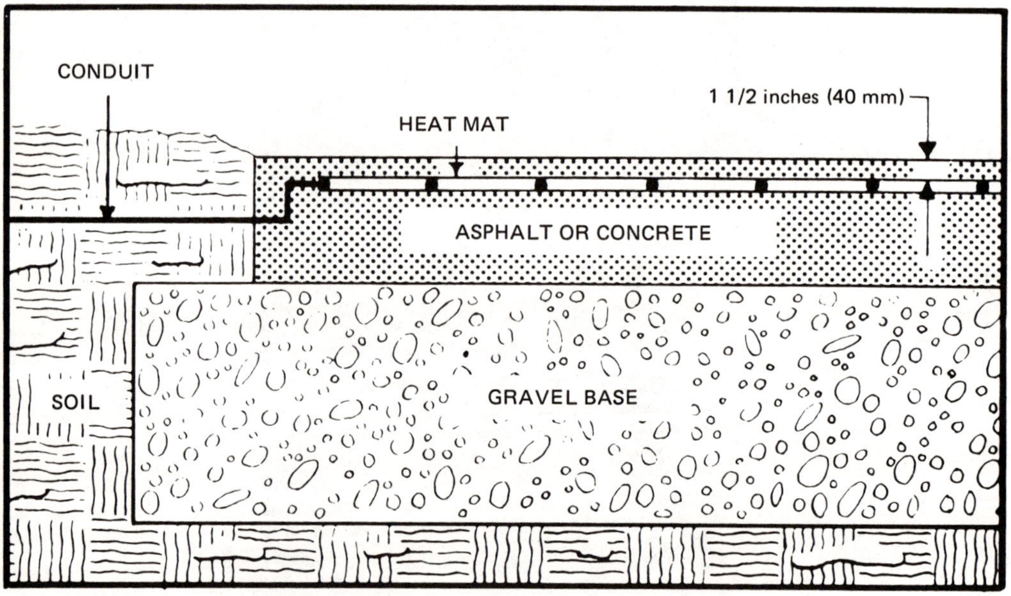

NOTE: FOR ASPHALT INSTALLATIONS, PLACE BITUMINOUS BINDER ON BASE COURSE BOTH UNDER MAT AND OVER MAT BEFORE PLACING FINAL COURSE.

Fig. 18-24 Installation of cable for driveway snow-melting use.

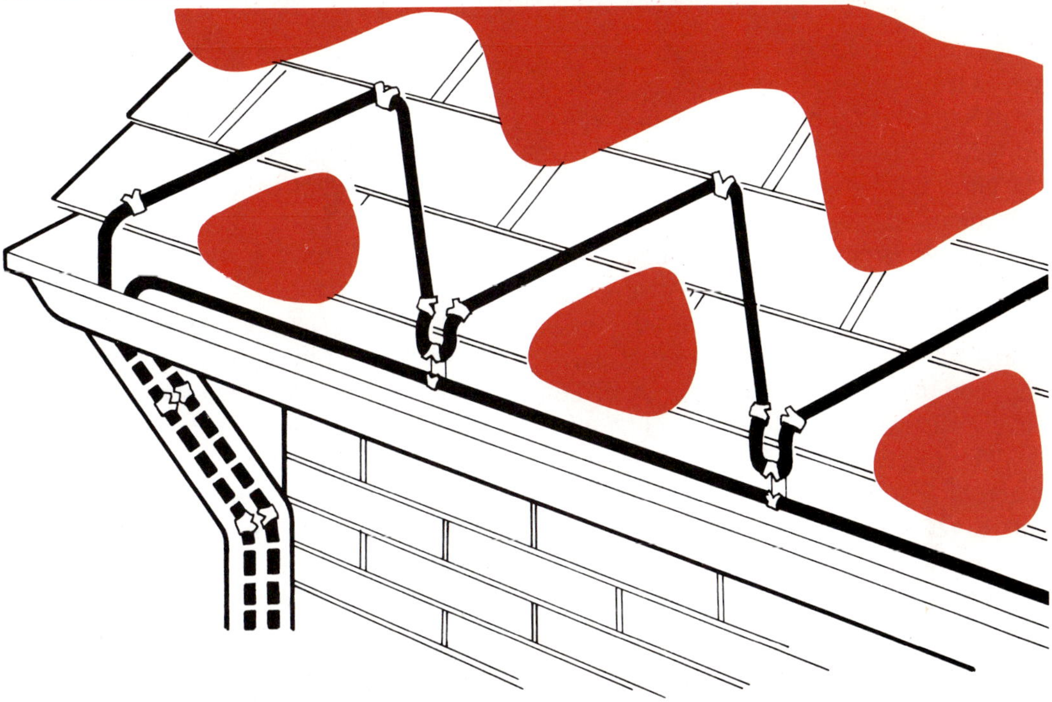

Fig. 18-25 Eavestrough heating cable installation.

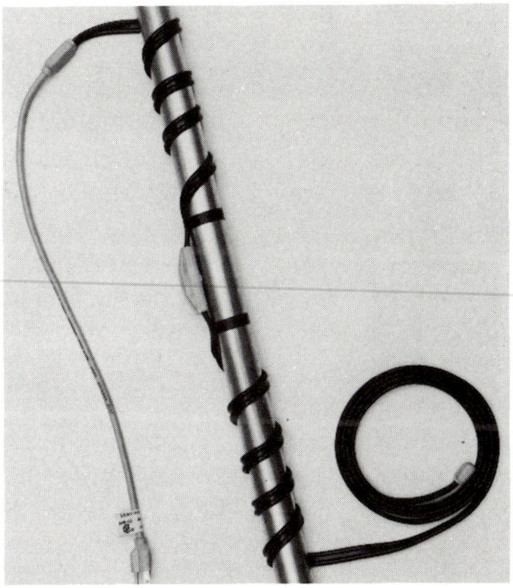

Fig. 18-26 Pipe-heating cable with thermostat.

RESIDENTIAL COOLING SYSTEMS

Generally, throughout most of the United States, a heating system is an essential part of any residence but, for year-round comfort, the cooling season must also be considered.

A complete system would include heating, humidifying and air cleaning for the heating season, and cooling, dehumidifying, and air cleaning for the cooling season. Distribution is also an essential part of the complete air-conditioning system.

Only the major considerations in cooling will be discussed. However, with this background information, the student will be able to recognize the elements of a year-round air-conditioning system, and understand the basic principles of its operation.

UNITS FOR THE COOLING SEASON

A unit for the cooling season may be only a window-type, self-contained, electrical plug-in air conditioner. It may be rated to operate from a single-phase, 120-V wall receptacle, or a single-phase, 240-V special-circuit outlet. These special electrical circuits may be installed in a particular location for the window-type, plug-in units, or they may be installed so as to operate from a switched controlled outlet. A switched controlled outlet may allow for the selection of either a baseboard heating unit or a plug-in air conditioner. A complex system of central heating and cooling, using several individual components and often a common ducting system for distribution of heating and cooling the air, is commonly used.

CHILLED-WATER COOLING COIL

A simple method of reducing the indoor temperature is the chilled water coil method, shown in Figure 18-27. This method can be used satisfactorily when a natural source of cold water (35°F to 55°F, 2°C to 13°C) is available. The cold water is pumped through the coil over which room air is circulated. The water discharged from the coil may go to a dry well, storm sewer, swimming pool, or lawn. Water cooled in a water chiller, as part of a mechanical-refrigeration system, is also used as the cooling agent in a chilled-water coil. Chilled water of this type, however, is recirculated through the water coil and is not discharged from the system.

When air is passed over the chilled-water coil, the room air is cooled because it gives up some of its heat to the water flowing through the coil. The discharged water is found to be 10°F to 15°F warmer than the supply water. This rise in water temperature through the coil is caused by the transfer of heat from the room air to the water in the coil. This principle is the basis of any cooling system. Cold is not created; heat is merely transferred from one place to another. A simple control, consisting of a low-voltage cooling thermostat and a relay operating from the electrical source, as shown in Figure 18-28, can be used for regulating the system cooling.

EVAPORATOR COIL

In any residential air-conditioning system using mechanical refrigeration, heat is removed from the room air with an evap-

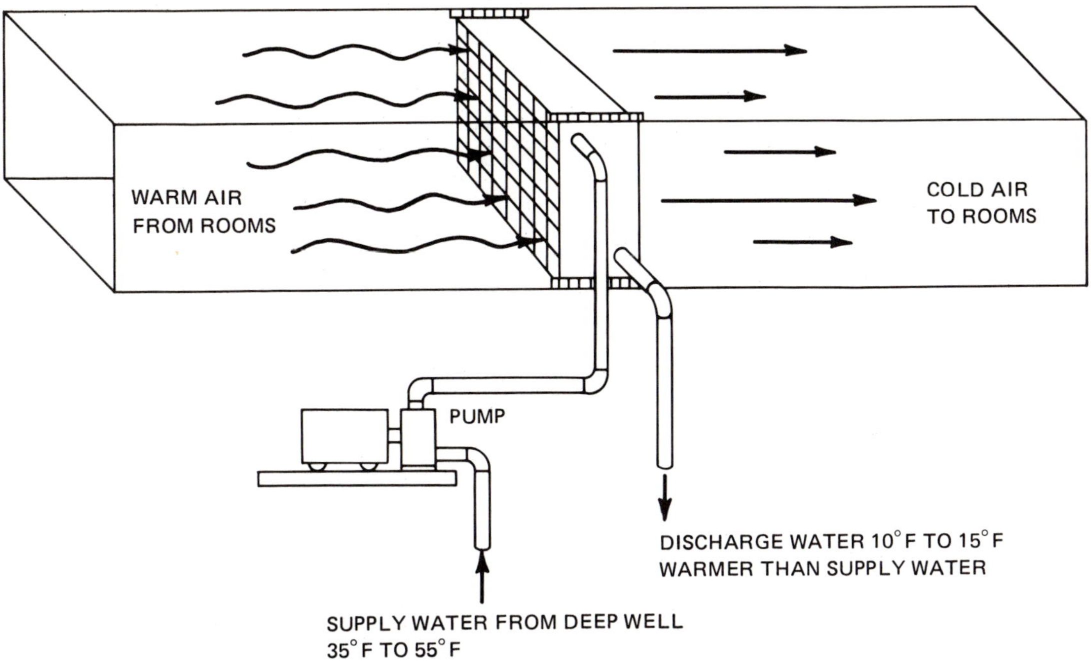

Fig. 18-27 Cold-water cooling with a chilled-water coil.

orator coil. The coil can be mounted in a moving air stream, as shown in Figure 18-29. With this method, the air is cooled by coming into actual contact with the evaporator coil.

SPLIT SYSTEM

A split system is usually used when cooling is applied to a steam heating, gas heating, hot-water heating, or electric duct heating system, Figure 18-30. Once the cooling equipment has been added to a cen-

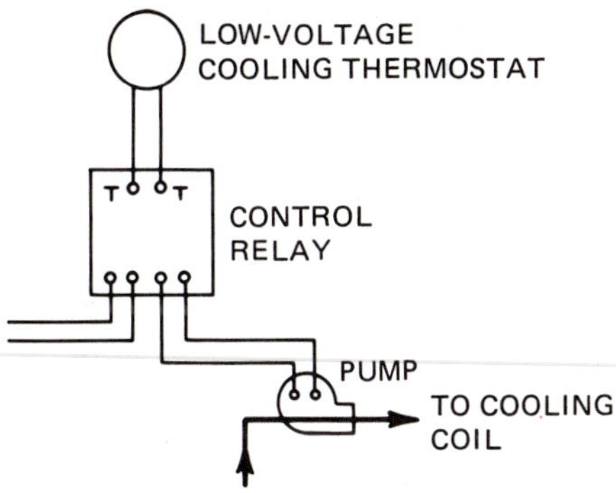

Fig. 18-28 Simple diagram for controlling cold-water cooling coil.

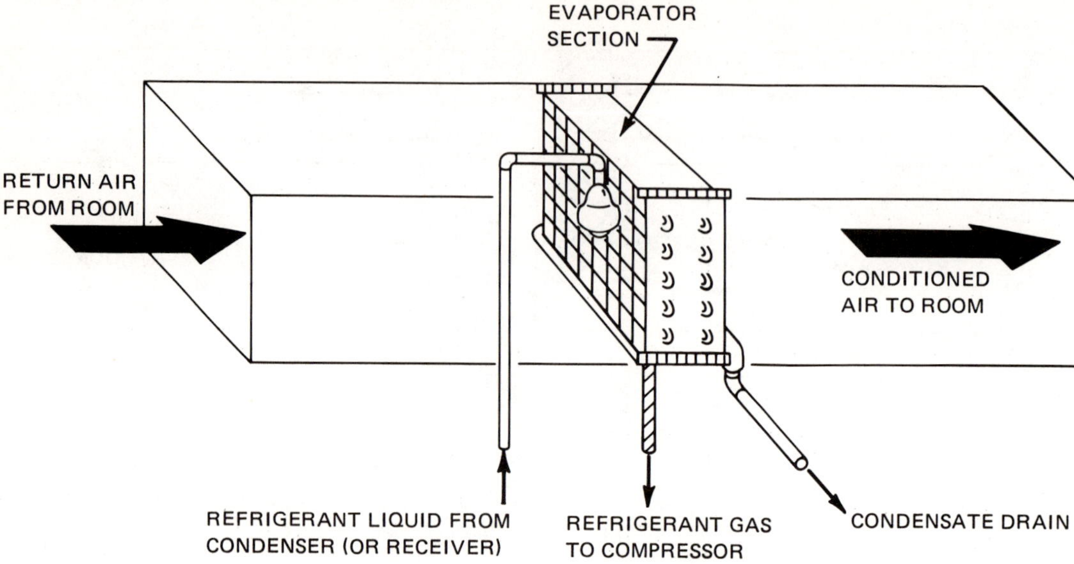

Fig. 18-29 Evaporator coil in air stream.

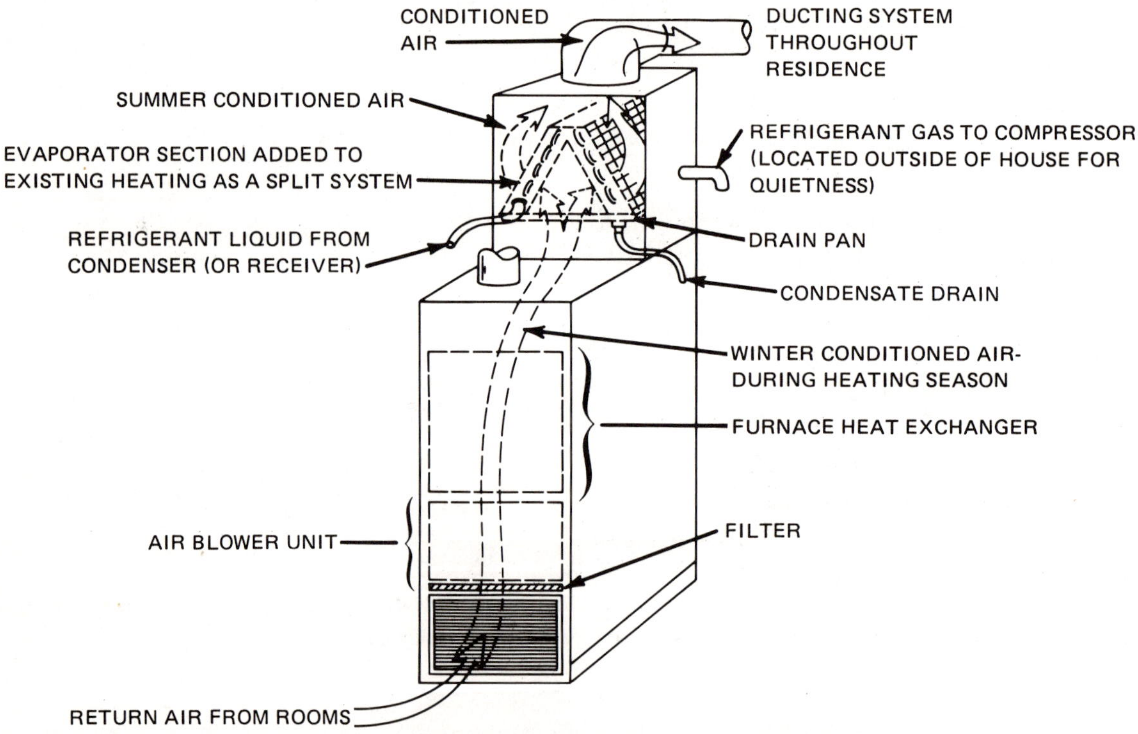

Fig. 18-30 A split system with cooling coil added to existing heating.

tral heating plant, the operation and controls are practically the same as a complete assembly year-round central air-conditioning package.

SYSTEM CHANGEOVER FROM HEATING TO COOLING

When heating and cooling plants are installed in a residence, there must be some method of switching over from one system to the other. Basically, there are three types of system changeover.

1. *Manual changeover* – the homeowner must go to the cooling or heating equipment to perform one or more operations.
2. *Semiautomatic changeover* – the homeowner must decide when the switchover must be made, but can carry out all functions of this changeover at the thermostat location.
3. *Automatic changeover* – the change from heating to cooling and back again is done automatically by the thermostat. The homeowner takes no part in the changeover. Often, manual switches are included with the automatic changeover. These switches allow the homeowner to shut down the cooling equipment, or both heating and cooling, and to turn on the circulating blower independently of the automatic controls.

HEAT PUMPS

In recent years, the heat pump has gained recognition as a method of providing year-round air conditioning. The heat pump uses the same basic components as the mechanical refrigeration system. During the cooling process, the operation of a heat pump is identical to the mechanical refrigeration method. This cooling process is shown in Figure 18-31. To heat a house the cycle is simply reversed, as shown in Figure 18-32. The heat pump usually operates

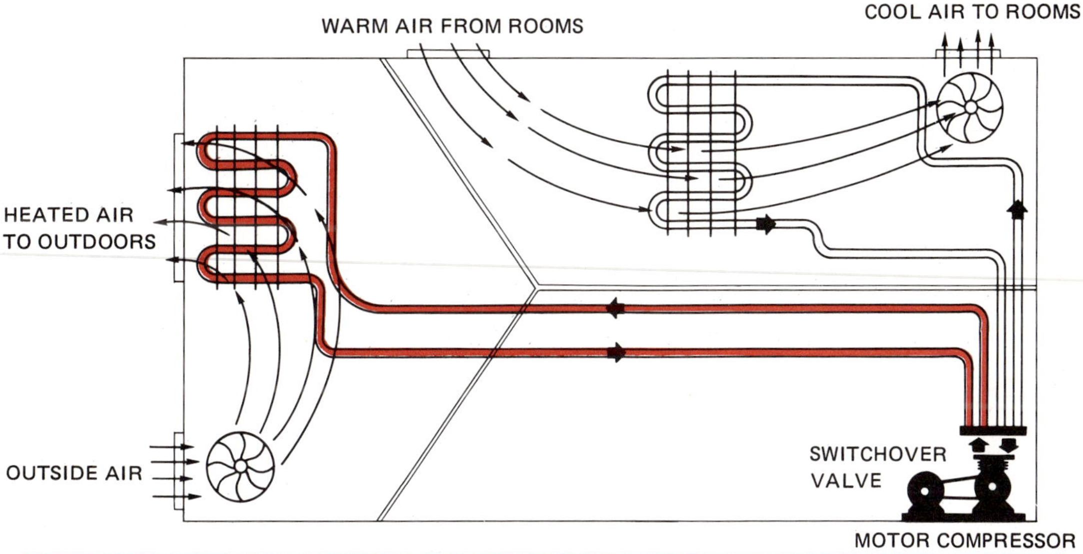

Fig. 18-31 Heat pump operating to cool a house.

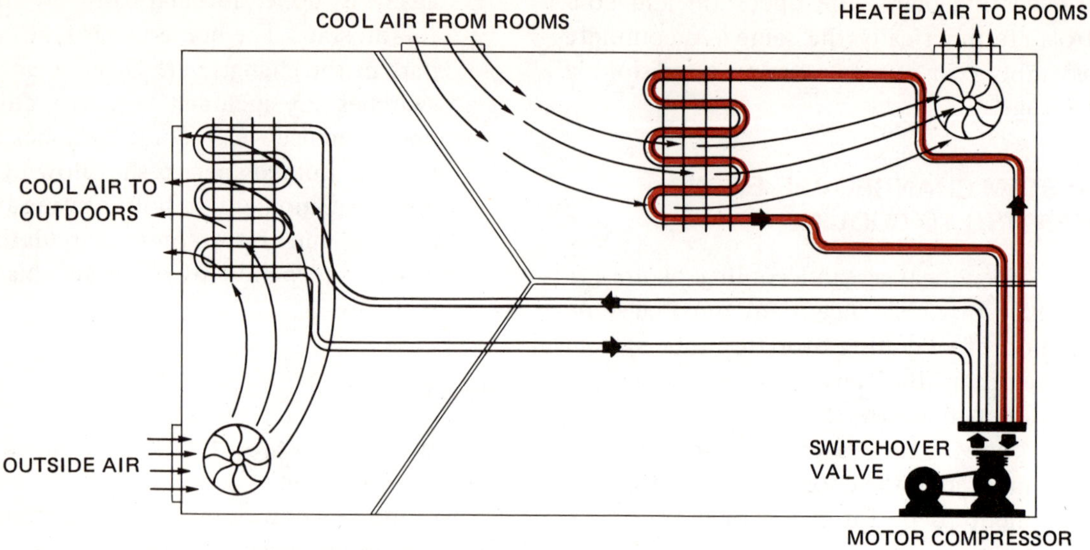

Fig. 18-32 Heat pump operating to heat a house.

satisfactorily down to approximately 40°F (4°C). At these lower temperatures, to obtain capacity on some heat pumps it may be necessary to add auxiliary heating equipment, usually in the form of electric-resistance strip heaters.

REFINEMENTS

Many refinements and an unlimited number of variations in equipment and control systems are continually developing. Electronic filtering and the use of solar energy are fast gaining prominence as efficient, energy-saving systems.

REVIEW

1. List six advantages that electric heating has over other heating systems.
2. What is the main purpose of insulation?
3. What are the minimum insulation requirements for an electrically heated house?
4. Explain what is meant by the R designation of an insulation.
5. List the four types of insulation that are suitable for electrically heated houses.
6. What is an air-vapor barrier? Where is it installed?
7. Why is it important to ventilate an attic?
8. Describe the general construction of a baseboard heater.
9. Why are baseboard heaters made in two wattage densities?

10. List three factors that must be taken into account when installing baseboard heaters.
11. Describe radiant-heating cable.
12. List three precautions that must be taken when installing radiant-heating ceiling cable.
13. List four major residential heat-loss areas.
14. List the advantages and disadvantages of aluminum-framed and wood-framed storm windows.
15. Why are heaters usually placed under windows?
16. What is the advantage of placing a forced-air heater unit in an entrance hall or foyer?
17. Why are baseboard heaters equipped with cut-out relays?
18. What precaution should be taken when insulating basement walls of a home that is likely to have water entering the basement?
19. Name one major advantage of using a forced-air electric furnace to heat a house instead of baseboard heaters.
20. Why must outside design temperature be taken into consideration when calculating a heat loss?
21. Would dehumidifying of the air be a part of the complete cooling or heating system?
22. Explain how a chilled-water coil cools the air inside the residence.
23. Can a split-system air conditioner be added to the existing heating equipment and share the same ducting?
24. Explain how a mechanical refrigeration evaporator coil cools the air in the residence.
25. What is the basic operation of a heat pump in an air-conditioning unit?

Chapter 19

Discharge Lighting Sources

FLUORESCENT LIGHTING

Fluorescent lighting was first developed during the 1930s. Its principle is simple enough, but it took years of research before it became the highly developed tube found in modern lighting fixtures.

TYPES OF FLUORESCENT LAMPS

Fluorescent lamps are available in a wide selection of lengths, tubular diameters, operating types, colors, shapes, and bases.

Lengths

4-watt lamps = 6 inches (152 mm)
6-watt lamps = 9 inches (229 mm)
8 watt lamps = 12 inches (305 mm)
13-watt lamps = 21 inches (533 mm)
15-watt lamps = 18 inches (457 mm)
20-watt lamps = 24 inches (610 mm)
30-watt lamps = 36 inches (914 mm)
40-watt lamps = 48 inches (1.22 m)
85-watt lamps = 72 inches (1.83 m)
110-watt lamps = 96 inches (2.44 m)

Circline Fluorescent Lamps

FC8T9/CW: 22 W, 9/8 inches, 8 1/2 inches outside diameter, cool white lamp

FC12T10/CW: 32 W, 10/8 inches, 12 inches outside diameter, cool white lamp.

FC16T10/WWX: 40 W, 10/8 inches, 16 inches outside diameter, warm white deluxe lamp.

Bases

4-W to 13-W lamps use the miniature bipin base.

14-W to 40-W lamps use the medium bipin base.

T-17, 60-inch length, 40-W through 90-W lamps use the mogul bipin base.

Slimline lamps, 25 W through 75 W use the single-pin base.

High-output (HO type) lamps and power-groove lamps use the recessed double-contact base.

ADVANTAGES OF FLUORESCENT LIGHTING

Fluorescent lamps last approximately 20 times as long as standard incandescent lamps. The higher initial cost of the fluorescent tube is more than offset by its long life. As a result, maintenance costs are much lower, because fluorescent lamps do not need to be replaced as often as standard incandescent lamps. Modern 40-W fluorescent tubes have an average life expectancy of 20 000 hours. The average incandescent bulb has a life expectancy of 1000 hours.

Fluorescent tubes produce more than 5 times as much light per watt of electricity consumed than the incandescent bulb. A 48-inch (1.22 m), 40-W fluorescent tube produces nearly as much light as a 150-W incandescent bulb. The fluorescent tube also keeps its brightness for a longer period of time.

Fluorescent tubes produce less heat than incandescent bulbs of the same size (wattage output). A large incandescent bulb will burn someone who is removing it while it is still in operation. In contrast, most fluorescent tubes can be handled safely, regardless of how long they have been in operation. When a building is air conditioned, fluorescent tubes have the advantage of providing adequate light without placing as large a heat load on the air conditioner as do incandescent bulbs of the same light output. In addition, fluorescent tubes produce more uniform light over a given area because of their length; there is no single bright spot as there is with incandescent lamps.

Note: Fluorescent tube life is greatly influenced by the number of times the fixture is turned on and off. The current that surges through the lamp circuit when the fixture is turned on tends to shorten the life of the lamp. A lamp that is turned on and off frequently will need to be replaced far sooner than a lamp that is allowed to operate for a number of hours between starts. Because of this, lights in many industrial plants are allowed to remain on during lunch hours, coffee breaks, and shift changes. As the cost of electrical energy continues to increase, a more realistic compromise has been reached between tube life and energy costs. It is now considered more economical to turn the lamps off when the off period is expected to be more than 20 minutes in duration.

DISADVANTAGES OF FLUORESCENT LIGHTING

In some places, such as clothing and fabric stores or meat markets, color is critical and the design of the lighting system is very important. Fluorescent lights give off many blue and green tones, but are low in reds and yellows. This prevents a customer from seeing the true color of the product. Walking outside to natural light is not always practical so, in the types of stores mentioned, both incandescent and fluorescent lights are often used to help bring out true colors. A fluorescent tube for "color critical" areas has now been developed, however. The "D" daylight tube can often serve this purpose.

Using discharge lamps over rotating machinery can sometimes be *dangerous.* The 60-Hz ac supply system of fluorescent fixtures makes the light flicker at a very high speed. This flickering is almost invisible to the naked eye, except when the light falls directly on rotating machinery. If the machine is rotating at a speed close to the speed at which the light is flickering, the machine will appear to be at a standstill. This *stroboscopic effect* deceives the human eye. Cases are known where an operator has absentmindedly reached in to touch a rotating part, thinking that the machine has come to a standstill. To some extent, this effect has been overcome by using two lamps in a fixture (in a lead-lag circuitry); the current to one lamp leading, while the other lamp has a lagging current.

FLUORESCENT TUBE PARTS

The heart of the fluorescent fixture is the tube itself. The tube has several components, Figure 19-1.

Glass Tube. The tube provides an airtight enclosure in which the mercury, gas, and phosphor can function.

Bases. Bases are the ends of the tube that connect the lamp to the electrical circuit. Several pin configurations are available.

Cathodes. Cathodes are small, oxide-coated filaments which heat up and emit electrons into the tube.

Mercury. Droplets of mercury are placed into the tube. They vaporize during the operation of the lamp and emit (give off) ultraviolet energy.

Filling Gas. A small amount of highly purified argon gas is placed in the tube. This gas *ionizes* (the producing of electrical conductivity in gases) when sufficient voltage is applied. This allows current to flow readily through the tube.

Phosphor Coating. All of the light energy produced by the mercury is ultraviolet and invisible to the naked eye. The phosphor coating reacts to the ultraviolet rays and turns this energy into visible light.

FLUORESCENT TUBE COLORS

Fluorescent lamps are available in a range of strong colors and in several different "whites."

The saturated colors – red, pink, gold, green and blue – are used for decorative effects, while the whites are used for both decorative and general lighting purposes.

White fluorescent lamps are designed to combine three important elements in lighting effects:

1. Efficiency – most light per dollar,
2. Color-rendering properties – the ability to bring out the beauty of colored materials and objects.
3. "Whiteness" – their appearance in relation to either natural outdoor daylight or the traditional artificial illumination such as filament lamps.

Lamp Designation

CW: Cool white – combines high efficiency with reasonably good color rendition. It is the most widely used fluorescent lamp color in factories, offices, and schools. It blends well with natural daylight.

WW: Warm white – provides the highest efficiency in white fluorescent lamps. It emphasizes orange, yellow and yellow-green at the expense of other colors. It is generally used where highest efficiency is more important than color rendition.

CWX: Deluxe cool white – most closely simulates the appearance and color-rendering properties of natural daylight. It is widely used in stores such as supermarkets, florists, apparel shops and other places where excellent color rendition of natural daylight is needed. It is also used in factories and office installations where best appearance of colors is important.

WWX: Deluxe warm white – sometimes referred to as home-lite, this lamp simulates the warm effects of fila-

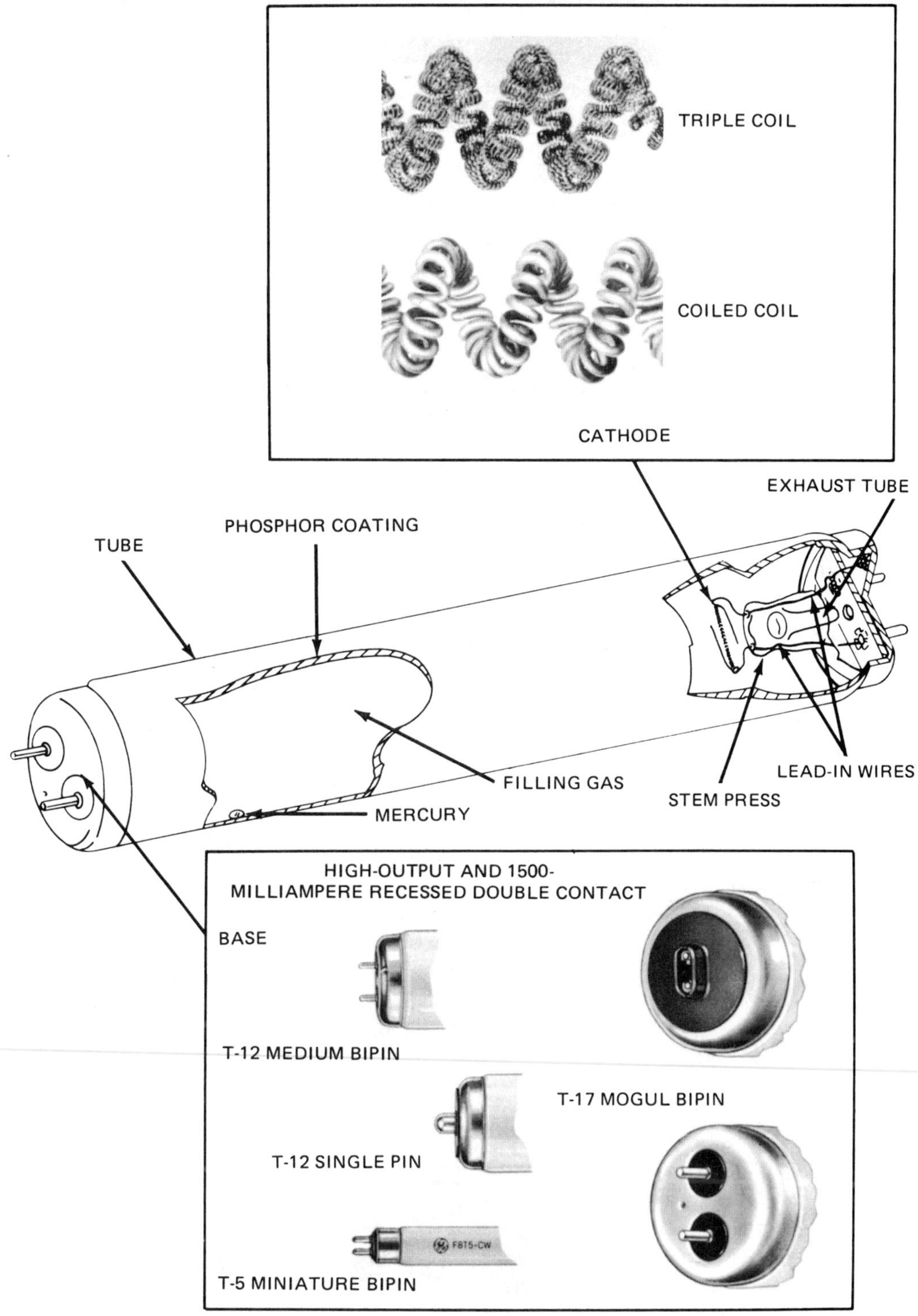

Fig. 19-1 Fluorescent tube and components.

ment lighting in both "whiteness" and color rendering. It is usually the first choice in residences, restaurants, beauty parlors, department stores, bakeries and other places where "homelike" lighting effects are desired.

D: Daylight – soft white and white lamps are still available for replacement purposes in existing installations and for new installations where their appearance or color-rendering properties are particularly suitable.

Lamp Ordering Abbreviations. For example, F4T5/CW, where

F = fluorescent lamp
4 = nominal lamp watts
T5 = tubular shape, 5/8 inch in diameter
T8 = 8/8 inch in diameter
T12 = 12/8 or 1 1/2 inches in diameter
CW = cool white color
WW = warm white
CWX = cool white deluxe
WWX = warm white deluxe (home-lite)
D = daylight
SW = soft white
B = blue
G = green
GO = gold
PK = pink
R = red

THE STARTER

Some fluorescent fixtures (preheat type) need a small starting mechanism to establish an electron emission from the cathode (filament), Figure 19-2.

Current enters the starter through one of the contact pins. The neon gas in the glass enclosure provides a high-resistance path from one contact to the other for a small amount of current. This causes the gas to glow. Heat from the glowing gas warms the bimetal strip, which then bends and closes the contacts. This action creates a low-resistance path through the starter, closing the series circuit applying the supply voltage through the lamp filament and the reactor (ballast). Once the contacts have closed, the gas ceases to glow, the bimetal strip cools, and the contacts open. All this takes about two or three seconds.

The capacitor in the starter controls the amount of arcing between the contacts when they open and close. This prolongs the life of the starter, and helps to eliminate radio interference which can be caused by the opening and closing of the contacts.

THE BALLAST

Mercury droplets in the tube vaporize during the heating operation of the lamp.

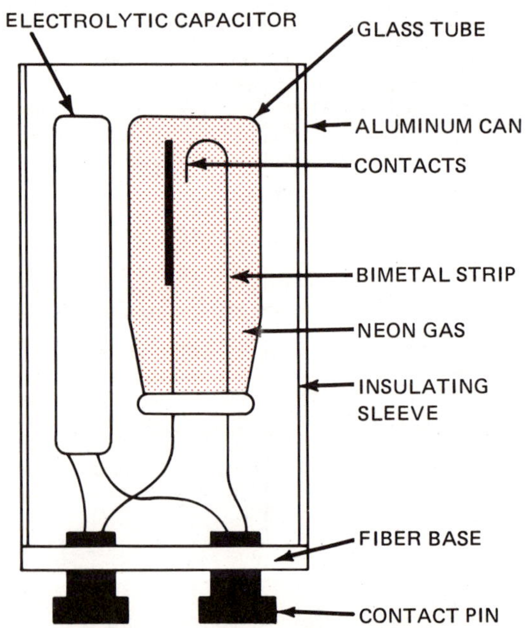

Fig. 19-2 Typical fluorescent starter.

Current then flows through the tube more and more easily. This ionization process, with its low-resistance path, could allow the current flow to increase to the point where the tube would destroy itself. The ballast controls and regulates this current flow in much the same way that ballast in a ship controls the stability of the vessel. The *ballast* is a coil of wire wound on a laminated steel core.

OPERATION OF THE FIXTURE

When a fluorescent lamp is switched on, current flows through the ballast, filaments, and starter, Figure 19-3. The high-resistance neon gas in the starter glows (heats up) and bends the bimetal contacts until they touch. This closing of the contacts provides a low-resistance path, and the lamp filaments heat up quickly as a result of the extra current flow. Heating of the filaments causes electrons to be emitted into the tube. The neon gas in the starter stops glowing, and the bimetal contacts cool and open the low-resistance path.

The increase in current flow through the filaments also causes a strong magnetic field around the ballast. When the bimetal contacts open, the high-resistance path is reestablished. This causes the collapse of the strong magnetic field around the ballast coils. As the magnetic field collapses, it induces a high-voltage kick into the ballast coils. This voltage is high enough to strike an arc across the tube, using the emitted electrons, argon gas, and mercury vapor as a current path.

Once an arc is established across the tube, most of the circuit current flows through the tube. Not enough current flows to the starter to reestablish a starting cycle in the starter's neon gas.

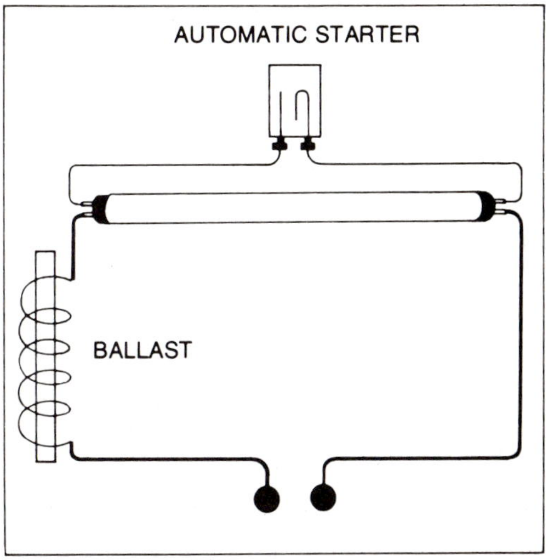

Fig. 19-3 Circuit for operation of one preheat-type lamp.

As mentioned before, current flow in the tube tends to increase. When the current flow increases, a stronger magnetic field is established around the ballast. This magnetic field produces (induces) a voltage in the ballast, which flows in the opposite direction to the applied voltage from the source. The more the current tries to increase in the tube, the more the ballast tries to hold back the applied voltage and current. The current flow in the fixture quickly stabilizes itself. The ballast is sometimes referred to as a "reactor" because of its operation.

When the ultraviolet light energy from the mercury vapor strikes the phosphor coating on the inside of the tube, a cool, comfortable, visible light is produced. As the tube nears the end of its life expectancy, an oxide coating from the filaments gradually appears on the ends of the tube. These darkened areas indicate that the tube is about ready to be replaced.

A tube that is flickering but not starting because of age should be replaced before the starter and/or ballast are damaged by the continuous starting currents. A defective starter may cause the lamp's filament to heat up, but not light the lamp; or it may not allow the filaments to become energized.

RAPID-START FIXTURES

The rapid-start type of fluorescent fixture does not need a separate starting device. Approximately 4 V is continuously supplied to the filaments of the tube by special heater windings in the ballast while the tube is in operation. The constantly heated filaments emit a steady stream of electrons, and this easily allows the ballast voltage to strike an arc across the tube.

Rapid-start fixtures, which were developed after the preheat and instant-start types, take advantage of the voltage that exists between the cathodes and the metal frame of the fixture. For this reason, all rapid-start fixtures must be grounded. Otherwise, in cool weather they will often fail to start.

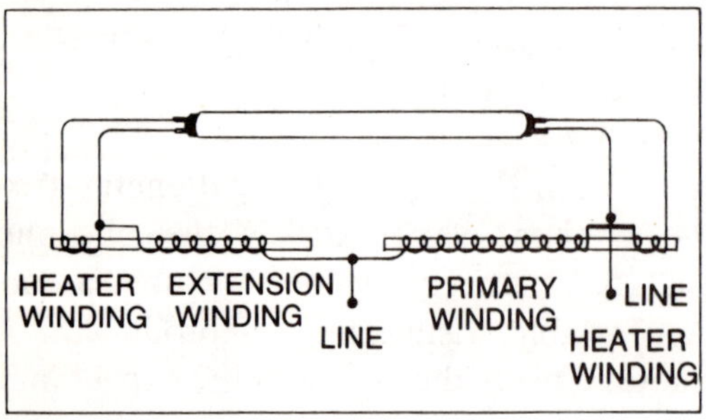

Fig. 19-4 Single-tube, rapid-start circuit. (Rapid-start ballast supplies small voltage continuously to cathodes.)

As a result of the continuous heating of the cathodes (filaments) a lower open-circuit voltage can be used. This reduces the physical size and weight of the ballast. NEC *Article 410* covers the construction and installation of lighting fixtures. Lighting equipment having an open-circuit voltage of more than 1000 volts is not permitted to be installed in residential occupancies.

Figure 19-4 shows a single-tube, rapid-start circuit. Two-lamp units are also available, Figure 19-5. The two-lamp fixture starts one tube slightly ahead of the other, with the help of a capacitor in the ballast.

Many industrial plants with a large number of electrical motors have a poor *power factor.* This simply means that they receive large quantities of electrical energy and require large conductors and control equipment, but waste much of the power. (This is like paying for only two flavors of ice cream from a three-flavor brick, because you want only two of the flavors. The cost of the wasted third flavor would then have

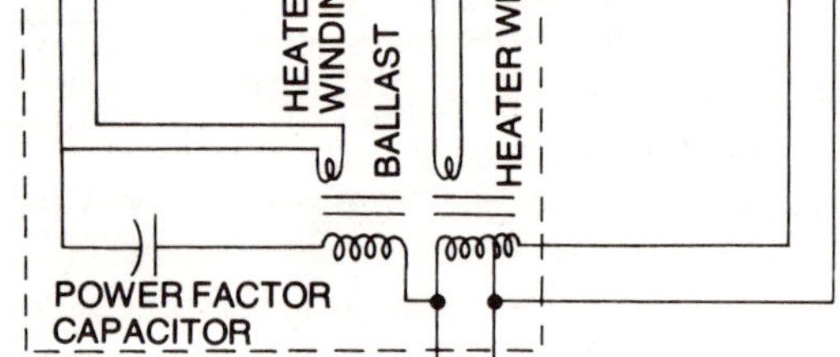

Fig. 19-5 Two-lamp rapid-start circuit.

to be covered by the supplier.) The capacitor in fluorescent fixture ballasts helps to overcome this problem. For this reason, industries with a poor power factor often save a lot on electrical bills by installing fluorescent fixtures.

BALLAST OVERHEATING

In the process of limiting tube currents and providing starting voltages, the ballast tends to warm up. As a ballast ages, the laminated steel core of the transformer often loosens, vibrates (producing a humming sound), and penetrates the insulation on the windings. At times, this will short-circuit some of the windings, causing an excess current flow in the remaining windings. As a result of this breakdown in insulation, a damaging amount of heat is produced in the ballast. Manufacturers now equip ballasts with a thermal protector (cut-out) to open the ballast circuit automatically when the heat reaches a dangerous level. This integral ballast protection is required by the NEC for all fluorescent fixtures installed indoors, except those using simple reactance ballasts. Ballasts are rated by the manufacturer to operate at temperatures up to 90°C (194°F).

Ambient temperature, the temperature of the surrounding air, also plays an important role in the life span of the ballast. Figure 19-6 shows the life expectancy of a ballast under various temperature conditions. Figure 19-7 shows a typical rapid-start ballast and its construction.

INSTANT-START FIXTURES

The tubes for instant-start fixtures are generally of the single-pin type, and no other type of tube can be substituted for them. At one time a bipin tube was made which is still found in some lighting installations.

An instant-start fixture does not require preheating. It is started by creating a higher ballast output voltage with a step-up transformer in the larger ballast. The lamp cathodes must be very strong to withstand this sudden and forceful starting technique.

Instant-start fixtures start their lamps in sequence, much like that of the rapid-start unit, Figure 19-8. Because of this higher starting voltage, many of the tube sockets are built so that they open the ballast circuit when a tube is removed from the fixture. This design reduces size, weight, wattage loss, and cost.

Instant-start lamps are made in lengths ranging from 24 inches to 96 inches (600 mm to 2.44 m).

METRIC LAMPS

A new, energy-saving lamp, produced in a length of 46 inches (1168 mm), operates

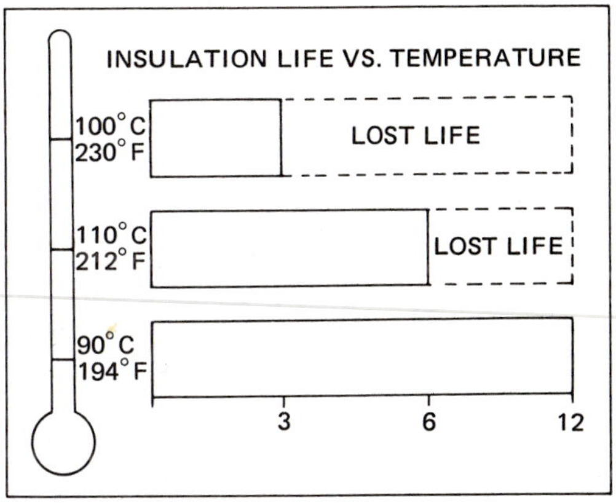

Fig. 19-6 Insulation life versus temperature. (Generally accepted "lost life rule" is "for every 10°C, or 50°F, increase in temperature, insulation life is cut in half.")

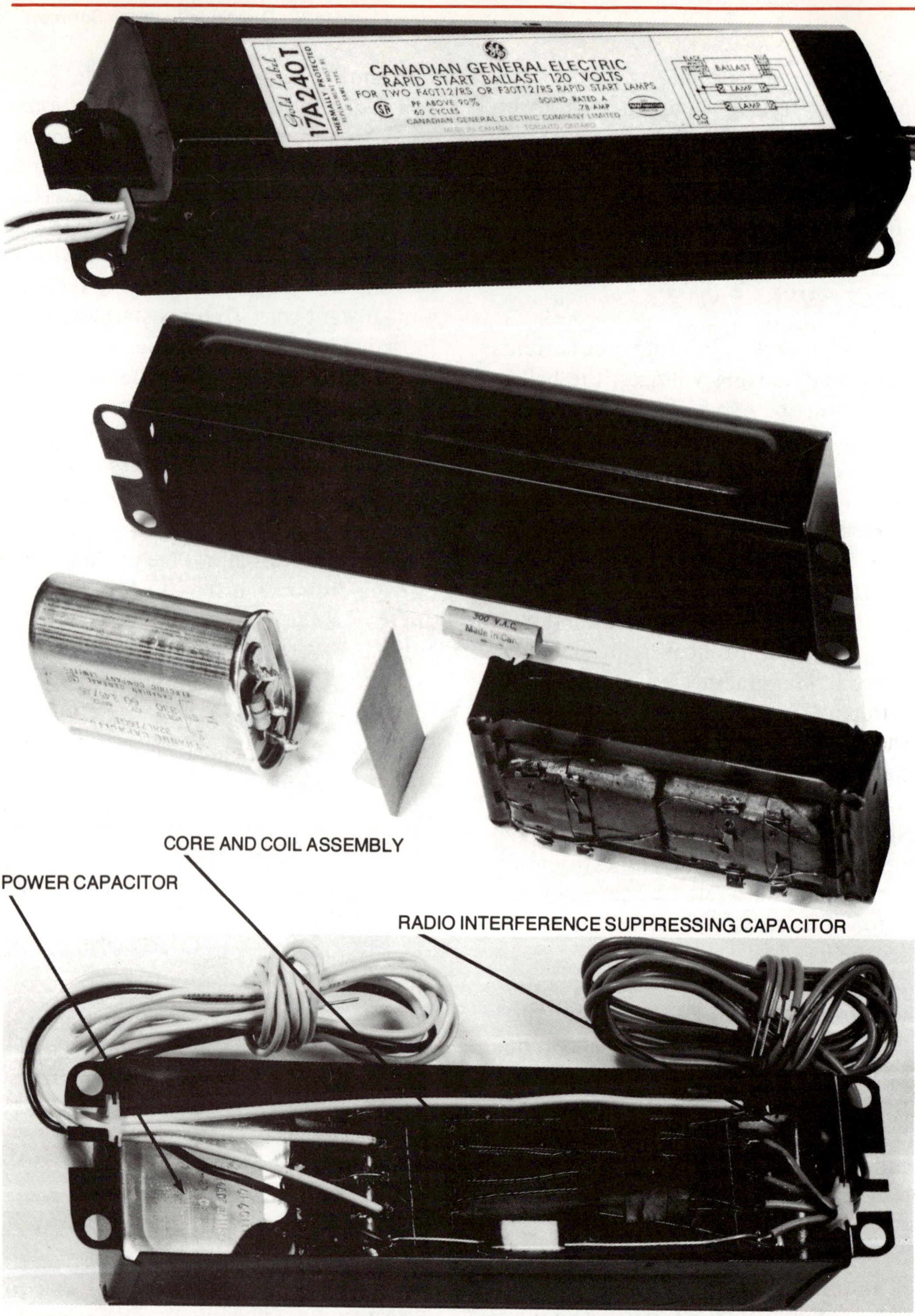

Fig. 19-7 Rapid-start ballast and its construction.

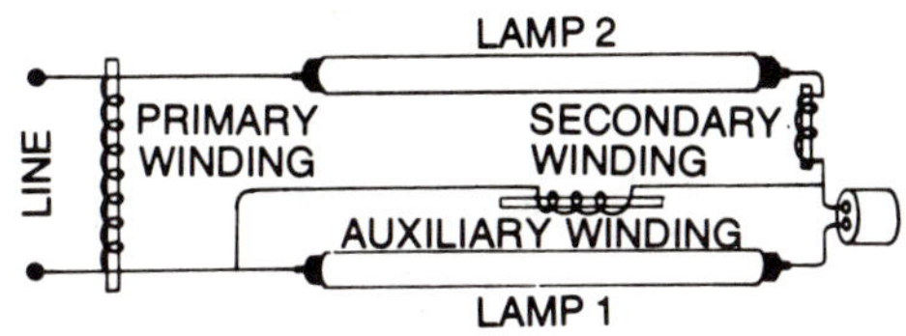

Fig. 19-8 Typical two-lamp rapid-start circuit.

satisfactorily with the F40 (40-W lamp) ballasts available. Light output and lamp life remain about the same. These metric lamps, as they are called, fit into the previously dimensioned 48-inch (1219 mm) fixtures with a small adapter at one end of the lamp to make up the difference in length. The new fixtures produced for use with these lamps are referred to as 48-inch (1219-mm) units. "SuperSaver" and "Thrift/Mate" are one manufacturer's trade names for the new energy-saving lamps.

POWER-GROOVE LAMPS

During the late 1950s, a fluorescent tube was developed that used indentations, or grooves, on *one* side of the glass tube to increase lighting output. Later research showed that grooving *both* sides of the tube produced an even higher lighting output. (See Figures 19-9 and 19-10.)

The grooves in the tube bring the phosphor-coated glass closer to the arc stream. This squeezing of the arc results in more visible light energy. In the power-groove tube, the arc stream must follow a wavy path as it travels the length of the tube. (If the arc is straightened out, there are approximately 9 feet (2.7 m) of arc contained in the 8-foot (2.4-m) tube. This increase and concentration of the arc length allows the tube to produce more light than a conventional fluorescent tube.)

The name *power groove* was given to this tube by one manufacturer. It is a very useful lighting source for industrial applications where the ceiling height is not too great (low bay), Figure 19-11.

HIGH-INTENSITY DISCHARGE (HID) LAMPS

The term *high-intensity discharge* (HID) describes a wide variety of lamps. However, all HID lamps are alike in one way: they produce light from a gaseous-discharge arc tube and operate at a pressure above that found in regular fluorescent lamps.

HID lamps were first introduced in 1934. They can be divided into three major categories: the mercury-vapor lamp, the metal-halide lamp, and the high-pressure sodium lamp.

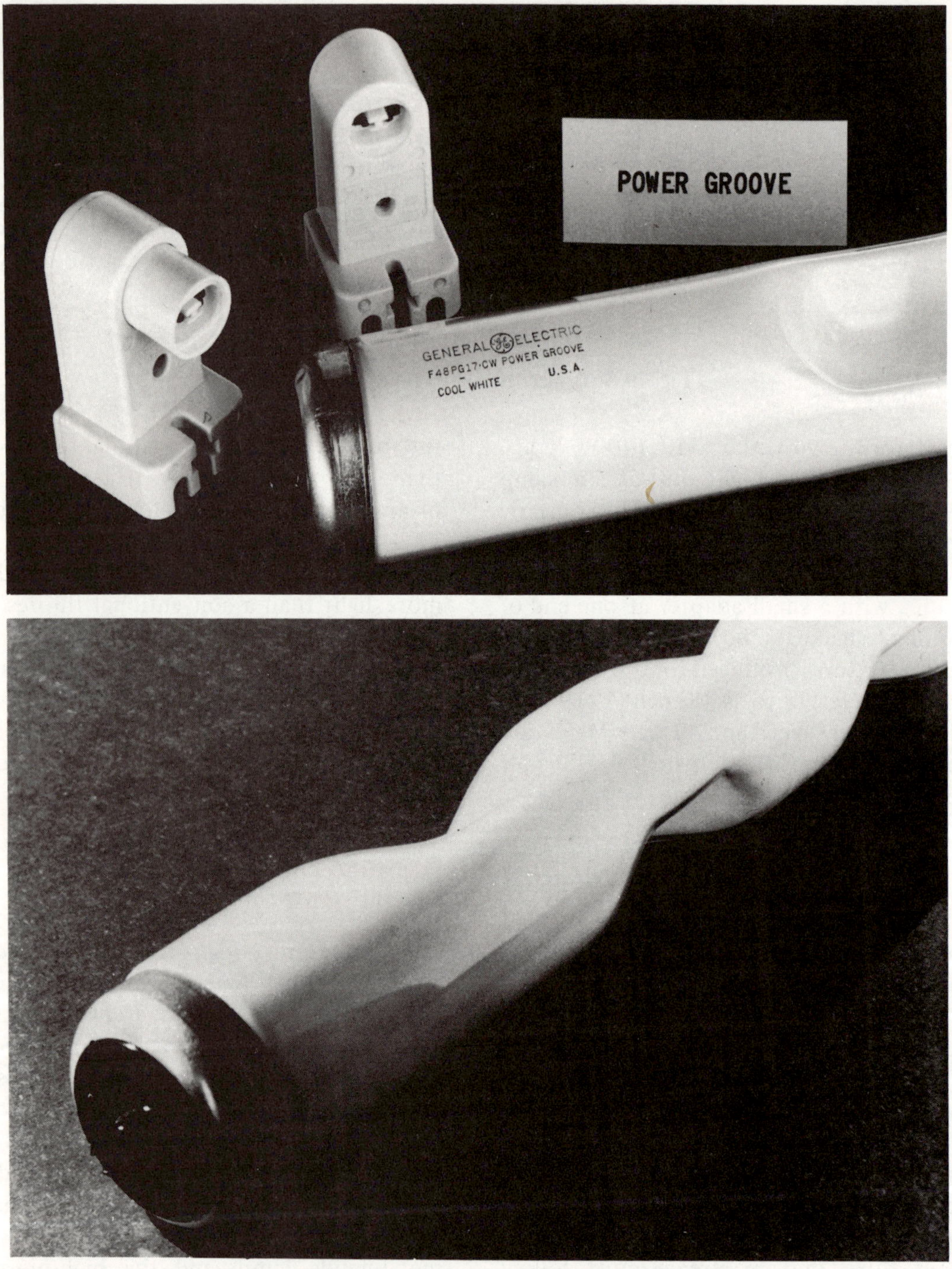

Fig. 19-9 Power groove fluorescent tubes and sockets.

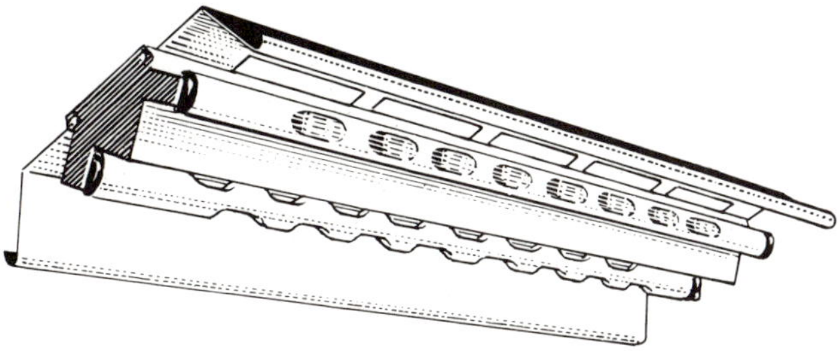

Fig. 19-10 Power-groove fluorescent fixture for industrial applications

MERCURY-VAPOR LAMPS

In a standard, low-pressure fluorescent tube, most of the light energy is in the ultraviolet range. Phosphor-coated tubes must be used to produce visible light. The higher-pressure mercury-vapor tube produces visible light directly. (See Figures 19-12 and 19-13.)

Some mercury-vapor lamps have a phosphor coating on the inside of the outer bulb. This coating reacts to the ultraviolet energy produced by the lamp and modifies the color of the light output.

The mercury-vapor lamp is produced in sizes ranging from 50-W to 1000-W output. Its average life expectancy is more than 24 000 hours, which makes it an ideal lighting source where replacing lamps is both costly and time consuming. Mercury-vapor lamps are used for street and road lighting, area lighting (for example, parking lots), and industrial lighting in factories and warehouses.

The mercury-vapor lamp has many components (as can be seen in Figure 19-12).

Base. The mogul screw-base used on most mercury-vapor lamps connects the lamp to the ballast and external circuits. Letters

Fig. 19-11 A power-groove lighting installation.

(matching the months of the year) and numbers are imprinted in the base. This system helps to keep a record of lamp life.

Starting Resistor. The starting resistor is a tiny, heat-withstanding resistor that limits current in the starting circuit to a safe value.

Starting Electrodes. The arc is established between the main electrodes and starting electrodes, ionizing the argon gas and helping to strike the main arc.

Main Electrodes. The main electrodes are made of a double layer of tungsten wire and coated with rare-earth oxides. They act as terminal points for the main arc.

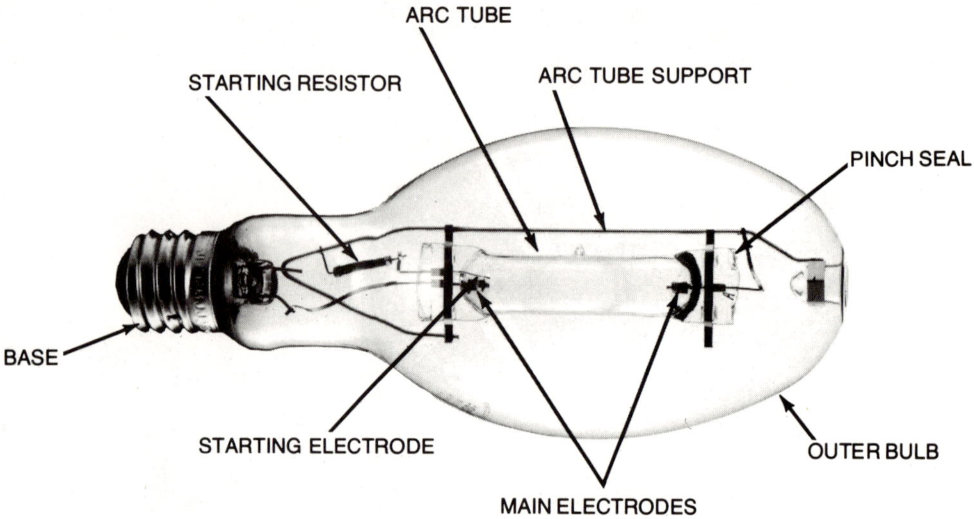

Fig. 19-12 Components of the clear mercury-vapor lamp (400 W).

Fig. 19-13 Typical phosphor-coated mercury-vapor lamp (1000 W).

Arc Tube Support. The arc tube support is a polished metal frame that serves two purposes: it supports the arc tube and it conducts current to the upper main electrode.

Arc Tube. The arc tube, a pure quartz tube, is called the heart of the mercury lamp. It contains a precise quantity of mercury and a small amount of argon gas. Some manufacturers coat the ends of this tube (around the electrodes) with platinum. This is to make sure that the tube will start in cold weather.

Pinch Seal. Pinch seals are used on the ends of the arc tube to prevent both the escape of argon and the entry of nitrogen. (Nitrogen is used between the arc tube and the outer tube.)

Outer Bulb. Heat- and weather-resistant glass is used to protect the internal parts and maintain a nearly constant arc tube temperature. Maintaining a high arc tube temperature is important if the lamp is to operate efficiently. Outer bulbs are sometimes phosphor coated, and they filter ultraviolet light energy.

How a Mercury-vapor Lamp Produces Light. The operation of the mercury-vapor lamp begins with the arc tube. One end of the tube has a starting electrode beside the main electrode. Increased starting voltage from the ballast (approximately double the line voltage) strikes an arc between these two electrodes, using the argon gas as a current path. The argon ionizes (breaks up) and spreads rapidly throughout the tube. As the ionized argon reaches the main electrode at the opposite end of the tube, the light-producing arc is formed.

The small starting resistor, approximately 40 000 ohms (40 kilohms), limits the arc current during starting. As a direct result of its high resistance, current flow quickly shifts to the lower resistance, main arc stream. Once the main arc has been established, current flow along its path is approximately 1000 times greater than the flow between the starting electrodes.

Heat from the main arc continues to vaporize the mercury in the arc tube for several minutes after starting. As more and more mercury is vaporized, current flow increases. Only the stabilizing, current-limiting feature of the ballast (similar to a fluorescent ballast) prevents the lamp from destroying itself.

Light energy is produced by the ionized argon gas particles colliding with the mercury atoms. As electrons in the mercury atom are jarred out of orbit and replaced by electrons from a nearby atom, radiation is given off. The color of light (wave length) produced depends upon which ring of the orbiting electrons has been hit by the colliding particles of argon. High pressures in the arc tube are responsible for deeper penetration than in the fluorescent tube. This results in more visible and less ultraviolet light energy.

Horizontal Operation of Mercury-vapor Lamps. Mercury-vapor lamps are slightly more efficient when operated while in a vertical position. Horizontal operation reduces the lamp's efficiency slightly because the arc will float upwards in the arc tube.

Restarting Mercury-vapor Lamps. If the power supply to a mercury-vapor lamp is interrupted, the arc will become extinguished and it will not restart for several minutes. This is because sufficient pressure will have built up in the tube during operation, thus preventing the arc from reestablishing itself immediately. Once the tube has cooled and the pressure lowered, the arc will restart automatically as usual. The ballast voltage is not high enough to restart the lamp until it cools and the pressure in the arc tube decreases.

Lamp life is shortened by continuous starting. If a lamp is allowed to operate for long periods of time, its life span will be much longer.

Mercury-vapor Ballast. Although bulbs that operate without ballasting have been designed, most mercury-vapor lamps need a ballast. The ballast does for HID lamps what it does for fluorescent lamps. Line voltage is boosted slightly, lamp current is steadied, and the power factor is corrected.

Figures 19-14, 19-15, and 19-16 show a typical ballast circuit, a cutaway view of an outdoor weatherproof unit, and an indoor model with the same circuitry as the outdoor unit.

Mercury-vapor Applications. For many years mercury-vapor lighting was used to replace incandescent lighting systems over roadways. More recent lamp developments

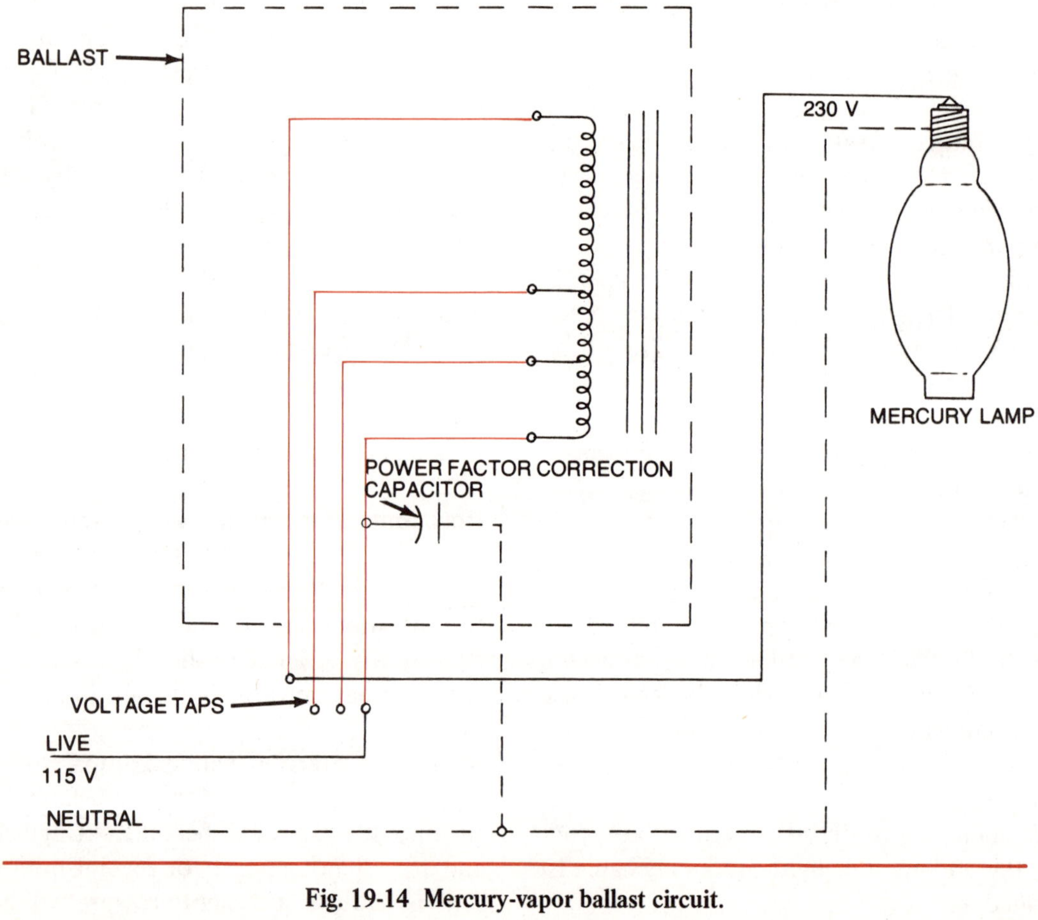

Fig. 19-14 Mercury-vapor ballast circuit.

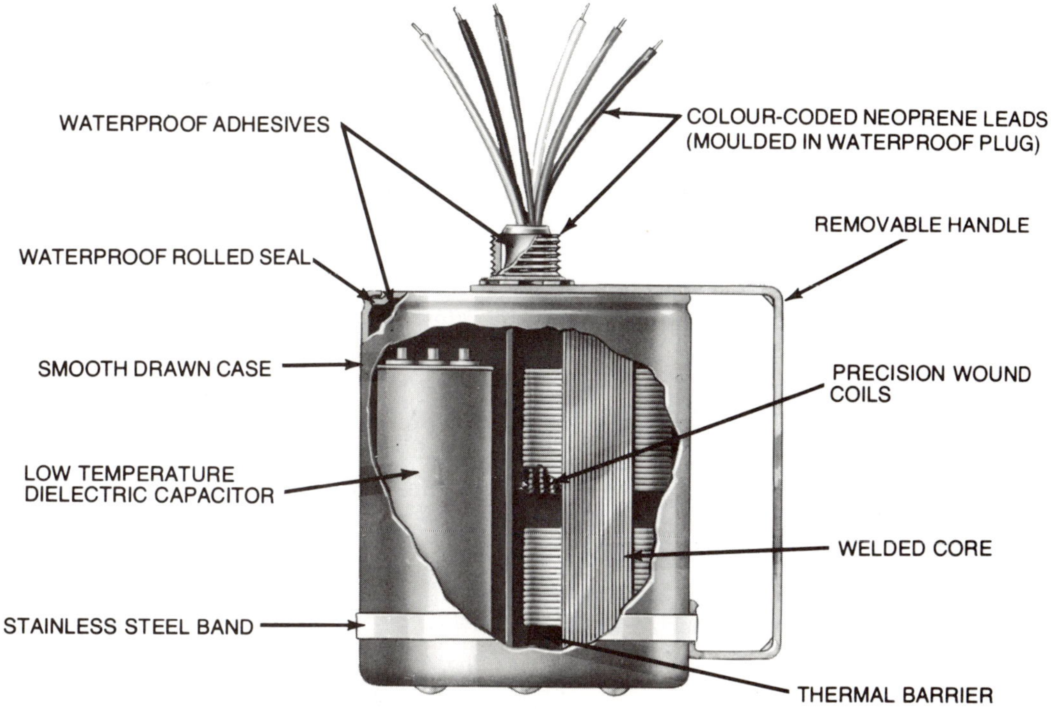

Fig. 19-15 Cutaway view of an outdoor weatherproof mercury-vapor ballast.

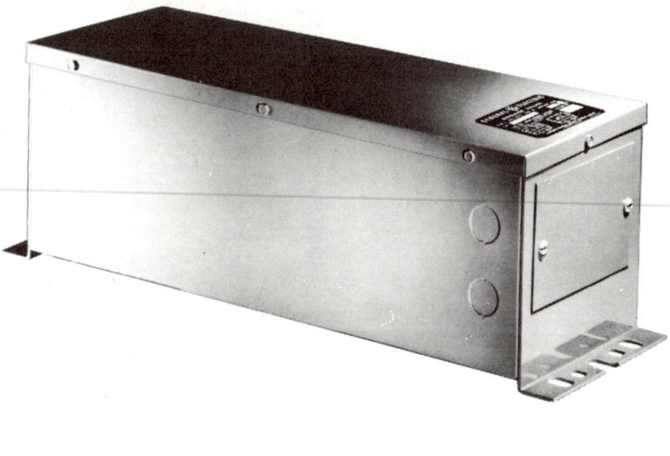

Fig. 19-16 Indoor mercury-vapor ballast.

are, in turn, replacing mercury-vapor lighting in these areas. However, mercury-vapor lighting is still used in shopping malls, commercial buildings, and safety and security systems where dusk-to-dawn lighting is required.

Figure 19-17 shows a section of roadway where two types of installations compare the difference in street light output. Figures 19-18, 19-19, and 19-20 show mercury-vapor lighting used for area lighting (for example, in parking lots) and decorative lighting.

Figures 19-21 and 19-22 show a 250-W lamp used for area lighting around a swimming pool. Figures 19-23 and 19-24 show 1000-W lamps and fixtures used for a large parking lot. Figure 19-25 shows a typical decorative, residential-street lighting unit as it is used to brighten the area around a driveway.

Applications requiring instant lighting cannot be served by mercury-vapor lamps. This is, of course, because of the time delay needed for the lamp to reach full brilliancy.

HID LAMP SIZES AND SHAPES

HID lamps are made in many shapes and sizes, Figure 19-26. HID lamps are known by various trade names, such as *Metalarc, Super Metalarc, Safeline, Brite-Lite, Lumalux,* and *Unalux.*

METAL-HALIDE LAMPS

During the early 1960s, experiments with other metals and combinations of metals produced a lamp with better light radiation characteristics than the mercury-vapor lamp. The problem was to find metals that could be easily vaporized and still re-

NOTE: 500-W INCANDESCENT LUMINAIRES. MOUNTING HEIGHT 25′ (7.6 m); POLES 80′ (24 m) STAGGERED.

NOTE: 400-W LUCALOX LUMINAIRES MOUNTING HEIGHT 35′ (11 m); POLES 50′ (15 m) STAGGERED.

Fig. 19-17 Difference in light output on 42-foot (12.8-m) roadway width.

Fig. 19-18 Typical lighting fixture suitable for use with mercury-vapor and other high-intensity discharge light sources.

Fig. 19-19 Mercury-vapor lighting for outdoor decorative use.

Fig. 19-20 Mercury-vapor and incandescent lighting used in combination for outdoor area lighting.

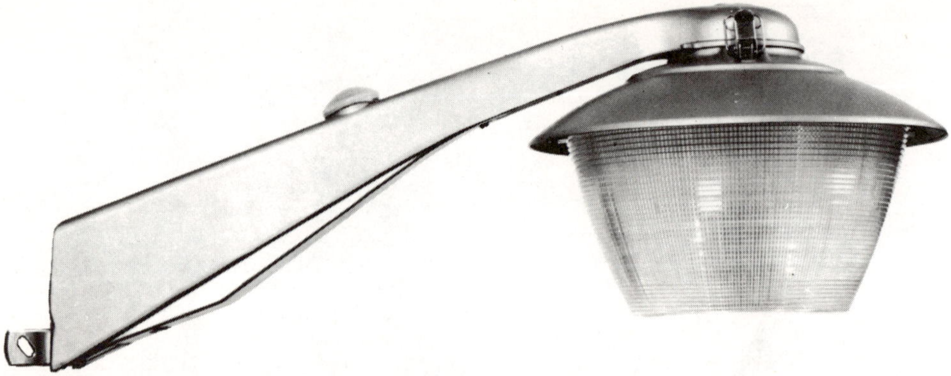

Fig. 19-21 Typical 250-W mercury-vapor, metal-halide, and high-pressure sodium light fixture (luminaire).

Fig. 19-22 Outdoor area lighting by 250-W lamp unit (luminaire).

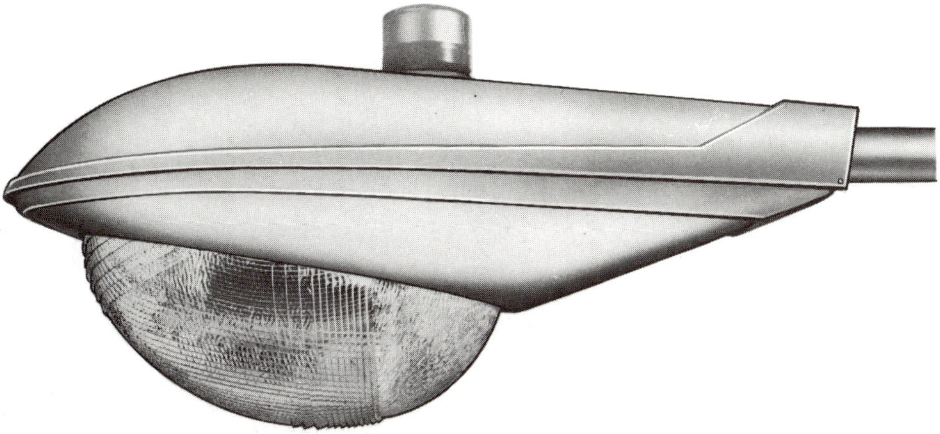

Fig. 19-23 Typical 400-W and 1000-W high-intensity discharge fixture for use in roadway lighting. This "cobra head" unit is equipped with a photoelectric control unit.

Fig. 19-24 Outdoor area lighting by 100-W mercury-vapor lamps.

Fig. 19-25 Decorative, residential-street lighting unit.

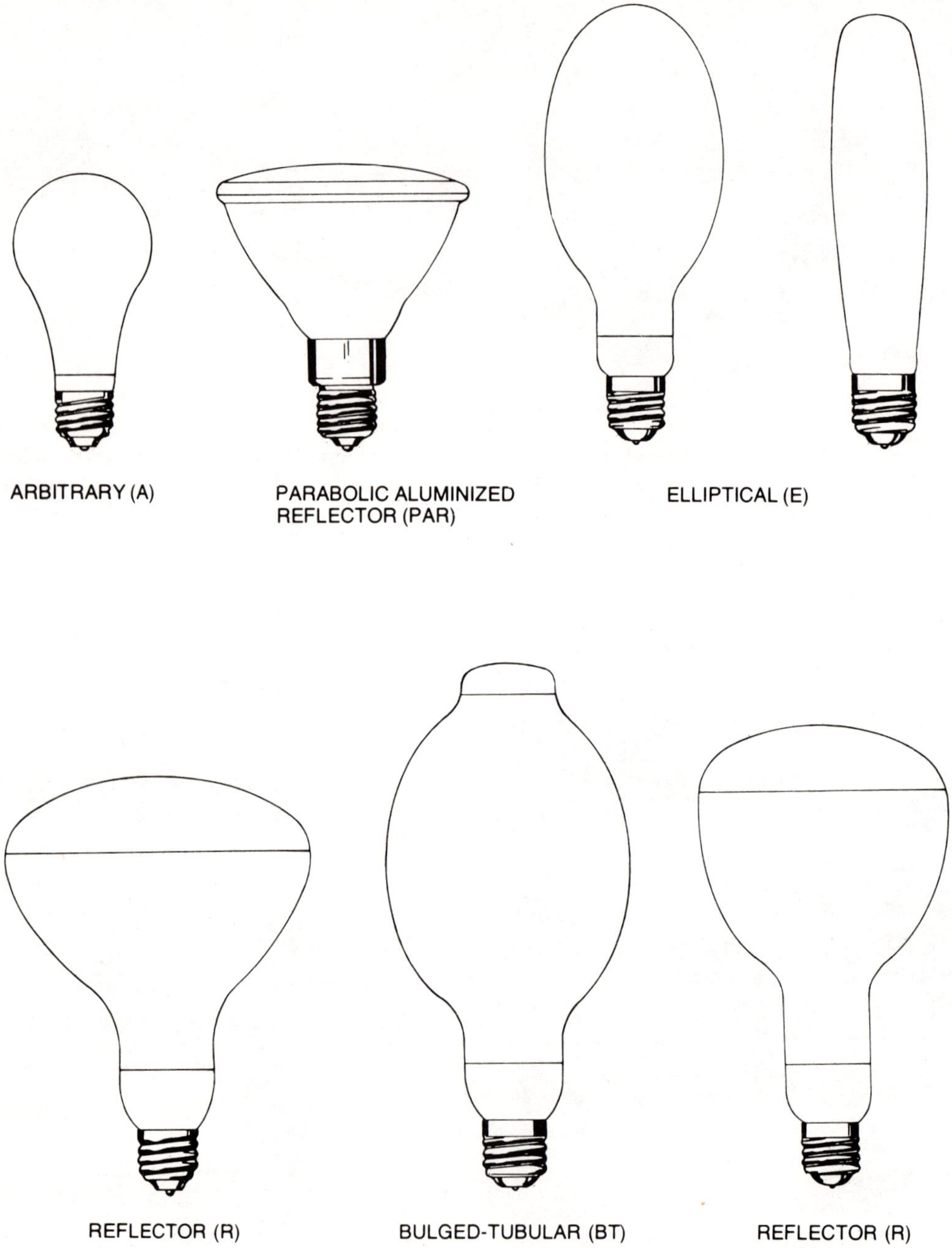

Fig. 19-26 Various HID lamp shapes.

Fig. 19-27 Typical 400-W metal-halide lamp.

main chemically stable. This was finally solved by adding metals in the form of their halide salts (usually iodides) to the basic mercury arc tube.

Today, lamps use iodides of sodium, thallium, and indium along with the mercury in the arc tube. The result is a metal-halide lamp that produces 50 percent more light than a mercury lamp, along with greatly improved color. (See Figures 19-27, 19-28, and 19-29.)

How a Metal-halide Lamp Produces Light. The operation of the metal-halide lamp is similar to that of the mercury lamp, and most mercury lamp fixtures will readily

Fig. 19-28 Typical 100-watt metal-halide lamp.

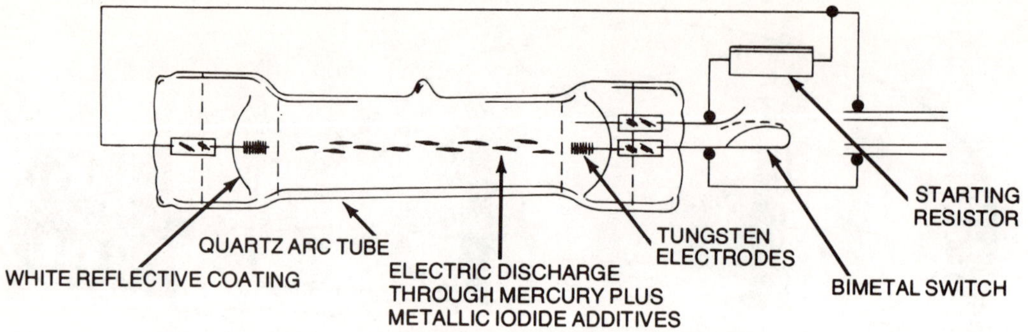

Fig. 19-29 Metal-halide lamp internal components.

accept the halide lamps. A metal-halide ballast must be used, however, because the lamp requires a higher voltage. The fused quartz arc tube of a metal-halide lamp is slightly smaller than that of the mercury lamp, and has a special coating on the ends to maintain proper electrode temperature during operation. A small bimetal switch is built into the lamp to short out the starting contacts during operation. This helps to prolong electrode life.

Metal-halide lamps maintain about 40 percent more light throughout their useful life than do the mercury-vapor lamps. This averages out to 15 000 hours, rated at 10-hour operation per start.

Metal-halide Lamp Applications. The clear, heat-resistant glass bulbs need no phosphor coating. They provide a better color than mercury-vapor lamps, and are used extensively to light sports stadiums where their color and light output is compatible with the requirements of color TV cameras. Made in 175-W to 1500-W sizes, metal-halide lamps are also used for stores and other commercial applications. Due to their higher efficiency, metal-halide lamps are gradually replacing mercury-vapor lamps as a lighting source.

Figure 19-30 shows a typical halide fixture. It is used for ball-park lighting and other area-lighting systems with many of the HID lamps. The circuit supplying this fixture requires a ballast.

Figure 19-31 shows a modern industrial fixture design. The ballast is built directly over the reflector on this unit.

Figure 19-32 shows a double-lamp unit. This unit is very useful for industrial lighting applications.

HIGH-PRESSURE SODIUM LAMPS

The high-pressure sodium lamp is quite different from other HID lamps, Figure 19-33. It is much simpler in design, because of the tremendous research and development that went into the production of materials used in its construction. It is regarded as the most efficient source of white light artificially produced. The 50-W, 70-W, 100-W, 150-W, 200-W, 250-W, 400-W, and 1000-W lamps put out approximately 50 percent more light than either mercury-vapor lamps or metal-halide lamps of the same wattage ratings, Figure 19-34.

Fig. 19-30 Typical P-1000 floodlight with trunnion-type mount. For use with mercury-vapor and metal-halide lamps.

Fig. 19-31 Metal-halide lamp indoor ballast and reflector unit.

The ceramic material used for the arc tube is one reason for the success of the high-pressure sodium lamp. A special method for sealing off the tube, which allows it to contain the high-pressure sodium discharges, was also developed. The ceramic is made of translucent aluminum oxide and was developed specially for the high-pressure sodium lamp. One trade name for this ceramic material is *Lucalox*.

Like many ceramics, Lucalox can withstand very high operating temperatures (1300°C, or 2372°F). Unlike many other ceramics, it is virtually free of tiny pores. This allows a high percentage (92 percent) of visible light to pass through. It contains few if any impurities, which makes it highly resistive to the corrosive effects of hot sodium. Quartz, on the other hand, deteriorates rapidly when exposed to sodium.

The ceramic end caps are joined to the arc tube in such a way that they will maintain the seal during the expansion and contraction (heating and cooling) cycles of the tube. The oxide-embedded electrodes and the end caps are also designed to withstand the corrosive effects of hot sodium.

The tube is filled with a mixture of xenon and mercury as well as sodium. A special circuit, consisting of a small electronic board, produces a high-voltage pulse (2500 V) for about 1 millionth of a second (1 microsecond) during each half of the alternating-current cycle. This voltage pulse is strong enough to ionize the xenon across the main electrode gap, making starting electrodes unnecessary. Once the arc has been established, the high-voltage pulsing is discontinued and normal ballast voltage maintains the arc.

High-pressure sodium lamps reach full brilliancy in a shorter period of time than mercury-vapor lamps or metal-halide lamps. They also may be restarted without waiting

Fig. 19-32 Metal-halide lamp indoor ballast and double reflector unit.

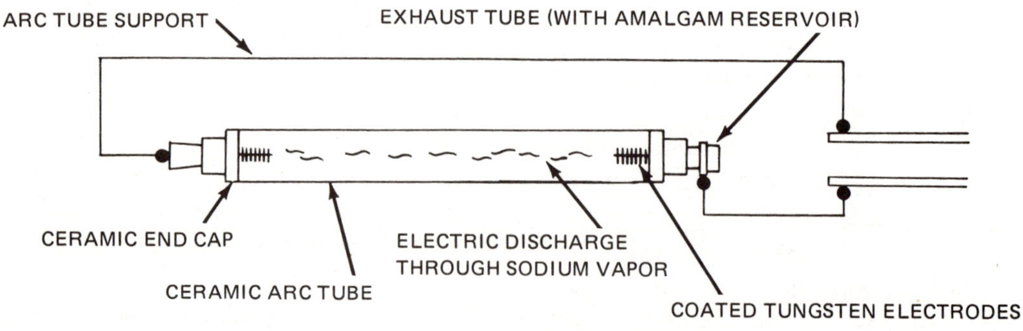

Fig. 19-33 High-pressure sodium lamp construction.

for the tube pressure to drop.

Color changes in the lamp can be seen as the sodium becomes fully vaporized. A dim, bluish-white light is given off as the xenon ionizes. This changes quickly to a brighter mercury-blue glow as temperature rises in the tube. A yellow shade takes over from the blue, showing that the sodium has reached low-pressure temperature. As temperature and pressure in the tube reach normal operating levels, the color changes to the white light seen during operation. The arc temperature is over 2000°C (3632°F) when the lamp is at normal operating temperature.

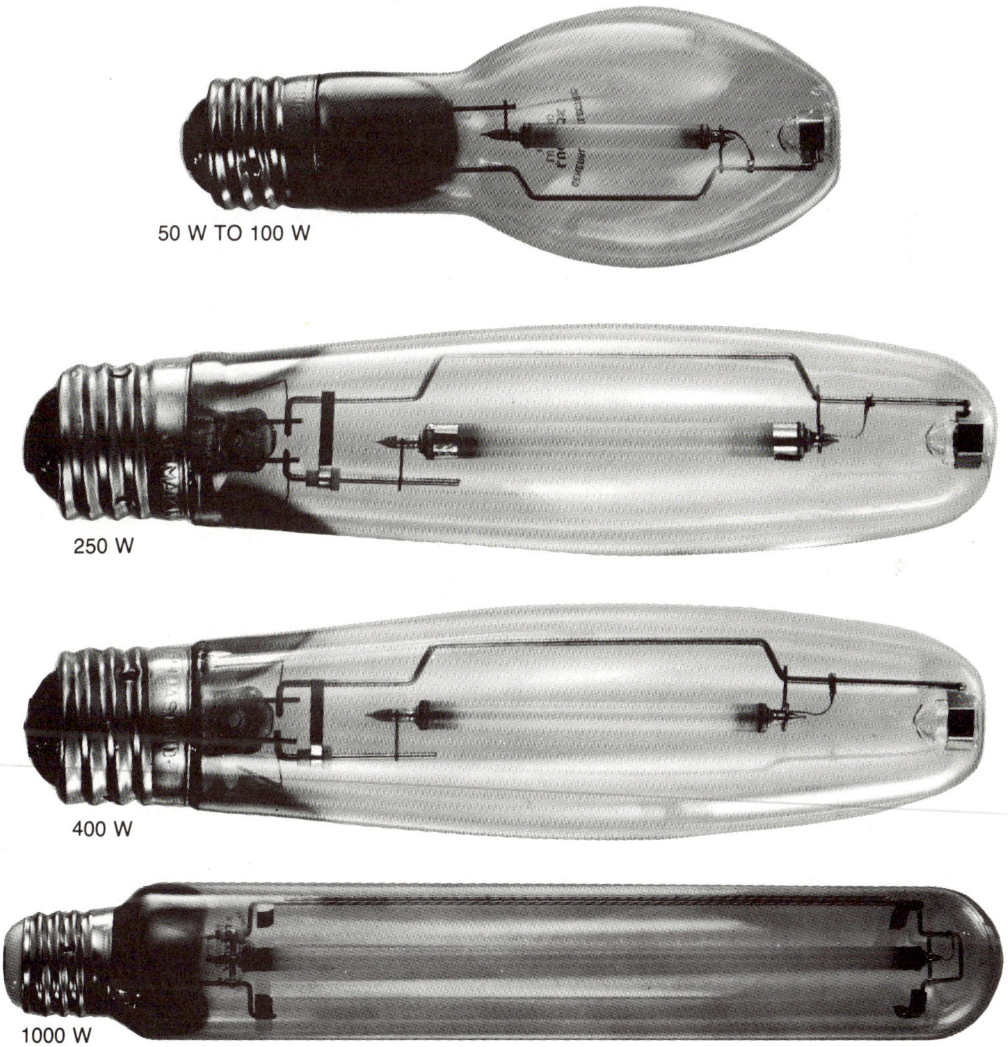

Fig. 19-34 High-pressure sodium lamps.

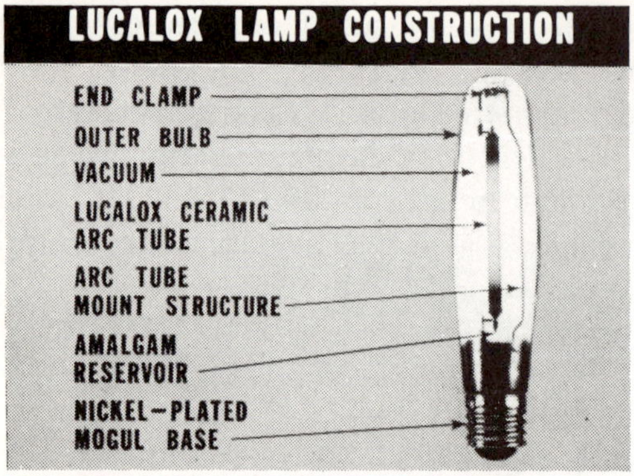

Fig. 19-35 New high-pressure sodium lamp, suitable for use in any burning positon.

At one time, Lucalox lamps were made in two different models for base-up or base-down operation. Each could operate in a nearly horizontal position. The reason for this was that the excess sodium mixture collected at the coolest point in the arc tube. A special reservoir was fitted to the arc tube at the coolest end of the lamp to collect it. In base-up lamps, the reservoir was placed at the end farthest from the base. In base-down units, the reservoir was placed near the base of the lamp.

Modern Lucalox lamps are universal burning (in any position) and have an external amalgam reservoir located at one end of the arc tube. This special reservoir contains the excess sodium amalgam, keeping it away from the arc stream, and thereby extending the life span of the lamp. These lamps are available in both clear and diffuse-coated versions to accommodate the various light distribution and luminaire requirements. The clear lamps provide the best optical control of the light energy, while the diffuse-coated lamps provide a smoother, but less bright, light in low-mounted, decorative applications. Figure 19-35 illustrates the new high-pressure sodium lamp.

Fig. 19-36 High-pressure sodium lamps floodlight the outside of a building.

High-pressure Sodium Lamp Applications. The highly efficient high-pressure sodium lamps produce more light per watt of electrical energy consumed than most other forms of lighting. They are used to floodlight the outside of buildings, Figure 19-36, to floodlight playing fields, Figure 19-37, and to illuminate the inside of manufacturing and repair buildings having high ceilings (high bays) as shown in Figures 19-38, 19-39, and 19-40. Warehouse and storage areas, as seen in Figure 19-41, make excellent use of these light sources as well. The tremendous light output of this form of lighting makes it suitable for highway lighting installations, as seen in Figure 19-42.

Fig. 19-37 High-pressure sodium lamps floodlight a playing field.

Fig. 19-38 High-bay lighting installation using metal-halide and high-pressure sodium lamps.

Fig. 19-39 High-bay lighting at a boiler assembly plant.

Fig. 19-40 High-bay lighting in a repair shop.

MULTIVAPOR LAMPS

A new series of high-efficiency, multivapor lamps has been produced that provides users with 50 percent more light over the life of the lamp than the older mercury-vapor lamps. These lamps will operate with approximately 80 percent of the mercury-vapor ballasts now in use, indoors or outdoors, and function in all burning positions. The lamps, rated at 400 W, are produced in clear and phosphored glass units. The clear lamp provides good color and is very useful where good light control and cutoff are important considerations in lighting. The phosphor-coated lamp provides even better color when a softer, diffused light source is needed.

Multivapor lamps, which operate on a mixture of metal halides rather than just one, such as the mercury or sodium lamps, are suitable for use in all weather conditions where the temperature is −18°C (22°F) or above. The lamps require between two and four minutes to reach full brilliancy, and will restart in 10 to 15 minutes after shutoff.

This type of lamp is recommended for use by energy-conscious users of lighting systems where funds are not available for complete system changeover to newer light forms. The lamps have an average rated life span of 15 000 hours based upon 10-hour operation per start.

In addition to being used to modernize the less efficient mercury-vapor lamp systems, multivapor lamps are suitable for installation in new systems. Universal (any position) burning, greatly increased light output, and lower initial cost are just a few of the reasons for selecting this type of lamp. It is highly acceptable for use in floodlighting buildings, merchandise displays, and playing fields.

Fig. 19-41 Warehouse storage area lighting system.

Standard multivapor lamps are used for new-equipment installations, while I-line (Interchangeable) lamps are used as replacements for the mercury-vapor lamp units.

TUNGSTEN-HALOGEN LAMPS

The tungsten-halogen lamp is not a member of the metallic vapor/arc family, but it is often confused with the vapor/arc lamps.

Like other incandescent lamps, the tungsten-halogen lamp has a tungsten filament. It produces light energy by passing current through this filament, causing it to glow brightly. Argon and a small amount of halogen gas are combined under relatively high pressure within the quartz filament tube to produce a brighter, whiter light, Figure 19-43.

A unique cleaning cycle is responsible for the long life and continued high output

Fig. 19-42 HIghway lighting system with Lucalox lamps.

of this lamp. As the tungsten filament reaches operating temperature, small particles are boiled off (emitted) into the tube. The halogen gas picks up these particles and returns them to the filament. This circulating process keeps the tube wall clear, prevents deterioration of the filament, and maintains high output and color rendition during the life span of the lamp.

Tungsten-halogen lamps are available in a linear arrangement of tube and filament, Figure 19-44A, and in standard, screw-base sizes, Figure 19-44B. The screw-base lamps can replace standard incandescent lamps without changing either the fixture or the wiring. Take care when handling lamps with an exposed filament tube. Perspiration from the hands will erode the quartz tube. Wearing cloth gloves will prevent damage to the tube during installation.

Most tungsten-halogen lamps are made for specific uses, such as for stage, studio,

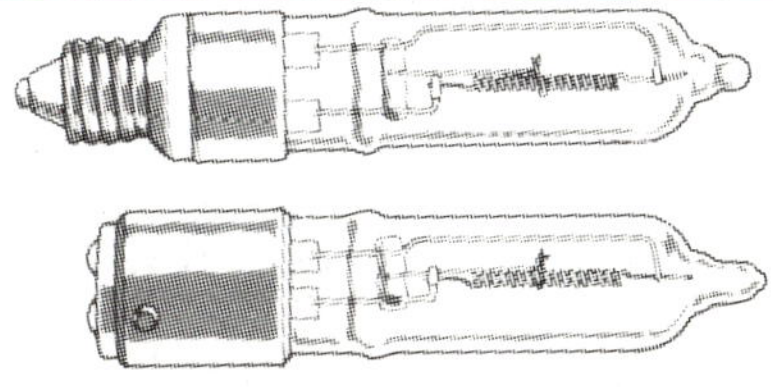

Fig. 19-43 Typical single-ended quartz halogen lamps.

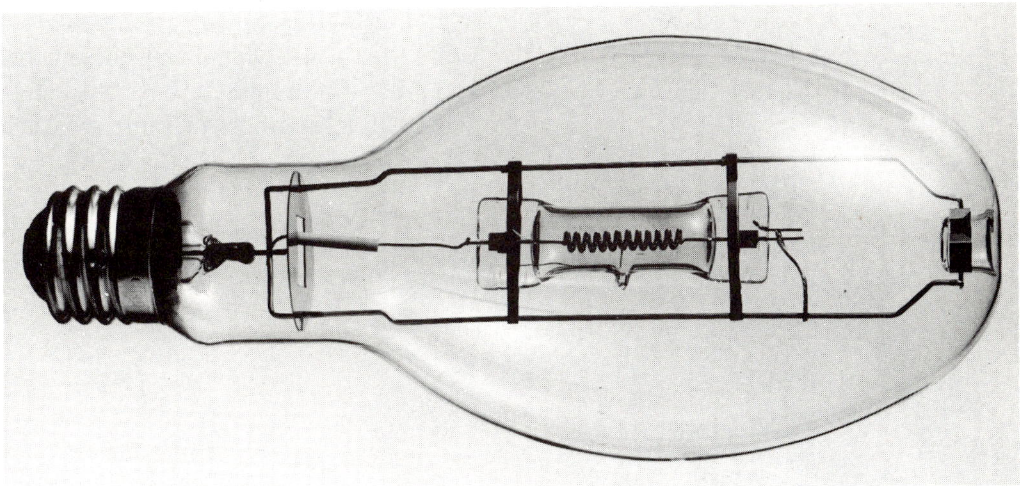

MOGUL BASE LAMP (FOR AREA LIGHTING)

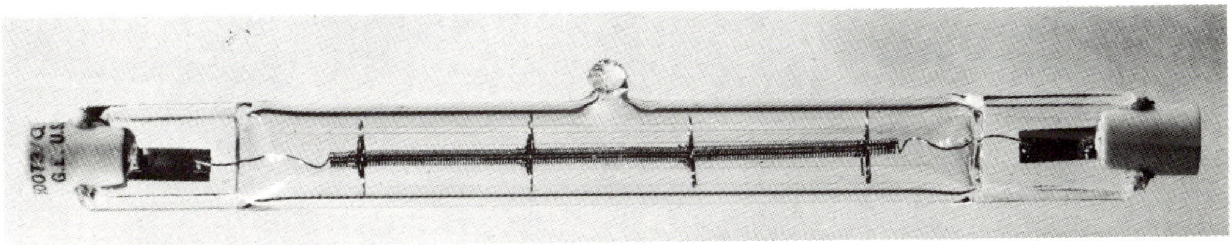

LINEAR QUARTZ FILAMENT TUBE (FOR FLOODLIGHTS)

Fig. 19-44A Tungsten-halogen lamps (linear arrangement of tube and filament.)

Fig. 19-44B Tungsten-halogen lamp (screw-base type).

and photographic lighting where control of color and direction are required. They are also used for decorative lighting around buildings. Figures 19-45 and 19-46 show a typical floodlight fixture and lamp, respectively.

ENERGY-SAVING LAMPS

Conservation-conscious manufacturers are now producing types of fluorescent tubes that provide nearly as much light output as conventional tubes, but reduce power consumption between 14 and 20 percent. This can add up to considerable savings over the period of a year. These new lamps are able to fit into existing fixtures, providing complete interchangeability with existing tube sizes and wattage ratings. No loss in average lamp life occurs when using these tubes.

Many commercial users of lighting, where large numbers of units are operating

Fig. 19-45 A quartz-flood floodlight fixture.

Fig. 19-46 Typical quartz halogen flood lamp.

for extended periods of time, are updating their systems to the newer, more efficient forms of lighting such as the high-intensity discharge (HID) lamps. Some companies have claimed complete recovery of their changeover costs within several years. The more efficient the light source, the more energy (and money) that can be saved. Table 19-1 illustrates the differences in lamp efficiency for the various types of light sources available. The cost of lighting can be compared for the various forms of lighting, as shown in Table 19-2.

LAMP MAINTENANCE

Lamps are rated by the manufacturers to give the user an idea of the expected average life from the lamp installed. Fluorescent lamps, for instance, are rated at 20 000 hours' average life. If the lamps were to burn 20 hours a day for 6 days a week, it would take more than 3 years for the lamps to burn out. In large industrial or commercial lighting layouts, the cost of labor for lamp replacement often outweighs the cost of the actual lamp. Manufacturers recommend that the lamps be replaced when they have been burning for 75 percent of their rated life span.

On the average, only 15 percent of the lamps will have actually burned out at the 75 percent life-span level, and so replacement costs will not be too extensive. Once the lamps reach 75 percent of their life span, the remaining lamps (85 percent of them) can be expected to burn out in fairly rapid succession. The labor costs from there on can be quite extensive.

If all the lamps are replaced at the rated 75 percent life-span level, and the fixtures cleaned and washed to increase reflection, the amount of light will be increased considerably and maintenance costs kept within reasonable limits.

Some high-intensity discharge lamps carry warnings against skin burns and eye inflammation from short-wave radiation. Use adequate shielding when installing them if the outer envelope is broken.

TABLE 19-1 Newer, more efficient light sources provide more light at a lower rate of power consumption.

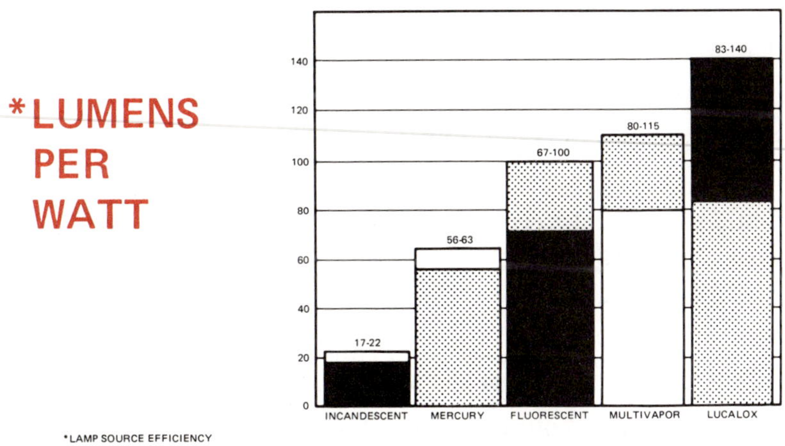

TABLE 19-2 Cost to illuminate 2000 yd^2 to 540 lx for a year (lx is the symbol for lux).

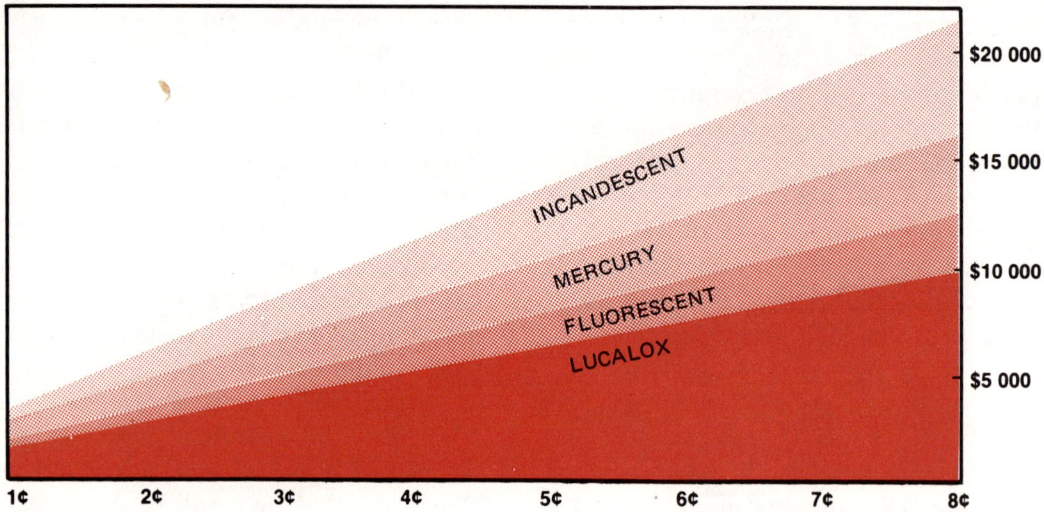

Power rate per kW-h

REVIEW

1. List three main advantages that fluorescent lighting has over incandescent lighting.
2. List the disadvantages of fluorescent lighting.
3. List and explain the use of the parts of the fluorescent tube.
4. Explain how a starter operates in a preheat lamp.
5. What are the two functions of the ballast? How does it work?
6. Why must rapid-start fixtures be grounded?
7. What effect does fluorescent lighting have on power factor in industrial plants?
8. What are manufacturers now doing to prevent damage from ballast overheating? What does the NEC require?
9. What change has been made in the design of tube sockets for instant-start fixtures? Why?
10. List and explain the use of the parts of a mercury-vapor lamp.
11. Explain briefly in your own words how a mercury-vapor lamp produces light.

12. What are the disadvantages of mercury-vapor lighting?
13. List four areas of use for mercury-vapor lighting.
14. How does a metal-halide lamp differ from a mercury vapor lamp?
15. What are the advantages of metal-halide lamps?
16. What special material was developed for use with the high-pressure sodium lamp?
17. What are the advantages of high-pressure sodium lamps?
18. Describe the color sequence of a high-pressure sodium lamp during its warm-up period.
19. To what family of lamps does the tungsten-halogen lamp belong?
20. Explain briefly how a quartz-halogen lamp works.
21. What are the advantages of the tungsten-halogen lamp?
22. What type and size of metric lamp is available?
23. What two methods can be used to conserve energy when using a lighting system?
24. At what point in a fluorescent lamp's rated life span should it be replaced? Explain why.
25. What type of lighting source can be used to replace mercury-vapor lamps for improved lighting costs and energy conservation?

Chapter 20

Motor Control

THE NEED FOR SPECIAL CONTROL EQUIPMENT

Electrical motors are used for a wide variety of residential, commercial, and industrial operations. Because of this, guidelines are necessary. This chapter deals with the need for motor control and some of the more common control systems.

Electrical motors operate on magnetism. The amount of current needed to create the magnetism depends upon the size and design of the motor.

Nearly all motors tend to draw much more current during the starting period (starting current) than when rotating at operating speed (running current). Motors are rated in horsepower (watts). *Note:* 1 horsepower (hp) = 746 watts (W) or 0.746 kilowatts (kW). The higher the rating of the motor, the higher the starting and running currents will be.

Every motor tends to produce a counter-voltage and a counter current within its windings. This generator action produces a voltage and current that flows opposite to the applied current. This may appear to be a problem but, in fact, the generated current flow helps to control the flow of incoming current through the motor. Manufacturers design motors with this current in mind. Proper control of incoming current extends the life span of the motor. Excessive input current will severely damage or burn the motor's windings.

One main factor in determining the amount of generated voltage and current in the motor is its speed. If the load placed on a motor reduces the speed, less generated current will be developed and more applied current will flow. That is, the greater the load on the motor, the slower it will rotate and the more applied current will flow through its windings. This is why a motor requires more current during the starting period. (Too great a load on the motor will cause the motor to stall.)

As the speed picks up, the generated voltage within the motor gradually increases. However, the instant the starting switch is closed, there is no generated voltage and the applied current is very high. If the motor is jammed or prevented from rotating in any way, it will burn out very quickly.

The rotating part of the motor is often called the *rotor.* When the rotor fails to turn, the excessive applied current is called the *locked-rotor current.* Most electrical motors have a starting current that is three to five times greater than the normal running current.

MOTOR CONTROL SWITCHES

Two basic methods for starting motors are: across-the-line starting, in which full-line voltage is applied to the motor, and reduced-voltage starting. This section discusses only the across-the-line starting method.

Any across-the-line switch or control device used to start (and stop) a motor must be able to withstand a high inrush of current during the starting period. Switch contacts must close rapidly, make a sure connection, and prevent arc damage to the switch. Since another important function of the switch is to open the circuit, allowing the motor to stop, any across-the-line switch or control device must also be able to open the circuit under locked-rotor conditions. (See NEC *Article 430*.) The switch or control device must have a very strong spring action to open the contacts quickly; otherwise, there may be considerable arc damage to the switch.

Sometimes a control switch does not have the proper components and is not suitable for the job. As a result, arc damage is so severe that the switch is destroyed. For this reason, experienced electricians usually do not stand in front of a switch when it is being operated. The safest way to stand is to keep the face and body off to one side and use the left hand to operate the switch. Although severe arcing does not happen often, the *danger* of serious damage to the switch and injury to the operator is always present. (The motor disconnect switch with the externally operated handle helps to eliminate this danger.)

Motor starting current and running current vary with the motor's horsepower. As the horsepower increases, the starting and running currents also increase. Any switch used to control a motor, whether in a basement workshop or for industrial duty, should have a horsepower rating marked on it. This indicates that the switch is rated for starting and stopping the motor safely. (See NEC *Article 430*.)

LOCATION OF CONTROL DEVICES

NEC *Article 430* requires that a controller be within sight from the motor and the driven machinery. This allows the operator to see that there is no danger to equipment or persons before starting the motor. Where the controller is not in sight from the motor and driven machinery, either a manually operable motor disconnect switch must be located in sight from the motor, or else the controller disconnecting means must be capable of being locked in the open position. These requirements make repairs or maintenance to the motor and driven machinery a safer operation.

OVERCURRENT PROTECTION

As with any other electrical circuit, motor circuits must be protected from overcurrent conditions; otherwise, a locked-rotor condition or a short circuit will damage both the wiring and control devices. NEC *Article 240* lists the standard ratings of fuses or circuit breakers that are readily available. (See Table 20-1 for maximum allowable ratings.)

As a general rule, overcurrent protection (running protection) should not exceed 300 percent of the motor's running current, which is listed as the full-load current on the nameplate of the motor. High starting currents are responsible for fusing the motor at a value above its rated current. NEC *Tables 430-147, 430-148, 430-149,* and *430-150* give more detailed figures for the various types of motors.

Type of Motor	Percent of Full-Load Current			
	Nontime Delay Fuse	Dual Element (Time-Delay) Fuse	Instantaneous Trip Breaker	* Inverse Time Breaker
Single-phase, all types				
No code letter	300	175	700	250
All ac single-phase and polyphase squirrel-cage and synchronous motors† with full-voltage, resistor or reactor starting:				
No code letter	300	175	700	250
Code letter F to V	300	175	700	250
Code letter B to E	250	175	700	200
Code letter A	150	150	700	150
All ac squirrel-cage and synchronous motors† with autotransformer starting:				
Not more than 30 amps				
No code letter	250	175	700	200
More than 30 amps				
No code letter	200	175	700	200
Code letter F to V	250	175	700	200
Code letter B to E	200	175	700	200
Code letter A	150	150	700	150
High-reactance squirrel-cage				
Not more than 30 amps				
No code letter	250	175	700	250
More than 30 amps				
No code letter	200	175	700	200
Wound-rotor —				
No code letter	150	150	700	150
Direct-current (constant voltage)				
No more than 50 hp				
No code letter	150	150	250	150
More than 50 hp				
No code letter	150	150	175	150

For explanation of Code Letter Marking, see Table 430-7(b).

For certain exceptions to the values specified, see Sections 430-52 through 430-54.

* The values given in the last column also cover the ratings of nonadjustable inverse time types of circuit breakers that may be modified as in Section 430-52.

† Synchronous motors of the low-torque, low-speed type (usually 450 rpm or lower), such as are used to drive reciprocating compressors, pumps, etc. that start unloaded, do not require a fuse rating or circuit-breaker setting in excess of 200 percent of full-load current.

TABLE 20-1 Maximum Rating or Setting of Motor Branch-Circuit Short-Circuit and Ground-Fault Protective Devices

MOTOR CONDUCTOR SIZES

The National Electrical Code lists ampacity ratings for conductors supplying current to a single motor. (See NEC *Table 310-16,* as shown in Table 5-2.) Motor feeders and motor branch circuits require conductors with ratings equal to 125 percent of the motor's full-load current. Motors that are operated for short periods can be supplied with conductors that have a somewhat lower current rating. Table 20-2 shows how to determine conductor sizes.

When conductors are permitted to supply two or more motors on the same circuit, conductor ampacity can be determined by adding the full-load currents of all the motors in the circuit. Then, 25 percent of the largest motor's full-load current is added to the total. This gives the conductor ampacity. (See NEC *Section 430-53.*)

THERMAL OVERLOAD (OL) RELAY PROTECTION

Often a motor is loaded beyond its designed capacity. Motors in wood- and metal-cutting machines, pumps, hoists, and fans are examples of where overloading can occur. Overloading slows down the motor, which results in less generated voltage and an increase of input current. For example, when a saw is used, if the board is damp or the cut is too deep, the motor will be overloaded and slow down. The current flow in the windings will increase and heat the motor beyond its design temperature. A jammed pump or an extra-heavy load on a hoist will have the same effect on a motor. Most electrical motors have a type of cooling fan built in, but the fan becomes less and less effective as the motor's speed is reduced. If nothing is done, there may be permanent damage to the motor's windings and expensive repairs will be needed.

TABLE 20-2 Determining conductor sizes for motors for different requirements of service.

Classification of Service	Percentage of Nameplate Current Rating of Motor			
	5 Min Rating	15 Min Rating	30 and 60 Min Rating	Continuous Rating
Short-time Duty. Operating valves, raising or lowering rolls, etc.	110	120	150	
Intermittent Duty. Freight and passenger elevators, tool heads, pumps, drawbridges, turntables, etc.	85	85	90	140
Periodic Duty. Rolls, ore- and coal-handling machines, etc.	85	90	95	140
Varying Duty	110	120	150	200

It is human nature to try for "one more cut" or "one last load." Knowing this, manufacturers have designed a heat-sensitive mechanical device. When a motor has been overloaded for a certain length of time, this device, called a *thermal overload relay*, opens the circuit. (See NEC *Section 430, Part C.*) The thermal overload relay does not cause "nuisance tripping," however, by opening the circuit every time there is a brief and/or minor overload. The relay opens the circuit *only* when there is a prolonged overload condition.

Modern three-phase motor control units require three overload relay units to provide full protection for the motor. On a common-block overload unit, three heater units are installed with a mechanical linkage to operate a single set of overload (OL) contacts. Any one of the heater units is able to open the contacts and turn the control circuit off if there is an overload condition on the motor. This method of overload protection has replaced an older system that made use of two overload heaters, each containing its own set of contacts to open the circuit under overload conditions.

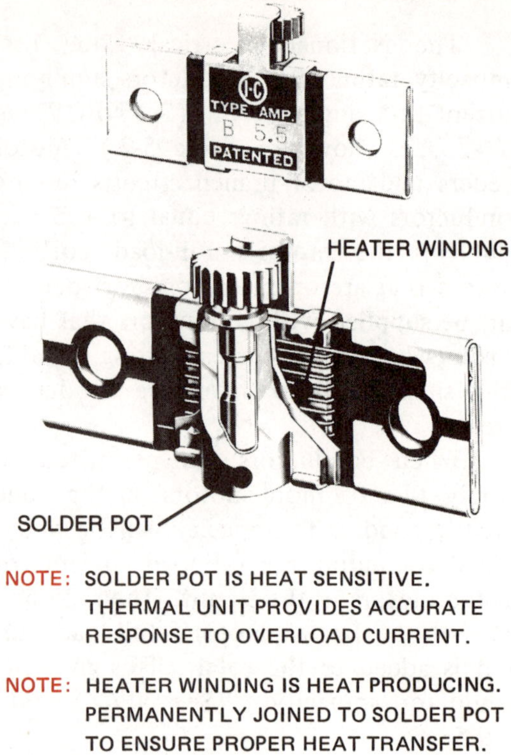

Fig. 20-1 Solder-pot thermal overload relay unit (front and cutaway views).

Operation of the Thermal Overload Relay. The thermal overload relay has several parts. Each part passes on its energy to the next part.

In the solder-pot style of overload relay, the heater part is connected to the motor in series with the supply conductor. Under normal load conditions, the heater remains at a moderate temperature. As the load increases beyond the motor's designed capacity, more current flows and the temperature of the heater rises. Within a short period of time, the heater melts the solder in the solder pot, Figure 20-1. This allows the small ratchet wheel to rotate, triggering the mechanical part of the relay, Figure 20-2. The movement of the spring-loaded mechanical section cuts off current to the motor in one of two ways.

In the manual motor starter, the action is a simple mechanical operation. The main switching mechanism of the starter is released and allowed to spring or trip to an open position by the operation of the overload relay. This opens the main contacts in the circuit delivering current to the motor, and prevents further operation of the motor. The manual motor starter must be reset by

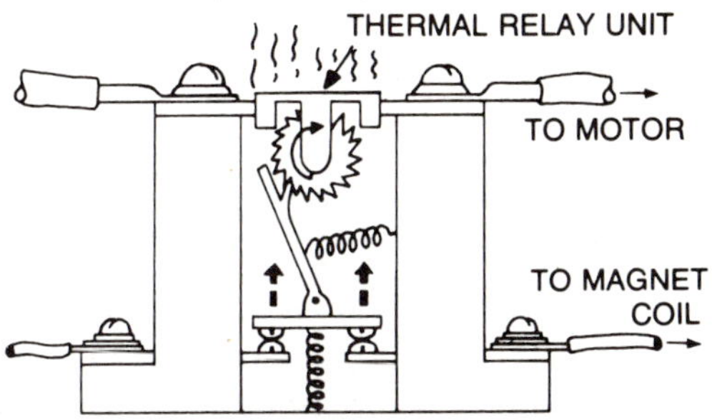

DRAWING SHOWS OPERATION OF MELTING ALLOY OVERLOAD RELAY. AS HEAT MELTS ALLOY, RATCHET WHEEL IS FREE TO TURN — SPRING THEN PUSHES CONTACTS OPEN.

Fig. 20-2 Solder-pot style overload relay mechanism.

the operator once the motor and OL relay have cooled down.

The magnetic motor starter uses the OL relay action in a slightly different manner. The mechanical action of the relay, set in motion by the heating of the relay's solder pot, opens a set of contacts which are connected in series with the control circuit of the starter's operating coil. When these contacts open, they cut off the current flow to the magnetic coil of the starter. As the coil loses its magnetism, it in turn releases the main contacts from their closed position. The contacts are spring-loaded open, and current flow to the motor is cut off. The OL relay unit must be reset by the operator once the relay and the motor have cooled down, but only the control circuit is reestablished by the resetting action. A separate start button must be operated to get the motor running again.

Many manufacturers produce the solder-pot type of OL relay for use in their motor control equipment. However, two other types of relay are used in various motor control units.

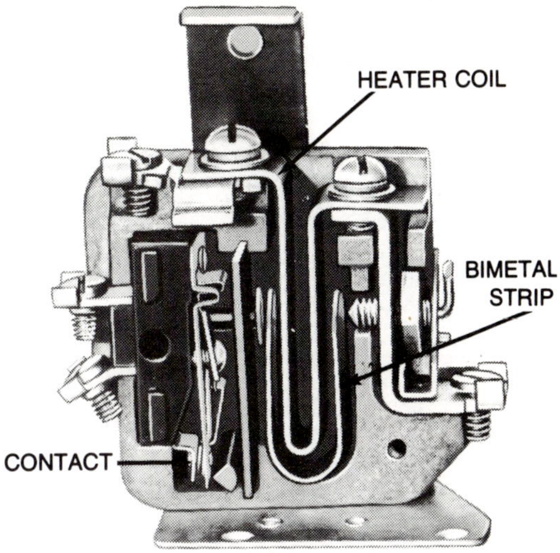

Fig. 20-3 Bimetallic overload relay unit.

Bimetallic thermal overload relays make use of a simple, U-shaped strip of metal (fabricated from two different types of metal) which bends or deflects when heated by the heater portion of the relay, Figure 20-3. This bending action opens a set of contacts connected in series with the control coil circuit or, in a manual starter, unlatches the start button, and prevents further operation of the motor. In many cases, these relays are adjustable over a range of 85 to 115 percent of nominal heater ratings. Another feature of the bimetallic magnetic starter is its ability to be set in a manual or automatic restart position by the operator. In the manual position it behaves much like the relays discussed previously. Once set in the automatic position, the relay restores power to the circuit as soon as it cools down to the proper level. Although this is quite convenient if the relay is installed in an inaccessible place, it can also cause several problems.

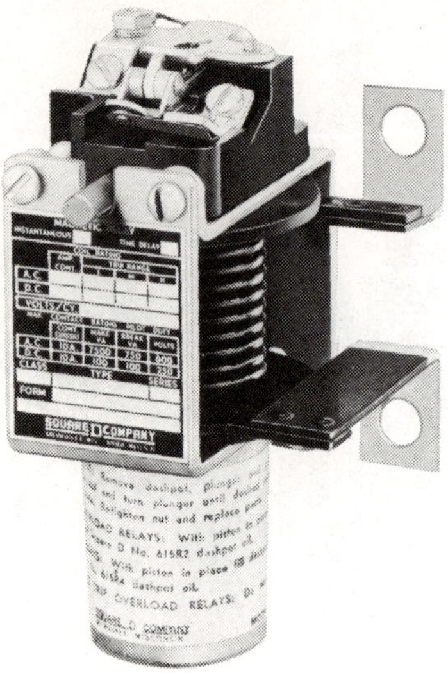

Fig. 20-4 Magnetic overload relay.

If the cause of the overload in the motor has not been removed, the relay allows the motor to restart each time it cools down. The motor will eventually be damaged (or an electrical fire may occur) as a result of the continuous inrush currents from the many restarts.

A more *serious* problem for the operator is the fact that the relay allows the motor to restart as soon as it cools down, even if the operator is working on the machine at the time. When using the automatic setting, great care must be taken to open the main switch for the motor. To prevent injury to the operator, the motor must *not* be allowed to restart while investigation or repairs are in progress.

Magnetic overload relays have a movable, magnetic core inside a coil carrying current to the motor, Figure 20-4. The coil is connected in series with the supply line to the motor. Normal operation of the motor allows the core to rest in such a position that the contacts of the relay are in a closed position. When an overload condition occurs in the motor, more current is drawn by the motor and through the coil. This makes the magnetic attraction of the coil stronger than normal, and it draws the core farther into the relay.

As the core is pulled into the relay, it opens the contacts in the relay that are used in the magnetic coil circuit to turn the motor starter off. To prevent nuisance tripping, movement of the core is slowed by the action of a piston-in-oil unit attached to the core. This *dash pot,* as it is called, works like an automobile shock absorber and prevents rapid movement of the core. The time taken for the core to move open the contacts is adjustable by a screw opening or closing an oil bypass. The amount of current required to draw the core into the coil (tripping current) is adjusted by a threaded rod, which positions the core inside the magnetic coil.

The magnetic OL relay is often used to protect motors that have unusual duty cycles or require longer than normal periods of time to reach operating speed.

Determining Overload Relay Size. The size of thermal overload relay used is determined by the current requirements of the motor. The time it takes to open the circuit is determined by the extent of the overload condition.

Method 1: Under normal operating conditions, motors can handle 125 percent of their rated full-load amperes (nameplate current). Currents larger than this are likely to damage the motor if allowed to continue for more than a few minutes. The overload relay should be capable of opening the circuit when current exceeds

115 percent to 125 percent of the full-load amperes. (See NEC *Article 430, Part C.*) For example, if a motor needs 8 A to operate, and the OL is rated at 125 percent, the overload relay should be capable of handling 10 A (125% X 8 A). This slight, extra allowance prevents nuisance tripping during minor overloads.

Method 2: Manufacturers usually place an overload relay selection chart inside the covers of control switches, such as the chart shown in Figure 20-5. The manufacturer's nameplate on the motor will show the full-load amperes.

Refer to the chart in Figure 20-5. For example, if the motor needs 8 A of current, look for the 8 A listing under the heading *Motor Full-load Current.* The category 7.58–8.15 is the correct listing. The heater number for this manufacturer's model is shown under the heading *Thermal Unit Number* opposite 7.58–8.15. This heater number is the manufacturer's catalog number. Use it when placing an order but remember that, although it may be quite close to the 125 percent calculation, the heater number is not a current rating.

Fractional Kilowatt Motor			
Motor Full-load Current	Thermal Unit Number	Motor Full-load Current	Thermal Unit Number
0.41 — 0.44 A	0.49	3.71 — 4.07 A	5.30
0.45 — 0.49 A	0.54	4.08 — 4.32 A	5.78
0.50 — 0.53 A	0.59	4.33 — 4.90 A	6.20
0.54 — 0.58 A	0.65	4.91 — 5.35 A	6.99
0.59 — 0.65 A	0.71	5.36 — 5.85 A	7.65
0.66 — 0.71 A	0.78	5.86 — 6.41 A	8.38
0.72 — 0.78 A	0.86	6.42 — 6.79 A	9.25
0.79 — 0.85 A	0.95	6.80 — 7.57 A	9.85
0.86 — 0.96 A	1.02	7.58 — 8.15 A	11.0
0.97 — 1.04 A	1.16	8.16 — 8.98 A	11.9
1.05 — 1.16 A	1.25	8.99 — 9.67 A	13.2
1.17 — 1.29 A	1.39	9.68 — 9.95 A	14.1
1.30 — 1.37 A	1.54	9.96 — 10.8 A	14.8
1.38 — 1.47 A	1.63	10.9 — 12.1 A	16.2
1.48 — 1.56 A	1.75	12.2 — 13.1 A	17.9
1.57 — 1.65 A	1.86	13.2 — 13.9 A	19.8
1.66 — 1.79 A	1.99	14.0 — 15.0 A	21.3
1.80 — 1.95 A	2.15	15.1 — 16.0 A	25.2
1.96 — 2.15 A	2.31	....	......
2.16 — 2.38 A	2.57	....	......
2.39 — 2.75 A	2.81	....	......
2.76 — 2.84 A	3.61	....	......
2.85 — 3.06 A	3.95	....	......
3.07 — 3.45 A	4.32	....	......
3.46 — 3.70 A	4.79	....	......

Fig. 20-5 Example of overload relay selection chart. This chart may also be used with motors rated in horsepower.

Restarting after an Overload. To restart a manual motor starter equipped with a thermal overload device, move the operating handle (toggle) from the central trip, or reset, position. (The overload condition will have moved the toggle handle from the "on" to the central trip position.) The toggle handle *must* be moved to the "off" position to reset the spring-loaded mechanical section of the relay. The control switch can then be turned *on,* once the relay has had ample time to cool.

The solder pot in the thermal relay will require several minutes to cool off before the solder solidifies and grips the ratchet wheel. This allows a cooling-out period for the motor before it is placed back into service. The mechanical section of the relay will *not* reset until the solder has cooled. The solder will not grip the ratchet wheel if attempts are made to reset the relay before the solder has cooled. If damaged, the thermoheater of the relay will then have to be removed from the motor control, heated with a match to remelt the solder, and allowed to cool properly before resetting. It can then be put back into the motor control and used to protect the motor. If a motor fault causes the OL relay to open again, locate and correct the fault before resetting.

MANUAL MOTOR CONTROL SWITCHES

Figure 20-6 shows a compact, manual motor control switch for fractional-horsepower motors, which is an across-the-line type of starter. The overload relay is on the left-hand side (marked A2.57). This type of switch is referred to by such names as *motor watchman switch,* or *motor toggle switch.* This unit fits into a special enclosure and can control most fractional horsepower (kW), single-phase motors, Figure 20-7. It is ideal for use in home workshops and in industry, and on sump pumps and swimming pool filter pumps.

Three-phase systems require 3-pole switches capable of opening all three live conductors of the circuit simultaneously. Figure 20-8 shows this type of compact manual switch and pilot light. The pilot light shows whether or not the switch is on and the circuit alive. This type of switch does *not* provide overload protection.

Figure 20-9 shows a more sophisticated, heavier duty, three-phase switch *with* thermal overload protection. Three relays are visible on this switch, one for each live conductor. The unit shown can control up to 600-V motors with ratings up to 10 hp (7.5 kW).

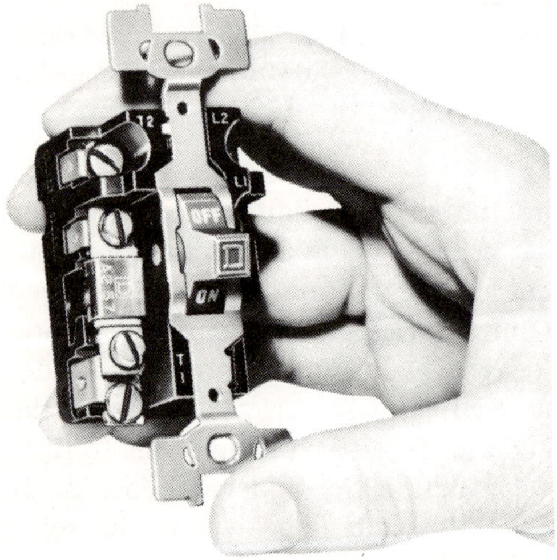

Fig. 20-6 A manual motor control switch for fractional hp motors.

Fig. 20-7 A single-phase fractional horsepower (kW) motor control switch with thermal overload relay and locking cover.

MAGNETIC MOTOR CONTROL

The magnetic motor controller is another type of across-the-line motor starter. This very useful type of starter uses an electromagnet to activate the motor control switch.

The magnetic motor controller has three main advantages. The first is the low-voltage protection feature of the unit. This simply means that a reduction, or loss, of line voltage allows the electromagnet to disengage and open the motor circuit. Motors that continue to operate when there has been a reduction in line voltage often overheat, causing damage to the motor's windings. Once the line voltage is returned to full value, the magnetic motor controller will *not* reactivate on its own. An operator must press the start button to reenergize the electromagnet. This prevents accidents when using machinery such as saws, presses, and conveyors, because there is no way of accurately predicting when line voltage will be restored. If the machine were able to start up without the starter being reset, operators working on or near the machine could be seriously injured.

The second advantage of magnetic motor control is that a great variety of activating devices are available. Mechanical activating devices in such styles as push buttons, limit switches, float switches, pressure switches, and foot controls are a few of the control stations on the market.

The magnetic motor controller is extremely versatile and can perform a great number of operations. It also can be located anywhere required for convenience or safety, and it can have any number of smaller control stations in the circuit to activate the starter. In contrast, a manual

Fig. 20-8 Typical 3-pole, three-phase manual motor starter (with pilot light).

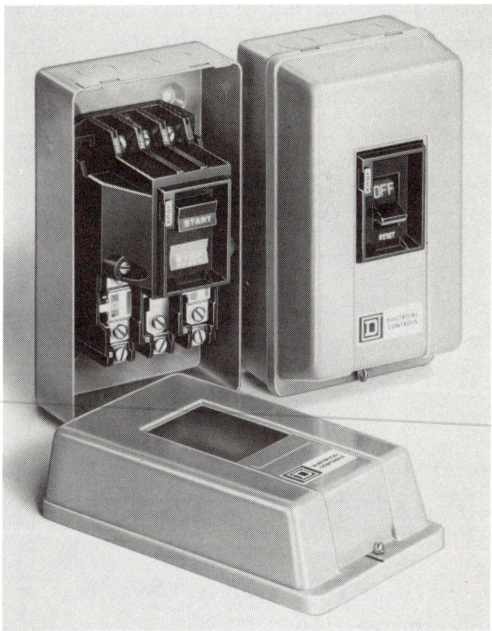

Fig. 20-9 Typical three-phase manual control switch with thermal overload protection.

motor starter must be placed within easy reach of the operator. If the starter is bulky, if damaging liquids or vapors are present, or if multiple control points are required, placing the starter near the operator may present a problem.

A third major advantage of magnetic motor control is the use of low voltage in the control circuits. Service and maintenance to the control circuit is less dangerous if 115 V is used to control a 575-V motor circuit. Controllers with this feature use a *step-down transformer* in the controller to arrive at the proper voltage for the control circuit. Figure 20-10 shows a combination protection unit with a magnetic starter. Some units may contain a control transformer.

Magnetic Motor Starter Sizes. Magnetic motor starters and control devices are divided into seven major groups, according to the power rating of the motors they can control safely. These groups are:

Fig. 20-10 A combination starter in a NEMA-1 enclosure.

HORSEPOWER (kW) SIZE	VOLTAGE – HORSEPOWER			
	200–208	220–240	440–480	550–600
00	1 1/2	1 1/2	2	2
0	3	3	5	5
1	7 1/2	7 1/2	10	10
2	10	15	25	25
3	25	30	50	50
4	40	50	100	100
5	75	100	200	200

Note: To change to kW ratings, multiply hp by 0.746.

Each motor controller must be protected by a short-circuit, ground-fault device. Either fuses or circuit breakers can be used. The motor feeder conductors must have an overload device. If a fuse system is used, take care to choose a horsepower-rated disconnect switch that can open and close the circuit safely. (See Chapter 17.)

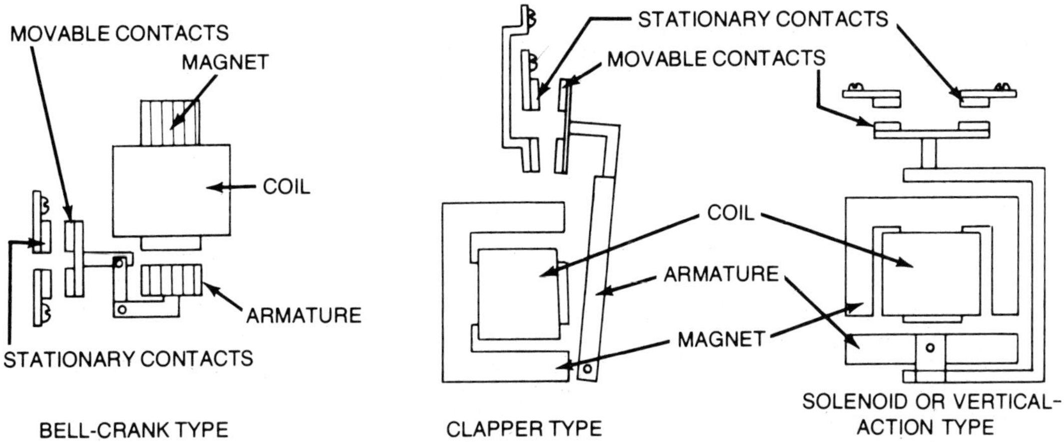

Fig. 20-11 Three basic methods for operating magnetic motor starter contacts.

Operation of the Magnetic Motor Starter. Three basic methods are used for opening and closing the contacts of a magnetic motor starter, Figure 20-11. When current is passed through the coil, the magnet attracts the armature and the movable contacts are brought into connection with the stationary contacts.

The pulsating effect of alternating current causes several problems. The armature and magnet assemblies tend to heat up as a result of the pulsating magnetic field around the main coil. To reduce this heating effect, called *hysteresis loss,* the armature and magnet are made of many thin layers of steel (laminations).

The coil loses its strength for an instant each time the alternating current falls to zero (120 times per second in a 60-Hz system). The result is a constant, high-speed attraction and release of the armature by the magnet. This is a damaging effect and can be heard as an annoying buzz or hum from the controller.

Extensive wear and heat on the magnet's pole faces can also be expected. Small copper rings, called *shading coils,* are embedded in the face of the ac magnet to counteract and reduce this undesirable effect, Figure 20-12. The pulsating magnetic field around the main coil induces (generates) a voltage and current in the shading coils. This induced current allows the shading coils to produce their own magnetic field, which helps to attract and hold the armature during the times that the main coil is weakened.

A magnetic starter is made up of two electrical circuits. The *main,* or *power,* circuit, which has line terminals, main contacts, overload heaters, and motor terminals, is used to supply current to the motor. This circuit normally comes prewired within the starter. The *control* circuit, which has an electromagnet, overload contacts, auxiliary (or maintaining) contacts, and an assortment of control stations, is used to activate the motor starter and ensure its safe operation. The control circuit must be wired to the starter in the field.

Both the main and control circuits can be represented in the form of schematic

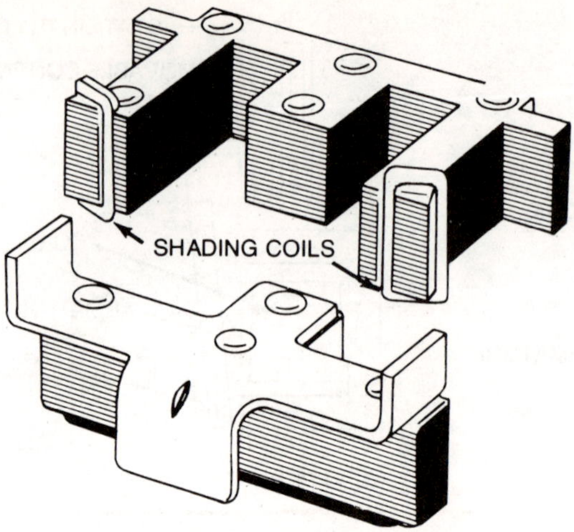

Fig. 20-12 Magnet and armature assembly of a starter.

wiring diagrams. Special symbols are used to represent each component in the circuit, as shown in Figure 20-13.

Figure 20-14 shows a typical three-phase magnetic motor starter (with pilot light). Figure 20-15 shows the wiring diagram for this controller. The main power circuit is outlined in black; the control circuit is marked "R" for red. (*Note:* the control circuit uses phase voltage.) This type of diagram, which shows each part in its correct location, is useful when wiring or troubleshooting a controller.

Figure 20-16 shows an elementary schematic diagram of the same controller. Although the parts are not in their true locations, this type of diagram gives an easily understood picture of the circuit, and is most helpful when troubleshooting more complicated types of controllers. When a complex control system is to be developed, electricians often make a schematic diagram of the control circuit only. Figure 20-17 shows a variety of these types of diagrams. When wiring the two circuits (power and control), electricians find it best to complete the power circuit first, and then the control circuitry.

HOLDING, OR MAINTAINING, CONTACTS

Magnetic motor starters *must* be able to open the power circuit to a motor whenever 1) an overload relay indicates a dangerous condition, 2) an undervoltage condition develops or, 3) any form of control switch indicates that operation of the motor should be stopped. Some control stations use a set of start contacts that remain closed only as long as the operator keeps a finger on the start button. The moment the button is released, a spring forces the contacts open again. This deenergizes the coil in the starter and allows the motor to stop.

A set of *holding,* or *maintaining,* contacts (sometimes called *sealing* contacts) are mounted on the same contact carrier bar as the main power circuit contacts. Therefore, the maintaining contacts open and close together with the main power circuit contacts. They are electrically connected in *parallel* with the start button contacts and keep the coil energized once the start button has been released. Maintaining contacts are represented by a normally open contact symbol (NO) in both schematic and wiring diagrams. Often, the

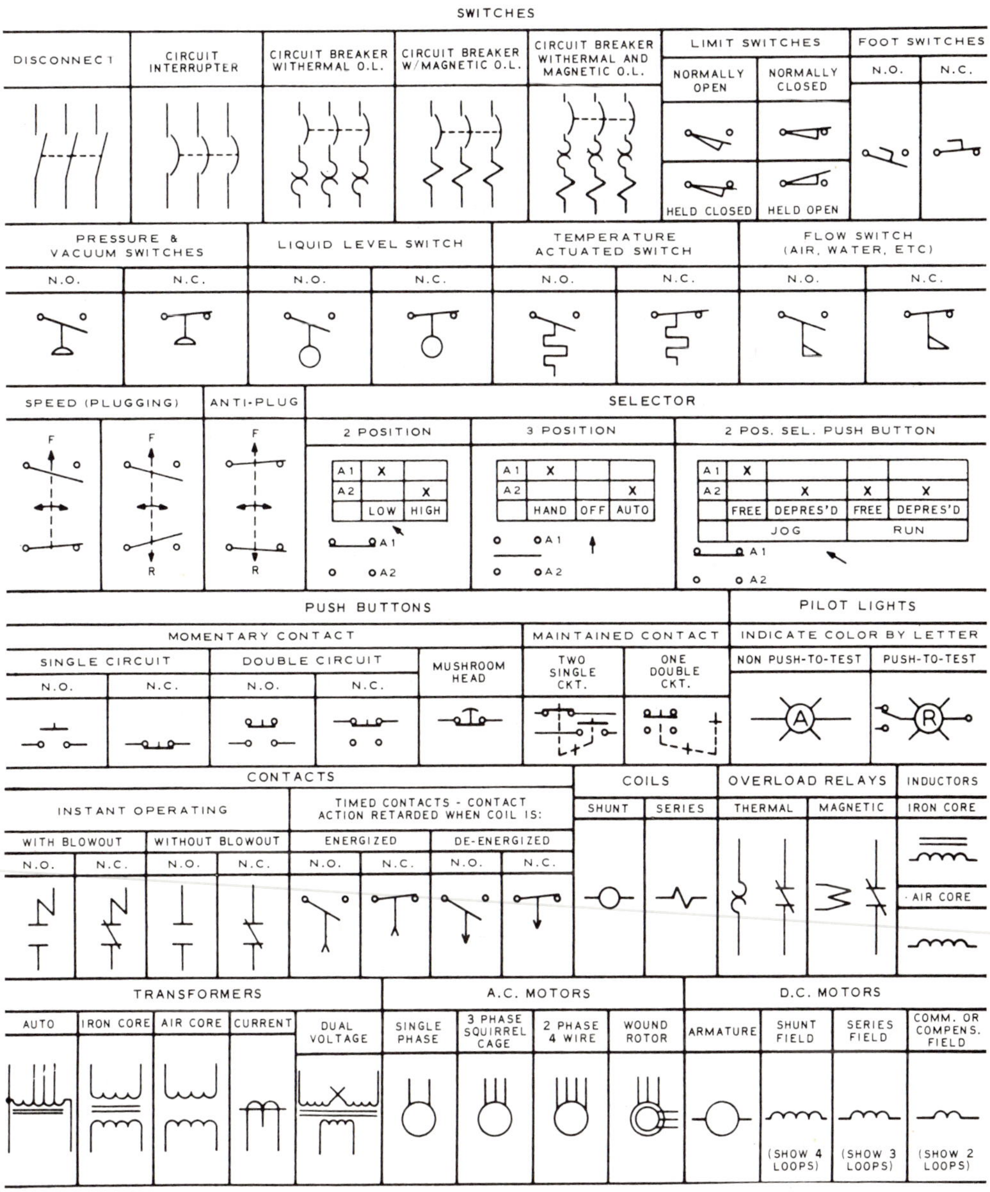

Fig. 20-13 Standard elementary diagram symbols.

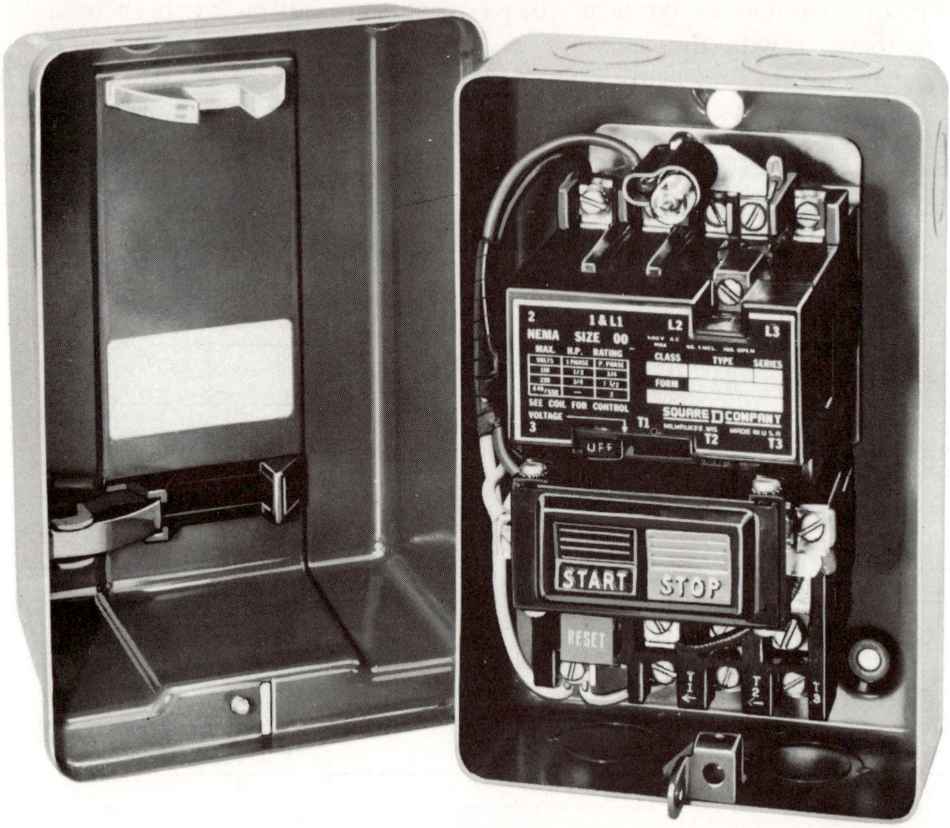

Fig. 20-14 A self-contained, three-phase magnetic motor starter (with pilot light).

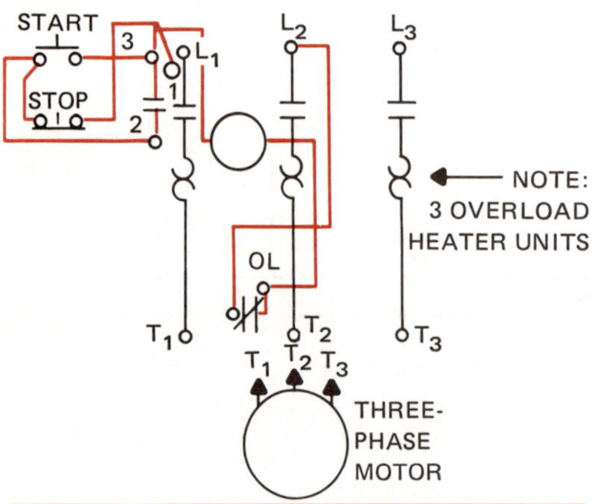

Fig. 20-15 Wiring diagram for a typical three-phase magnetic motor starter.

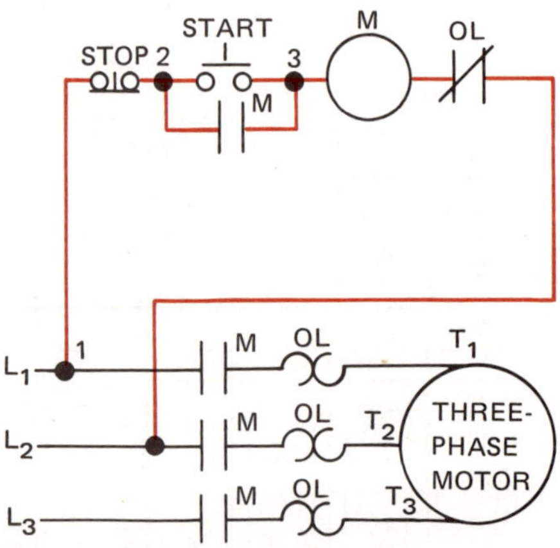

Fig. 20-16 Elementary schematic diagram for a typical three-phase magnetic motor starter.

letter M will appear beside the contacts to show that they are the maintaining contacts. (See Figures 20-15 and 20-16.)

CONTROL STATIONS

Figure 20-14 shows a self-contained motor controller with the start/stop control station mounted on the front of the unit. When multiple-point control is necessary, a remote push button station is used. Other types of control stations are available, such as limit switches to control the up and down movement of an electric door, float switches to control the level of liquid in a tank, or dual-action pump control switches. (See Figures 20-18, 20-19, 20-20, and 20-21.)

Sometimes it is necessary to bypass the automatic control device and operate the

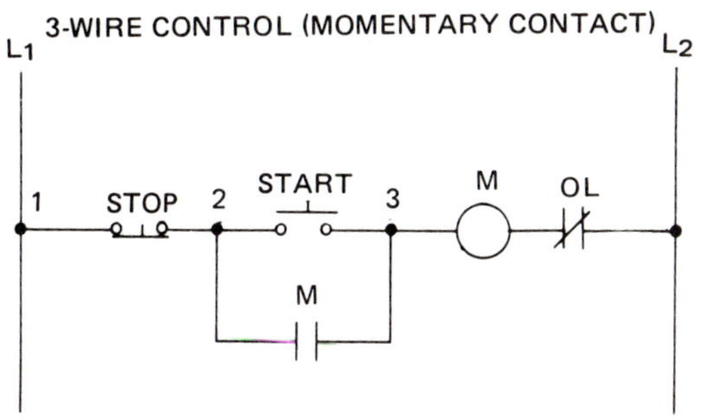

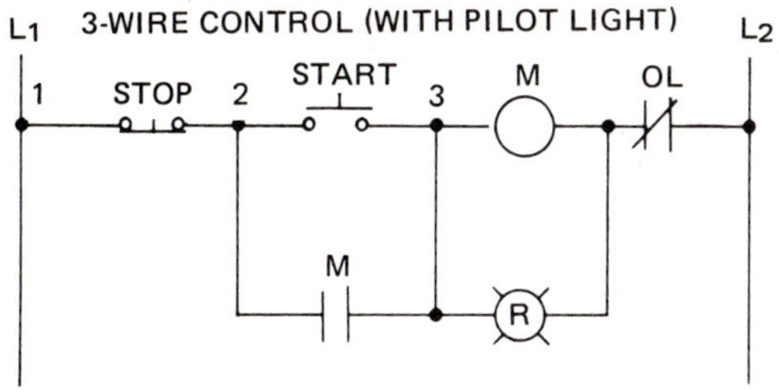

NOTE: PILOT LIGHT CAN BE WIRED IN PARALLEL WITH STARTER COIL TO SHOW WHEN STARTER IS ENERGIZED AND MOTOR IS RUNNING.

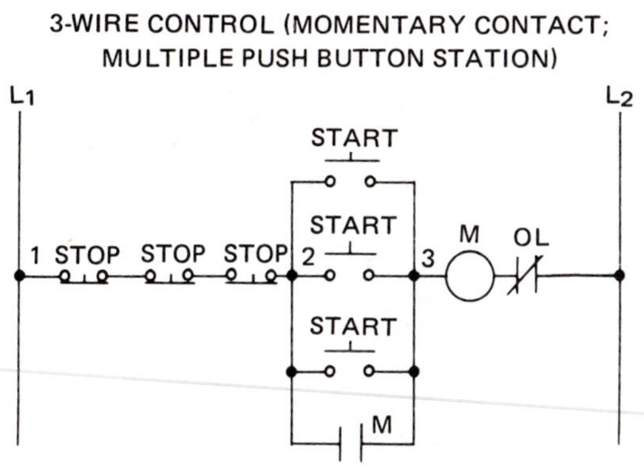

NOTE:
WHERE A MOTOR MUST BE STARTED AND STOPPED FROM MORE THAN ONE LOCATION, ANY NUMBER OF START AND STOP PUSH BUTTONS MAY BE WIRED TOGETHER.. ALSO, IT IS POSSIBLE TO USE ONLY ONE START/STOP STATION IN COMBINATION WITH SEVERAL STOP BUTTONS AT DIFFERENT LOCATIONS FOR USE IN EMERGENCIES.

Fig. 20-17 Schematic diagrams for typical motor control circuits.

Fig. 20-18 Remote push button start/stop control station.

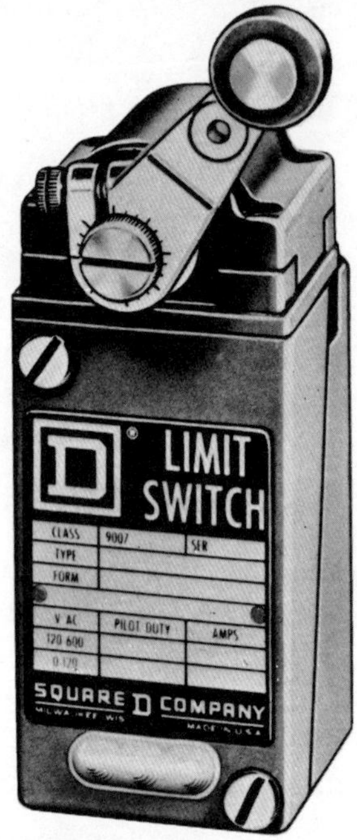

Fig. 20-19 Limit switch.

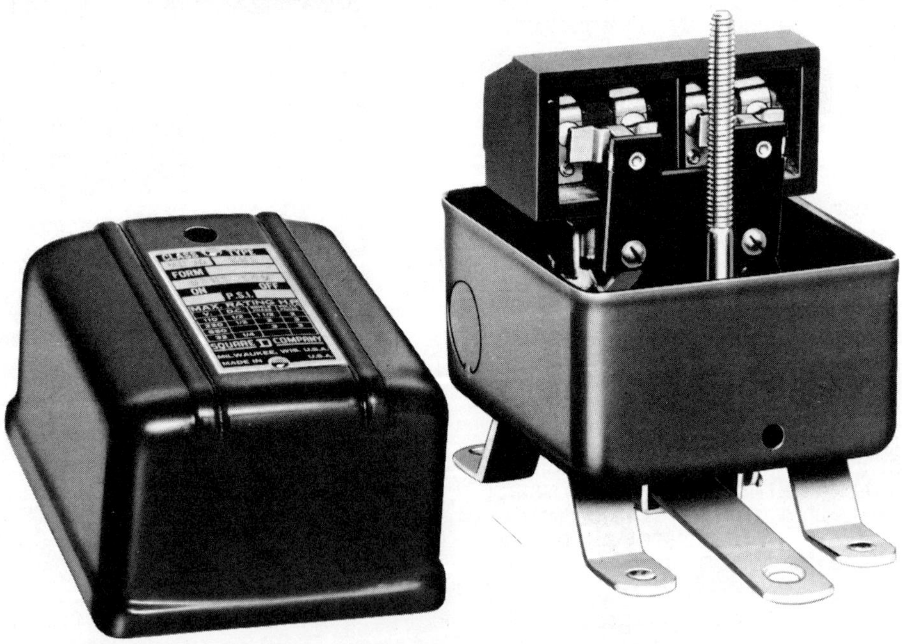

Fig. 20-20 Float switch.

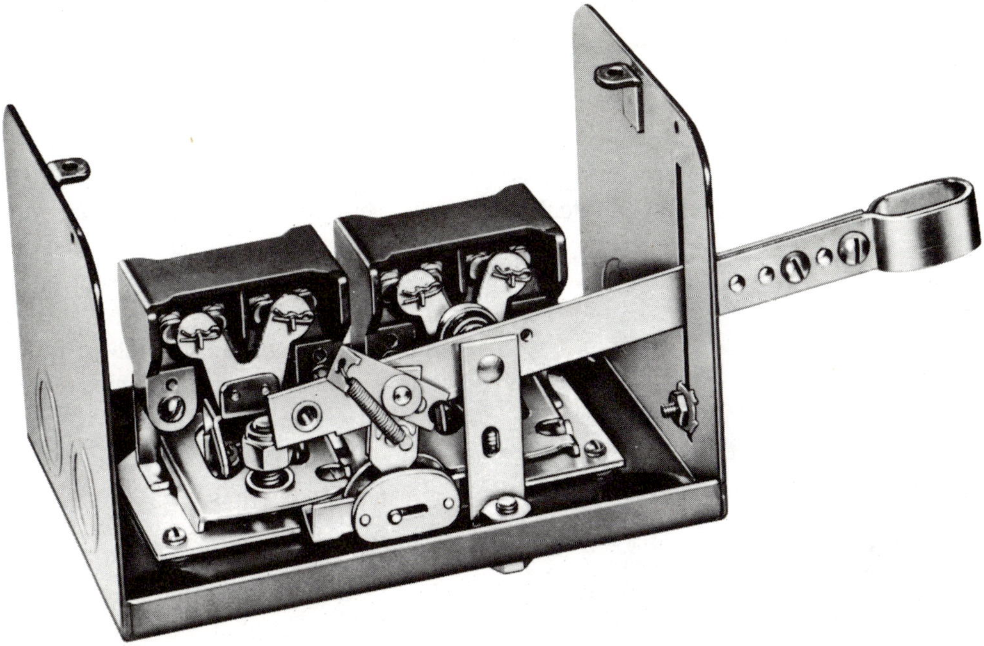

Fig. 20-21 Dual-action pump control switch.

controller manually to fill a tank or start a pump. A push button station with a selector switch is used with the controller to provide both manual and automatic control functions, Figure 20-22. This selector switch is often referred to as a *hand-off-automatic* (HOA) switch.

Occasionally, an operator desires to control the length of time a motor operates or the length of time before a second control circuit is energized. A pneumatic timing controller is used for this purpose, Figure 20-23. To calibrate the time cycle, an adjusting screw at the top of the unit regulates the flow of air (or oil) in and out of a compression chamber. This type of control device is very useful in reduced-voltage starting units for large motors to control the length of time of the starting to running sequence.

The NEC requires three overload relays (one per phase) on a three-phase motor controller, Figure 20-24. This is because a three-phase motor will continue to operate on any two of its three phases if one should fail. It will not restart on two phases but, if one phase is disconnected, it can handle approximately 60 percent of its rated power. A motor that is putting out less than its rated power will soon heat up from the overload, and its windings will be damaged if it is not disconnected from the line. An overload relay on each phase ensures that two or more phases will indicate overload conditions and remove the motor from the line.

Overload relays on a magnetic starter function in a way very similar to those on manual motor starters. The usual thermal element, solder pot, and ratchet wheel are used. The spring-loaded mechanical section of the relay, however, does not open the power circuit of the controller directly. Instead, when the ratchet wheel is allowed to rotate, a set of normally closed contacts (NC) are forced open by the mechanical section. These contacts are connected in series with the control circuit. Once they have been opened, the magnetic holding coil is deenergized, and the armature falls back to open the main power circuit contacts. As in the case of the manual starter, a reset button in the controller must be pressed (after the overload relay has cooled) to reactivate the mechanical section of the relay and close the overload contacts. These contacts are shown on schematic and wiring diagrams by the normally closed contact symbol (NC) with the letters OL beside it.

Fig. 20-22 Push button control station for manual/automatic control.

JOGGING CIRCUITS

Work must be positioned carefully on metal-cutting or forming machines before operation. Large machines often have motor-driven tables or material platforms. These motors are activated several times in

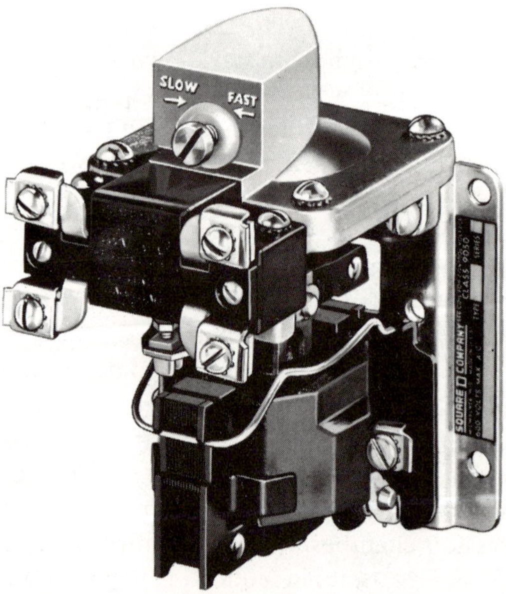

Fig. 20-23 Motor control unit pneumatic timer.

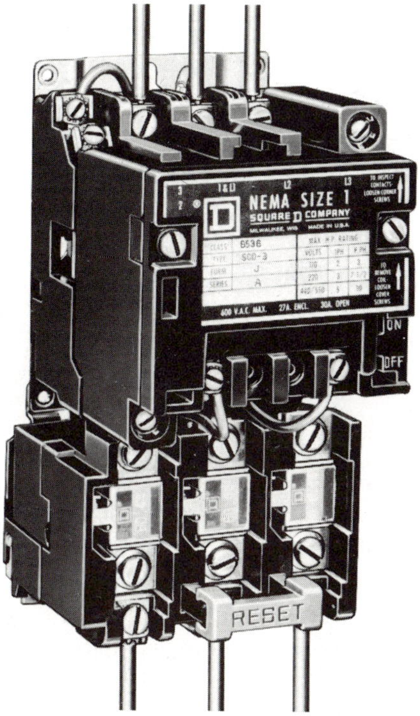

Fig. 20-24 Magnetic motor controller with three overload relays.

quick succession to position the material accurately. This is called *jogging* (or *inching*). The main power contacts must open and deenergize the motor as soon as the operator releases the jog button.

Two basic types of circuits are used for jogging: the *push button* jog and the *selector jog*.

Push Button Jog. The push button jog uses a standard start/stop control circuit with a 4-terminal (double circuit) push button for the jogging operation, Figure 20-25. The upper contacts of the jog button open the regular current path, while the bottom contacts close a bypass circuit around the maintaining contacts. With the normal current path interrupted by the jog button, the maintaining contacts cannot keep the coil energized.

One danger exists with this type of circuit. An operator in a hurry may release the jog button by allowing a finger to slide off. The spring in the jog button might close the upper contacts quickly, before the armature and power contacts have had a chance to open the power circuit. If this

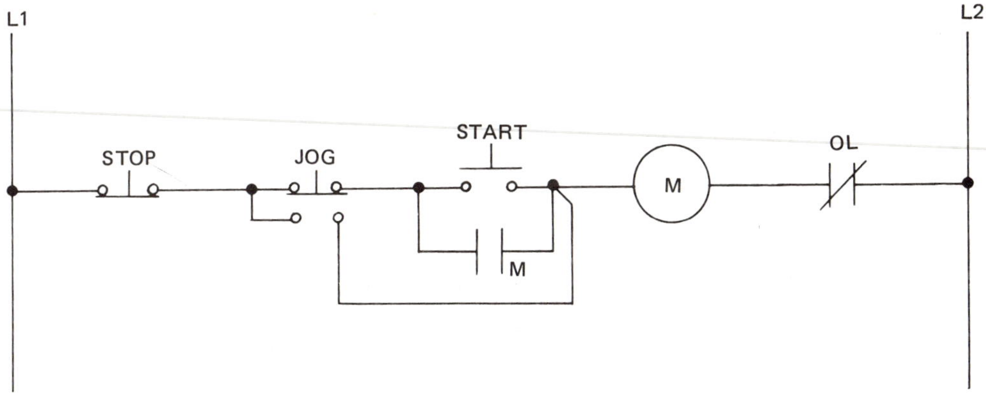

Fig. 20-25 Start/stop/jog push button control circuit.

happens, the maintaining contacts will still be closed and the coil may remain energized, keeping the motor operating. To stop the motor, the stop button must be pressed. Although this situation does not happen often, when it does, the operator can be *injured* if the motor fails to stop when the jog button is released.

Selector Jog. The selector jog is a safer form of jogging circuit than the push button type, Figure 20-26. The selector switch jog circuit eliminates the danger of an operator releasing the jog button too quickly. This circuit places a single-pole selector switch in series with the maintaining contacts. When the switch is closed (run position), the circuit operates normally. When the switch is open (jog position), the maintaining contacts cannot keep the coil circuit energized. The power circuit opens as soon as the operator releases the start button.

Specialized jogging circuits that make use of a control relay are also produced, Figure 20-27.

If a motor circuit is to be jogged continuously (more than five times per minute), controllers must be derated. This simply means that either motors smaller than the nameplate rating of the controller must be used or a larger controller must be installed. For instance, if excessive jogging or reversing occurs using a Size 1 starter, a Size 2 starter must be used. The constant exposure to high inrush currents (motor starting current) will soon damage the contacts of the controller if jogging operations are carried on continuously.

REDUCED-VOLTAGE CONTROL CIRCUIT

Many three-phase motors operate on 440 V and 550 V. Voltage at this level can cause painful *injury* to an operator or service person if contact is made with live equipment. At this voltage level, it is not uncommon to control the motor starting equipment with a reduced-voltage control system. This system makes it safer for

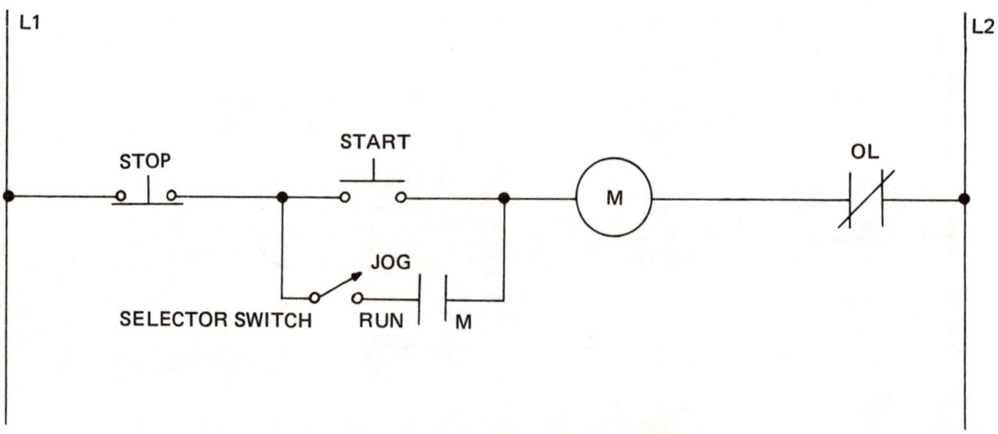

Fig. 20-26 Start/stop/selector jog control circuit.

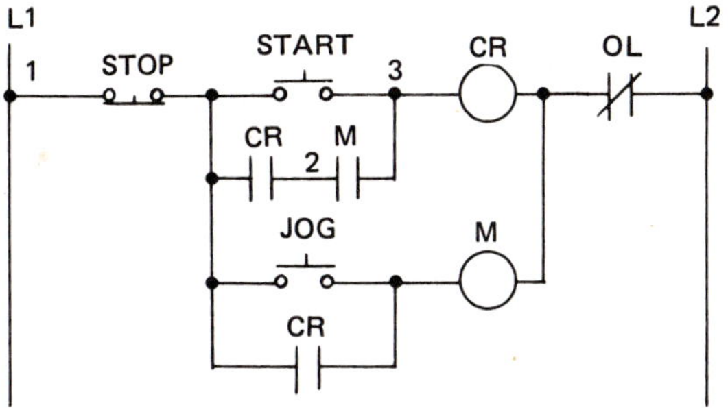

NOTE: PRESSING THE START BUTTON ENERGIZES THE CONTROL RELAY, WHICH IN TURN ENERGIZES THE STARTER COIL. THE NORMALLY OPEN STARTER INTERLOCK AND RELAY CONTACT THEN FORM A HOLDING CIRCUIT AROUND THE START BUTTON. PRESSING THE JOG BUTTON ENERGIZES THE STARTER COIL INDEPENDENTLY OF THE RELAY. NO HOLDING CIRCUIT FORMS, AND SO JOGGING CAN BE OBTAINED.

Fig. 20-27 Jog circuit with control relay.

service personnel to work on the control circuit, and simplifies the actual equipment in the control circuit, since there is less danger of a flashover or arc. Two basic methods are used to connect this type of reduced-voltage control circuit, as shown in Figures 20-28 and 20-29.

Figure 20-28 shows the use of a control circuit transformer to reduce the line voltage to a safer level (120 V). With this method, both motor and control voltage can be cut off by the same disconnect switch used to supply the motor circuit.

Figure 20-29 shows the use of a separate source of supply for the control voltage. This system is useful when direct current (battery) or other separate source is needed to control the motor starter. Windows or natural light sources are unavailable in some work areas. If the lighting system should fail, machine operators are left in a dark room with all the machines running. This can be very *dangerous.* By connecting the control circuit source to the lighting panel in such an area, a loss of lighting voltage will also shut off operation of the machines, making it safer for operators to move about in the dark area. In such work areas, it is best to install an emergency (secondary) lighting system.

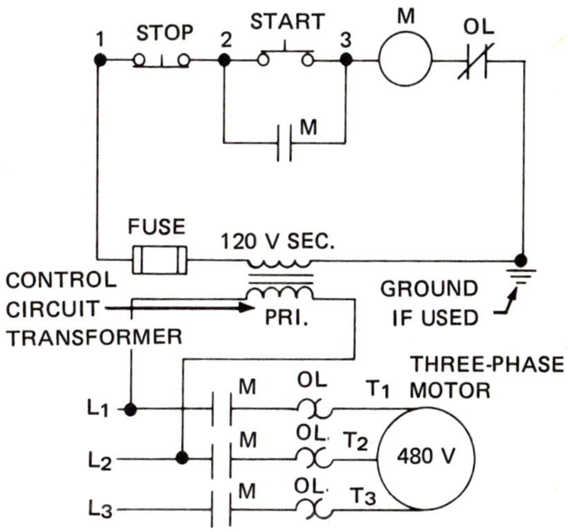

Fig. 20-28 Typical reduced-voltage control circuit using a source common to motor and control circuit.

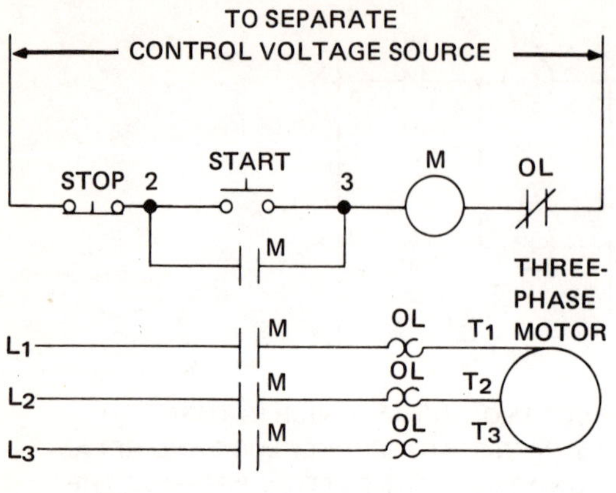

Fig. 20-29 Typical reduced-voltage control circuit using a separate voltage source.

While 440 V and 550 V are mentioned in particular at the start of this section, supply voltages of many levels can be controlled by a reduced-voltage system. The higher the supply voltage, the more likely it is that a control circuit using reduced voltage will be required.

REVERSING CONTROLLERS

Interchanging any two of the leads to a three-phase motor will cause it to run in the reverse direction. Single-phase motors tend to be more complicated, and often need connection changes within the motor itself. For this reason, only three-phase controllers are discussed in this section.

A three-phase reversing starter consists of two contactors enclosed in the same cabinet. Only one set of overload relays is used, however, since both forward and reverse control circuits are interconnected.

When reversing a motor, it is very important to make sure that both contactors are not energized at the same time. This action would cause a short circuit, since the line conductors are reversed on one contactor. Prevention of this occurrence is accomplished by a system called *interlocking*. Mechanical and electrical interlocking systems are available in most reversing controllers.

The *mechanical* interlock uses a system of levers to prevent the armature of the reversing contactor from engaging when the forward contactor's armature is in operation (and vice versa). Figure 20-30 shows a schematic diagram of the control circuit for this type of unit.

The *electrical* interlock makes use of two double-contact (4-terminal) push buttons, Figure 20-31. When the forward button is pressed, the upper contacts open the reverse coil circuit. Even if the reverse coil is energized, the control circuit will be broken. The power circuit is opened as soon as the forward button is pressed.

The electrical interlock circuit allows the motor to be reversed simply by pressing the reverse button. The mechanical interlock system requires that the stop button be pressed, thus disengaging the forward contactor before the reversing contactor can be brought into service.

Auxiliary contact blocks actuated by the armature of the starter assembly are also used for electrical interlocks. The control circuitry for the reversing coil is passed through the normally closed contacts of the forward starter, and vice versa.

Figure 20-32 is a schematic diagram for a typical reversing control circuit, showing the location of limit switches (if used). The dotted lines between the upper and lower rows of contacts show which normally open and normally closed contacts operate together.

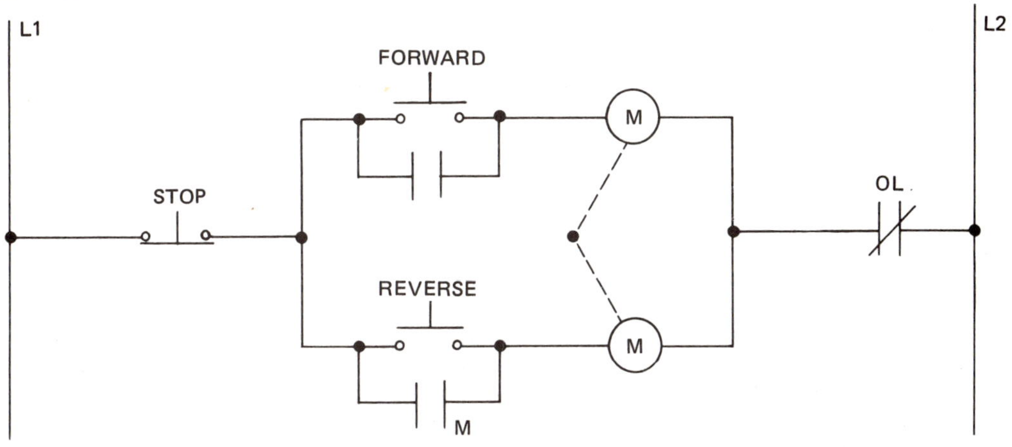

Fig. 20-30 Mechanical interlock forward/reverse/stop control circuit.

Figures 20-33 and 20-34 show complete wiring diagrams for a reversing controller. An extra set of contacts (numbers 6 and 7 in Figure 20-32) have been installed to perform as an electrical interlock. Operation of either the forward or reverse contactors will open the matching interlocking contact. This makes it necessary to press the stop button before the motor's direction of rotation can be changed.

Take care when reversing large motors. The sudden jar of direct reversal can damage the machine or equipment the motor is driving. High inrush currents can cause damage to both the motor and the controller if the motor is reversed without allowing enough

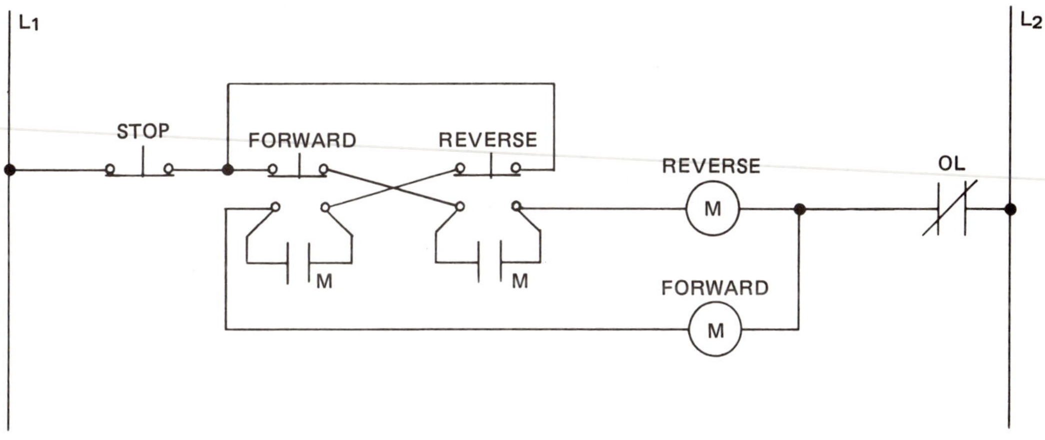

Fig. 20-31 Electrical interlock forward/reverse/stop control circuit.

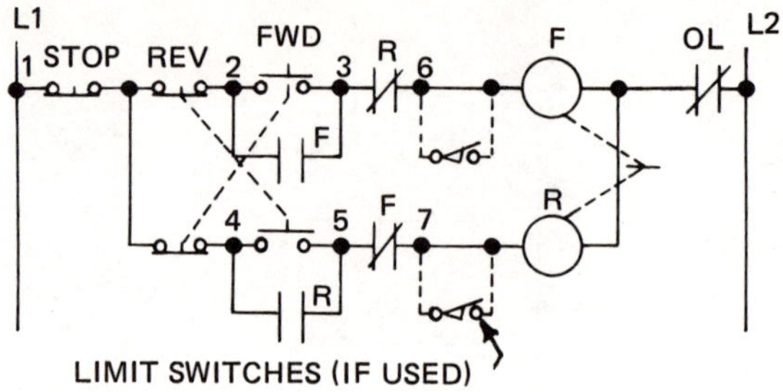

NOTE: 3-WIRE CONTROL OF A REVERSING STARTER IS POSSIBLE WITH THIS FORWARD/REVERSE/STOP PUSH BUTTON STATION. LIMIT SWITCHES CAN BE ADDED TO STOP THE MOTOR AT A CERTAIN POINT IN EITHER DIRECTION. JUMPER 6 TO COIL AND JUMPER 7 TO COIL MUST THEN BE REMOVED.

Fig. 20-32 Schematic diagram showing typical reversing control circuit (with limit switches).

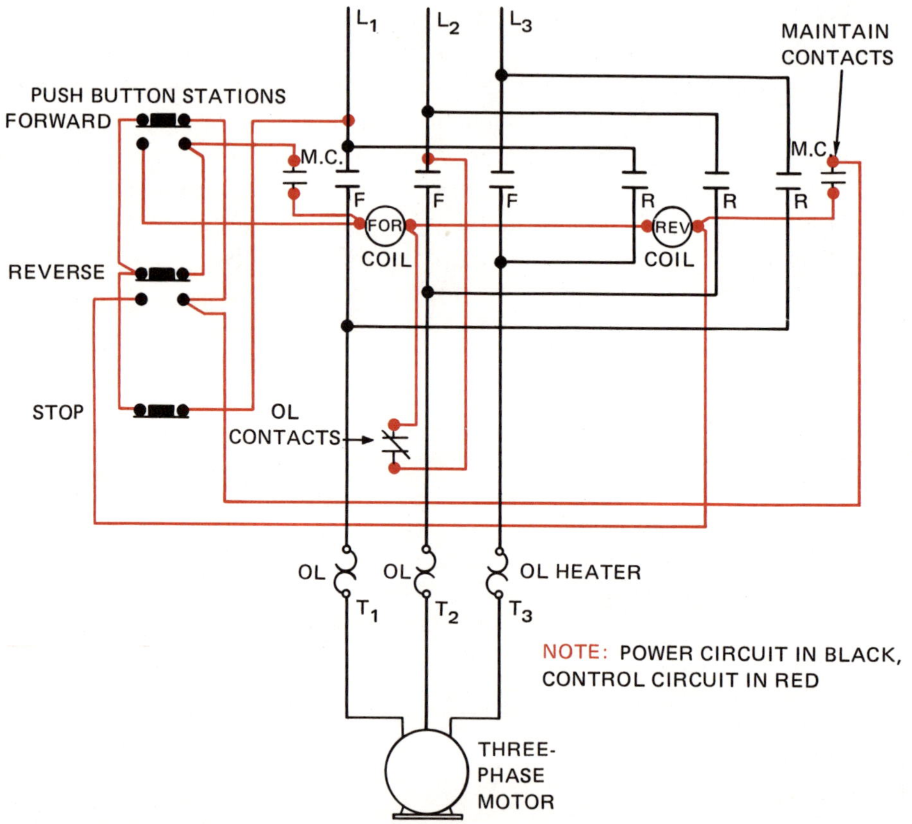

Fig. 20-33 Typical three-phase, forward/reverse/stop wiring diagram.

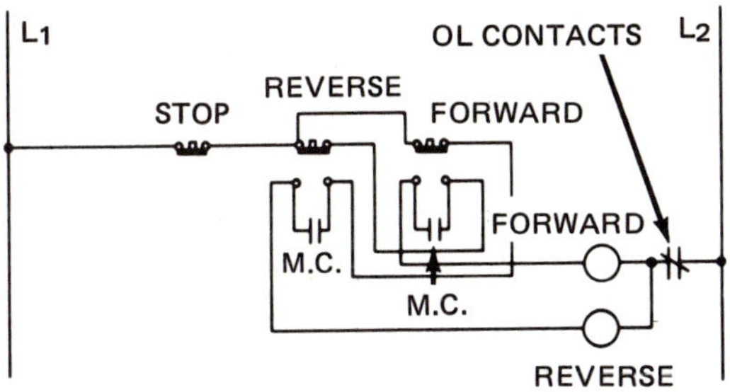

Fig. 20-34 Alternate schematic diagram, forward/reverse/stop control circuit.

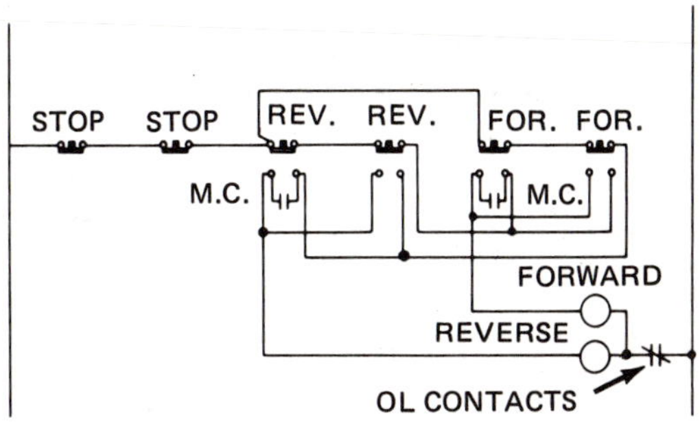

Fig. 20-35 Schematic diagram showing multiple push button control for a reversing starter.

time for the speed of the motor to decrease.

Multiple push button stations can be used with the reversing controller. Figure 20-35 shows a schematic diagram of two forward/reverse/stop stations.

DUAL-SPEED MOTOR CONTROL

Some three-phase motors are designed with a dual set of windings to provide two separate speed ranges, Figures 20-36 and 20-37. If a few minor changes are made to it, the standard reversing controller can be used to operate these motors. Since the dual-speed, three-phase motor has multiple windings, overload relays (normally of different heater values) are required in both high- and low-speed circuits to provide adequate protection for each speed range. There is no reversal of the line conductors between the two contactors, since the motor must rotate in the same direction for each speed range.

The push button control station and its electrical connections are similar to those used with the reversing controller. *High* and *low* markings are used on the push button station instead of *forward* and *reverse*.

REDUCED-VOLTAGE STARTERS

When motors are started with full-line voltage, high (locked rotor) currents and maximum torque can be expected. Under full-line voltage conditions, large motors often develop enough torque to damage belts, gears, and other driveline components. Starting current is often high enough to endanger the controller contacts. It also can create a voltage disturbance on the line, causing interference with other motors, lighting, or other electrical equipment. For these reasons, a starting system with less than line voltage is used.

Reduced-voltage starters are made in Size 2 and larger. They are of two types: the *primary resistor* and the *autotransformer.*

Primary Resistor Starter. The primary resistor starter connects a set of resistors in series with the line, which reduces line voltage for a predetermined length of time (3 seconds to 15 seconds). A timing relay activates the controller automatically for the second state. The motor-starting process begins with the pressing of the start push button. This activates the line contacts coil, allowing current to flow through the primary

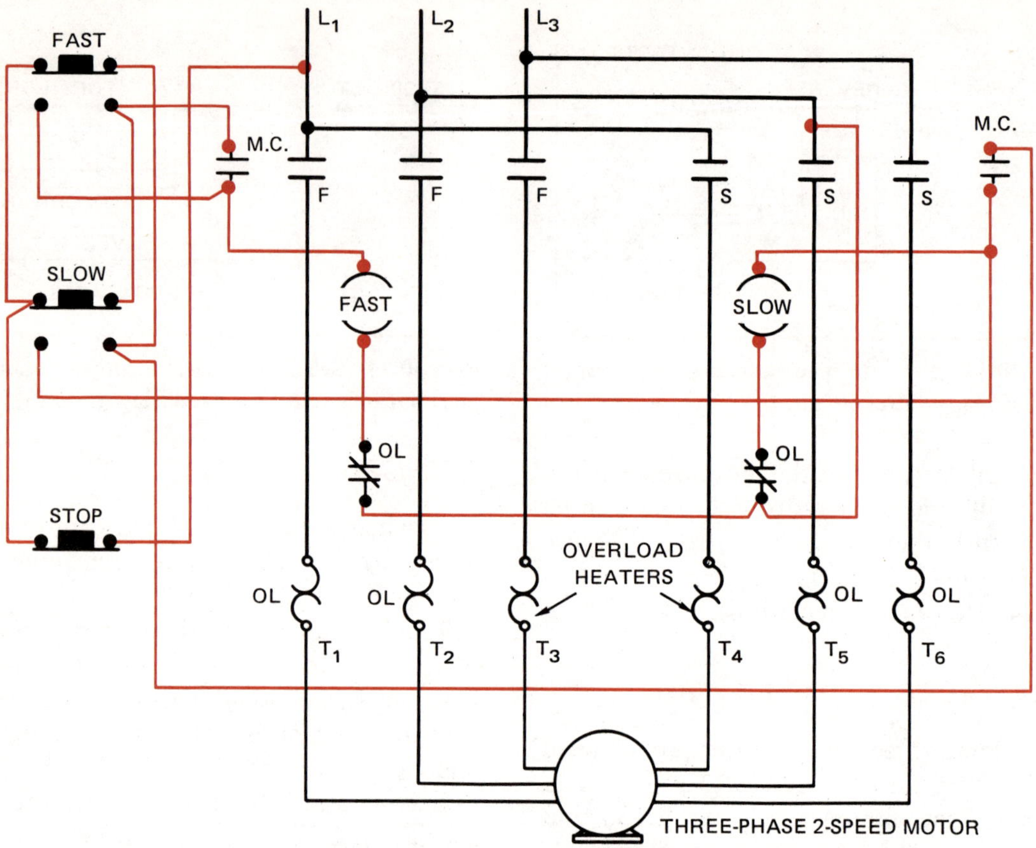

Fig. 20-36 Typical dual-speed, three-phase motor control circuit (dual winding motor).

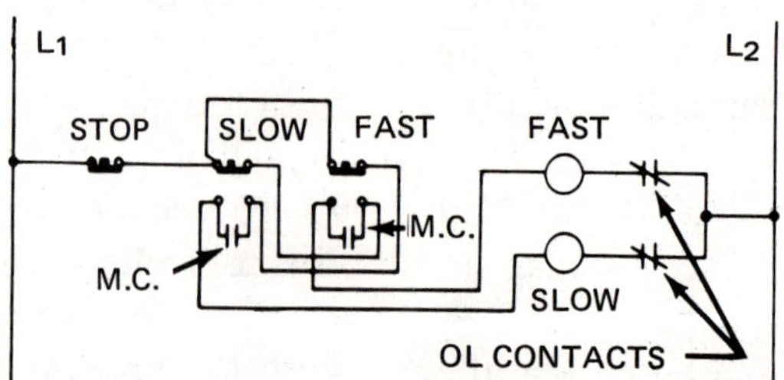

Fig. 20-37 Schematic diagram showing control circuit for a two-speed (dual winding) motor.

starting resistors, and into the motor. The primary starting resistors reduce the line voltage to a level that allows the motor to start without drawing an excessive amount of starting current.

A timing relay is activated at the same time as the line contacts coil. This relay can be preset by the operator to close its contacts at a predetermined time. When the contacts of the timing relay close, they activate the bypass contacts coil. This in turn closes the resistor bypass contacts and current now flows around the primary resistors to the motor. Full source voltage is now applied to the motor, thus no voltage is lost or reduced by this path to the motor, and full speed is soon reached by the motor. The motor can be stopped simply by pressing the *stop* push button.

This type of starting gives smooth acceleration to full operating speed without loss of speed during the changeover cycle.

Figure 20-38 shows a wiring diagram for the primary resistor-type of motor starter.

Autotransformer Starter. The autotransformer starter, which is often called a *compensator,* uses a set of single-winding (autotransformer), step-down transformers to reduce the line voltage. As with the primary resistor starter, partial voltage is supplied to the motor on starting, with a timer relay calibrated to activate the second stage of the operation.

The starting sequence of the autotransformer starter is set in motion by the pressing of the start push button. This

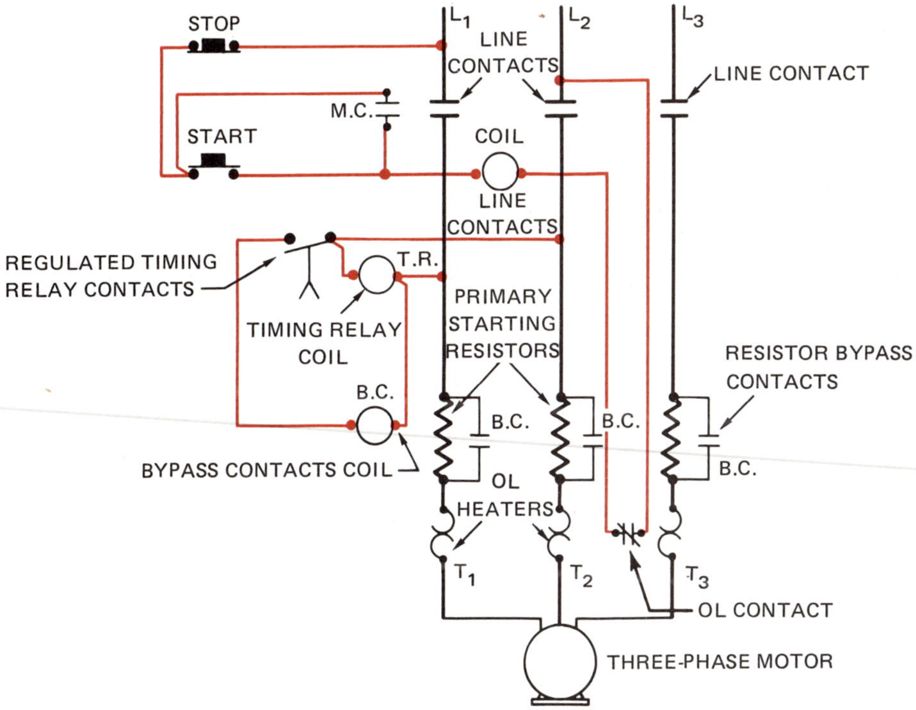

Fig. 20-38 Typical primary resistor, reduced-voltage, motor starter wiring diagram.

activates three electromechanical devices at the same time. (A) The timing relay is set for its predetermined time and begins its countdown. The *maintain* contacts in this controller are activated by the timing relay and keep the control circuit energized once the start button has been released. (B) The *start* contacts coil is activated and in turn closes the starting contacts. (C) The *transformer* contacts coil is also energized and closes the transformer contacts, allowing current to flow into the three-phase transformer circuit (wye connected). As can be seen in the wiring diagram for a typical autotransformer starter, Figure 20-39, the starting contacts receive their voltage and current from approximately midpoint on the transformer. This starting voltage is likely to be around 65 percent of the full-line voltage available. This is enough voltage to start the motor with a moderate amount of starting current, thus preventing dangerous current surges.

When the timing relay reaches the end of its countdown (5 seconds to 15 seconds) it automatically opens the starting and transformer contacts by deenergizing the coils controlling those contacts. At the

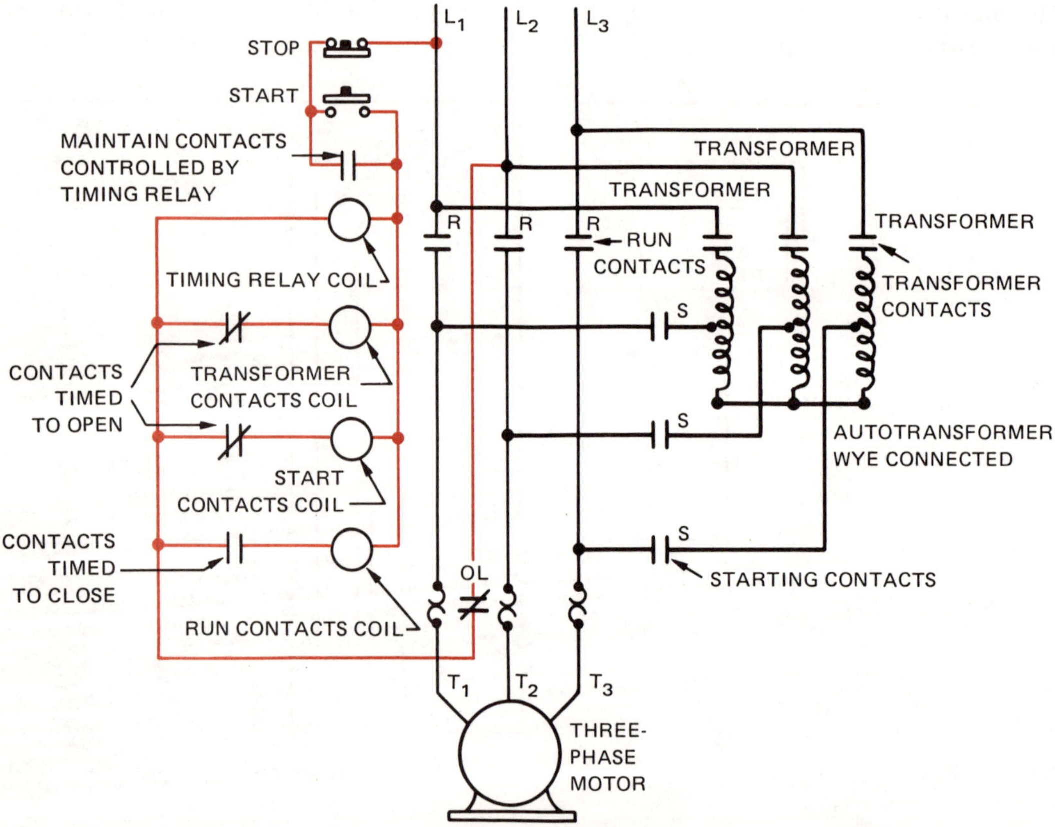

Fig. 20-39 Typical autotransformer, reduced-voltage, motor starter diagram.

same time, the run contact coil is activated and the run contacts are closed. Current is now allowed to travel straight through to the motor instead of detouring through the transformer circuit. Full-line voltage is impressed on the motor, and it soon reaches its full operating speed. Operation can be brought to a halt by pressing the *stop* push button.

More starting torque is available with the autotransformer type of controller than with the primary resistor type, but acceleration to full speed is not as smooth. The open-circuit condition during changeover from starting to full-line voltage can cause a current surge for an instant.

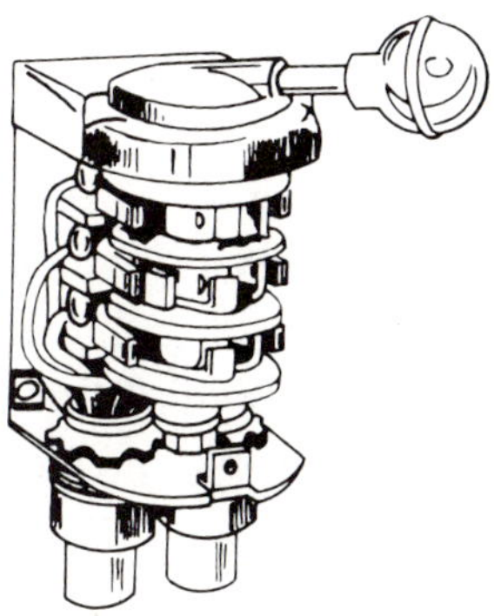

Fig. 20-40 Typical 3-position reversing drum switch.

REVERSING SINGLE-PHASE MOTORS

Single-phase motors (split-phase type) are usually used in home workshops and light industry. They have two independent windings, called *start* and *run.*

The smaller-gauge wire start winding is connected to the circuit during the starting period *only*. It determines the direction of the motor's rotation.

The larger, heavier-gauge run winding is connected to the line at all times that the motor is operating. It keeps the motor running.

To reverse the split-phase motor, the start winding or run winding leads must be interchanged (preferably the start winding). A mechanical reversing unit, called a *drum switch,* can be used to do this, Figure 20-40.

Figure 20-41 shows the internal connections of the three-position reversing drum switch, and Figure 20-42 shows a typical wiring diagram for the split-phase motor with drum switch. If a drum switch is used for reversing the motor, it may be necessary to take the motor apart to gain access to the start winding leads.

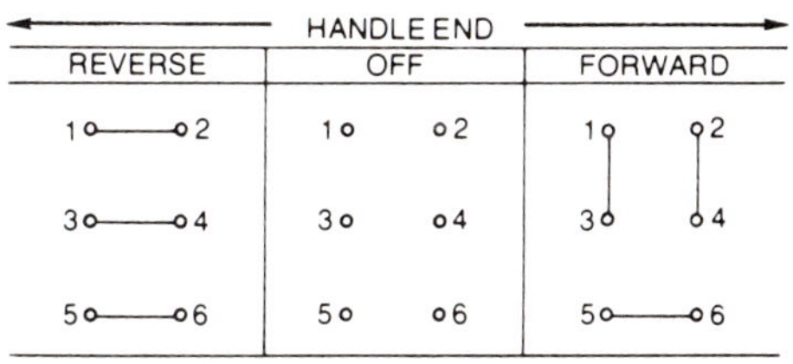

Fig. 20-41 Internal connections for a 3-position reversing drum switch.

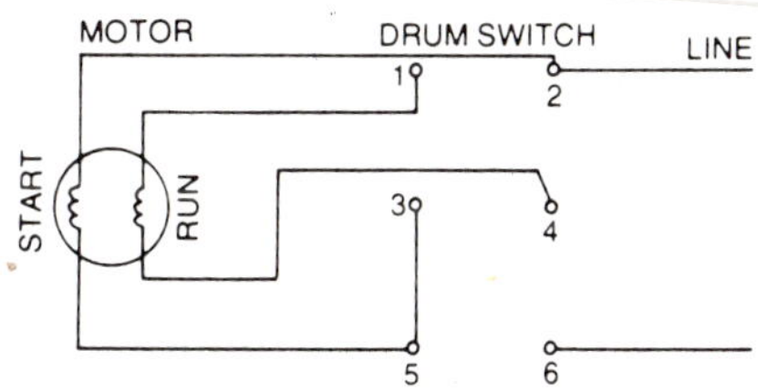

Fig. 20-42 Wiring diagram showing a split-phase motor (with drum switch) reversing.

DIRECT-CURRENT MOTOR CONTROL

Direct-current motors are used far less often than alternating-current motors, and they need special starting equipment.

As with large ac motors, high starting currents are also a problem with large dc motors. The starting controllers used have a tapped resistor to raise the motor's speed gradually without excessive starting current.

Direct-current motors of the compound type have two field windings, called *series* and *shunt* (parallel) fields.

To provide starting torque, the shunt field receives line voltage at all times. The armature and series field, however, has a variable resistor connected in series. Figure 20-43 shows a 3-point (three connections to the starter) reduced-voltage starter with no-field release.

As the starting arm is moved gradually across the face of the starter, series field and armature circuit resistance is decreased gradually. This exposes these windings to more and more voltage, until full-operating speed is reached. A holding coil keeps the starting arm in the *run* position until the operator returns it to the *off* position.

If the shunt-field circuit is opened or damaged in any way, the holding coil, which is connected in series with the shunt field, releases the spring-loaded starting arm and the motor shuts off. This *no-field release* protection is very important for dc motors. Without it, an extremely *dangerous* speed will be reached quickly if the shunt field is disconnected. Large dc motors can speed up to the point at which the armature windings are thrown out of their retainers (by centrifugal force) and the motor destroyed. The direct-current controller can be used to regulate the starting of a shunt motor, which has no series field, by placing the A^2 side of the armature directly on the armature terminal of the controller.

Figure 20-44 shows the wiring for another type of manual motor starter, the 4-point (four connections to the starter) motor starter. These controllers also provide no-voltage protection, which prevents the motor from restarting by itself if a power failure has caused the holding coil to release the starting arm.

Successful starting of the motor with a manual starter depends entirely upon the operator. If the starting arm is moved too quickly, strong inrush currents will damage the controller and/or the motor. An automatic motor starter to prevent such human error is available. The operator simply presses a start button, which in turn activates a solenoid (electromagnet). A *solenoid* is another form of timer relay, which controls the speed at which the contacts of the controller close. Closing of the contacts gradually eliminates the starting resistance, and the motor accelerates smoothly.

Figure 20-45 shows a wiring diagram for an automatic dc motor starter.

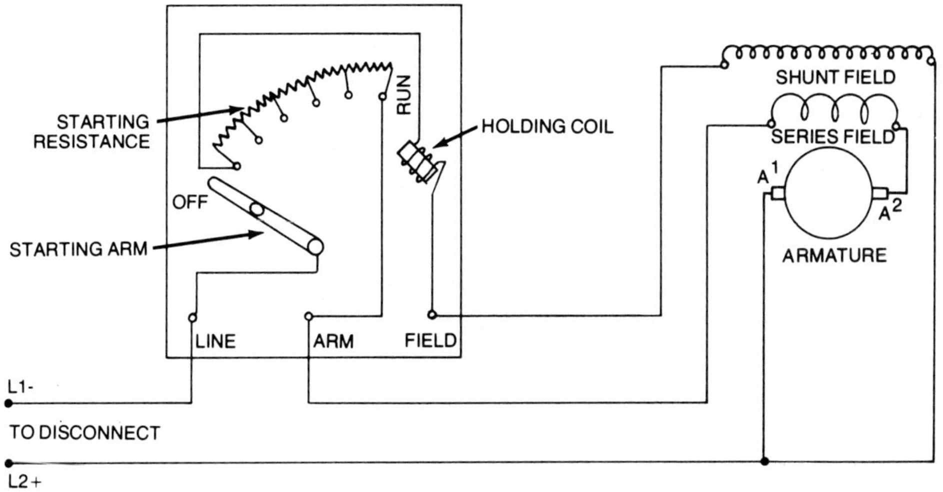

Fig. 20-43 Typical 3-point, manual dc motor starter with no-field release.

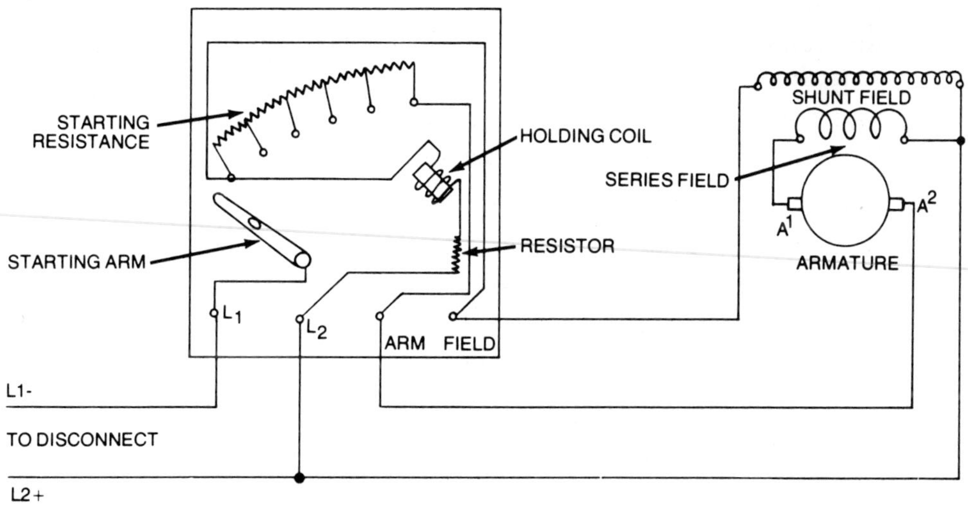

Fig. 20-44 Typical 4-point, manual dc motor starter with no-voltage protection.

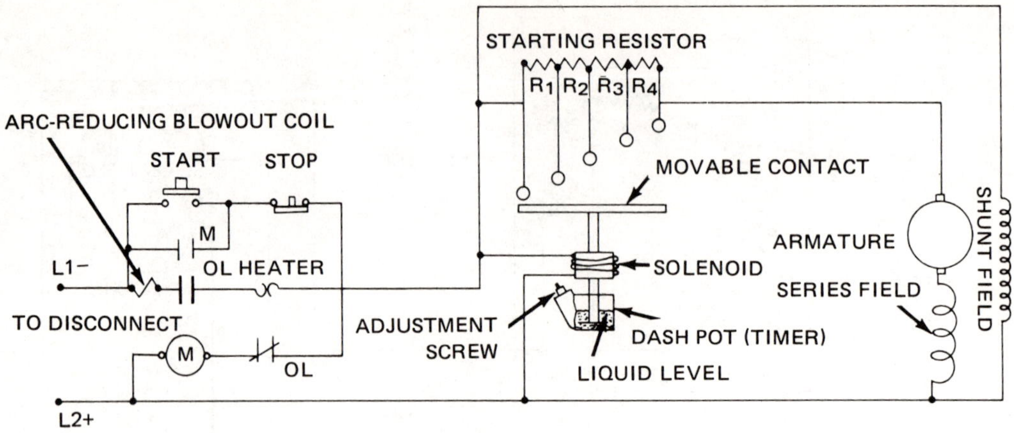

Fig. 20-45 Wiring diagram showing automatic dc motor starter.

REVIEW

1. Why do motors need special control switches?
2. How does the voltage generated in a motor affect the input current?
3. Define *locked-rotor current.*
4. What two basic methods are used for starting motors? What are the advantages of each method?
5. Why are motor starting switches rated in horsepower (watts or kilowatts)?
6. According to the National Electrical Code, where should control switches be located?
7. How is the fuse size for a motor determined? What effect does the fuse size have on the type of disconnect switch installed?
8. Explain how motor conductor sizes are determined.
9. What three types of relay are used to give thermal overload protection? Explain briefly how each works.
10. What two factors determine the size of the overload relay? Explain in your own words two methods for determining relay size for a motor.
11. What precaution must be taken when restarting a motor after an overload has tripped the relay?
12. List the three main advantages of magnetic motor control.
13. List the seven sizes, or classes, of magnetic motor starters and the motors each can control safely.

14. What is the purpose of the shading coil in a starter?
15. What two circuits make up a magnetic motor starter?
16. Draw a schematic diagram of the control circuit used when four start/stop stations are to be connected to the magnetic starter.
17. What is the purpose of the holding (maintaining) contacts? How do they work?
18. List three types of control stations that may be used with magnetic motor starters.
19. Why are three overload relays required in a three-phase starter?
20. How does an overload relay used with a magnetic motor starter differ from a relay used on a manual starter?
21. What is the purpose of the jogging circuit?
22. What danger exists when using a push button jog circuit?
23. Explain how a three-phase motor is reversed.
24. What is the purpose of the interlock on a reversing controller?
25. What might happen if a large motor is reversed too quickly?
26. What is the difference between a dual-speed motor controller and a standard reversing controller?
27. What is the purpose of the timer relay in a reduced-voltage starter?
28. Explain briefly how the primary resistor starter and the autotransformer work.
29. What are the advantages of using the two types of reduced-voltage starters?
30. Explain briefly how a single-phase motor can be reversed.
31. What problem do both dc and ac motors have when starting?
32. Define *no-field release.*
33. Why is no-field release protection important when using a dc motor?
34. Define *no-voltage protection.*
35. Why is an automatic starter safer than a manual motor starter when controlling dc motors?

Chapter 21

Fastening Devices

USES OF FASTENING DEVICES

As in most building construction areas, the electrical trade relies heavily on fastening devices to mount boxes and panels, to support conduits and cables, and to secure the many fittings associated with the trade. Fastening devices are available in many forms for the support of electrical equipment on wood, steel, and all types of masonry surfaces. Due to the variety of materials that the fasteners must penetrate, they are designed to accommodate either a wood- or a machine-type screw. To make the best use of the fastening systems available, an understanding of wood screw and machine screw features is desirable.

SCREW FASTENERS

The following parts of a screw fastener are important to the installer and will be discussed: head, driving configuration, neck and/or shoulder, shank and body, thread, and point. It should be noted that with screws given numbered size designations, the higher the number, the larger the physical size of the screw.

Head Design. The enlarged, preformed shape on one end of the screw is known as the *head.* Heads are shaped to meet the many requirements of the electrical industry, as well as others, Figure 21-1. The oval and flat-undercut heads are designed to allow a semiflush or flush fit with the surface of the object being held, (for example, switch or receptacle cover plates). The other types of heads are for use where it is desirable to have the entire head of the screw on the surface being held (for example, the cover on an octagon box or square box). The head of any threaded fastener is the *bearing surface* and supports the load.

Driving Configuration. To meet the many installation situations that are encountered, many types of tools and/or drivers are available for the screw fasteners. Figure 21-2 illustrates the various driving configurations produced in the heads of screws.

The *slotted* type of driving configuration is an old standard, designed for the common screwdriver found in most home or shop areas. It is produced in many sizes to meet the requirements of the various screw sizes.

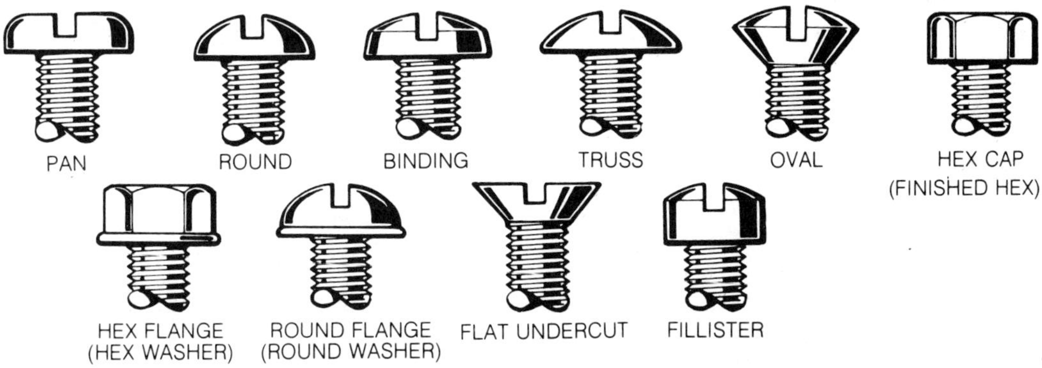

Fig. 21-1 Common head designs for screw fasteners.

The electrician's tool kit would not be complete without a set of *square recess* (commonly known as *Robertson head*) screwdrivers. This form of driving configuration is patented under many trade names, including *Scrulox.* It is produced in six sizes ranging from No. 4 (largest) to No. 00 (smallest).

A color-coded handle indicates the size of the driver. The No. 4 black-handled driver is used for No. 16 gauge screws and larger. No. 3 black is somewhat more common and is used for No. 12 and No. 14 gauge screws. No. 2 red drivers install No. 8, No. 9, and No. 10 gauge screws. No. 1 green is used for No. 5, No. 6, and No. 7 gauge screws. No. 0 yellow installs gauges No. 3 and No. 4. The No. 00 orange-handled driver is used for the very small No. 1 and No. 2 gauge screws.

The square recess design has two very desirable features. The square recess for the driver is tapered slightly and causes the screw to "cling" once it has been placed on the tip of the driver. This allows the installer to use one hand on the driver/screw combination, leaving the other hand free to support the equipment being installed. Heat treating (hardening) along with proper recess design produces an excellent driving

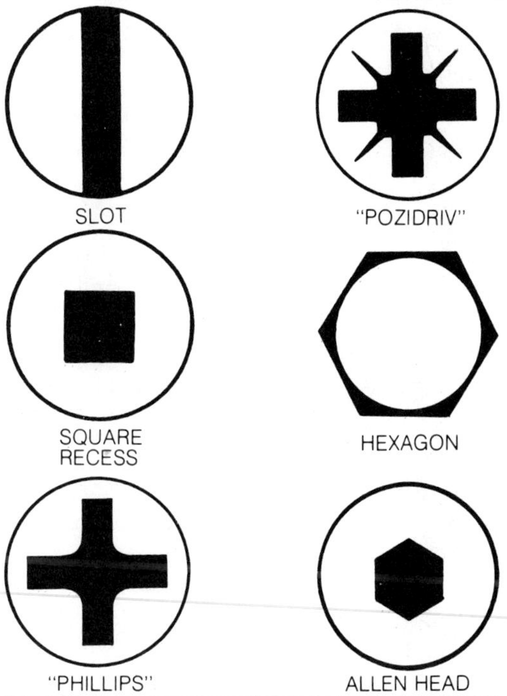

Fig. 21-2 Common head configurations for screw fasteners.

configuration, thus reducing the frequency of "cam-out." *Cam-out* is the action between driver and recess that causes the driver to disengage from the recess in the head. The square recess driving configuration is very suitable for use with mechanical or power-operated screwdrivers.

Phillips-type screws and screwdrivers are used extensively for appliance assembly. This type of driving configuration permits the use of air- or electric-powered driving tools, thus speeding up assembly-line procedures. Phillips-head screwdrivers are available in sizes similar to square recess drivers, but they are seldom color coded. The *Pozidriv* unit is designed primarily for use with power-driven tools.

Hexagon-shaped heads are found on most of the larger machine screws, and a wrench is required to tighten them securely without damage to the head. Socket wrenches or specially designed nutdrivers (screwdriver handle and shaft with a socket at the tip) are very useful in hard-to-reach areas.

Hexagon recess heads, frequently referred to as *Allen head* screws, make use of an L-shaped wrench or key to tighten them. They are frequently used as "grub screws" to hold a motor pulley to the shaft and in the assembly of mechanical devices or frameworks. The fastening tools, known as *Allen keys,* are available in sets of different sizes to cover the broad range of screws equipped with this type of driving configuration.

Neck and/or Shoulder. The area directly under the head of a screw is frequently given a special design treatment. Figure 21-3 illustrates a number of *neck* and *shoulder* designs used to prevent rotation of the screw while a nut is being tightened on the bolt. This special design feature usually appears on screws that rely on the tightening of a nut to secure them, rather than on some form of driver configuration in the head.

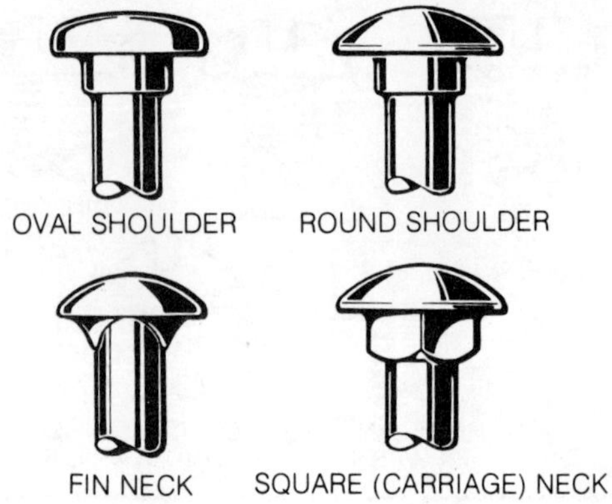

Fig. 21-3 Typical neck and shoulder designs for screw fasteners.

Units of this design are very useful in areas where access to both ends of the screw is impossible or (for security reasons) to prevent removal of the screw from the outside (for example, on the hinges of a material/supply box).

Shank and Body. The length of the screw from under the head to the tip is known as the *shank.* Any unthreaded portion of the shank is called the *body* of the screw. Long screw fasteners often have thread only at the tip area, leaving a long body. Shorter screws are frequently threaded over the entire length of the shank, leaving no body at all. Production time and costs are reduced, and stronger upper portions of the screw are produced, when only the necessary amount of thread is formed on the shank.

Thread Types. A variety of thread types is available, each designed to fasten securely into a particular building material. Some of the more common thread types are as follows.

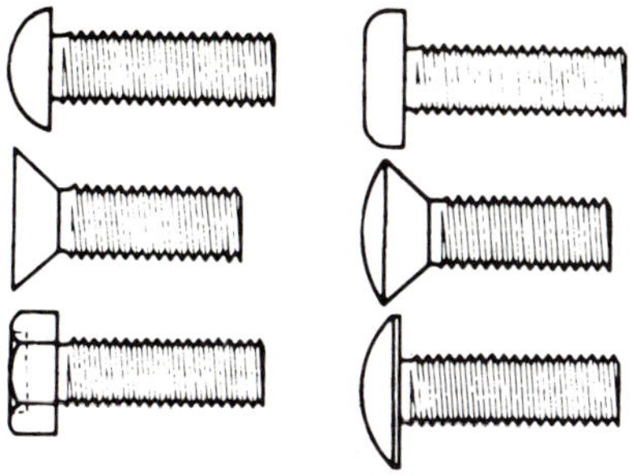

Fig. 21-4 Typical small-diameter (Gauge No.) machine screws.

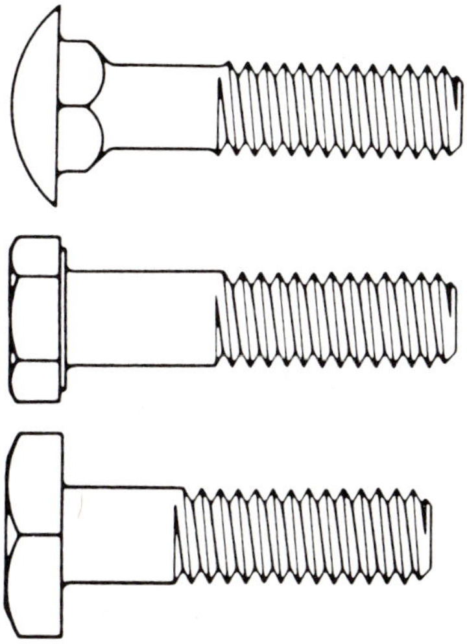

Fig. 21-5 Typical large-diameter machine bolts.

Machine screws. Machine screws are bolt-and-nut units designed primarily to join metal to a variety of other materials. They are produced in many thicknesses and thread-pitches, depending upon the required amount of support strength and/or compression between surfaces. The small-diameter machine screws (having a gauge number) are available with a variety of head shapes and designs, as shown in Figure 21-4. Large-diameter units, frequently called *bolts,* are produced in three basic head shapes, as shown in Figure 21-5.

Most machine screw and/or bolt sizes are produced with either coarse or fine threads. The coarse-thread bolt installs faster, since the nut advances along the bolt *(thread pitch)* a greater distance for each complete turn. Some manufacturers of electrical panelboards use coarse-thread bolts for quick fastening of the panel covers. Fine-thread units require more turns of the nut to tighten them, but excellent compression is obtained between the surfaces being joined. Table 21-1 compares common, unified (coarse- and fine-thread) screw sizes with their metric equivalents. *Caution:* Never mismatch metric screws with Imperial (inches) nuts or tapped holes. Instead of producing a metric screw that exactly matches its Imperial measure (inches) counterpart, producing awkward dimensions (decimals), manufacturers have established a new set of popular metric sizes. The new screw sizes have simplified metric dimensions that are reasonably close to the inch units they will replace. Table 21-2 offers a comparison between the metric and the English dimensions for fasteners.

For proper installation, machine screws require either a clearance hole to be drilled or a prethreaded hole in the materials being joined. The internal threads of the hole or nut being used to secure the assembly must match the external threads of the bolt or machine screw to prevent damage to either part. Typical machine screw nuts are shown in Figure 21-6.

Self-tapping screws. Self-tapping screws are made of low-carbon, heat-treated (hardened) steel, and are available in five basic types.

TABLE 21-1 Comparing common coarse- and fine-thread screw sizes.

COMMON MACHINE SCREW SIZES

Figures in this table are based on *Unified* screw thread sizes as established by the American National Standards Institute.

Unified Screw Threads				Metric Screw Threads
Coarse		Fine		
no.	tpi.	no.	tpi.	Screw Thread
2	56	2	64	M2.2×0.45
3	48	3	56	M2.5×0.45
4	40	4	48	M3×0.5
5	40	5	44	
6	32	6	40	M3.5×0.6
8	32	8	36	M4×0.7
10	24	10	32	M4.5×0.75
12	24	12	28	M5×0.8
diam.	tpi.	diam.	tpi.	
1/4	20	1/4	28	M6×1
5/16	18	5/16	24	M7×1
3/8	16	3/8	24	M8×1.25
7/16	14	7/16	20	M10×1.5
1/2	13	1/2	20	M12×1.75
9/16	12	9/16	18	M14×2
5/8	11	5/8	18	M16×2

Example: to convert a 12-24 machine screw to metric, select M5 × 0.8.
Caution: never mismatch metric screws with Imperial (inches) nuts or tapped holes.

1. *Thread-forming* screws, Figure 21-7, reshape the material in the pilot hole and do not remove any of the surrounding material. They provide an excellent fit and fast assembly when joining metal to metal.
2. *Thread-cutting* screws are designed with cutting edges and chip cavities to remove material as they are being installed, Figure 21-8. They are used in thick, brittle, or granular materials where thread-forming screws are not suitable.
3. *High-performance thread-forming* screws, Figure 21-9, engage approximately 30 percent more efficiently than ordinary self-tapping screws, and 60 percent more efficiently than machine screws in tapped holes. Figure 20-10 illustrates how the tight-fitting, thread-forming screw with its thread clearance fits the material much more securely than the tapped hole and machine screw. Thread stripping and screw breakage are virtually eliminated by the well-designed projections (spaced 120° apart at the tip) which form an accurate mating thread in the screw-supporting material. These screws are made for assembly-line production.
4. *High-performance thread-cutting* screws, Figure 21-11, are particularly suited for use with brittle mate-

TABLE 21-2 Comparison guide for popular metric and inch screw fastener sizes.

DIAMETERS

The range of diameters listed in metric standards is from M1.6 to M100. Wherever possible designers are asked to use the stock and preferred sizes listed here.

Metric diameter	Nearest Inch diameter
M2	#2
M2.5	#3
M3	#5
M3.5	#6
M4	#8
M5	3/16''
M6	1/4''
M8	5/16''
M10	3/8''
M12	7/16''
	1/2''
M14	9/16''
M16	5/8''
M20	3/4''
M24	1''
M30	1 1/8''
M36	1 1/4''
	1 3/8''
M42	1 1/2''
M48	1 3/4''
	2''
M56	2 1/4''
M64	2 1/2''
M72	2 3/4''
M80	3''
M90	3 1/2''
M100	4''

LENGTHS

The metric lengths given are the preferred lengths and these should be used wherever possible. Some of the shorter lengths will not be available in the larger diameters.

If lengths over 200 mm are required then increments of 20 mm should be used and lengths over 300 mm increments of 25 mm should be used.

Metric length	Nearest Inch length
10 mm	3/8''
12 mm	1/2''
16 mm	5/8''
20 mm	3/4''
25 mm	1''
30 mm	1 1/4''
35 mm	1 3/8''
40 mm	1 1/2''
45 mm	1 3/4''
50 mm	2''
55 mm	2 1/4''
60 mm	
65 mm	2 1/2''
70 mm	2 3/4''
75 mm	3''
80 mm	3 1/4''
90 mm	3 1/2''
100 mm	4''
110 mm	4 1/2''
120 mm	4 3/4''
130 mm	5''
140 mm	5 1/2''
150 mm	6''
160 mm	6 1/4''
170 mm	6 1/2''
180 mm	7''
190 mm	7 1/2''
200 mm	8''

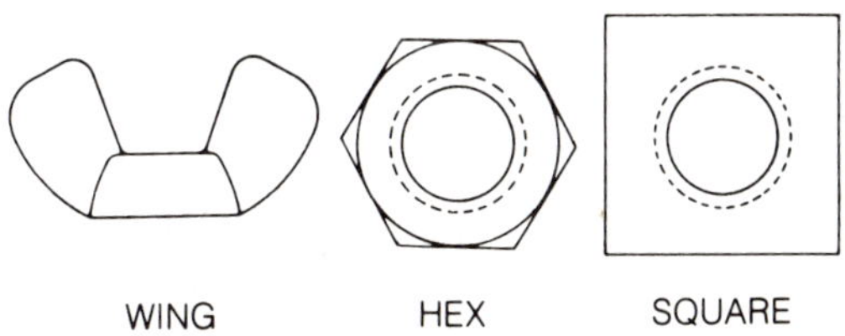

Fig. 21-6 Machine screw nuts.

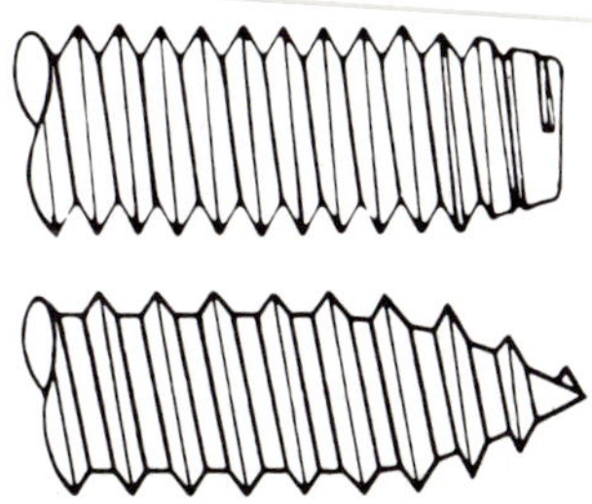

Fig. 21-7 Thread-forming screws.

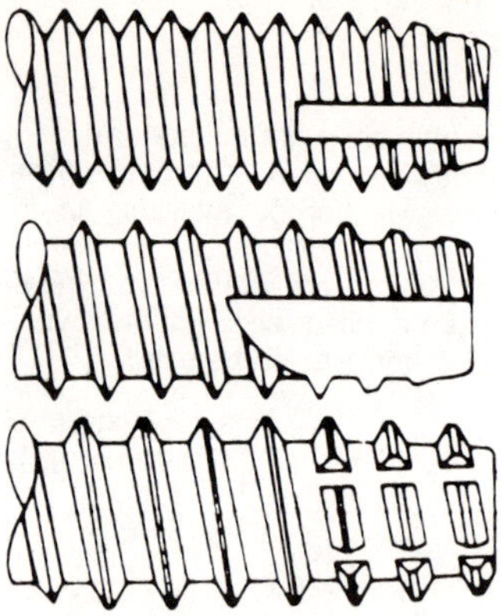

Fig. 21-8 Thread-cutting screws.

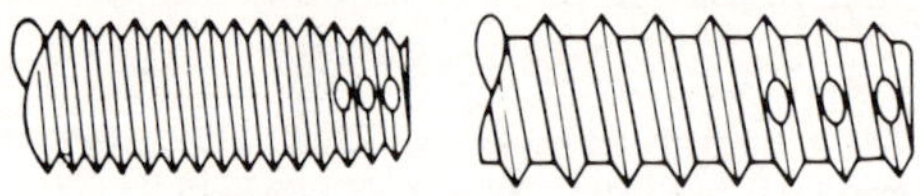

Fig. 21-9 High-performance thread-forming screws.

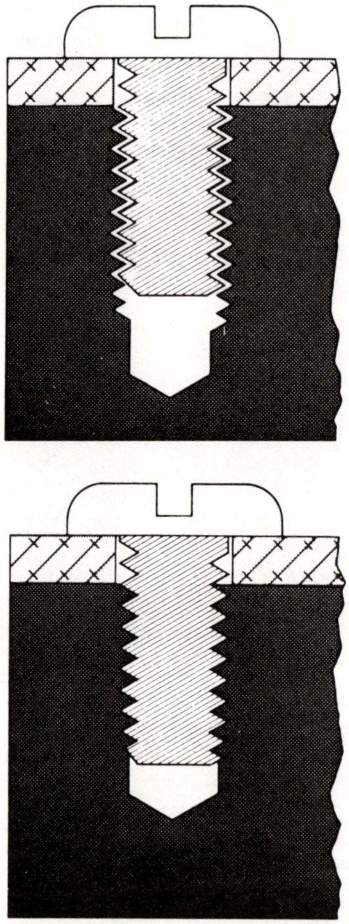

Fig. 21-10 Machine screw installation (top) and typical thread-forming screw installation (bottom).

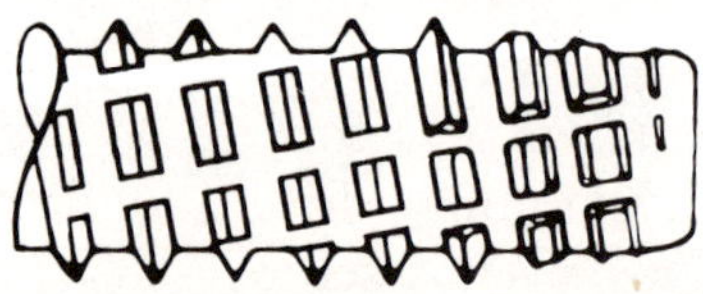

Fig. 21-11 High-performance thread-cutting screw.

rials. Chips are quickly removed, reducing stripping or cracking hazards. These screws provide quick and easy assembly.

5. *Metallic-drive* screws, Figure 21-12, are designed for permanent fastenings. They are pressure driven (punch press or hammer) into the holding material and are well suited for attaching nameplates, covers, and so forth.

Wood screws. Wood screws are used in many different areas of the construction field, and are available in four basic thread configurations.

1. *Single-lead* (thread) tapered wood screws have been in use for many years, Figure 21-13. The tapered neck and shank design, however, caused frequent splitting problems when installed in wood. The amount of torque required to install the screw increases greatly as it penetrates deeper into the wood. As a direct result of this increase in torque, more stress is placed both on the screwdriver and on the installer.

Fig. 21-12 Metallic-drive screw.

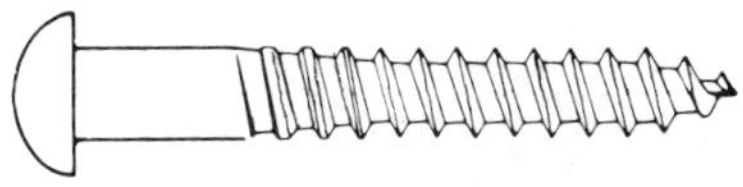

Fig. 21-3 Single-thread, tapered wood screw.

2. The *double-lead* (thread), *fast-spiral-thread* screw is a more recent development. Produced under various trade names (such as *Kwixin* and *Twin-fast*), this type of screw has a neck and body thickness that is less than the thread diameter. This feature virtually eliminates the splitting problem and reduces driver/installer stress significantly. The sides of the double-threaded shank are parallel, providing greater holding power and faster penetration. Figure 21-14 shows a comparison between single- and double-lead wood screws. The sharp point on this screw reduces the need for predrilling and permits easier starting of the screw.
3. Fastening to particle board, plastic, and similar materials often presents a powdering or stripping problem in the material to which the screw is being fastened. A specially designed screw, with wider thread spacing, sharper and deeper threads, and a well-designed self-starting point is available. One manufacturer has named this product the *Lo-root* screw. It provides approximately 70 percent greater thread engagement with the material. Figure 21-15 illustrates this thread type and the eight-point driver configuration that makes this screw convenient for use with power screwdrivers.

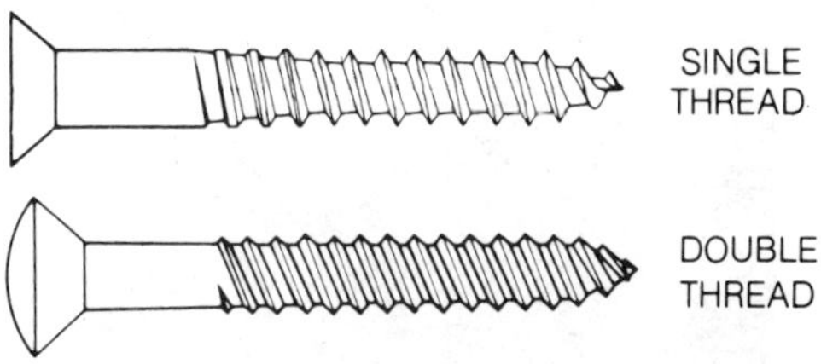

Fig. 21-14 Comparison between single-thread and double-thread screws.

Fig. 21-15 Typical Lo-root screw and driver configuration.

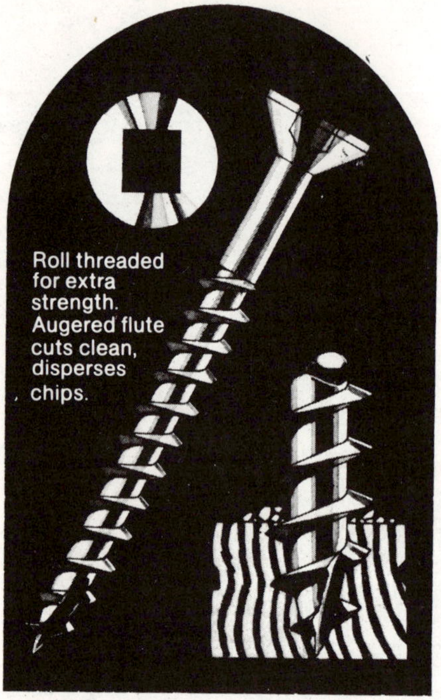

Fig. 21-16 Typical augered flute-tip screw and driver configuration.

4. A wood screw for use in hardwoods and kiln-dried woods is available with a special *augered flute* in the tip that cuts and clears the wood fibers during installation. This eliminates the need for predrilled holes and speeds up installation considerably. Figure 21-16 illustrates this type of screw and its driver configuration. It is a fast-starting screw, suitable for use with power drivers and is produced by one manufacturer under the trade name *Candril* screw.

Length of Screws. The length of a wood screw is determined by the distance between the head and tip of the screw, Figure 21-17. Machine and sheet metal (self-tapping)

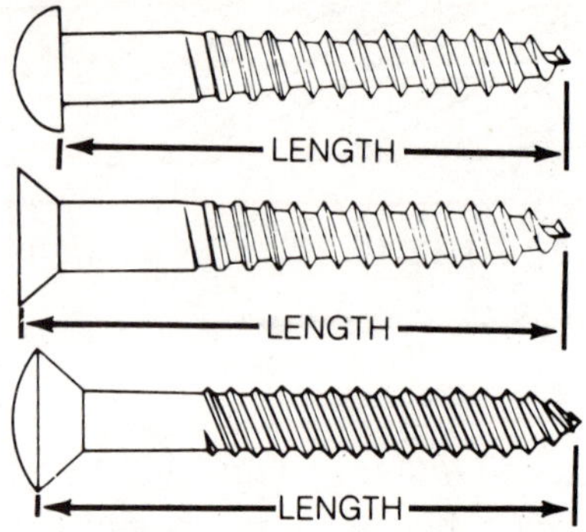

Fig. 21-17 Method of determining screw length.

screws are sized in a like manner. Screws of each thread type are produced in various lengths to meet construction needs and are available in both Imperial and metric measurements. Round-headed wood screws do not measure full length, but may be from 1/16 inch to 3/16 inch short. For example, a No. 4 by 1/2-inch wood screw measures 7/16 inch in length; a No. 20 by 2-inch wood screw measures about 1 7/8 inches under the head.

Screw Diameter. The thickness, or diameter, of a screw is indicated by a gauge number ranging from 0 to 24. The higher the gauge number, the thicker the screw. Large, heavy-duty screw-thread units are available, Figure 21-18. These *Lag bolts,* as they are called, are used extensively in expansion-type masonry fasteners, and are available in standard machine bolt diameters and lengths.

Point Design. The tip or point of a screw is designed to perform a variety of operations. Figure 21-19 illustrates a selection of point designs. *Gimlet* and *pinch* points are used for the penetration of material. *Conical, pilot* (dog), and *flat* (plain) points are used

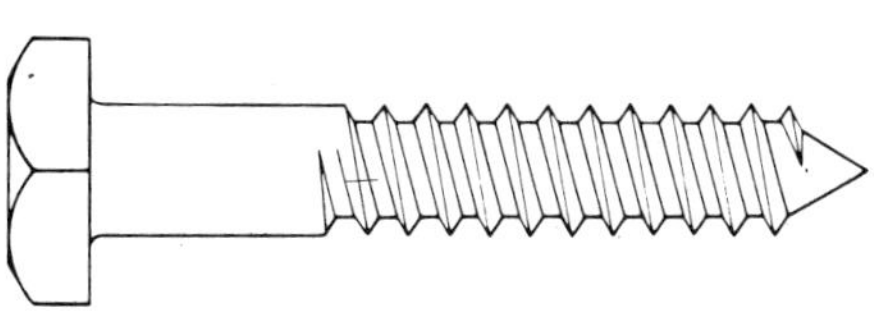

Fig. 21-18 Typical Lag screw.

to assist in the alignment of the parts being secured, while *spherical, header,* and *chamfer* points ease the insertion and starting of the screw thread.

Screw Construction Materials. Many types of metals and alloys are used to produce the great variety of screws available. Some of the most commonly used metals are discussed.

Aluminum is used to produce a screw suitable for use with other aluminum products where chemical (galvanic) action between the screw and the material could take place. If other than aluminum screws were to be used, severe damage could result to the material. Aluminum screws do not have a great deal of strength, and care must be taken during installation not to twist them off.

Brass is used a great deal where corrosion from the elements (such as rain, snow, air, and saltwater) could be a problem. It is easily plated (with chrome, for instance) by the manufacturer for further corrosion resistance and/or improved appearance. Silicone bronze is similar to brass in strength and corrosion resistance.

Monel, a nickel alloy, and stainless steel are used where strength and extra corrosion resistance are required. Steel is by far the metal most commonly used in the manufacture of screws. It is strong and easily worked, and can be plated for resistance to corrosion.

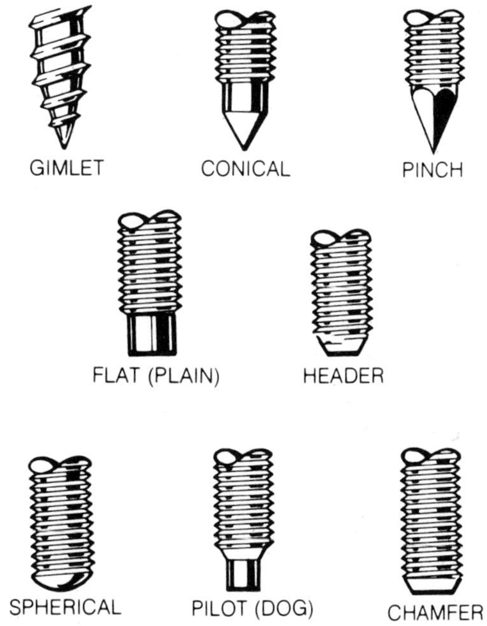

Fig. 21-19 Various point designs for screws.

Zinc or cadmium plating is applied over the steel to protect it from corrosion. Cadmium is the better of the two coatings.

Bolt Strength. Steel hex-head machine bolts, or *cap screws* as they are often called, are produced in a number of different tensile strengths, or S.A.E. (Society of Automotive Engineers) grades. The grades range from Grade 1 through Grade 8, and the user is advised of the potential holding power of each grade of bolt. A simple pattern of ridges on the head forms a grade identification mark, indicating the S.A.E. rating. Table 21-3 illustrates typical markings, along with size, load, strength, and hardness ratings.

TABLE 21-3 Typical markings and mechanical properties of hex head cap screws.

STRENGTH GRADES

Metric (ISO 898) | **Inch** (SAE J429)

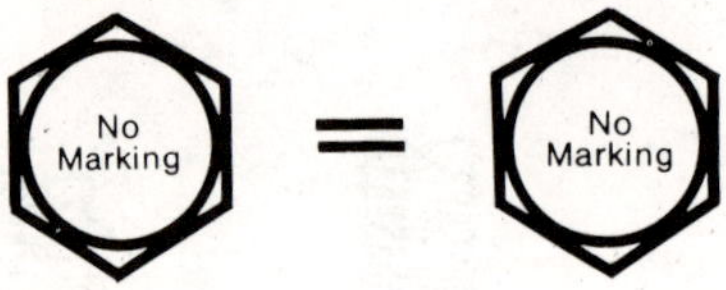

Grade: 4.8 (4.6, 5.8)
Tensile: 420 MPa

Grade: 2
Tensile: 60.000 psi

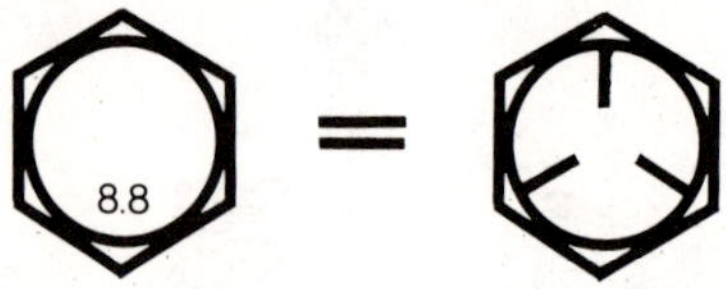

Grade: 8.8
Tensile: 830 MPa

Grade: 5
Tensile: 120.000 psi

NOT IN GENERAL USE

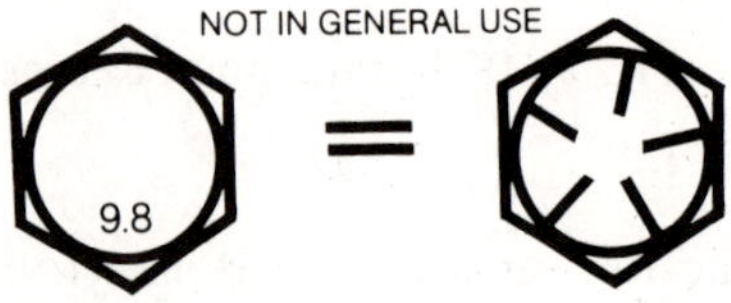

Grade: 9.8 (to 16 mm only)
Tensile: 900 MPa

Grade: 7
Tensile: 133.000 psi

Metric (ISO 898) | **Inch** (SAE J429)

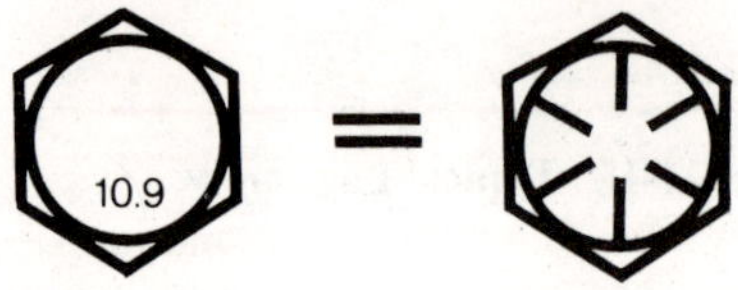

Grade: 10.9
Tensile: 1040 MPa

Grade: 8
Tensile: 150.000 psi

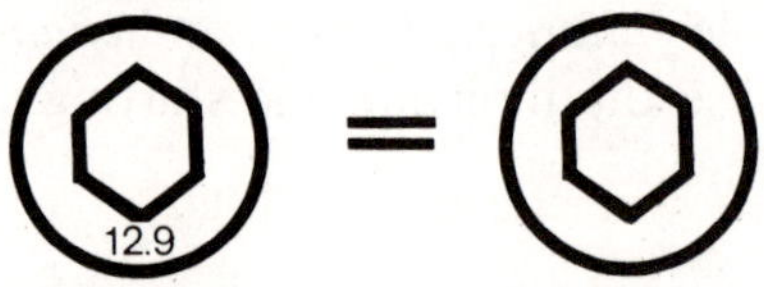

Grade: 12.9
Tensile: 1220 MPa (176.900 psi)
Note: Metric Hex Socket Cap Screws are available in lower strength grades (8.8, 10.9) and marked accordingly.

Grade: ASTM A574
Tensile: 170.000 psi
Note: Generally not marked

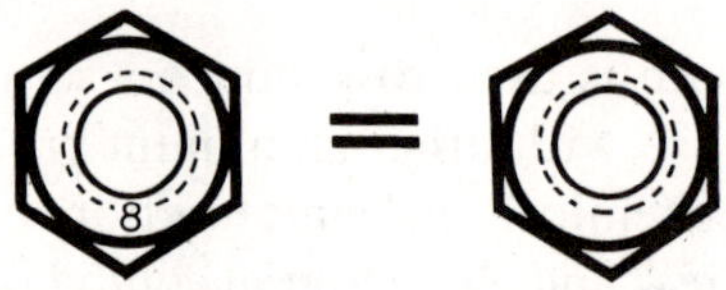

Grade: Corresponding to bolt
Marking: Grade 8 and higher
Note: Always use equal or higher nut grade for assembly with respective bolt grade.

Grade: Corresponding to bolt
Marking: Nuts for use with bolts SAE Grade 8 and higher only are marked (REF SAE J995 or ASTM A563).

Quantity	**Metric Symbol**	**Conversion (Approx.)**
Stress	MPa (megapascal)	1 MPa = 145 psi 7 MPa = 1000 psi
Force	N (newton)	1 N = 3.6 ozf 4.4 N = 1 lbf
Torque	N·m (newton metre)	1 N·m = 9 in/lb 1.4 N·m = 1 ft/lb 1 N·m = 0.75 ft/lb

NOTE: The term kilogram meters (kg·m) may be encountered and is comparable to 10 N·m (2% deviation). (Use: kg·m = 10 N·m)

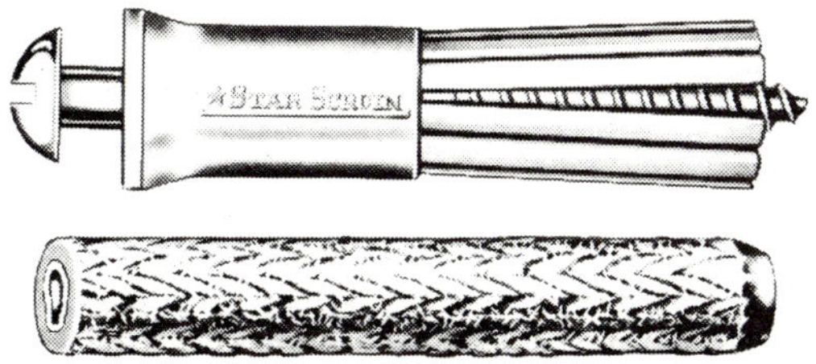

Fig. 21-20 Screw anchors; lead (top) and jute fiber (bottom).

Grade 1 and Grade 2 bolts can be used for the simple mounting and/or fastening of equipment where strain and load are not severe. The mid- and high-grade units are used for heavy loads where stress and strain are greater, thus preventing bolt and thread failure. As the grade number increases, the carbon steel of the bolt is subjected to a more intensive heat/quench (hardening) process.

MASONRY FASTENERS

Due to the extensive use of masonry materials (such as concrete and brick) in both old and new buildings, a greater variety of fastening devices has been developed for use by the construction trades. Much of the electrical equipment installed must be fastened to or supported on masonry surfaces, with one or more of the following fastening systems. Note that wood inserts are *never* used in masonry or concrete. Anchors are *never* installed in vertical mortar joints, but *may* be installed in horizontal mortar joints.

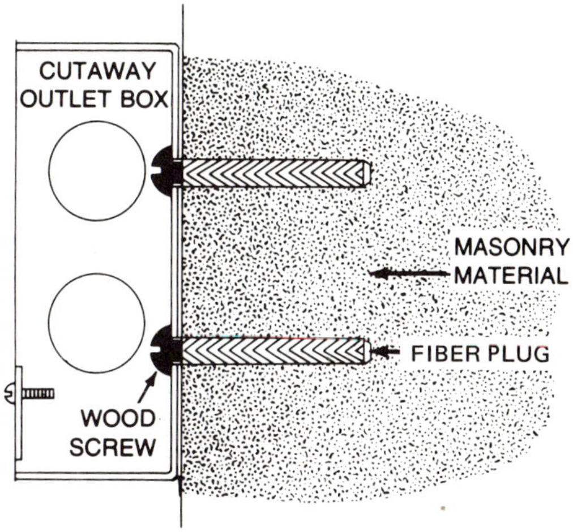

Fig. 21-21 Fiber screw-anchor installation.

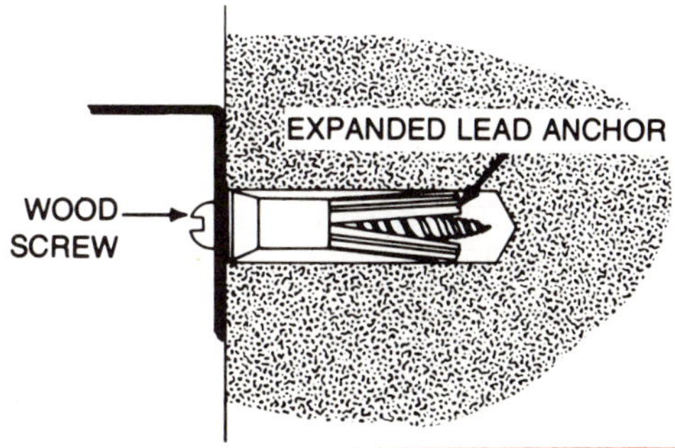

Fig. 21-22 Lead screw-anchor installation.

Screw Anchor. Screw anchors are available in jute fiber, lead, nylon, and plastic. This common fastening device is inserted into a predrilled hole in the masonry surface. A wood screw or self-tapping screw is then used to complete the fastening system. Figure 21-20 illustrates two of the more common screw anchors. Plastic inserts are used and installed similarly to the jute fiber screw anchor. The anchor used should be as long as the threaded shank of the wood screw (approximately 2/3 of the screw length). Screw anchors are also made in a variety of widths, to coincide with the gauge diameters of screws. Figures 21-21 and 21-22 illustrate typical fiber and lead screw-anchor installations.

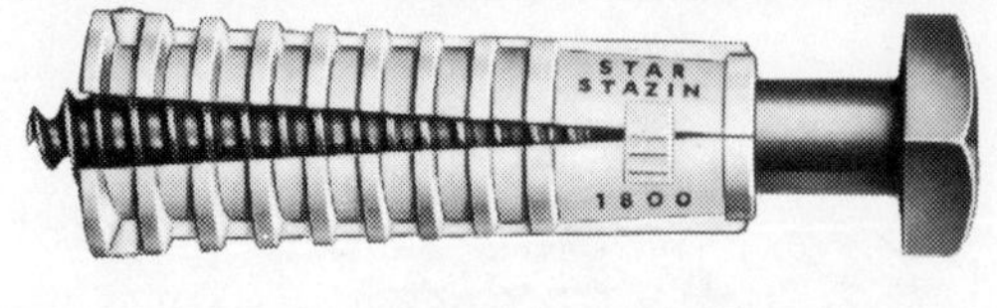

Fig. 21-23 Lag bolt expansion shield.

For a more rugged installation, a metal alloy expansion *(Lag)* shield is available, Figure 21-23. These units have been designed to accept the various-sized Lag bolts and are very useful where a larger support device is required for heavy loads.

Expansion Shields. Expansion shields are primarily designed for use with machine bolts. These zinc alloy units are produced in various lengths and diameters, with internal threads in common machine screw/bolt sizes. A hole of the correct diameter, as indicated on the expansion shield, must be drilled into the masonry for the full depth of the shield. The shield is then inserted and the machine bolt threaded into position. As the bolt is tightened, the threaded section of the shield is drawn towards the mounting surface, expanding the body of the shield against the sides of the predrilled hole. Care must be taken to use this shield in solid or firm masonry material as great pressure can be exerted on the sides of the hole as the shield is expanded. Cracking of the masonry and shield pullout can result if this precaution is not taken. Figure 21-24 illustrates two types of zinc alloy expansion shields.

Self-drilling Shield. A hardened steel expansion shield is produced that is capable of drilling its own hole, Figure 21-25. The body of the shield has a break-off, tapered end which is held in a power hammering device during installation. Once the hardened teeth at the tip of the shield have drilled the hole, a tapered steel plug is inserted into the tip. The shield assembly is then placed back in the masonry hole. Further pressure from the power hammer drives the tapered steel plug up into the shield, expanding the cutting tips and widening the bottom of the hole. This type of shield has extraordinary holding power due to the increased hole and shield diameter at the base of the hole. The tapered break-off neck of the shield can be removed easily once the shield has been fully installed. A bolt of a higher S.A.E. grade can be used with these shields, since the internal threads of the shield are also formed of hardened steel. Figures 21-26 and 21-27 illustrate the use of expansion shields and self-drilling shields.

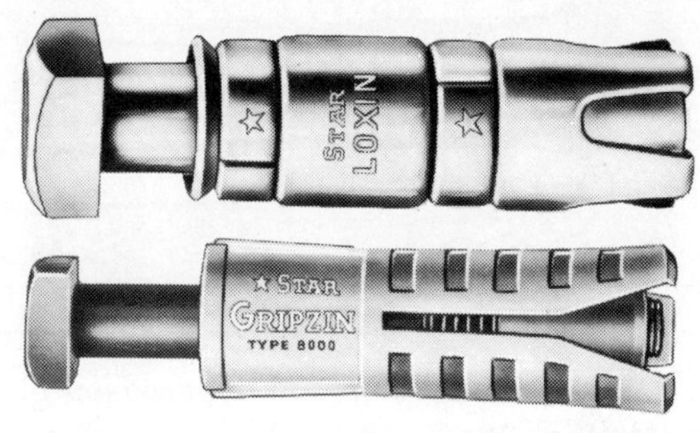

Fig. 21-24 Zinc alloy expansion shields.

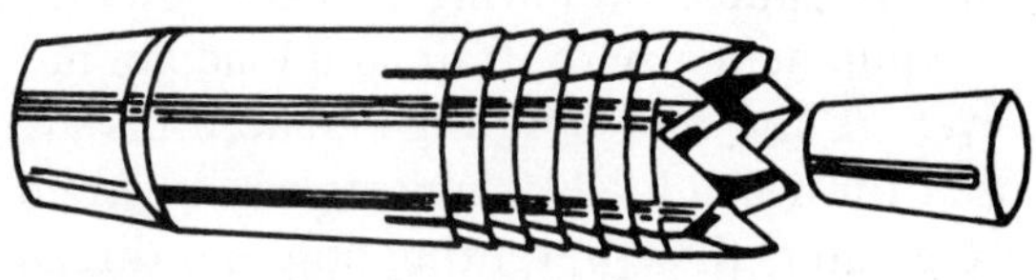

Fig. 21-25 Hardened steel, self-drilling, expansion shield and plug.

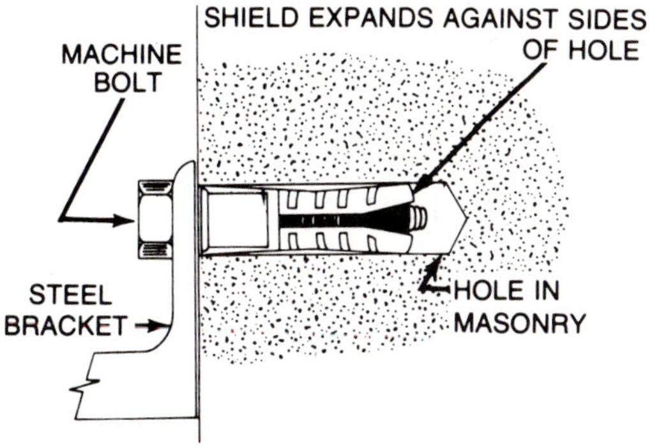

Fig. 21-26 Installation of expansion shield, zinc alloy type.

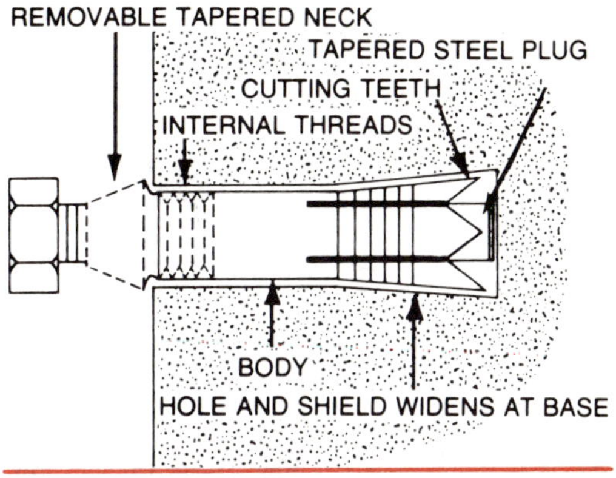

Fig. 21-27 Hardened steel, self-drilling expansion shield installation.

Lead Sleeve Anchor. The lead sleeve anchor is a simplified form of expansion shield, Figure 21-28. These units are placed in a predrilled hole and set into their final expanded position by a setting tool, Figure 21-29. (Make sure that the expansion end is placed into the drilled hole first.) Blows from a hammer provide the necessary force to expand the lead anchor into irregularities in the sides of the hole, Figure 21-30. An internally threaded, zinc-alloy cone assists in the expansion and accepts standard machine screws/bolts in a variety of sizes. The drive-pin's pilot (dog) ensures that no thread damage will occur. Sleeve anchors do not provide as much holding power as other types of fasteners, but are widely accepted for use in positioning machinery on a concrete floor. The anchor illustrated in Figure 21-30 uses a machine bolt instead of a threaded cone. The bolt is left protruding from the floor; the machine is then put into position and a standard nut placed on the threaded end of the bolt to tighten it in place.

Fig. 21-28 Lead sleeve anchors.

Fig. 21-29 Lead sleeve setting tool.

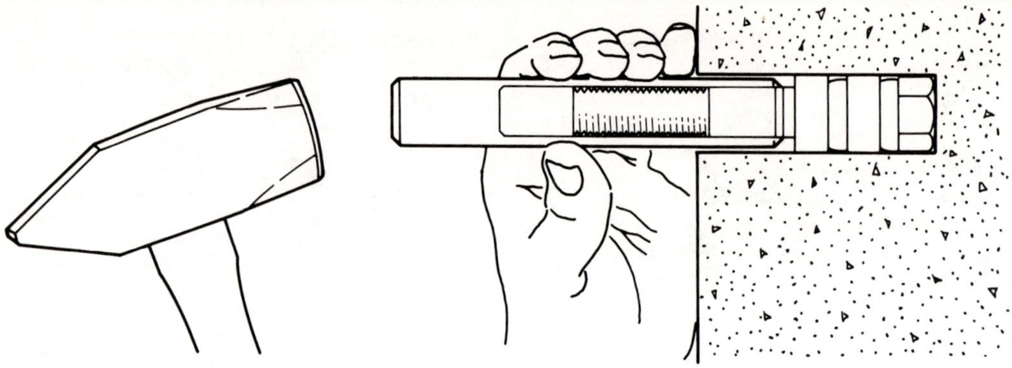

Fig. 21-30 Installation of a lead sleeve anchor and bolt.

Drive-in Anchors. Drive-in anchors are a basic type of expansion shield that is secured in the masonry material by a driven nail or pin. A hole must be predrilled in the mounting surface; the anchor is then placed through the device being supported and inserted into the masonry hole. A series of blows from a hammer drives the nail or pin into the shield, thus expanding the back portion of the shield against the sides of the hole. The shield itself supports the load with this type of fastener, since it passes through the device being supported. One example of this type of unit is illustrated in Figure 21-31. These units are normally produced in small diameters, 3/16 inch to 1/2 inch (5 mm to 13 mm), for light-duty applications.

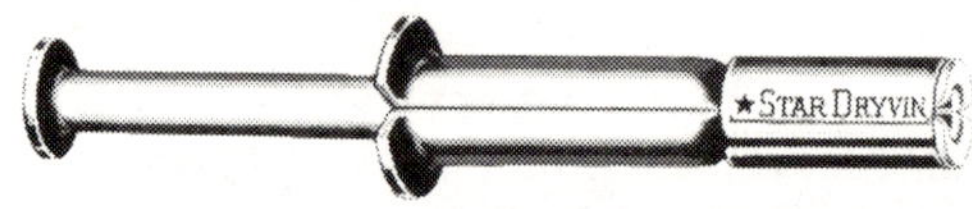

Fig. 21-31 Drive-in anchor.

DRILLING DEVICES

Hammer-driven Drills. In the past, holes in masonry surfaces were made with the use of a hand-held, hammer-driven, masonry drill. This manual drill is still used by installers, especially in areas where electrical power or air power is not readily available. The drill comes in two parts, as shown in Figures 21-32A and 21-32B. The twist drill holder has a rubber grip assembly to lessen the hammering vibrations during drilling and to protect the user's hand. The drill bits are formed of high-quality carbon steel. They are hardened, tempered, and sharpened at the tip to produce a relatively quick hole in concrete, brick, or stone surfaces. The shank is tapered at the end to fit easily into the drill holder. It is recommended that the drill unit be held perpendicular to the surface and rotated by the operator during the drilling process to prevent binding and eventual breakage of the drill bit in the hole. The drill bits are produced in diameter sizes up to 9/16 inch (15 mm) for use with small expansion-type shields.

Large holes also can be produced with a four-point drill, available in diameters up to 1 1/2 inch (40 mm). The four-point (Star) drills are also made in lengths of 15 inches (380 mm) and 18 inches (460 mm) where a deep hole must be made, or access to the

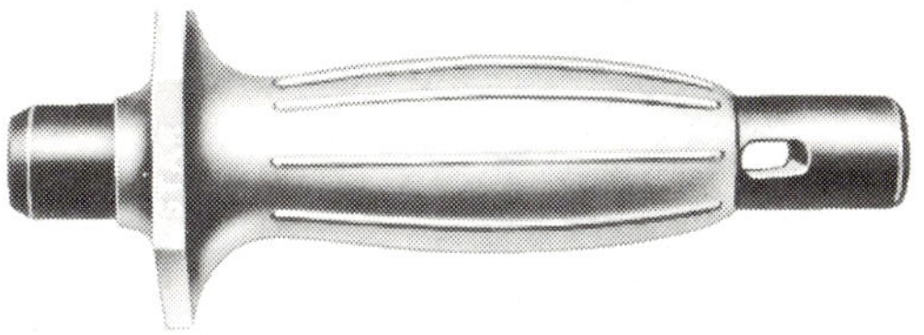

Fig. 21-32A Manual, rubber grip, masonry drill holder.

Fig. 21-32B Drill bit for masonry drill holder.

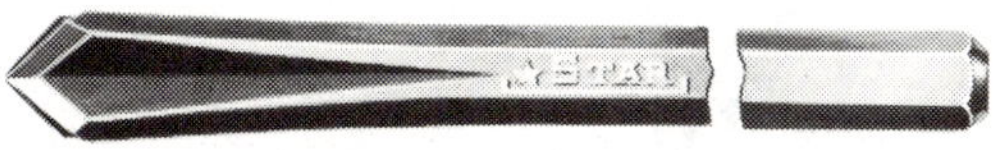

Fig. 21-33 Four-point masonry drill, hammer driven.

Fig. 21-34 Carbide-tipped, power-driven, masonry drill bits.

work surface is awkward. Figure 21-33 illustrates a typical four-point drilling device. A hammer is used to drive this drill into the masonry surface. This hardened-steel drill is for use on all masonry surfaces, and it should be rotated during use to remove rutted material and to prevent binding and/or jamming in the hole.

Power-driven Drills. Due to the popularity and availability of quality electric-drill motors, a masonry drill bit has been produced for use with these tools. Air-powered electric-drill motors can also use this type of bit for work on masonry surfaces. The shape of the bit closely resembles that of a standard steel drill bit. The cutting tip, however, is made from a small piece of carbide brazed (welded) into position. The carbide is an extremely hard metal, capable of cutting clean, fast, and accurate holes in most masonry surfaces. The hardness of the carbide tip makes it necessary to resharpen these bits on a special type of grinding wheel specially compounded for the purpose. Attempts to sharpen the bits on a standard grinding wheel will usually result in the wearing down or forming of grooves in the grinding wheel itself. The bits are produced in a variety of sizes, and examples of these are shown in Figure 21-34. The carbide bit is excellent for use with most of the small fastening devices. Figure 21-35 illustrates the use of carbide bits.

Large holes can be produced in masonry surfaces with a multilayer, carbide-tipped, spiral-flute drill bit. This drill bit has a hollow core, allowing clearance for chips and removal of larger pieces of the masonry material, Figure 21-36. The drill bits are available in assorted diameters and lengths to meet the needs of most large-hole installa-

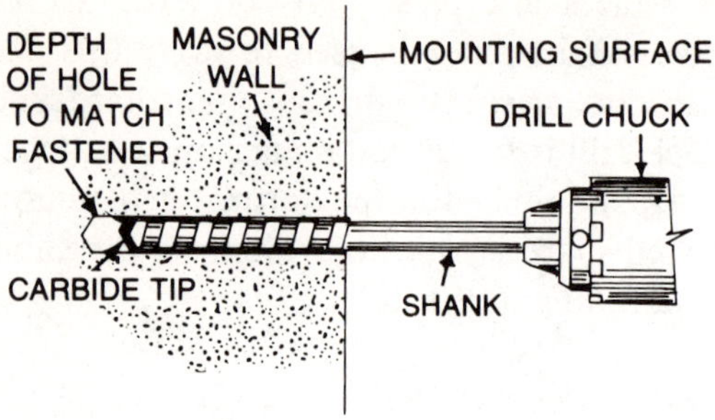

Fig. 21-35 Drilling a hole in masonry surfaces.

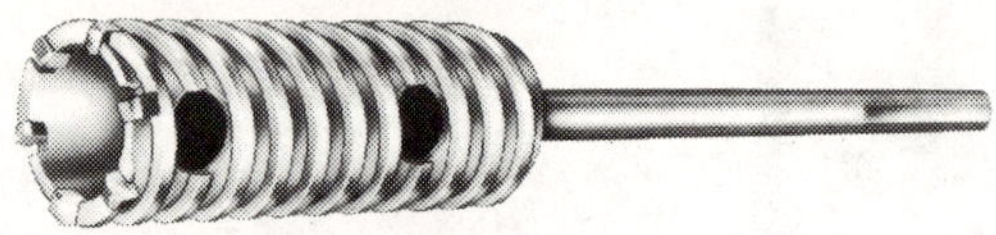

Fig. 21-36 Typical multilayer carbide-tipped masonry drill bit.

tions. Due to the cost of these drill bits, care should be taken during the drilling not to damage or remove the carbide cutting tips. Removal of the drill bit from the hole several times during the drilling operation will aid in chip clearance and cooling of the bit, and will prolong the life span of the bit.

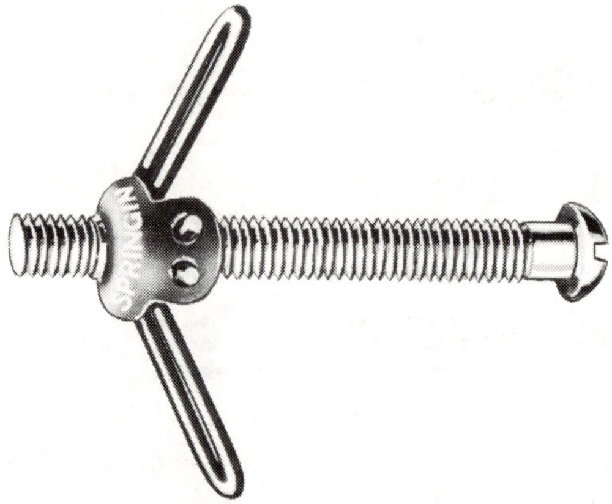

Fig. 21-37 Spring-wing toggle bolt.

Hollow-wall Fasteners. Many fastening operations are performed in residential buildings or other buildings where plaster materials are applied to the wall and ceiling areas. These surfaces are generally too thin and low in density to accommodate fasteners other than the small screw-type anchors. A series of mechanical fasteners is designed, however, to make use of the space behind these plaster or similar surfaces. Figure 21-37 illustrates a spring-wing, toggle-bolt fastener. The tempered steel wings are installed on the standard-thread machine screw after it has been passed through the device being mounted. The wings are then inserted into a predrilled hole in the mounting surface. Once clear of the back of the hole, the wings spring to an open position. Tightening of the machine screw draws the wings up against the inside surface of the mounting area, thus securing the mounted device. Figure 21-38 illustrates a spring-wing toggle-bolt installation. This type of fastener cannot be reused, since it is virtually impossible to remove the open, spring-wing portion from the space behind the wall. The machine screw, however, can normally be reclaimed and reused with a new wing unit. These units are available in a number of sizes and types for the mounting of equipment on hollow-wall surfaces.

Where it is necessary to remove and replace equipment on the mounting surface, a type of hollow-wall screw anchor that remains in place is used, Figure 21-39. The fastener fits into a predrilled hole in the surface, expands, and grips the back of the

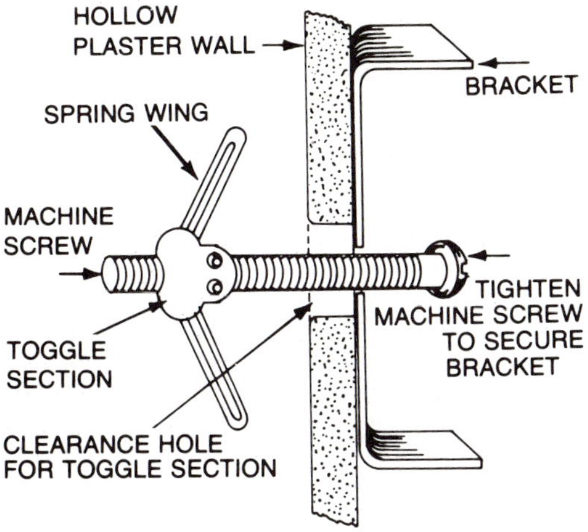

Fig. 21-38 Spring-wing, toggle-bolt installation.

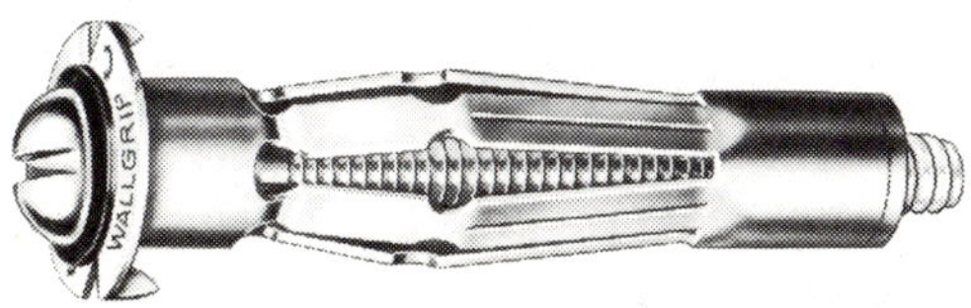

Fig. 21-39 Hollow-wall screw anchor.

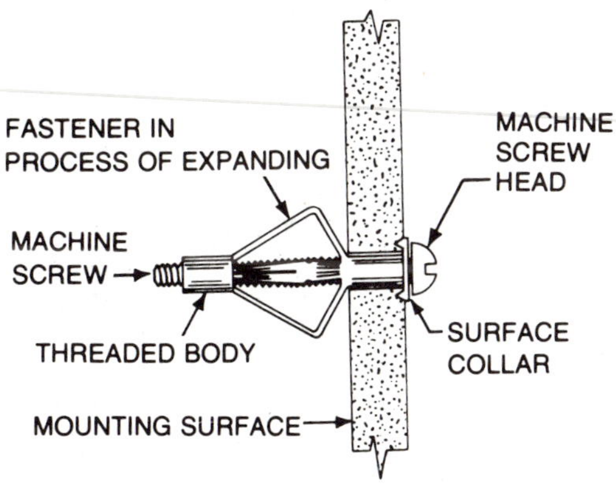

Fig. 21-40 Hollow-wall screw-anchor installation.

mounting surface when the machine screw is tightened. The machine screw can be removed and the complete fastener will remain in position for reuse. These units are produced in several sizes for use where a light load is to be supported. Figure 21-40 illustrates the installation of this fastening device.

POWDER-ACTUATED FASTENERS

The drilling of holes in masonry and/or steel surfaces is a tedious and often time-consuming operation. Labor versus fastener costs must be considered when choosing a fastening system. Gun-type powder-actuated fastening systems are available that speed up fastening time and greatly reduce the physical strain on the installer. These systems release a great deal of energy at the time of firing, making it necessary fully to train competent operators. Manufacturers of powder-actuated equipment are most willing to train and advise installers in the use of their product and its related equipment. In many areas, operators of this type of fastening equipment are tested, and issued operator's licenses upon successful completion of the training program, Figure 21-41. Powder-actuated tools are divided into two main categories: low- and high-velocity systems.

Low-velocity Equipment. Low-velocity tools operate at 300 ft/s (90 m/s) or less, and are suitable for fastening to most masonry or steel surfaces. They are the safest form of powder- (explosive-) actuated tool available, due in part to the fact that both piston and fastener must be accelerated by the powder charge. If at any time the fastener should pass through the receiving surface, its speed and energy will not be sufficient to endanger human life. These tools have a somewhat higher recoil than the high-velocity tools but are quieter in operation. Due to the low velocity of the fastener, the tool does not have a controlled fire angle. This feature is most necessary, however, on the high-velocity tools. (The tool must be flat against the surface being penetrated in order to fire the charge.) A leveling device is available as an option with the low-velocity tools and will provide fire-angle control, if desired. A well-designed buffer system prevents overtravel of the drive mechanism if and when improper charges are inserted. Figure 21-42 illustrates the mechanism of a typical low-velocity tool.

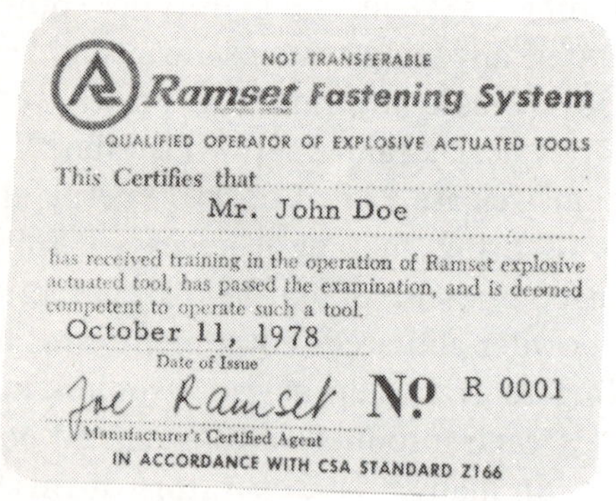
NOT TRANSFERABLE
Ramset Fastening System
QUALIFIED OPERATOR OF EXPLOSIVE ACTUATED TOOLS
This Certifies that
Mr. John Doe
has received training in the operation of Ramset explosive actuated tool, has passed the examination, and is deemed competent to operate such a tool.
October 11, 1978
Date of Issue
Joe Ramset Nº R 0001
Manufacturer's Certified Agent
IN ACCORDANCE WITH CSA STANDARD Z166

Fig. 21-41 Typical explosive-actuated tool operator's license.

Operation. The low-velocity tool discharges a powder cartridge inside the breach of the tool. This energy is then transferred to a special drive piston, which in turn forces the hardened-steel fastener into the mounting surface. Since the fastener is nearly in contact with the receiving surface prior to firing, it cannot reach a high speed. The energy from the piston is then used to force the fastener into the mounting surface.

Powder-charge Cartridges. The explosive used in low-velocity tools comes in the form of a brass, rim-fire cartridge. These cartridges are factory-loaded to precise power levels and are given both a number and a color code to indicate these levels. Both cartridge and box or container are equipped with the code markings. Figure 21-43 illustrates typical powder-charge cartridges.

Manufacturers have agreed upon a common coding system. However, cartridges from one manufacturer should not be used in another manufacturer's equipment, due to variations in design and tool capability. The cartridge coding system is as follows.

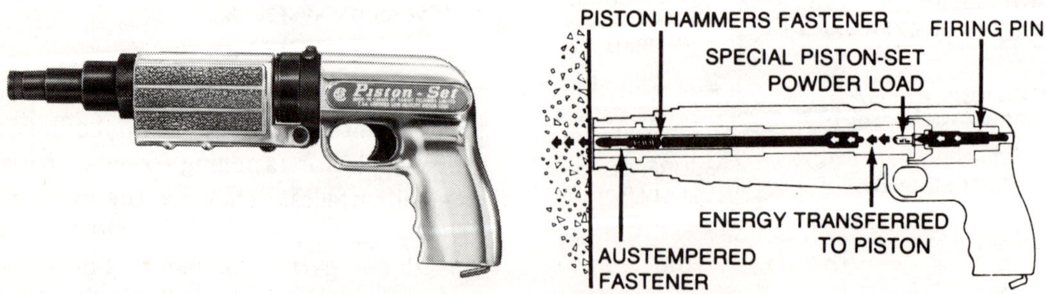

Fig. 21-42 Typical low-velocity, .22 caliber, piston-set tool and mechanism.

Fig. 21-43 Typical powder-charge cartridges. From left, .22 caliber, rim-fire, low-velocity cartridge; two types of .22 caliber rim-fire, high-velocity cartridges; .38 caliber center-fire, high-velocity cartridge.

Number	Color	Application
No. 1	gray	the lightest charge, for use in soft material
No. 2	brown	a light charge, for use in dense material
No. 3	green	a medium-duty charge, for use in solid masonry or mild steels
No. 4	yellow	a heavy charge, for use in steels or dense concrete and masonry surfaces

Fasteners. Four basic types of fasteners are available, as shown in Figure 21-44.

Drive Pins. Drive pins are used to nail materials directly to concrete, steel, and horizontal mortar joints. These fasteners are permanent, and are intended for use with equipment that does not need to be removed and remounted from time to time. They are available in assorted lengths and diameters.

Threaded Studs. Threaded studs are for use where washer and threaded-nut mounting is desired, such as to facilitate the removal and remounting of equipment. They are also ideal where adjustment or positioning

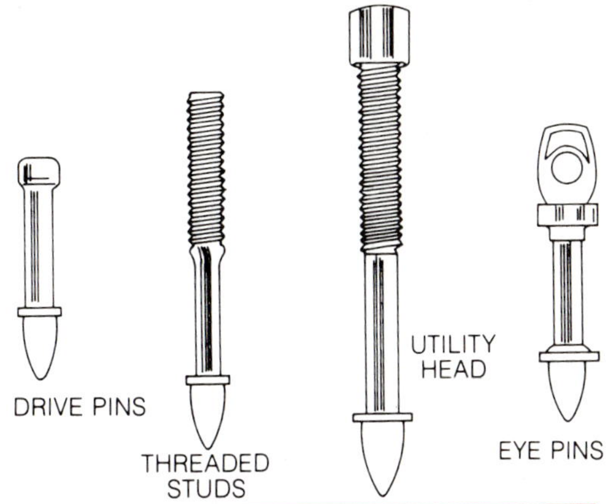

Fig. 21-44 The four basic types of powder-actuated fasteners.

of the equipment is necessary from time to time. Care should be taken not to overtighten the nut, thereby loosening the stud in the mounting-surface material. These versatile fasteners are produced in several bolt sizes, and various thread and stud lengths are available.

Utility Head. Utility-head fasteners have a round nut, and are generally recommended for fastening material 1 1/2 inches (40 mm) in thickness (or greater) to concrete and/or steel. This fastener has been designed for use with the high-velocity tools *only,* due to its mass and shank diameter.

Eye Pins. Eye pins are often used in suspended-ceiling installations and hanging-type lighting fixtures, and for attaching veneers or wire mesh to concrete. They are available in several sizes and lengths for use with high-velocity tools *only.* The piston of the low-velocity tool would crack or damage the eye section of this fastener, rendering it unsafe. Hanging-type lighting fixtures and suspended ceilings often require the extra

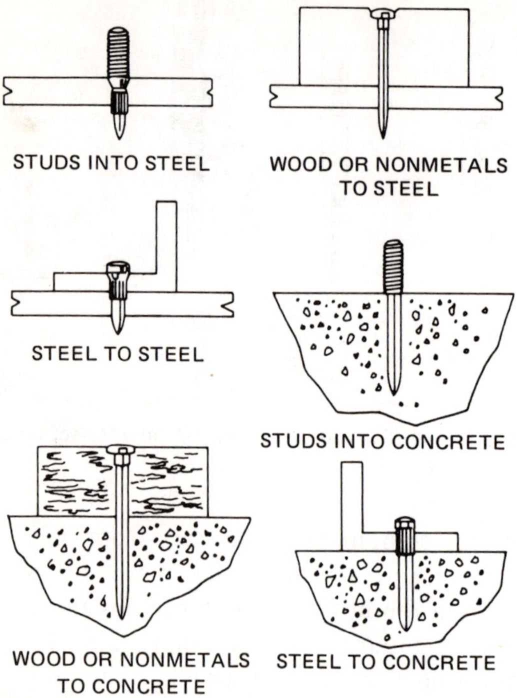

Fig. 21-45 Typical combinations of material to be secured by powder-actuated fasteners.

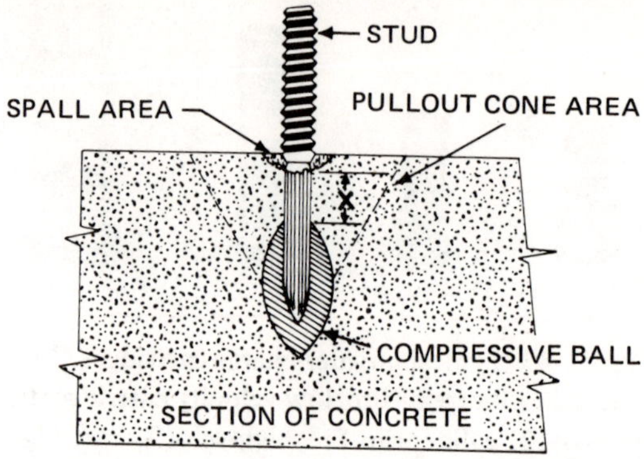

Fig. 21-46A Threaded stud in concrete.

holding strength achieved with the high-velocity tool. A specially designed clip and fastener is available for use with the low-velocity tool when the suspended load is not too great.

Fastening Surfaces. Figure 21-45 illustrates several combinations of fastening surfaces that can be secured or supported by powder-actuated fasteners.

When a stud or pin is fired into concrete (or similar masonry), a compressive bond or ball is formed around the point of the fastener. Since this area accounts for much of the holding power, the fastener should be driven into the concrete to a depth of approximately eight times the diameter of the fastener. Referring to Figure 21-46A, the maximum holding power is achieved when the strength of the concrete in area X is greater than the bond at the point. If penetration is not sufficient, the fastener may pull out under load, removing a cone-shaped section of the concrete as indicated by the dash lines in the figure.

Concrete requires about 28 days to reach its full compressive strength. Any fastener set in "green" concrete (less than four days old) will develop only half of its potential holding power, but will improve slightly as the concrete ages. It is recommended that the fasteners *not* be driven into any concrete that is less than four days old.

The small chips of concrete (known as *spall*) that break away from around the pin area do not lessen the holding power of the fastener. A *spall guard* on the fastening tool helps to eliminate this, however, and improves the appearance of the finished job, Figure 21-46B.

Any fastener driven into a hollow block must not be allowed to protrude through the block. The compressive bond at the point

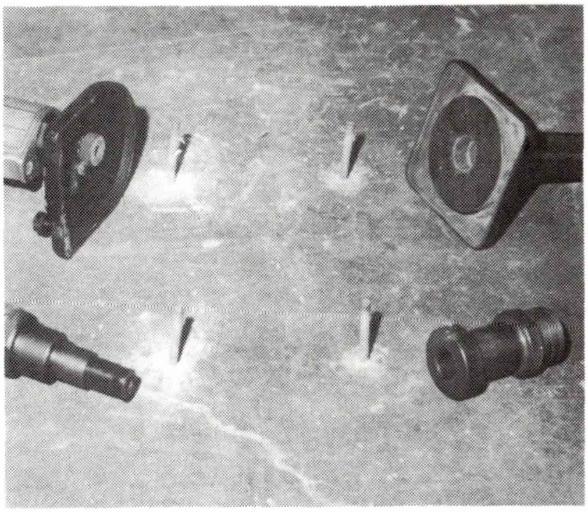

Fig. 21-46B Threaded studs driven into concrete without a spall guard (left) and with a spall guard (right).

is lost completely under these conditions and little or no holding power will be the result. Fasteners should not be driven into vertical mortar seams on any brick or block wall. Mortar in these seams tends to be less in quantity and definitely lacking in compressive quality. Only horizontal mortar seams provide sufficient quantity of material and adequate compressive holding power.

Care should be taken to drive several test pins into an out-of-the-way area of the mounting surface to determine proper cartridge strength and pin size. Use great care to ensure that the pin does not pass completely through the receiving surface and *endanger* someone on the other side. It is wise to start with the weakest charge and progress upward in strength until the correct combination is determined.

The user of a powder-actuated tool should be familiar with the minimum distance recommended between fasteners, minimum distance from the edge, and the minimum thickness of the concrete to provide safe installation and adequate holding power. Table 21-4 lists the important distances to be observed with various fasteners.

Fasteners of different diameters have different holding powers in concrete. Table 21-5 lists the holding power of pins or studs in average concrete.

Fastening to steel requires a technique somewhat different from fastening to concrete. Holding power in steel is determined by the ability of the steel to push back (elastically) against the shank of the fastener. Unlike masonry surfaces, the point of the fastener *must* pass completely through the steel, Figure 21-47. If the point does not pass completely through, the springlike action of the steel will exert pressure on the rounded portions of the tip. This pressure tends to force the fastener back out. Once the proper amount of penetration has been achieved, the holding power is determined by steel thickness and fastener diam-

TABLE 21-4

SHANK DIAMETER		FASTENER SPACING		EDGE TO PIN		SLAB THICKNESS	
inches	(mm)	inches	(mm)	inches	(mm)	inches	(mm)
0.762	3	16.51	65	19.05	75	16.51	65
1.016	4	19.05	75	19.05	75	16.51	65
1.1176	4.4	24.13	95	19.05	75	16.51	65
1.4224	5.6	27.94	110	38.10	150	25.40	100
1.6256	6.4	29.21	115	38.10	150	25.40	100

TABLE 21-5

SHANK DIAMETER		PENETRATION (8 x shank)		HOLDING POWER	
inches	(mm)	inches	(mm)	pounds	(kg)
1.016	4	7.62	30	1760	800
1.2192	4.8	10.16	40	2420	1100
1.4224	5.6	11.43	45	3520	1600
1.6256	6.4	12.70	50	5390	2450

eter. The thicker the steel and/or the larger the fastener, the more holding power is achieved, Table 21-6. On occasion, fasteners must be driven into thick steel where complete penetration is impossible. Lightweight loads can be safely supported on these fasteners, since 50 percent of the fastener's potential holding power can be achieved without full penetration. When in doubt about the fastener's ability to support a load, consult the manufacturer.

A straight spline-knurl on the shank of the fastener adds about 25 percent to the holding power and prevents the threaded-stud type of fastener from turning while the nut is being tightened. Care should be taken not to overtighten the nut, since tremendous pressure can be exerted by the turning effort, even with a short wrench.

Threaded fasteners for steel can be easily recognized by the knurled section on the shank and should not be used in masonry surfaces/materials. Some manufacturers' drive pins will not have a knurled section because no further tightening or nut application is required. Figure 21-48 compares concrete and steel fasteners.

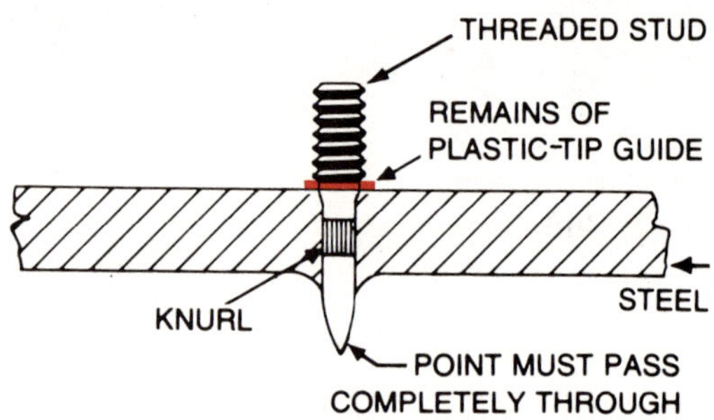

Fig. 21-47 Threaded stud in steel.

The holding power figures presented in Table 21-6 are actual. Therefore, it is recommended that a suitable margin of safety be allowed when choosing a fastener.

Many fasteners come equipped with a colored plastic-tip guide on the point of the fastener. The plastic tip guides the fastener

TABLE 21-6 Typical Holding Powers in Steel

1.016 inches (4 mm) SHANK DIAMETER				1.4224 inches (5.6 mm) SHANK DIAMETER			
THICKNESS		HOLDING POWER		THICKNESS		HOLDING POWER	
inches	(mm)	pounds	(kg)	inches	(mm)	pounds	(kg)
1.6256	6.4	1980	900	2.413	9.5	3520	1600
2.413	9.5	2640	1200	3.302	13	4620	2100
3.302	13	3190	1450	4.064	16	5610	2550
4.064	16	3740	1700	5.08	20	6600	3000

while in the barrel of the tool, assures straight entry into the receiving surface, and helps clear the residue of previous shots from the barrel. Another important function of the plastic tip is to act as a penetration guide. If most of the plastic tip is visible at the point of penetration after firing, penetration is insufficient. When no plastic is visible, too strong a charge has been used. Correct penetration will leave a small ring or flange of plastic where the fastener enters the steel, Figure 21-49. Once again, start with a light charge, and increase the charge strength until proper penetration is achieved. Pins should be driven not closer than 1/2 inch (13 mm) from the edge of steel. Fastening closer to the edge could cause a *dangerous* ricochet. When steel has been welded, there is frequently an increase in the hardness of the steel surrounding the weld. For this reason, fasteners should not be driven any closer than 2 inches (50 mm) from a welded area.

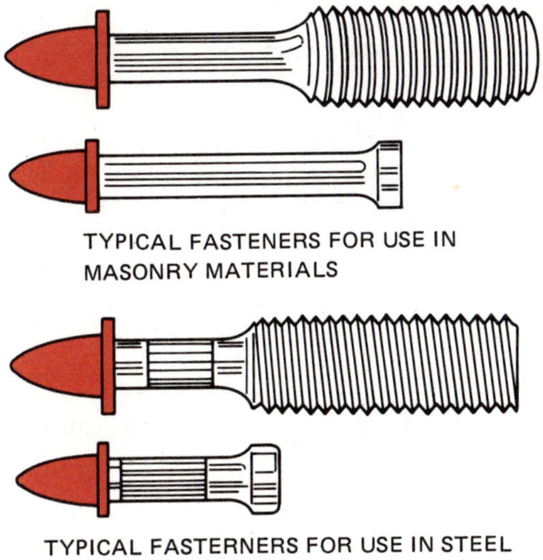

Fig. 21-48 Fasteners for masonry and steel.

High-velocity Equipment. High-velocity tools operate at 300 ft/s (90 m/s) and higher. They are suitable for use in dense materials where penetration is more difficult and/or increased holding power is required by using larger fasteners. The high-velocity tool is similar to a gun in that the fastener accelerates down the length of the barrel, striking the receiving surface with considerable force. The tools are designed in such a way that

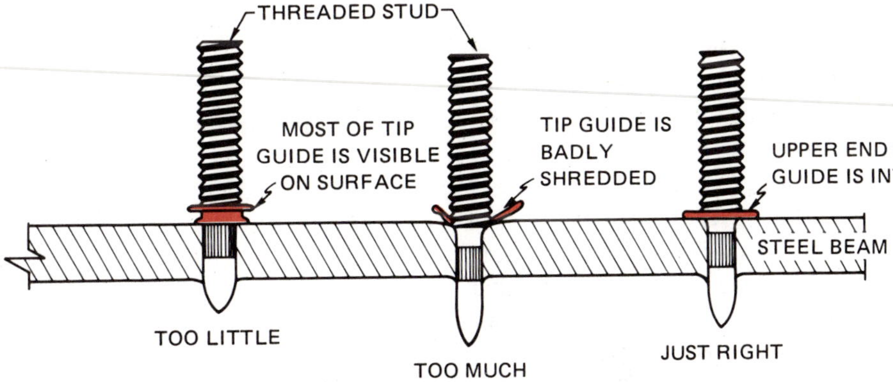

Fig. 21-49 The plastic-tip guide gives proof of proper penetration.

they cannot be fired unless pressed very firmly against the receiving surface. In this way the tool cannot be misused or accidentally discharged. The firing mechanism of the tool is also designed so that it will not function when the tool is held at an angle of 8° off the perpendicular to the receiving surface. Because of the fastener's speed and striking energy, care must be taken to use the manufacturer's recommended procedures, safety guards, and equipment, as well as common sense.

High-velocity tools, when used by trained, competent operators, perform many otherwise tedious tasks in a fraction of the time required by other fastening methods. *Danger* exists only when the tool or its related equipment is used improperly. Figure 21-50 illustrates a typical high-velocity tool mechanism.

Powder charges used with high-velocity tools often exceed the strength of those used in low-velocity units. The No. 3 green and No. 4 yellow cartridges used with the low-velocity equipment can be used in high-velocity tools. However, two additional charges – the No. 5 red (heavy duty) and the No. 6 purple (extra heavy duty) – are available for use in more demanding applications.

Most powder-actuated tools and fasteners are produced in a standard .22 caliber. Some manufacturers have developed an extremely powerful range of cartridges that are slightly longer and coded as *No. 7 long gray* and *No. 8 long brown.* These cartridges are for use in tools with a barrel capable of driving 3/8-inch (9.5-mm) diameter fasteners. A .38-caliber tool is available that accommodates a full range of 3/8-inch (9.5-mm) fasteners for use in steel or masonry materials. These units, however, are not as popular as the smaller, lighter-duty units in .22 caliber. Use of the heavy .38-caliber fastener is mainly limited to the suspension of heavy loads, and is often used in underwater firing applications. (See Figures 21-51, 21-52, and 21-53.)

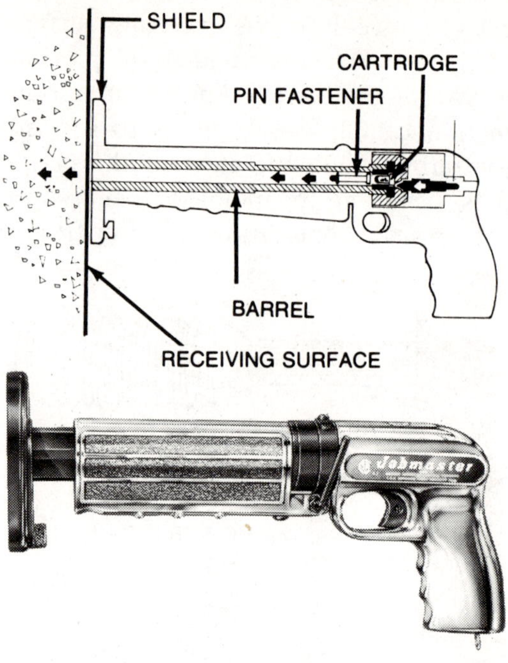

Fig. 21-50 Typical high-velocity, .22 caliber tool and mechanism.

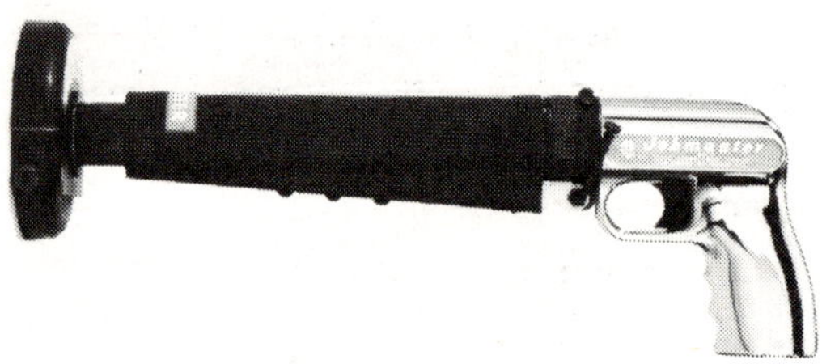

Fig. 21-51 Typical .38 caliber high-velocity tool.

Safety Equipment. Safety glasses should be worn by operators of any powder-actuated tool. In many cases, a hard hat is recommended to protect the operator from spall when firing in the area of the head and shoulders. Noise from the more powerful tools, especially in a confined area, makes ear protection advisable. Common sense and consideration of others working nearby greatly adds to the safe use of the versatile powder-actuated fastening systems. Manufacturers of the tools produce a variety of special guards and fastening aids to speed up the safe operation of their equipment.

Fig. 21-52 High-velocity tool for underwater use.

Safety Recommendations. The following is a list of safety recommendations well worth remembering.

- **Never** use a powder-actuated tool without having its operation and limitations explained to you.
- **Always** use the proper recommended safety shield on the tool.
- **Never** attempt to set a fastener through a predrilled hole in steel.
- **Always** try the weakest cartridge on the first shot. Progress to the next heavier load only when necessary.

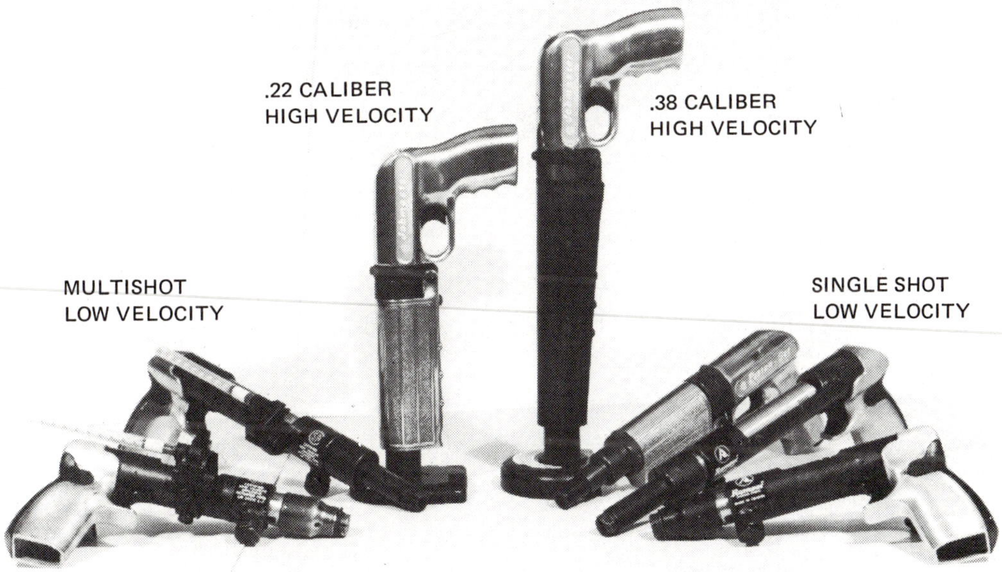

Fig. 21-53 Typical powder-actuated fastening tools.

Never fire a second fastener into the immediate area where a previous fastener has just failed.

Always check the receiving surface to make sure it will safely accept the intended fastener.

Never attempt to fasten into hard steel or near welds.

Always maintain proper, safe firing distances between fasteners and the edge of mounting surfaces.

Never attempt to modify any tool or to adapt pieces of one manufacturer's tool for use in another's.

Always check to see whether the barrel is clear before inserting a fresh cartridge or fastener.

Never carry fasteners or metal objects in the same pocket as the powder cartridges.

Always wear protective safety equipment when using these tools.

Never load a tool until you are ready to fire; never put a loaded tool away: an untrained person could fire it.

Always keep the tool in good shape. Clean it regularly, and have it checked by a manufacturer's representative periodically.

Never point the tool, loaded or unloaded, at yourself or another person.

Always keep the tool against the receiving surface for at least 15 seconds to 20 seconds if it should fail to fire when triggered.

REVIEW

1. List four pieces of electrical equipment that may require the use of fastening devices to assist in mounting them to masonry or similar surfaces.
2. What are the three main types of screw fasteners?
3. Name five driver configurations for screw fasteners.
4. List the different head types used for screw fasteners.
5. State three advantages of using square-recess *(Robertson head)* screwdrivers.
6. Where are hexagon *(Allen)* recess-type machine screws used?
7. What are the advantages of using self-tapping screws?
8. What is the difference between a standard single lead and a *Kwixin* wood screw?
9. How is the length of a wood screw determined?
10. What are the physical differences between a wood screw and a Lag bolt?
11. Name three different materials used for wood-screw construction, and state one use for each.
12. Why is "bolt strength" important when using a machine bolt?
13. Name four different materials used in the construction of masonry fasteners.

14. Outline the procedure for fastening an electrical box to a masonry wall with a masonry fastener and wood screw.
15. Describe four different methods of creating a hole in a masonry wall for an expansion-type fastening device.
16. Name two types of hollow wall fasteners and give one application of each.
17. State three advantages of a powder-actuated fastening system.
18. Why is a license necessary for operating powder-actuated fastening systems?
19. List the cartridge and charge identifications for low-velocity fastening systems.
20. Explain why a threaded stud is used to mount equipment, rather than a drive pin.
21. How does a threaded stud for steel differ from a masonry stud?
22. List three precautions to be taken when driving fasteners into steel surfaces.
23. What is the main difference between the gun mechanisms of a low-velocity tool and a high-velocity tool?
24. What safety equipment is needed by the operator of a powder-actuated tool?
25. What is the caliber of the powder cartridges used in high- and low-velocity tools?

ACKNOWLEDGMENTS

Amalgamated Electric Corporation, Ltd.
Bryant, Division of Westinghouse
Burndy Corporation
Canada Wire and Cable Co., Ltd.
Canadiam Chromalox Company, Ltd.
Canadian Electrical Code
Canadian General Electric Company, Ltd.
Canadian Johns-Manville Co., Ltd.
Commercial Enclosed Fuse Company
Efcor of Canada, Ltd.
Greenlee Tool Company
Ilsco of Canada
3M Company
Marr Electric
Ray C. Mullin, author ELECTRICAL WIRING – COMMERCIAL
Murray-Jensen Mfg, Ltd.
National Electrical Code
Ontario Electrical League
Ontario Hydro
Phillips Cables, Ltd.
Pyle-National of Canada, Ltd.
Pyrotenax of Canada, Ltd.
Ramset Fasteners, Ltd.
Ron Ridsdill
P.L. Robertson Company, Ltd.
Sangamo Company, Ltd.
Scepter Manufacturing Co., Ltd.
Robert Shaw Control
Robert L. Smith, author ELECTRICAL WIRING–COMMERCIAL
Smith and Stone
Square D Company
Star Fasteners, Ltd.
Steel Company of Canada, Ltd.
Wolberg Electrical Supply Co., Inc.

FOR REFERENCE

Building Wire and Corflex BW-78 Cables, The Modern Wiring Method, AS-67, Canada Wire and Cable.

Canadian Electrical Code, Part I, Canadian Standards Association. (See Also Electrical Safety Code for province of residence.)

Electrical and Electronic Drafting, Bruce A. Renton, General Publishing.

Electrical and Electronic Technology One, F.

Electrical Construction Wiring, Walter N. Alerich, American Technical Society.

Electrical Experiments, Frank J. Long, General Publishing.

Electricity and Electronics, Howard Gerrish, Goodheart-Willcox.

Electricity and Electronics – Basic, 4th Edition, William F. Steinberg and Walter B. Ford, American Technical Society.

Electricity – Electronics Dictionary, Howard H. Gerrish, Goodheart-Willcox.

Fluorescent Ballasts, Section 2, 7310, General Electric.

Fluorescent Lamps, TP-111, General Electric.

Fundamentals of Electricity – Basics of Electricity, Electronics, Controls, and Computers, 6th Edition, Robert G. Seippel, American Technical Society.

High Intensity Discharge Ballasts, Section 3, 5574, General Electric.

High Intensity Discharge Lamps, HID Lamps 11/71, General Electric.

House Wiring Simplified, Floyd M. Mix, Goodheart-Willcox.

How to Read Electrical Blueprints, 2nd Edition, Kennard C. Graham, American Technical Society.

Intermediate Electricity, Frank J. Long, General Publishing.

Introducing Electricity, Frank J. Long, General Publishing.

Medallion All-Electric Award Requirements, C273.1-1971, Planning Guide for Residential Electric Heating, Canadian Standards Association; Ontario Electrical League.

Mercury Lamps, TP-109, General Electric.

National Electrical Code 1981, National Fire Protection Association.

National Electrical Code and Blueprint Reading, 5th Edition, Kenneth Gebert, American Technical Society.

The Wireman's Handbook, WH.79, Canada Wire and Cable.

Index

A

Across-thc-line starting, 303
Adapters, 30
Air-vapor barrier, 240–241
Allen keys, 338
Aluminum-sheathed cable, 125–132
  corrugated, 125, 129
  covering, 126
  currents, 126–128
  electrical ratings, 126
  single versus multiconductor, 126
  smoothed-sheathed, 125, 129, 130
  supports, 132
  termination, 128
    preparation of cable for, 128, 131
Ambient temperature, 219, 275
American Wire Gauge (AWG), 48
Ampacity (ies), 51, 55–56, 305
  allowable, if insulators conductors, 55
Amperage, 8
Amperes, 5, 13
Annealing, 46, 220
Architectural plans, electrical symbols for, 83
Arcing, 12, 15
Arc-quenching material, 225
Armored bushed cable (ABC), 118–119
Armored cable (AC), 118–123
  accessories, 122
  construction of, 119
  electrical ratings, 118
  fishing, 122
  installation in wet areas, 122
  special connectors, 123
Armored cable (continued)
  supports, 120–122
  termination, 119–120
  trade names, 118–119
  uses of, 118
    prohibited, 118
Asbestos, 54
Attic ventilation, 241
Augered flute-tip screw, 344
Automotive Engineers (S.A.E.), Society of, 345

B

Ballast
  defined, 273
  overheating, 275
Baseboard heater(s), 241–242
  air convection in, 243
  installation of, 242
    hallway, 256
Basement heating, 246–247
Bearing surface, defined, 336
Blanket insulation. *See* Insulation, batts
Bolt strength, 345–346
Bolts, 339
Bonding jumper, 193
Buzzer tester, 16, 17

C

Cable clamps, in outlet boxes, 73
Cable connector, 87
  mechanical, 92–95
    preparation of cable for, 94–95
Cable ripper, 87
Cable wiring, nonmetallic-sheathed, 77–87
Cables, insulated, 49
Cam-out, 337
Canadex, 78
Candril screw, 344
Cap screws, 345
Caulking, 245
Circuit breakers, 197, 213–214
  thermomagnetic, 213
Circular mil area, 48
Cold flow, 46
Cold lead, 245
Compensator, 329
Compressed-air fishing, 168
Compression cable connectors, tubular, 95
Compression tool(s)
  effect of, on cable and lug, 99
  hand operated
    hydraulic, 95, 96
    mechanical, 95
  motor-driven hydraulic, 95, 97
Conductor(s), 45–56
  forms, 46, 48
    cable, 48
    cord, 48
    wire, 46
  importance of, 45
  insulation, 51, 54
    asbestos, 54
    cross link, 54
    glass over thermoplastic, 54
    neoprene, 54
    rubber and cotton, 54
    varnish, 54

Conductor(s), (continued)
materials, 45–46
aluminum, 45–46
steel, 46
overheating of, 220
service drop, 173
sizes, 48, 51
American Wire Gauge, 48
cross-sectional area, 48
Conduit(s)
flexible metal (flex), 163–165
intermediate metal (IMC), 150–151
liquidtight flexible, 165–166
rigid aluminum, 160–161
rigid PVC, 161–163
rigid (thickwall), 143–150
thinwall. *See* Electrical metallic tubing (EMT)
Conduit fill, 168
Conduit fishing, 167, 168
using compressed air, 168, 169
Conduit wiring, 142–170
applications, 142
installing conductors, 167–168
protective features, 142
sizes, 142
types of, 143
Condulet(s), 154
covers, 156
Controllers, reversing, 324–327
Convection of air, 242
Cooling systems, residential, 262
changeover from, to heating, 265
chilled-water cooling coil, 262
evaporator coil, 262–263
heat pumps, 265–266
refinements in, 266
split system, 263–265
units for, 262
Copper, hard-drawn, 46
Cord caps, heavy-duty, 58–59
Cord fittings, 58–61. *See also* Plug cap(s)
two main types, 58
Cords, insulated, for lighting fixtures and appliances, 50
Cross-link (X-link), 118
Current flow
control of, 11
three factors in, 8
Current-limiting fuse, 225–226
Cutting pliers, 87

D

Dash pot, 308
Dead air, 235
Demand factor, 212
Design conditions, outside
for localities in United States, 251
for localities throughout world, 252
Design data for selected locations in Canada, 250
Dies, 56
Dimmer switch(es), 24–26
high-low, 24, 26
infinite-control, 24, 26
rating, 15
Distribution transformer, 1–2
Drills
hammer-driven, 350–351
power-driven, 351–352
Drip loop, 181
Drive pins, 355
Drum switch, 331
Dryer cable, 81
Dual-element fuses, 231

E

Eavestrough heating cable installation, 258, 261
Eddy currents, 127
Edison-base fuse. *See* Plug fuse
Electric supply lines, 173
Electrical metallic tubing (EMT), 152–159
bending, 153
fittings for, 152, 153–154
electrical boxes for, 157–158
types of, 155, 156
PVC-jacketed, 152
termination, 154, 159
threading, 154
Electrolier switch(es), 22–24
three-circuit, 24, 25
alternate, 24
consecutive, 24
two-circuit, 22–24, 25
Electrolysis, 45
Energy-saving lamps, 275, 277, 298–299
Eye pins, 355–356

F

Fastening devices, 336–362
uses of, 336
Fastening surfaces, 356–359
Ferrule-contact cartridge fuses, 224
Fish tape, 167
securing conductors to, 168, 169
550V and 440V systems, 201–202, 203
grounding, 205
metering, 202–205
method of distribution, 202, 205
Flat-rate water heater system, 182–183
Flex. *See* Flexible metal conduit
Flexible metal conduit (flex), 163–165
applications, 164
cutting, 165
ground continuity, 164
supports, 164
terminating, 165
Fluorescent lamps. *See also specific type*
designation, 270–272
maintenance, 299
operation of, 273–274
ordering abbreviations, 272
types of, 268
Fluorescent lighting, 268
advantages of, 268–269
ballast, 272–273
disadvantages of, 269
instant-start, 275
power factor capacitor, 274
rapid-start, 274
starter, 272
Fluorescent tube(s)
colors, 270
parts, 270
Forced-air heating unit, 256, 257
Four-way switch control, wiring diagrams for, 86
Frequency, defined, 1
Fuse(s), 219–232
cartridge
ferrule-contact, 224
knife-blade, 224
one-time, 225
renewable-link, 226
replaceable-link (renewable), 224
code requirements, 226
continuous current rating, 221
defined, 220
dual-element, 231
Edison-base, 221–223
high-interrupting capacity (HIC), 227–231
high-rupture capacity (HRC), 227
interrupting capacity, 221
maximum rated voltage, 221
ratings, 221
plug, 221–223

G

GFCI. *See* Ground-fault circuit interrupter
Gooseneck bend, 183, 184
Ground fault, defined, 215
Ground-fault circuit interrupters (GFCI), 40–43, 215–217
  operation of, 216–217
Grounding rods, 193–195
  installation, 195
Grounding slot, 38

H

Hand-off automatic (HOZ) switch, 319
Hanger strips, 245
Heat generation, 219–220
Heat pumps, 265–266
Heat sink, 231
Heat-control switch(es), 106–115
  five-heat, double-pole, 106, 107
  infinite-heat switch, 110–112
    operation of, 112
  oven control, 112–113
  rotisserie, 115
  self-cleaning ovens, 113–115, 116
    continuous, 113
    pyrolytic, 114–115, 117
  series-parallel connections, 106
  seven-heat, 109–110
  three-heat
    double-pole, 106
    single-pole, 106
Heater
  baseboard, 241–242
  location of, 256
  size, determining, 255
  snow-melting, 259
Heating
  cost, 256
  electric
    advantages of, 234
    costs of, 234–235
Heat-loss areas, 245–246
Heat-loss calculations, 247–255
  through ceiling, 254–255
  through doors and windows, 253
  heat-retaining walls and ceilings, 247–248
  infiltration loss, 253–254
  total loss for room, 255
  through walls and ceilings, 248
Hex head machine bolts, 345, 356
Hickey, 146
High-intensity discharge (HID) lamps, 277
  sizes and shapes, 284
High-interrupting capacity (HIC) fuses, 227–231
  Class G, 230
  Class J, 228
  Class K, 228
  Class L, 229
  Class R, 230–231
  Class T, 229–230
High-pressure sodium lamps, 290–294
  applications, 294
High-rupture capacity (HRC) fuse, 227
  form I, 228
  form II, 228
High-voltage metering system, 208
Hi-temp clean switch, 115
Holding powers in steel, 358
Hollow wall fasteners, 252–253
Hysteresis loss, 313

I

Inching. *See* Jogging circuits
Industrial services, 200–217
  circuit breakers, 213–214
  demand meters, 211–212
  voltages, 200
Insulating disc, 59
Insulating putty, 101
Insulating tapes, 95–105
  colored, 104
  electrical coating, 103
  ethylene propylene rubber (EPR), 100
  fiberglass, 101
  friction, 96
  vinyl plastic, 96, 100
Insulation, 235–240
  air-vapor barrier and, 240–241
  attic, 241
  batts, 235–237
    installing and cutting, 239
  conductor, 51, 54
  pouring, 237–238
  rigid sheets, 238–239
  rolls, 237
  thermoplastic, 51, 54
  varnish, 54
Insulation tester, 138
Insulators, pull-chain, 33
Interlocking system, 324–327
  electrical, 324, 325
  mechanical, 324, 325
Intermediate metal conduit (IMC), 150–151
  construction, 151
  size, 151
  supports, 151
  uses, 151

J

Jogging circuits, 320–322
  with control relay, 323
  push button, 321–322
  selector, 322

K

Kilowatt-hour meter, 2
Knife-blade cartridge fuses, 224
Knockout cutters, 166–167
  hydraulic-powered, 167
Kwixin, 343

L

Lag bolts, 344
Lag shield, 348
Lamp(s)
  energy-saving, 275, 277, 298–299
  fluorescent. *See* Fluorescent lamps
  high-intensity discharge (HID), 277, 284
  high-pressure sodium, 290–294
  mercury-vapor, 279–284
  metal-halide, 284, 285–286
  metric, 275–277
  multivapor, 295–298
  power-groove, 277, 278
  quartz-halogen, 297, 298
  tungsten-halogen, 297, 298
Lampholder(s), 29–33
  circuit connection, 32
  construction, 30
  key type, 32
  keyless type, 32
  location of, 33
  screw-base, 29
  switch mechanisms, 32
Lighting sources, discharge, 268–299
Lighting-control switches, 11–26
Lighting-load calculations, 15
Line cord, 58
Line voltage, 8
Liquidtight flexible conduit, 165–166
  grounding, 166
Live wire(s)
  defined, 2
  fusing, 5
  switching, 2–4

Locked-rotor current, 303
Loom wire, 78
Loop system, and NMC installation, 78
Lo-root screw, 343
Lucalox, 291, 294
Lugs, 93

M

Magnesium oxide, 135
Magnetic motor controller, 311-312
  electrical circuits of, 313
  holding or maintaining contacts, 314
  operation of, 313-314
  sizes, 312
Magnetic motor starter. *See* Magnetic motor controller
Masonry fasteners, 347-350
  drive-in anchors, 350
  expansion shields, 348
  lead sleeve anchor, 349
  screw anchor, 347-348
  self-drilling shield, 348
Mercury-vapor lamps, 279-284
  applications, 282-284
  arc tube, 281
    support, 281
  ballast, 282
  base, 279
  electrodes, 280
  operation of, 281
  outer bulb, 281
  pinch seal, 281
  restarting, 281-282
  starting resistor, 279
Messenger wires, 21
Metal fatigue, of fuse link, 232
Metal-halide lamps, 284, 285-286
  applications, 290
  operation of, 289-290
Meter bases. *See* Meter socket
Meter socket(s), 181-182
Meters, 196-197
  demand, 211-212
  five-dial, 196, 197
Metric lamps, 275-277
Mineral-insulated (MI) cable, 135-140
  applications, 139
  development of, 135
  durability, 136
  features of, 135
  installation, 138-139
  making, 135
  sheath currents, 139
  termination, 136-138
Motor conductor sizes, 305
  determining, 305
Motor control, 302-333
  control stations and, 317-320
  devices, maximum rating or setting of, 304
  direct-current, 332
    series field, 332
    shunt (parallel) field, 332
  dual-speed, 327
  location of devices, 303
  magnetic, 311-312
  manual switches, 310
  need for, 302-303
  overcurrent protection, 303-304
  switches, 303
Motor toggle switch, 310
Motor watchman switch, 310
Multimeters, 16
Multivapor lamps, 295-298
Muscular freeze, 42

N

National Electrical Manufacturers Association (NEMA), 35, 39
Neoprene, 54
Neutral wire, 2, 8
  fusing, 5-6
  grounding, 6-7
No-field release, defined, 332
Nonmetallic cable (NMC), 118. *See also* Nonmetallic-sheathed cable
  accessories, 82, 87
Nonmetallic-sheathed cable, 77
  circuit diagram, 81-82
  concealed installations, protection for, 81
  construction, 77-78
  fastening, 79-80
  installation, 78-79
  residential applications, 81
  restrictions, 78
  supports, 79
  trade names, 78
  underground feeders, 80-81
  uses permitted, 78

O

Octagon concrete rings, 64, 65
Ohms, 56
Ohm's Law, 8
One-time cartridge fuses, 225
Outlet boxes, 63-76
  capacity, calculation of, 75
  concrete-masonry-tile, 72
Outlet boxes (continued)
    construction, 72
    covers, 72
    dimensions, 72
    methods of mounting, 73
  conductor capacity of, 74
  grounding, 73-74
  octagon, 63-65
    concrete rings, 64, 65
    covers, 64
    dimensions, 63
    knockouts, 65
    methods of mounting, 65
  pancake, 65
  requirements, 63
  round, 63
  sectional plaster, 68-70
    construction, 68
    covers, 69
    dimensions, 69
    methods of mounting, 70
  square, 66
    covers, 67
    dimensions, 61
    methods of mounting, 67
  utility, 70
    construction, 71
    covers, 71
    dimensions, 71
    methods of mounting, 71
Outlets, 34-42. *See also* Receptacles
Overhead supply system, residential, 2, 3
Overheating, 74
  damage from, 220
  protection, 257
Oxidation, 45

P

People protection, 215
Philex, 78
Pipe cutter, use of, for threading, 144
Pipe-heating cable with thermostat, 259, 261
Plaster ears, 40, 41
Plug cap(s), 58
  appliance attachment, 60
  connection, 59
  dead-front attachment, 59
  electrical ratings, 60
  grounded, 61
  male and female attachment, 58
Plug fuse, 221-223
  circuit fault indications, 222-223
  time-delay, 221

Polyphase (three-phase) system, 200–201
Polyvinyl chloride (PVC), 126, 152
Powder-actuated fasteners, 353–355
  cartridges for, 354–355
  high-velocity, 359–360
  low-velocity, 353–354
  operation of, 354
  safety equipment, 361
  safety recommendations, 361–362
Power-groove lamps, 277, 278
Pryout, 73

Q

Quartz-halogen lamp, 297, 298

R

Radiant-heating ceiling cable, 242–245
  installation of, 244
Receptacles, 34–43
  combinations, 35
  construction, 34–35
  direct current, 37
  duty ranges, 34
  grounding, 38
  heavy-duty, 34
  light-duty, 34
  NEMA, 39–40
  number of, on circuit, 34
  and plug shapes, 35–37
    crow's-foot, 36
    old-style two prong, 35
    tandem, 37
    twist-lock, 37
    "U" ground, 36, 37
  polarized, 40
  range and dryer, 37
  replacement of, 39
  split, 38–39
  uses, 34
  wiring diagrams with, 85
Reduced-voltage control circuit, 322–324
Reduced-voltage starter(s), 303, 327–331
  autotransformer, 329–331
  primary resistor, 327, 329
Relay, thermal overlead (OL). *See* Thermal overload relay
Renewable-link cartridge fuse, 226
Replaceable-link fuse, 224
Residential electric heating and cooling, 234–266
Residential service wiring, 173–198
  consumer's service, 173
  electrical utility, 173
  grounding, 190–196
    most common method of, 192
    precautions, 193
  indoor meter method, 175, 190
  inspection permit, 197
  installation techniques, 177
  outdoor meter, 187
  service boxes, 185–187
    location of, 185
    mounting, 185
  service mast, 178–181
    installation, 180, 183
  service-entrance panel, 176, 177
  size and capacity, 173, 176–177
  typical, 174
  underground, 185, 186
Resin splicing kits, 101, 103
Resistance, 8, 56
Rigid aluminum conduit, 160–161
  bending, 160
  installing, 160–161
  threading, 160
Rigid PVC conduit, 161–163
  applications, 161
  bending, 161–162
    heat sources for, 163
  conductor installation, 163
  expansion, 162
  joining and terminating, 161
  light weight advantage of, 161
  supports, 163
Rigid (thickwall) conduit, 143–150
  bending, 146–150
    45° offset bends, 146
    methods for, 147
    90° bend, 146, 157
    offset bend, 146
    outside corner bend, 146
    round saddle bend, 146
    square saddle bend, 146
  reaming, 144, 145
  threading, 144–146
Robertson head screwdriver, 337
Romex, 78
Root-mean-square (RMS) values, 221
Rotary switch, 22
Rotisserie, 115
Rotor, 302
Rubber-covered conductors, dimensions of, 52–53

S

Screw(s)
  augered flute-tip, 344
  candril, 344
  cap, 345
  construction material, 345
  diameter, 344
  length of, 344
  Lo-root, 343
  point design, 344–345
  self-tapping, 339–341
  sizes, 340, 341
  wood, 342–344
Screw base sizes, 29–30
  candelabra, 30
  intermediate, 30
  medium, 30, 31
  miniature, 30
  mogul, 29
Screw fasteners, 336–337
  driving configurations, 336–337
    Allen head, 338
    hexagon recess heads, 338
    hexagon-shaped, 338
    Phillips-type, 338
    slotted, 336
    square recess, 337
  head design, 336
  neck and shoulder, 338
  shank and body, 338
  thread type, 338
    machine screws, 339, 340
    self-tapping screws, 339–342
    wood screws, 342
Scrulox, 337
Sealing contacts, 314
Self-tapping screws, types of, 339–341
Series-lamp tester, 16
Series-parallel connections, 106
Service drop. *See* Overhead supply system, residential
Service ell, 184
  installation, 185
Service laterals. *See* Residential service wiring, underground
Service-entrance elbows, 183–185
Shading coils, 313
Short circuits, 225
Single-phase motors, reversing, 331
Single-pole switch, cable wiring diagrams for, 84
Skin effect, 56
Snap switches, ratings and uses of, 14–15
  AC general-use, 14

Snap switches (continued)
- AC-DC general use, 14–15
- CO/ALR, 15
- DC, 15
- horsepower, 15

Snow-melting heaters, 259
- installation of cable for, 260

Solderless connectors, 88–104
- compression, 90, 92
- development of, 88
- set-screw, 89–90
  - installing, 90, 92
- twist-on, 88–89, 90

Solenoid, 332
Spall, defined, 356
Spall guard, 356
Splice, taping, 104–105
Spring-wing toggle bolt, 352
Standpipe, 181
Step-down transformer, 312
Storm windows, aluminum versus wood framed, 245
Stroboscopic effect, of fluorescent light, 269
SuperSaver, 277
Switch(es)
- dimmer, 24–26
- double-pole, 18, 19
- dual-action pump control, 317, 319
- electrolier, 22–24
- equipment for testing, 16
- float, 317, 318
- flush-mounted, 11, 12
- four-way, 17, 21–22
- internal construction, 11–12, 13
- limit, 317, 318
- line side, 4
- non-indicating, 21
- operating mechanisms, 11, 12
- single-pole, 17–18

Switch(es) (continued)
- surface-mounted, 11
- three-way, 17, 20–21
- types of, 11
- uses, 16
- wiring symbols, 16, 17

Switch ratings, 13–14

T

Temporary service, 197, 198
Test-laboratory listing, 14
Thermal overload (OL) relay
- bimetallic, 307
- magnetic, 308
- operation of, 306–308
- protection, 305–306
- restarting after, 309–310
- size, determining, 308–309
- solder pot, 306, 307

Thermoplastic insulation, 51, 54
Thermoplastic-covered conductors, dimensions of, 52–53
Thermostat, location of, 256
Thickwall conduit. *See* Rigid (thickwall) conduit
Thinwall conduit. *See* Electrical metallic tubing (EMT)
Threaded studs, 355
Three-phase magnetic motor starter, 316
Three-phase voltage system. *See* Polyphase (three-phase) system
Three-way switch control, wiring diagrams for, 86
Three-wire distribution system, 1–10, 173
Thrift/Mate, 277
Transformer(s)
- distribution, 1–2
- step-down, 312

Travelers, 21
Trilite (three-way) switch, 22, 23
Tungsten-halogen lamps, 297, 298
Twin-fast, 343
208Y/120-V system, 205–206
- conductor and conduit sizes, 211
- grounding, 206
- installation, 207, 209
- service-entrance equipment for, 206

Type S fuse(s), 223

U

Unbalanced system, 7–10
Underground feeders (UF) cable, 80–81
Underground Service Entrance (USE) cable, 80
Underground supply system, residential, 2, 3. *See also* Residential service wiring, underground
Underwriter's knot, 59
Underwriters' Laboratories, Inc. (UL, Und. Lab.), 14
Utility head fasteners, 355

V

Varnish insulation, 54
Ventilating devices, 241
Voltage(s), 5, 13
- obtaining two, from three wires, 1, 2
- residential, 1
- single-phase, 1

W

Wire size uses, 51
Wires, insulated, common types of, 47
Wood screws, types of, 342–344